Phase Behavior of Petroleum Reservoir Fluids

Developed in conjunction with several oil companies using experimental data for real reservoir fluids, *Phase Behavior of Petroleum Reservoir Fluids* introduces industry standard methods for modeling the phase behavior of petroleum reservoir fluids at different stages in the process. Keeping mathematics to a minimum, this book discusses sampling, characterization, compositional analyses, and equations of state used to simulate various pressure–volume–temperature (PVT) properties of reservoir fluids. The Third Edition has been updated throughout.

- Reflects advances in equation of state modeling for reservoir fluids and CO_2-rich fluids
- Presents association models along with non-classical mixing rules for handling fluids with aqueous components
- Has an extended coverage of reservoir fluid communication, energy properties, and asphaltene precipitation
- Provides practical knowledge essential for achieving optimal design and cost-effective operations in a petroleum processing plant

This book offers engineers working in the energy sector a solid understanding of the phase behavior of the various fluids present in a petroleum reservoir.

Phase Behavior of Petroleum Reservoir Fluids

Third Edition

Karen Schou Pedersen, Peter Lindskou Christensen
and Jawad Azeem Shaikh

CRC CRC Press
Taylor & Francis Group
Boca Raton London New York

CRC Press is an imprint of the
Taylor & Francis Group, an **informa** business

Designed cover image: Karen Schou Pedersen, Peter Lindskou Christensen and Jawad Azeem Shaikh

Third edition published 2024
by CRC Press
2385 NW Executive Center Drive, Suite 320, Boca Raton FL 33431

and by CRC Press
4 Park Square, Milton Park, Abingdon, Oxon, OX14 4RN

CRC Press is an imprint of Taylor & Francis Group, LLC

© 2024 Taylor & Francis Group, LLC

First edition published by CRC Press 2007
Second edition published by CRC Press 2015

Library of Congress Cataloging-in-Publication Data
Names: Pedersen, K. S. (Karen Schou) author. | Christensen, Peter L. (Peter Lindskou) author. | Shaikh, Jawad Azeem, author.
Title: Phase behavior of petroleum reservoir fluids / Karen Schou Pedersen, Peter Lindskou Christensen and Jawad Azeem Shaikh.
Description: Third edition. | Boca Raton, FL : CRC Press, 2024. | Includes bibliographical references and index.
Identifiers: LCCN 2023031173 (print) | LCCN 2023031174 (ebook) | ISBN 9781138313811 (hbk) | ISBN 9781032642222 (pbk) | ISBN 9780429457418 (ebk)
Subjects: LCSH: Petroleum reserves—Mathematical models. | Petroleum—Geology—Mathematical models. | Phase rule and equilibrium. | Thermodynamic equilibrium. | Heavy oil. | Hydrocarbons.
Classification: LCC TN871 .P423 2024 (print) | LCC TN871 (ebook) | DDC 553.2/8—dc23/eng/20230826
LC record available at https://lccn.loc.gov/2023031173
LC ebook record available at https://lccn.loc.gov/2023031174

ISBN: 978-1-138-31381-1 (hbk)
ISBN: 978-1-032-64222-2 (pbk)
ISBN: 978-0-429-45741-8 (ebk)

DOI: 10.1201/9780429457418

Typeset in Times
by Apex CoVantage, LLC

Contents

Preface

New data and new models have become available since the previous edition of this book and contribute to an ever-improving understanding of the phase behavior of oil and gas mixtures. In that light, it is noteworthy that the van der Waals equation of state presented in 1873, with minor modifications, remains the standard in the petroleum industry for describing the properties and phase behavior of oils and gases. The equation expresses that the phase behavior in the entire gas and liquid range is determined by the location of the critical points of the individual components. Also, for a multicomponent reservoir fluid, the location of the critical point greatly influences the phase behavior of the fluid far away from the critical point. It is described in this edition how this observation can be used to simplify the determination of the optimum equation of state parameters. When the focus is on specific components or mixtures, other equations may be more accurate. The Span–Wagner equation was developed for pure carbon dioxide and the GERG-2008 equation for natural gas mixtures. For mixtures with aqueous components, the CPA equation is a good alternative to the non-classical mixing rules that have long been used to handle mixtures of water and hydrocarbon fluids.

The composition varies with the depth of a petroleum reservoir through the influence of gravity and a vertical temperature gradient. A recent observation is that for heavy oils, there is an additional contribution from viscosity.

Asphaltene precipitation from reservoir fluids has until recently not been fully understood, but thanks to extensive data for asphaltene precipitation from live oils becoming available, it has been possible to establish that an asphaltene phase can be described as a heavy liquid phase using a cubic equation of state. This has enabled the development of algorithms for the determination of a complete asphaltene phase diagram.

Wax data and modeling have long focused on the wax appearance point. Equally interesting is the amount of wax precipitated at lower temperatures, especially from live oils. We are pleased to be able to present that kind of data and would like to extend special thanks to Equinor for making such data available and for many years of fruitful collaboration with the authors of this book.

We are grateful for the many exciting tasks within phase equilibria we have had the opportunity to solve in collaboration with people in the oil industry, which have given substance and inspiration to this Third Edition of *Phase Behavior of Petroleum Reservoir Fluids*.

Karen Schou Pedersen
Peter Lindskou Christensen
Jawad Azeem Shaikh

Authors

Karen Schou Pedersen earned a PhD in liquid physics at the Technical University of Denmark. Her work with liquids at a fundamental level continued in positions as a research associate at the University of Edinburgh and at the nuclear research center Institute Laue-Langevin, Grenoble, France. She founded Calsep A/S in 1982, worked for Calsep as a senior principal consultant until 2022, and was responsible for software development and projects within reservoir fluid modeling and flow assurance. She is the author of more than 60 publications on the phase behavior of oil and gas mixtures. Her current position is Senior Principal Consultant at Kapexy Aps, a company specializing in reservoir fluid modeling.

Peter Lindskou Christensen earned a PhD at the Technical University of Denmark. In the 1980s, he was employed at Risø National Laboratories in Denmark, where he was responsible for PVT modeling for use in reservoir simulation studies. For a 10-year period in the 1990s, he was an associate professor at the Technical University of Denmark and taught thermodynamics, unit operations, and oil and gas technology. After leaving the university, he became a senior principal consultant at Calsep A/S, where he developed new techniques for matching experimental PVT data and was engaged in several fluid modeling projects. Peter is today a senior consultant at Kapexy Aps.

Jawad Azeem Shaikh earned an MSc in petroleum technology at the University of Pune in India. His current position is Sales Director and Principal Consultant of Calsep FZ LLC in Dubai where he has been responsible for several projects carried out in collaboration with the local oil industry within the design of enhanced oil recovery studies, equation of state modeling, and reservoir fluid communication. Before joining Calsep, he was an advanced studies supervisor for Core Laboratories International BV in Abu Dhabi. His job functions at Core Laboratories included technical support to field engineers to carry out fluid sampling as well as project coordination and data evaluation within PVT, compositional, and enhanced oil recovery projects. He has authored several papers and articles on sampling, PVT lab work, fluid communication, and equation of state modeling.

1 Petroleum Reservoir Fluids

1.1 RESERVOIR FLUID CONSTITUENTS

Petroleum reservoir fluids are multicomponent mixtures consisting primarily of hydrocarbons. Methane (CH_4) is the simplest of all hydrocarbons and also the most common component in petroleum reservoir fluids. Because methane contains one carbon atom, it is often referred to as C_1. Similarly, the term C_2 is used for ethane (C_2H_6), C_3 for propane (C_3H_8), and so on. Hydrocarbons with seven and more carbon atoms are called C_{7+} components, and the entity of all C_{7+} components is called the C_{7+} fraction. Petroleum reservoir fluids may contain hydrocarbons as heavy as C_{200}. A particular C_{7+} component will belong to one of the following component classes:

Paraffins: A paraffinic compound consists of hydrocarbon segments of the type C, CH, CH_2, or CH_3. The carbon atoms are connected by single bonds. Paraffins are divided into normal paraffins (*n*-paraffins) and iso-paraffins (i-paraffins). In an *n*-paraffin, the carbon atoms form straight chains, whereas an i-paraffin contains at least one side chain. Paraffins are sometimes also referred to as *alkanes*. Figure 1.1 shows the structure of methane (C_1), ethane (C_2), and *n*-hexane (nC_6), which are all examples of paraffinic compounds.

Naphthenes: These compounds are similar to paraffins in the sense that they are built of the same types of hydrocarbon segments, but they differ from paraffins in that they contain one or more cyclic structures. The segments in the ring structures (e.g., CH_2) are connected by single bonds. Most naphthenic ring structures contain six carbon atoms, but naphthenic compounds with either five or seven carbon atoms connected in ring structures are also common in petroleum reservoir fluids. Naphthenes are also called *cycloalkanes*. Cyclohexane and methyl cyclopentane shown in Figure 1.1 are examples of naphthenic components.

Aromatics: Similar to naphthenes, aromatics contain one or more cyclic structures, but the carbon atoms in an aromatic compound are connected by aromatic double bonds. Benzene (C_6H_6), the simplest aromatic component, is shown in Figure 1.1. Polycyclic aromatic compounds with two or more ring structures are also found in petroleum reservoir fluids. An example of the latter type of components is naphthalene ($C_{10}H_8$), whose structure is also shown in Figure 1.1.

The percentage of paraffinic (P), naphthenic (N), and aromatic (A) components in a reservoir fluid is often referred to as the PNA distribution.

Petroleum reservoir fluids may also contain inorganic compounds, of which nitrogen (N_2), carbon dioxide (CO_2), and hydrogen sulfide (H_2S) are the most common. Water (H_2O) is another important constituent of reservoir fluids. As water has limited miscibility with hydrocarbons and a higher density, most of the water in a reservoir is usually found in a separate water zone located beneath the gas and oil zones.

1.2 PROPERTIES OF RESERVOIR FLUID CONSTITUENTS

Table 1.1 shows selected physical properties of some constituents found in naturally occurring oil and gas mixtures. By comparing, for example, the normal boiling points, it is evident that hydrocarbons in petroleum reservoir fluids cover a wide range of component properties. At atmospheric pressure, pure methane will be in gaseous form at temperatures above the normal boiling point of

DOI: 10.1201/9780429457418-1

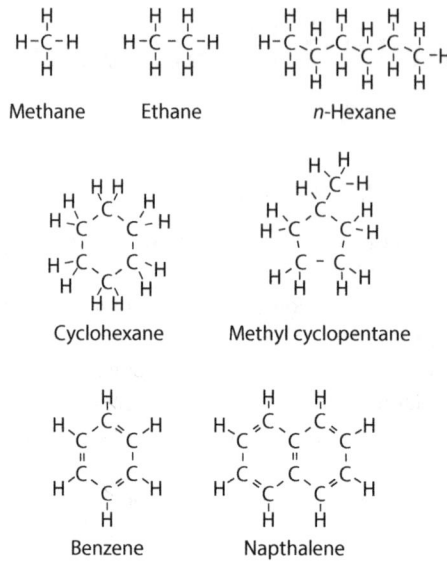

FIGURE 1.1   Molecular structures of some petroleum reservoir fluid constituents.

−161.6°C, whereas at the same pressure the temperature must be raised to 218.0°C before naphthalene evaporates. The properties of hydrocarbons with the same number of carbon atoms may also differ substantially. n-hexane (nC_6), methyl cyclopentane ($m\text{-}cC_5$), and benzene all contain six carbon atoms. However, the properties of these three components are quite different. For example, it can be seen in Table 1.1 that the density of nC_6 at atmospheric conditions is lower than that of $m\text{-}cC_5$, which component has a lower density than that of benzene. This suggests that densities of components with the same carbon number will increase in the order P → N → A. It is rare to see a measured PNA distribution, and, in the absence of experimental information about the predominant molecular structures, the trend in component densities may be used to give an idea about the distribution of P, N, and A components in a given C_{7+} fraction.

The pure component vapor pressures and critical points (CPs) are essential in calculations of component and mixture properties. The pure component vapor pressures are experimentally determined by measuring the corresponding values of temperature (T) and pressure (P) at which the substance undergoes a transition from liquid to gas. Figure 1.2 shows the vapor pressure curves of methane and benzene, both of which are common constituents of oil and gas mixtures. The vapor pressure curve ends at the CP, above which no liquid-to-gas phase transition can take place. The CP of methane is −82.6°C and 46.0 bar and that of benzene 289°C and 48.9 bar. The temperature at CP is called T_C and the pressure P_C.

As illustrated in the right-hand-side plot in Figure 1.3, the phase behavior of a pure component at a given temperature, T_1, may be studied by placing a fixed amount of this component in a cell at temperature T_1. The cell volume may be varied by moving the piston up and down. At position A, the cell contents are in the gaseous state. If the piston is moved downward, the volume will decrease, and the pressure increases. At position B, a liquid phase starts to form. By moving the piston further downward, the volume will further decrease, but the pressure remains constant until all gas is converted into liquid. This happens at position C. A further decrease in the cell volume will result in a rapidly increasing pressure. The left-hand-side curve in Figure 1.3 illustrates the phase changes when crossing a vapor pressure curve. A pure component can only exist in the form of two phases in equilibrium right at the vapor pressure curve. When the vapor pressure curve is reached, a conversion from either gas to liquid or liquid to gas will start. This phase

TABLE 1.1
Physical Properties of Common Petroleum Reservoir Fluid Constituents

Component	Formula	Molecular Weight (g/mol)	Melting Point (°C)	Normal Boiling Point (°C)	Critical Temperature (°C)	Critical Pressure (bar)	Acentric Factor	Density (g/cm³) at 1 atm and 20°C
Inorganics								
Nitrogen	N_2	28.013	−209.9	−195.8	147.0	33.9	0.040	—
Carbon dioxide	CO_2	44.010	−56.6	−78.5	31.1	73.8	0.225	—
Hydrogen sulfide	H_2S	34.080	−83.6	59.7	100.1	89.4	0.100	—
Paraffins								
Methane	CH_4	16.043	−182.5	−161.6	82.6	46.0	0.008	—
Ethane	C_2H_6	30.070	−183.3	−87.6	32.3	48.8	0.098	—
Propane	C_3H_8	44.094	−187.7	−42.1	96.7	42.5	0.152	—
Iso-butane	C_4H_{10}	58.124	−159.6	−11.8	135.0	36.5	0.176	—
n-butane	C_4H_{10}	58.124	−138.4	−0.5	152.1	38.0	0.193	—
Iso-pentane	C_5H_{12}	72.151	−159.9	27.9	187.3	33.8	0.227	0.620
n-pentane	C_5H_{12}	72.151	−129.8	36.1	196.4	33.7	0.251	0.626
n-hexane	C_6H_{14}	86.178	−95.1	68.8	234.3	29.7	0.296	0.659
Iso-octane	C_8H_{18}	114.232	−109.2	117.7	286.5	24.8	0.378	0.702 (16°C)
n-Decane	$C_{10}H_{22}$	142.286	−29.7	174.2	344.6	21.2	0.489	0.730
Naphthenes								
Cyclopentane	C_5H_{10}	70.135	−93.9	49.3	238.6	45.1	0.196	0.745
Methyl cyclopentane	C_6H_{12}	84.162	−142.5	71.9	259.6	37.8	0.231	0.754 (16°C)
Cyclohexane	C_6H_{12}	84.162	6.5	80.7	280.4	40.7	0.212	0.779
Aromatics								
Benzene	C_6H_6	78.114	5.6	80.1	289.0	48.9	0.212	0.885 (16°C)
Toluene	C_7H_8	92.141	−95.2	110.7	318.7	41.0	0.263	0.867
o-xylene	C_8H_{10}	106.168	−25.2	144.5	357.2	37.3	0.310	0.880
Naphthalene	$C_{10}H_8$	128.174	80.4	218.0	475.3	40.5	0.302	0.971 (90°C)

Source: Data from Reid, R.C., Prausnitz, J.M., and Sherwood, T.K. *The Properties of Gases and Liquids*, McGraw-Hill, New York, 1977.

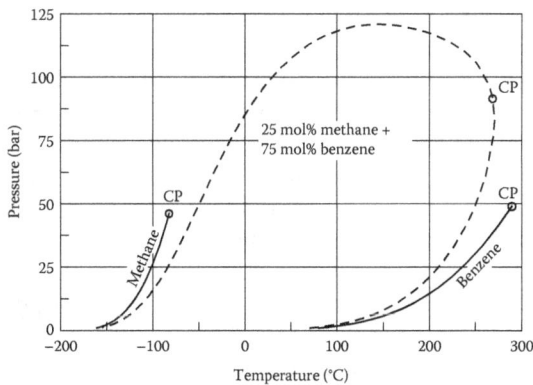

FIGURE 1.2 Vapor pressure curves of methane and benzene (full-drawn line). Phase envelope (dashed line) of a mixture of 25 mol% methane and 75 mol% benzene calculated using the Soave–Redlich–Kwong equation of state as presented in Chapter 4. CP stands for critical point.

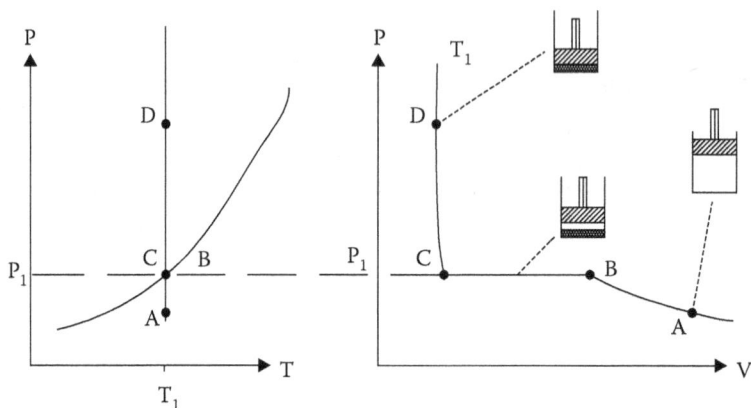

FIGURE 1.3 Pure component phase behavior in PT (left) and PV (right) diagrams.

transition is associated with volumetric changes at constant T and P. At the point B, the component is said to be at its dew point or in the form of a saturated gas. At position C, the component is at its bubble point or in the form of a saturated liquid. At position A, the state is undersaturated gas, and at D, it is undersaturated liquid.

Another important property is the acentric factor, ω, as defined by Pitzer (1955):

$$\omega = -1 - \log_{10}\left(\frac{P^{sat}}{P_c}\right)_{T=0.7T_c} \tag{1.1}$$

where P^{sat} stands for vapor pressure (or saturation pressure). The idea behind this definition is outlined in Figure 1.4. A plot of the logarithm of the reduced pure component vapor pressure, $P_r^{sat} = P^{sat}/P_c$, against the reciprocal of the reduced temperature, $T_r = T/T_c$, will for most pure substances give an approximately straight line. Figure 1.4 shows plots of $\log_{10} P_r^{sat}$ versus $1/T_r$ for argon (Ar) and n-decane (nC$_{10}$). For $T_r = 0.7$ ($1/T_r = 1.43$), $\log_{10} P_r^{sat} = -1.0$ for argon and -1.489 for nC$_{10}$. Argon is used as a reference and assigned an acentric factor of 0. In general, the acentric factor of a component equals the difference between $(\log_{10}P_r^{sat})_{T_r=0.7}$ for argon and for the actual substance. With this definition, the acentric factor of nC$_{10}$ equals $[-1 - (-1.489)] = 0.489$, which is consistent with the acentric factor given for nC$_{10}$ in Table 1.1.

The acentric factor got its name because the acentric factor of n-paraffins increases with carbon number. Methane (C$_1$) has an acentric factor of 0.008, ethane (C$_2$) 0.098, propane (C$_3$) 0.152, and so on. With increasing carbon number, molecules of this component class get more elongated (less spherical). More fundamentally, the acentric factor can be seen as a measure of the curvature of the pure component vapor pressure curve. Figure 1.5 shows the vapor pressure curves of three hypothetical substances, all having the same critical temperature and pressure as nC$_{10}$ (344.5°C and 21.1 bar), whereas the acentric factors are 0.0, 0.5, and 1.0, respectively (the acentric factor of nC$_{10}$ is 0.489). With the CP locked, the vapor pressure curves are bound to end at the same point, whereas the bend on the curve is determined by the acentric factor. For an acentric factor of 1.0, the vapor pressure curve is relatively flat at low temperatures and then increases steeply when approaching the critical temperature. If the acentric factor is lower, a more even increase is seen in vapor pressure with temperature. The vapor pressure curves in Figure 1.5 have been calculated using the Peng–Robinson equation of state presented in Chapter 4.

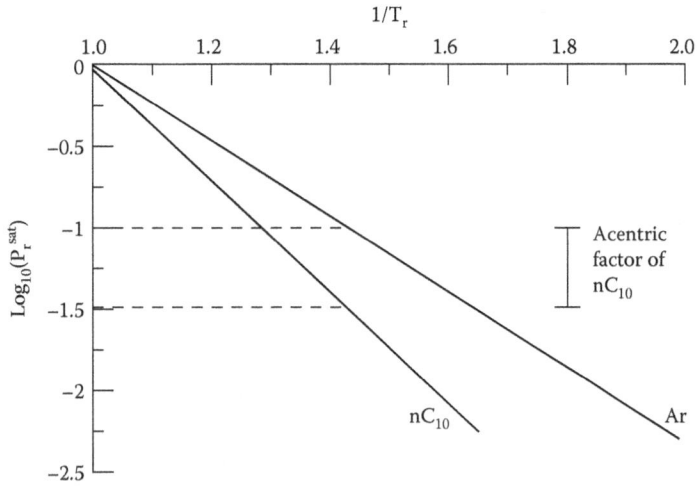

FIGURE 1.4 Acentric factor of nC_{10} from vapor pressure curves of Ar and nC_{10}. P_r^{sat} is the reduced saturation point (P^{sat}/P_c), and T_r the reduced temperature (T/T_c).

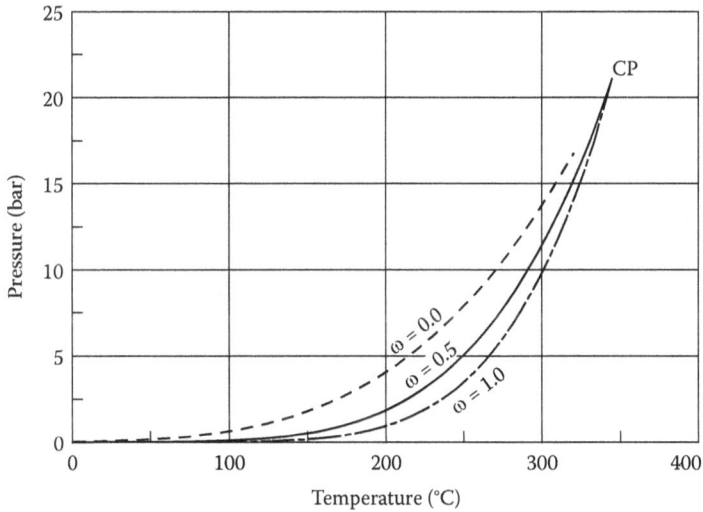

FIGURE 1.5 Vapor pressure curves of components with the same T_c and P_c as nC_{10} and acentric factors of 0.0, 0.5, and 1.0.

1.3 PHASE ENVELOPES

Petroleum reservoir fluids are multicomponent mixtures; therefore, it is of much interest to find a mixture equivalent of the pure component vapor pressure curve. With two or more components present, the two-phase region is not restricted to a single line in a PT diagram. As is illustrated in Figure 1.2 for a mixture of 25 mol% methane and 75 mol% benzene, the two-phase region for a mixture forms a closed area in P and T. The line surrounding this area is called the *phase envelope*.

Figure 1.6 shows the phase envelope for a natural gas mixture with the composition given in Table 1.2. The phase envelope consists of a dew point branch and a bubble point branch meeting at the CP of the mixture. On the dew point branch, the mixture is in gaseous form and in equilibrium with an incipient amount of liquid. Under these conditions, the gas (or vapor) is said to be saturated.

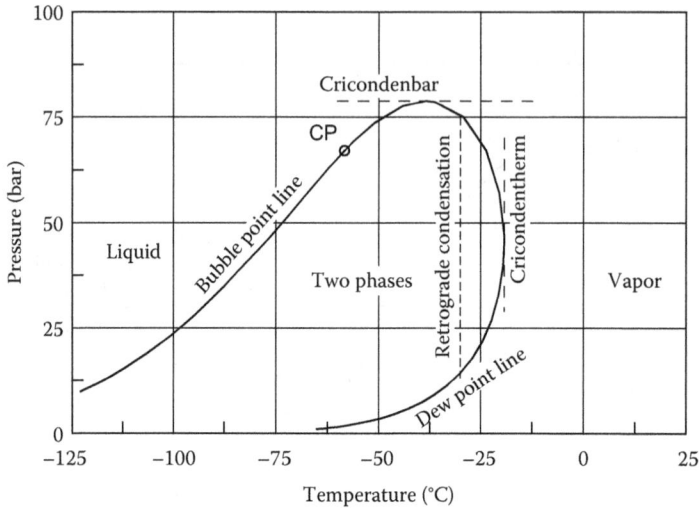

FIGURE 1.6 Phase envelope of natural gas in Table 1.2. CP stands for critical point. The phase envelope has been calculated using the Soave–Redlich–Kwong equation of state as presented in Chapter 4.

TABLE 1.2

Composition of Natural Gas Mixtures

Component	Mole Percentage
N_2	0.340
CO_2	0.840
C_1	90.400
C_2	5.199
C_3	2.060
iC_4	0.360
nC_4	0.550
iC_5	0.140
nC_5	0.097
C_6	0.014

Note: Phase envelope is shown in Figures 1.6 and 1.7.

At higher temperatures and the same pressure, there is no liquid present. By contrast, the gas may take up liquid components without liquid precipitation taking place. The gas is therefore said to be undersaturated. On the bubble point branch, the mixture is in liquid form and in equilibrium with an incipient amount of gas, and the liquid is said to be saturated. At lower temperatures and the same pressure, the liquid (or oil) is undersaturated. At the CP, two identical phases are in equilibrium, both having a composition equal to the overall composition. At temperatures close to the critical temperature and pressures above the critical pressure, there is only one phase present, but it can be difficult to tell whether it is a gas or a liquid. The term *super-critical fluid* is often used. Phase identification in the super-critical region is discussed in more detail in Chapter 6. The highest pressure at which two phases can exist is called the *cricondenbar*, and the highest temperature with two phases present is called the *cricondentherm*.

The phenomenon called *retrograde condensation* is illustrated in Figure 1.6 as a dashed vertical line at T = −30°C. At this temperature, the mixture is in gaseous form at pressures above the upper dew point pressure, that is, at pressures above approximately 75 bar. At lower pressures, the mixture will split into two phases, a gas and a liquid. Liquid formation taking place as the result of falling pressure is called retrograde condensation. If the pressure at a constant temperature is decreased to below the lower dew point pressure of approximately 15 bar, the liquid phase will disappear, and the entire mixture will be in gaseous form again.

1.4 CLASSIFICATION OF PETROLEUM RESERVOIR FLUIDS

Petroleum reservoir fluids may be divided into:

Natural gas mixtures
Gas condensate mixtures
Near-critical mixtures or volatile oils
Black oils
Heavy oils

Tables 1.2–1.7 give examples of each of these fluid types. The various fluid types are distinguished by the position of the mixture's critical temperature relative to the reservoir temperature.

TABLE 1.3
Composition of Gas Condensate Mixtures

Component	Mole Percentage	Molecular Weight	Density (g/cm³) at 1 atm and 15°C
N_2	0.53	—	—
CO_2	3.30	—	—
C_1	72.98	—	—
C_2	7.68	—	—
C_3	4.10	—	—
iC_4	0.70	—	—
nC_4	1.42	—	—
iC_5	0.54	—	—
nC_5	0.67	—	—
C_6	0.85	—	—
C_7	1.33	91.3	0.746
C_8	1.33	104.1	0.768
C_9	0.78	118.8	0.790
C_{10}	0.61	136	0.787
C_{11}	0.42	150	0.793
C_{12}	0.33	164	0.804
C_{13}	0.42	179	0.817
C_{14}	0.24	188	0.830
C_{15}	0.30	204	0.835
C_{16}	0.17	216	0.843
C_{17}	0.21	236	0.837
C_{18}	0.15	253	0.840
C_{19}	0.15	270	0.850
C_{20+}	0.80	391	0.877

Note: Phase envelope is shown in Figure 1.7.

TABLE 1.4
Composition of Near-Critical Mixtures

Component	Mole Percentage	Molecular Weight	Density (g/cm³) at 1 atm and 15°C
N_2	0.46	—	—
CO_2	3.36	—	—
C_1	62.36	—	—
C_2	8.90	—	—
C_3	5.31	—	—
iC_4	0.92	—	—
nC_4	2.08	—	—
iC_5	0.73	—	—
nC_5	0.85	—	—
C_6	1.05	—	—
C_7	1.85	95	0.733
C_8	1.75	106	0.756
C_9	1.40	121	0.772
C_{10}	1.07	135	0.791
C_{11}	0.84	150	0.795
C_{12}	0.76	164	0.809
C_{13}	0.75	177	0.825
C_{14}	0.64	190	0.835
C_{15}	0.58	201	0.841
C_{16}	0.50	214	0.847
C_{17}	0.42	232	0.843
C_{18}	0.42	248	0.846
C_{19}	0.37	256	0.858
C_{20+}	2.63	406	0.897

Note: Phase envelope is shown in Figure 1.7.

TABLE 1.5
Composition of Black Oil Mixtures

Component	Mole Percentage	Molecular Weight	Density (g/cm³) at 1 atm and 15°C
N_2	0.04	—	—
CO_2	0.69	—	—
C_1	39.24	—	—
C_2	1.59	—	—
C_3	0.25	—	—
iC_4	0.11	—	—
nC_4	0.10	—	—
iC_5	0.11	—	—
nC_5	0.03	—	—
C_6	0.20	—	—
C_7	0.69	85.2	0.769
C_8	1.31	104.8	0.769
C_9	0.75	121.5	0.765
C_{10+}	54.89	322.0	0.936

Note: Phase envelope is shown in Figure 1.7.

TABLE 1.6
Composition of Near-Critical Chinese Reservoir Fluids

Component	Mole Percentage	Molecular Weight	Density (g/cm³) at 1 atm and 15°C
N_2	3.912	—	—
CO_2	0.750	—	—
C_1	70.203	—	—
C_2	9.220	—	—
C_3	2.759	—	—
iC_4	0.662	—	—
nC_4	0.981	—	—
iC_5	0.402	—	—
nC_5	0.422	—	—
C_6	0.816	—	—
C_{7+}	9.873	192.8	0.8030

Note: The phase diagram is tabulated in Table 1.7 and shown in Figure 1.8.

TABLE 1.7
Composition of Biodegraded Heavy Oil. The Shown Molecular Weight and Density Are of the Total C_{7+} Fraction

Component	Mole Percentage	Molecular Weight	Density (g/cm³) at 1 atm and 15°C
N_2	0.600		
CO_2	0.445		
C_1	32.302		
C_2	0.186		
C_3	0.064		
iC_4	0.048		
nC_4	0.032		
$C_{7(+)}$	0.027	412.3	0.9678
C_8	0.048		
C_9	0.170		
C_{10}	0.441		
C_{11}	0.979		
C_{12}	1.485		
C_{13}	2.086		
C_{14}	2.476		
C_{15}	2.982		
C_{16}	2.721		
C_{17}	2.777		
C_{18}	2.878		
C_{19}	2.844		
C_{20}	2.664		
C_{21}	2.261		

(Continued)

TABLE 1.7   (*Continued*)

Composition of Biodegraded Heavy Oil. The Shown Molecular Weight and Density Are of the Total C$_{7+}$ Fraction

Component	Mole Percentage	Molecular Weight	Density (g/cm³) at 1 atm and 15°C
C$_{22}$	2.191		
C$_{23}$	2.053		
C$_{24}$	2.012		
C$_{25}$	1.803		
C$_{26}$	1.719		
C$_{27}$	1.657		
C$_{28}$	1.630		
C$_{29}$	1.554		
C$_{30}$	1.440		
C$_{31}$	1.399		
C$_{32}$	1.320		
C$_{33}$	1.224		
C$_{34}$	0.980		
C$_{35}$	0.915		
C$_{36+}$	17.586		

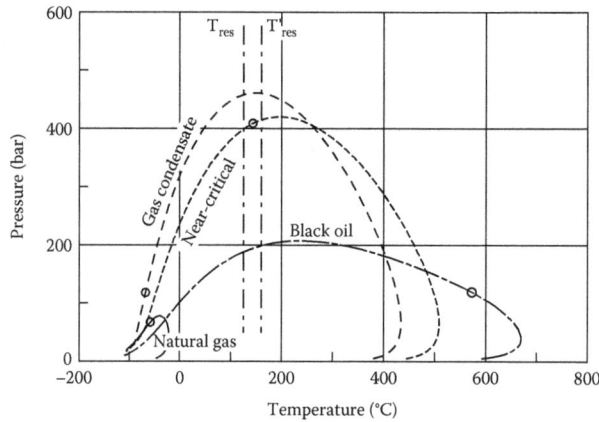

FIGURE 1.7   Phase envelopes of various types of reservoir fluids. The molar compositions are shown in Table 1.2 (natural gas), Table 1.3 (gas condensate), Table 1.4 (near-critical fluid), and Table 1.5 (black oil). The phase envelopes have been calculated using the Peng–Robinson equation of state presented in Chapter 4. The reservoir fluid compositions have been characterized using the procedure of Pedersen et al. presented in Chapter 5.

This is illustrated in Figure 1.7. During production of oil and gas from a reservoir, the temperature remains approximately constant at the initial reservoir temperature, T^{res}, whereas the pressure decreases as material is removed from the reservoir. For a natural gas, this pressure decrease will have no impact on the number of phases. The gas will remain single phase at all pressures. For a gas condensate, a decreasing pressure will, at some stage, lead to the formation of a second phase. This happens when the pressure reaches the dew point branch at temperature T^{res}. The second phase formed will be a liquid phase with a higher density than that of the original phase.

For a near-critical mixture, a pressure decrease will also, at some stage, lead to the formation of a second phase. If the reservoir temperature is T^{res}, as shown in Figure 1.7, the second phase will be a gas phase, because the point at which the phase envelope is reached is located at the bubble point branch. Such a mixture is classified as a volatile oil. If the reservoir temperature is slightly higher, as indicated by T'_{res} in Figure 1.7, the entry into the two-phase region will take place in the dew point branch, and the mixture is classified as a gas condensate. Near-critical reservoir fluids are mixtures with critical temperatures close to the reservoir temperature. Inside the phase envelope, the compositions and properties of the gas and liquid phases are similar.

Finally, for black oils and heavy oils, entry into the two-phase region at the reservoir temperature will always take place at the bubble point branch, and, accordingly, the new phase formed is a gas.

Figure 1.8 shows a close-up of the near-critical region of a Chinese reservoir fluid (Yang et al. 1997), whose composition is given in Table 1.6. The liquid volume percentages plotted in Figure 1.8 are tabulated in Table 1.8. Figure 1.8 illustrates that the relative volumetric amounts of gas and

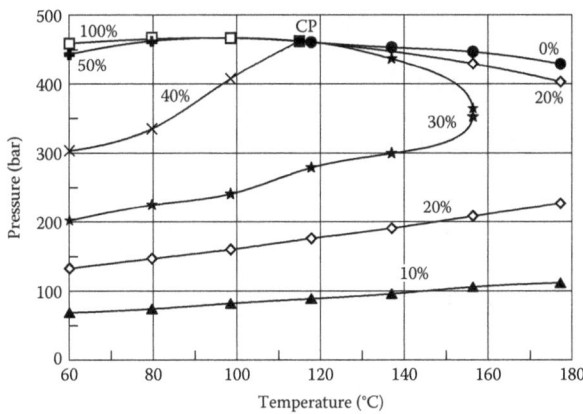

FIGURE 1.8 Near-critical part of phase envelope for the composition in Table 1.6 (Yang et al. 1997). The values stated are liquid volume percentages. The data points are tabulated in Table 1.8. CP stands for critical point.

TABLE 1.8
Pressure Tabulation Points for the Near-Critical Part of the Phase Envelope for Chinese Reservoir Fluid

Temperature (°C)	Pressures (in bar) at Specified Liquid Volume Percentages						
	0%	10%	20%	30%	40%	50%	100%
60.1	—	68.6	132.5	201.8	303.2	443.2	458.9
79.7	—	74.0	146.7	224.5	334.8	462.4	465.3
98.5	—	81.9	159.9	240.6	407.2	—	466.8
117.8	460.7	88.7	175.8	278.6	—	—	—
137.0	453.1	95.8	190.5	299.2, 436.4	—	—	—
156.4	446.6	105.7	208.2, 429.2	352.0, 364.0	—	—	—
177.2	—	111.7	226.6, 402.8	—	—	—	—

Source: Data from Yang et al., Phase behavior of near-critical reservoir fluid mixture, *Fluid Phase Equilib.* 128, 183–197, 1997.

Note: The molar composition is shown in Table 1.6. The critical point has been measured at 115°C and 462 bar. The results are plotted in Figure 1.7.

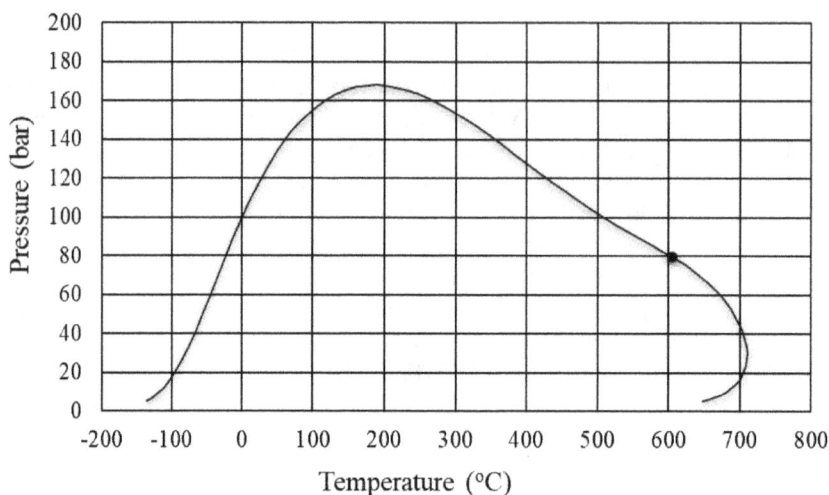

FIGURE 1.9 Phase envelope of the heavy biodegraded oil in Table 1.7. The phase envelope has been calculated using the Soave–Redlich–Kwong equation of state presented in Chapter 4. The fluid composition has been characterized using the heavy oil characterization procedure presented in Chapter 5.

liquid change rapidly with pressure and temperature in the vicinity of the CP. For example, at a temperature of around 100°C, only a marginal change in pressure is needed to change the liquid-phase amount from 50 vol% to 100 vol%.

The heavy oil in Table 1.7 is an example of a biodegraded oil. It lacks C_5–C_6 components and has low contents of C_7–~C_{15} components. Microorganisms have broken down the n-paraffins in that carbon number range. Figure 1.9 shows the simulated phase envelope of the heavy oil in Table 1.7. Because of the relatively low content of gaseous components, the bubble point pressures are relatively low. The high C_{7+} molecular weight signals a significant concentration of C_{100+} components, which makes the dew point temperatures higher than those of the black oil, as can be seen by comparing with Figure 1.7. Biodegraded oils are further dealt with in Section 5.7.1.

1.5 SHALE FLUIDS

Shale oil and gas are hydrocarbon reservoir fluids found trapped within shale formations. Oil shale has been known for centuries and has in the United States since the early twentieth century occasionally been mentioned as a potential additional energy resource. Extensive development of shale fluids started in the United States around 2000 and has since then increased almost exponentially. Shale fluids belong to the group of unconventional resources, but not because they chemically deviate from other petroleum reservoir fluids. They differ by being contained in very tight formations, so unconventional production strategies, including fracking, must be applied. Getting a representative sample is a challenge, and the produced fluid will often be depleted because of the pressure drop (draw down) in the near-well bore area. Major shale reservoirs in North America are Eagle Ford in Texas and Bakken in Montana and North Dakota and spreading into Canada. Other shale reservoirs that could potentially be recovered are located in Russia, China, Argentina, Libya, and Australia. Table 1.9 shows a shale gas condensate fluid from Argentina.

TABLE 1.9
Composition of Shale Gas Condensate Fluid

Component	Mole Percentage	Molecular Weight	Density (g/cm³) at 1 atm and 15°C
N_2	0.387		
CO_2	1.033		
C_1	62.955		
C_2	12.575		
C_3	6.768		
iC_4	1.428		
nC_4	2.658		
iC_5	0.955		
nC_5	0.996		
C_6	1.439		
C_7	1.707	96.7	0.709
C_8	2.186	109.4	0.730
C_9	1.205	122.1	0.750
C_{10}	0.736	134.8	0.767
C_{11}	0.523	147.4	0.783
C_{12}	0.403	160.1	0.797
C_{13}	0.365	172.8	0.810
C_{14}	0.297	185.5	0.822
C_{15}	0.252	198.2	0.833
C_{16}	0.200	210.9	0.843
C_{17}	0.188	223.6	0.851
C_{18}	0.138	236.3	0.859
C_{19}	0.121	248.9	0.866
C_{20+}	0.485	313.6	0.892

Source: Cismondi, M., Tassin, N.G., Canel, C. Rabasedas, F., and Gilardone, C., PVT experimental and modeling study of some shale reservoir fluids from Argentina, *Braz. J. Chem. Eng.* 35, 313–326, 2018.

REFERENCES

Cismondi, M., Tassin, N.G., Canel, C., Rabasedas, F., and Gilardone, C., PVT experimental and modeling study of some shale reservoir fluids from Argentina, *Braz. J. Chem. Eng.* 35, 313–326, 2018.

Pitzer, K.S., Volumetric and thermodynamic properties of fluids. I. Theoretical basis and virial coefficients, *J. Am. Chem. Soc.* 77, 3427–3433, 1955.

Reid, R.C., Prausnitz, J.M., and Sherwood, T.K., *The Properties of Gases and Liquids*, McGraw-Hill, New York, 1977.

Yang, T., Chen, W.-D., and Guo, T., Phase behavior of near-critical reservoir fluid mixture, *Fluid Phase Equilib.* 128, 183–197, 1997.

2 Sampling, Quality Control, and Compositional Analyses

The quality of an experimental pressure-volume-temperature (PVT) study depends heavily on the quality of the samples collected from the field. A PVT study will only reflect the reservoir fluid if performed on fluid samples that are representative of the fluid at down-hole conditions. Figure 2.1 shows the workflow from the time a fluid sample is taken until an equation of state (EoS) model is developed, which matches the observed fluid behavior and can be used as input to compositional reservoir, process, and flow simulations.

2.1 FLUID SAMPLING

To collect representative samples, the flow into the well must be single phase. That will only be the case if the saturation pressure of the reservoir fluid is lower than the reservoir pressure. It must further be taken into consideration that the bottom hole flowing pressure (BHFP) will in general be lower than the reservoir pressure. It is therefore a further requirement that the saturation pressure be lower than the BHFP at which the sample was taken.

The samples can be

Bottom hole samples (subsurface)
Separator samples (surface)
Wellhead samples

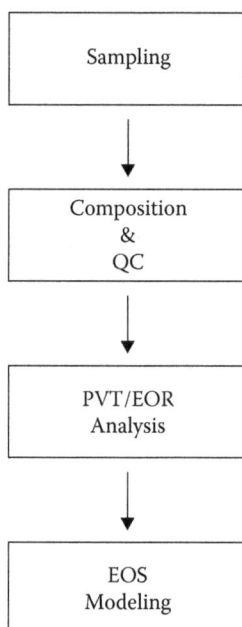

FIGURE 2.1   Workflow for reservoir fluid PVT study.

                                            DOI: 10.1201/9780429457418-2

Bottom hole sampling (subsurface) is well suited for undersaturated oil reservoirs, that is, oil reservoirs with a reservoir pressure exceeding the saturation pressure. A bottom hole sample is taken by lowering a single-phase sampler into the well using wire line technology, as sketched in Figure 2.2, and making sure the samples are captured at the desired depth. Subsurface sampling is appropriate for black oils, volatile oils, and dry gases. Because the samples are taken at or close to the reservoir pressure, it is the preferred technique for samples to undergo an asphaltene study. At a lower sampling pressure, asphaltenes might precipitate, and that would cause irreversible damage of the samples.

For gas condensate fluids, a bottom hole sample may be "contaminated" by liquid dropping out higher up in the well where the pressure is lower and the fluid has split into two phases. It is therefore advisable for gas condensate type of fluids to carry out surface or separator sampling. As illustrated in Figure 2.3, the reservoir fluid is let through a separator operating at a particular temperature and pressure. Separator oil and gas are sampled at the same time. This pair of separator samples is then recombined to the producing gas/oil ratio (GOR) to give a representative reservoir

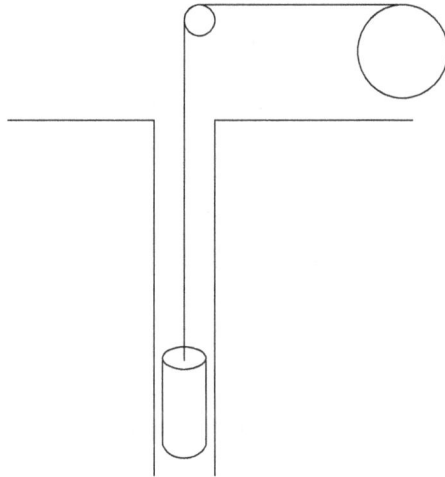

FIGURE 2.2 Bottom hole sampling technique. A single-phase sample cylinder is lowered into the well using wire line technology.

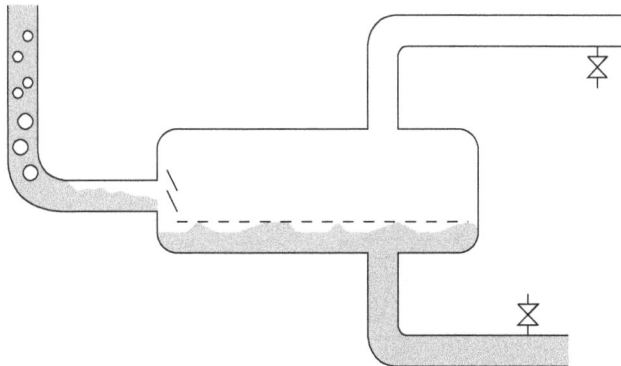

FIGURE 2.3 Separator sampling. Samples are taken at the gas and liquid outlets.

fluid composition. Separator sampling is applicable for black oils, volatile oils, gas condensates, wet gases, and dry gases.

Separator sampling is further recommended for reservoirs that have been in production long enough to have the reservoir pressure drop to below the saturation pressure of the original reservoir fluid (depleted reservoirs). For such reservoirs, bottom hole sampling is unlikely to provide a representative sample.

Wellhead sampling, as sketched in Figure 2.4, is the preferred method for any fluid that is single phase at the wellhead conditions. When that condition is fulfilled, wellhead sampling is a very reliable and cost-effective method of sampling.

No matter which sampling technique is used, it is important to condition the well properly. Well conditioning has the purpose of cleaning the well and reducing the pressure drop (drawdown) from the reservoir to the wellbore to a minimum. The area with a pressure drop is called the drainage area and is illustrated in Figure 2.5. Omitting well conditioning can result in two-phase flow in the

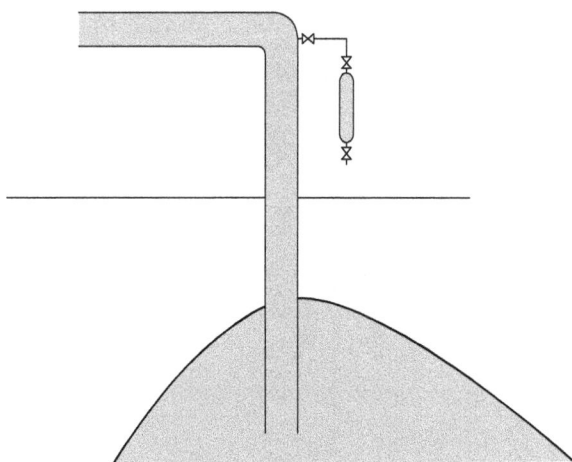

FIGURE 2.4   Wellhead sampling. A single-phase sample is taken at wellhead conditions.

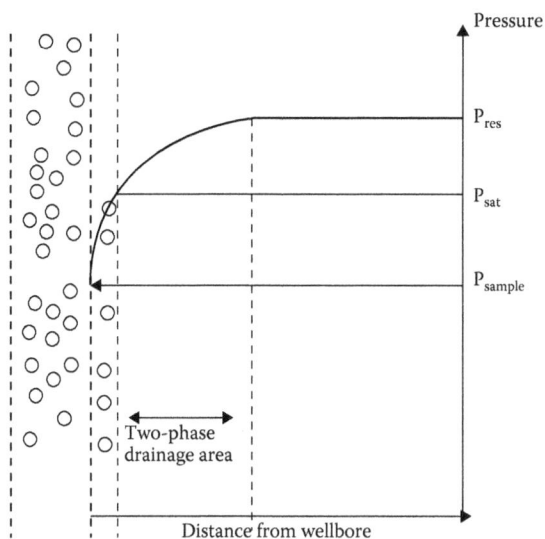

FIGURE 2.5   Drainage area near wellbore.

wellbore or near the perforations. The fluid entering into the wellbore will in that case be different from the original reservoir fluid in the reservoir outside the drainage area. Well conditioning is particularly important when the reservoir pressure is approaching the saturation pressure.

If bottom hole samples are to be collected, well conditioning is to be performed as follows:

Start off with a high flow rate to clean up the well. This will almost certainly increase the drainage area, and free gas will form.

To afterwards reduce the drainage area, flow the fluid at the lowest possible flow rate for about 4 days.

Shut in the well for a period of 1 week to force the released gas into solution in the oil, thus raising the saturation pressure and obtaining a representative fluid in the wellbore.

Before collecting separator samples, it is important to measure the gas and liquid flow rates continuously. The flow rate should be stable and as low as possible.

2.2 QUALITY CONTROL OF FLUID SAMPLES

The compositional analyses and fluid studies are carried out by a PVT laboratory, which will initially perform a sample validation (quality control) to check whether the samples represent the reservoir fluid. The actual validation process depends on whether bottom hole, wellhead, or separator samples have been collected.

For all types of samples, the sample chamber is initially heated to approximately 90°C to redissolve any wax that may have precipitated during sampling or shipping.

2.2.1 BOTTOM HOLE/WELLHEAD SAMPLES

The quality control of a bottom hole oil sample is carried out as follows:

The preferred type of chamber to collect bottom hole and wellhead samples has a mechanism for retaining the pressure high enough to also keep the fluid single phase at temperatures lower than the reservoir temperature. At the PVT laboratory, the opening pressure of the sample cylinder at ambient temperature is recorded. As schematically shown in Figure 2.6, this can be done by attaching a pressure gauge to the bottom of the chamber and then opening the bottom valve of the chamber to record the pressure. The opening pressure must comply with the shutting pressure at the sample site.

To detect whether the sample contains water, approximately 5 ml of sample is removed from each side. If water comes out, the fluid displacement is continued until reservoir fluid comes out.

Approximately 10 ml of sample is displaced and released to atmospheric conditions. The densities of different flashed oil samples are compared to check for consistency. The density may possibly also be compared to the density of oil samples from neighboring wells. A sample of a higher density than other samples from the same field could be contaminated by water.

An initial gas chromatographic (GC) analysis is carried out to check whether the fluid sample is contaminated by base oil from drilling mud. This analysis is called a fingerprint analysis. The purpose is not to determine the fluid composition but to look for peaks in the chromatogram originating from oil contained in drilling mud. If a sample of the drilling fluid exists, the chromatogram of the reservoir fluid sample may be compared with that of the drilling fluid. Any similarity in peaks may indicate that the sample is contaminated. Section 2.4.1 has more on GC analyses and Section 2.7 more on mud-contaminated samples.

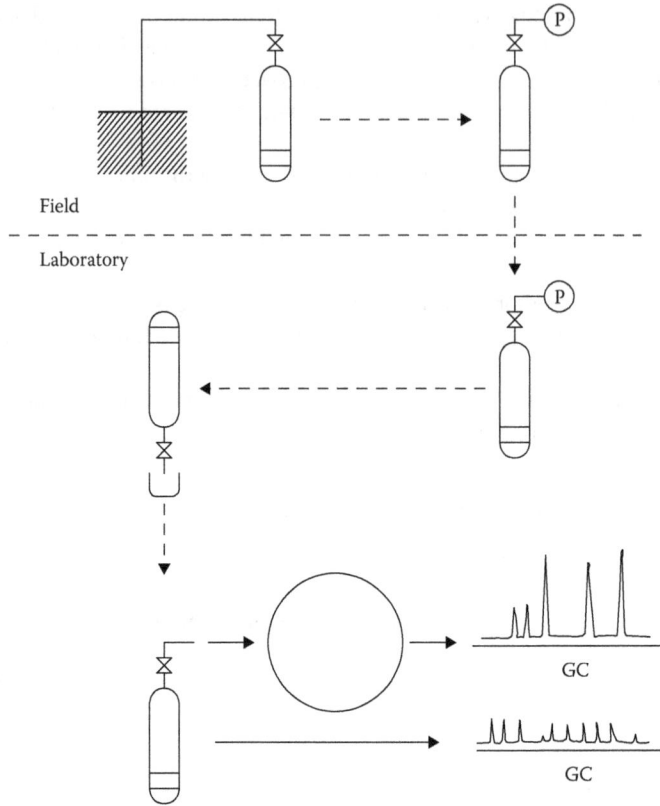

FIGURE 2.6 Quality control of bottom hole samples.

A partial constant composition expansion (CCE) test is performed to determine the saturation pressure of the fluid at the reservoir temperature. This test is performed by charging
approximately 50 ml of sample to a high-pressure PVT cell. After stabilization at the reservoir temperature, the sample pressure is reduced until the saturation pressure is observed
either visually or as a sharp change in the slope of the pressure–volume (PV) curve. CCE
experiments are further dealt with in Section 3.1.1. The saturation pressure may be known
from other sources or from other bottom hole samples taken from the same reservoir. If the
saturation point of the actual sample deviates significantly from the established saturation
point, the sample quality may be questionable.

The sample is flashed to atmospheric pressure and 15°C. The volumes of the flashed gas and
residual liquid are measured. The residual liquid is also called stock tank oil (STO). The
GOR at atmospheric conditions is called the single-stage GOR. The ratio of the oil volume
at reservoir and atmospheric conditions is called shrinkage. The measured GOR, shrinkage, and STO density should be consistent between samples and agree reasonably with
data for neighboring fields. The compositions of the evolved gas and liquid are analyzed
using GC (see Section 2.4.1) and checked for impurities from atmospheric air.

2.2.2 SEPARATOR SAMPLES

A separator sample consists of an oil and a gas sample taken from a separator, both at the same time.
The process the separator sample undergoes from field to compositional analysis in the PVT laboratory is shown schematically in Figure 2.7. The samples must be collected at equilibrium conditions

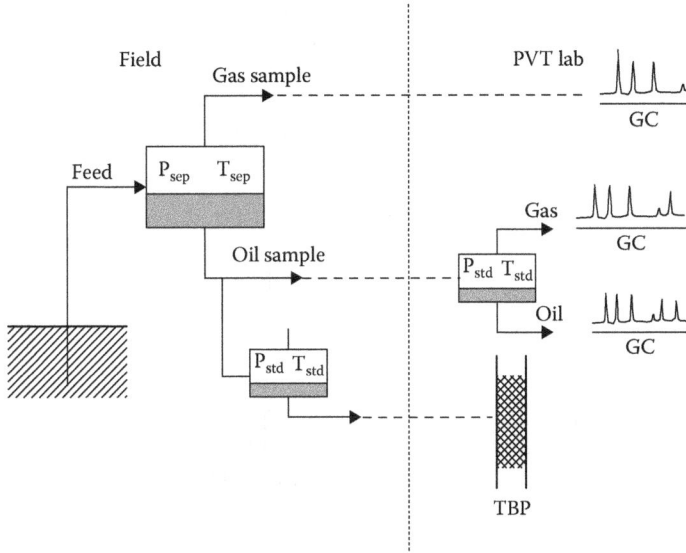

FIGURE 2.7 Handling of separator samples from field to compositional analysis.

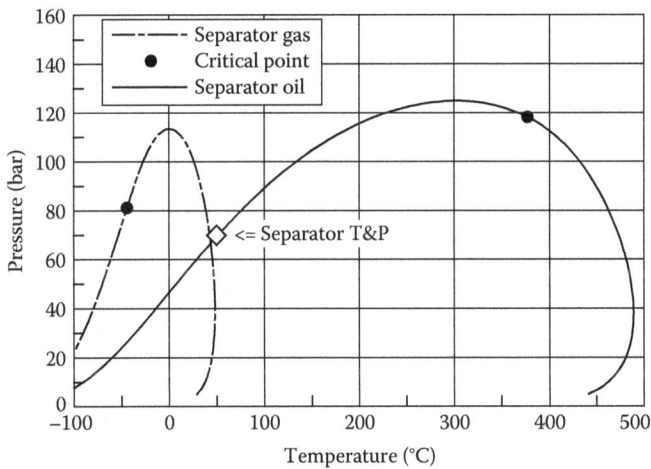

FIGURE 2.8 Phase envelopes of a separator gas and separator oil. The separator conditions of 50°C and 70 bar are marked on the figure.

at which the separator gas is at its dew point and the separator oil at its bubble point. As illustrated in Figure 2.8, this means that the phase envelopes of the separator gas and liquid will intersect at the separator conditions.

2.2.2.1 Quality Control of Separator Gas
Figure 2.9 shows how a separator gas is quality controlled (QCed).

The laboratory will measure the opening pressure to see whether it complies with the shutting pressure at the field. Because the sample cylinder is opened at a lower temperature than the separator temperature, the opening pressure will be lower than the shutting pressure.

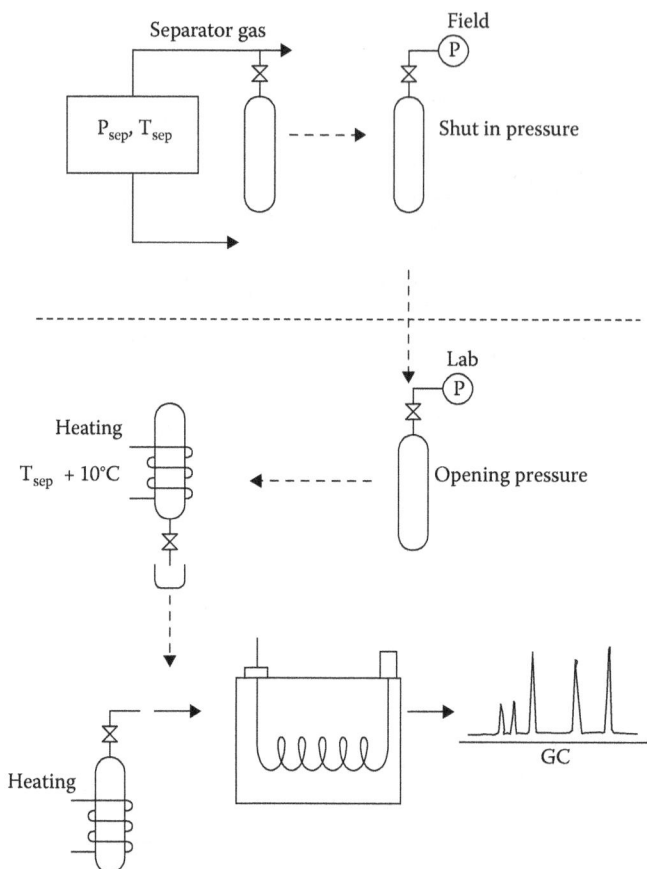

FIGURE 2.9   QC of separator gas sample.

A quantitative check of compliance between the shutting and opening pressure can be performed using a PVT simulation package. A pressure–temperature (PT) flash calculation at separator conditions will give the molar volume of the fluid in the sample chamber. The opening pressure can be found by a volume–temperature (VT) flash for the laboratory temperature. Flash calculations are further dealt with in Chapter 6.

If heated to above the separator temperature, the separator gas sample is not to contain any liquid. This is checked by heating the gas cylinder to a temperature at least 10°C above the separator temperature. If the cylinder contains any liquid at these conditions, it suggests either liquid carryover from the separator liquid or that the sampling temperature was incorrect. It could also be that the cylinder was not cleaned before taking the sample. Such a sample must be discarded.

If the separator gas passes the quality check, its composition is measured using GC, as further described in Section 2.4.1.

2.2.2.2   QC of Separator Liquid

A separator liquid is QCed as shown in Figure 2.10.

At the laboratory, it is checked whether the opening pressure complies with the shutting pressure at the field. This can be done by subsequent PT and VT flash calculations, as for a separator gas.

The separator liquid cylinder is heated to least 10°C above the separator temperature. To
detect whether the sample contains water, approximately 5 ml of sample is removed. If
water comes out, the fluid displacement is continued until reservoir fluid starts coming out.
If the water content is more than 10%, the fluid sample is discarded.

The separator liquid should be at its bubble point at the separator temperature. The bubble point is determined by a partial CCE test (not shown in Figure 2.10) carried out
at separator temperature. The partial CCE is performed as for a bottom hole sample
(Section 2.2.1).

The sample is flashed to atmospheric pressure and 15°C. The volumes of the flashed gas and
residual liquid (STO) are measured. The compositions of the evolved gas and liquid are
analyzed using GC (see Section 2.4.1).

When the reservoir fluid composition is established, it is possible to carry out a PT flash
calculation at separator conditions using a PVT simulation package. The gas and oil compositions from the flash should ideally be identical to those of the sampled separator gas
and liquid. It is often found to be more convenient to compare the K-factors than phase
compositions. The K-factor of component i is the ratio of the mole fraction of component
i in the separator gas (y_i) and the mole fraction of component i in the separator liquid (x_i).
A plot of the simulated K-factors against the experimental ones should give a straight line
expressed as $y = x$. An example of such a plot is shown in Figure 2.11.

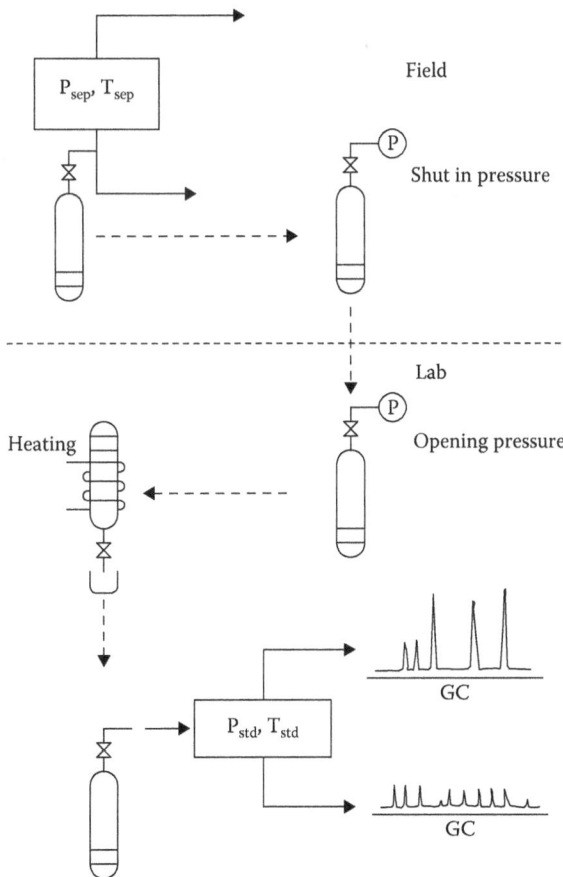

FIGURE 2.10 QC of separator liquid sample.

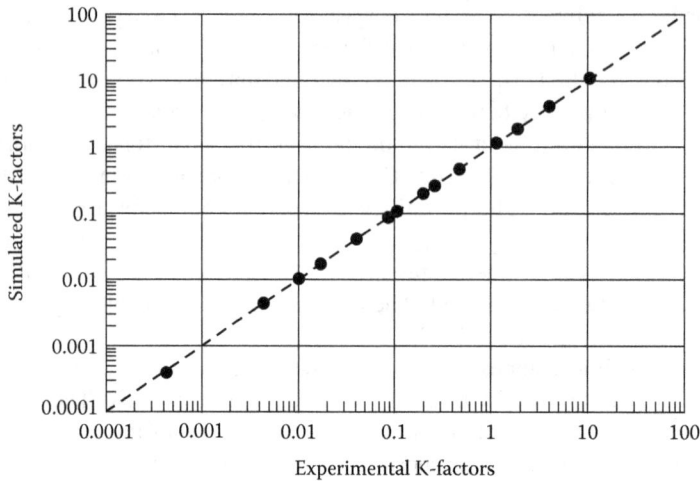

FIGURE 2.11 Simulated versus experimental K-factors for separator compositions used to check whether separator compositions are in equilibrium. The quality check is successful if an approximately straight line is seen expressed as $y = x$.

2.3 HOFFMANN PLOT

The Hoffmann plot (Hoffmann et al. 1953) is much used in oil industry as a method to validate whether gas and liquid separator samples are in equilibrium at separator conditions. As outlined in the following, the Hoffmann plot is based on simple and approximate correlations that are otherwise no longer used in the oil industry.

Table 2.1 shows the composition of a volatile oil reservoir fluid obtained by recombining the shown gas and liquid compositions from a separator operating at 69 bar and 65°C. The recombined reservoir fluid composition will only be representative of the reservoir fluid if the separator gas and separator liquid compositions are in thermodynamic equilibrium at separator conditions. The Hoffmann plot is intended to clarify whether that is the case.

It assumed that a linear relation exists for

$$ln\left(K_i \times P\right) \; vs. \; b_i\left(\frac{1}{T_{Bi}} - \frac{1}{T}\right) \tag{2.1}$$

where i is a component index and

$$b_i = \frac{ln\left(P_{ci}\right)}{\dfrac{1}{T_{Bi}} - \dfrac{1}{T_{ci}}} \tag{2.2}$$

The remaining quantities are

K K-factor (ratio of component mole fractions in gas and liquid)
P Pressure in atm
P_c Critical pressure in atm
T Temperature in K
T_B Boiling temperature at atmospheric pressure in K
T_c Critical temperature in K

TABLE 2.1

Recombined Reservoir Fluid and Gas and Liquid Compositions from a Separator at 69 bar and 65°C. The Recombined Fluid has a C_{7+} Molecular Weight of 181.5 and a C_{7+} Density of 0.824 g/cm³

Component	Recombined Reservoir Fluid Mol %	Separator Gas Mol%	Separator Liquid Mol%	Boiling Point Temperature K
N_2	1.775	2.320	0.226	77.4
CO_2	2.929	3.390	1.620	194.6
H_2S	11.450	10.669	13.667	213.5
C_1	61.314	76.443	18.340	111.6
C_2	2.707	2.873	2.235	184.6
C_3	2.145	1.786	3.166	231.1
iC_4	0.772	0.498	1.550	261.4
nC_4	1.521	0.851	3.426	272.7
iC_5	0.872	0.332	2.407	301.0
nC_5	0.872	0.289	2.527	309.2
C_6	1.470	0.268	4.885	341.9
C_7	1.666	0.134	5.871	
C_8	1.783	0.083	6.529	
C_9	1.520	0.040	5.609	
C_{10+}	7.204	0.025	27.944	

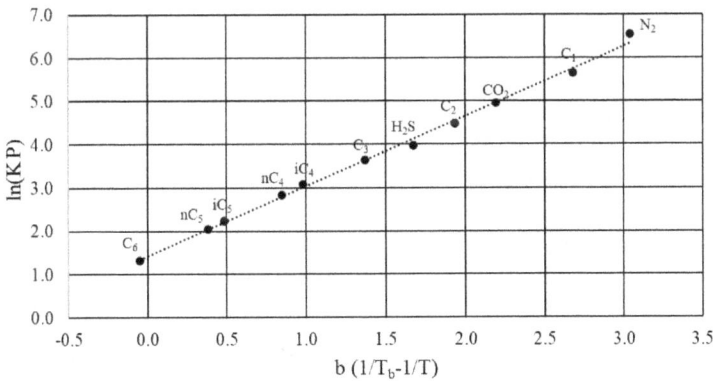

FIGURE 2.12 Hoffmann plot for the separator gas and oil compositions in Table 2.1. The separator conditions are 69 bar and 65°C.

A Hoffmann plot to C_6 for the separator gas and liquid compositions in Table 2.1 is shown in Figure 2.12. A perfect Hoffmann plot would have all the dots on the dashed line, which criterion is not completely fulfilled in Figure 2.12.

To arrive at the linearity in Equation 2.1, the pure component vapor pressure of component i is assumed to follow the simplified Antoine equation (Antoine 1888)

$$ln_{10}\left(P_i^{sat}\right) = a_i - \frac{b_i}{T} \tag{2.3}$$

where a_i and b_i are constants specific for component i.

At the normal boiling point of component i, T_{Bi}, the vapor pressure, equals atmospheric pressure (1 atm), giving

$$a_i = \frac{b_i}{T_{Bi}} \tag{2.4}$$

This allows the simplified Antoine equation to be rewritten to

$$ln_{10}\left(P_i^{sat}\right) = b_i\left(\frac{1}{T_{Bi}} - \frac{1}{T}\right) \tag{2.5}$$

At the critical point of component i, the vapor pressure equals P_{ci}, and the temperature is T_{ci}. This gives the following expression for b_i

$$b_i = \frac{ln_{10}\left(P_{ci}\right)}{\frac{1}{T_{Bi}} - \frac{1}{T_{ci}}} \tag{2.6}$$

For an ideal liquid mixture and an ideal gas, the K-factor of component i can be calculated from (Raoult 1886):

$$y_i P = x_i P_i^{sat} \Rightarrow P_i^{sat} = \frac{y_i}{x_i}P = K_i \times P \tag{2.7}$$

which means

$$ln\left(K_i \times P\right) = ln\left(P_i^{sat}\right) \tag{2.8}$$

An ideal mixture is one in which the interaction between two molecules of the same chemical species is the same as between two molecules of different chemical species.

Combining Equation 2.8 and Equation 2.5 gives

$$ln\left(K_i \times P\right) = b_i\left(\frac{1}{T_{Bi}} - \frac{1}{T}\right) \tag{2.9}$$

This is a special case of the Hoffmann relation, where the straight line assumed in the Hoffmann method follows the equation $y = x$. The right-hand side is independent of pressure, and a pressure correction is needed to arrive at the final Hoffmann relation.

Poynting (1905) introduced a pressure correction through what he called a modified saturation pressure, as defined in the following. It was later known as the pure component fugacity

$$ln\left(P_i^{sat,modified}\right) = Poynt_i \times ln\left(P_i^{sat}\varphi_i^{sat}\right) \tag{2.10}$$

φ_i^{sat} is the fugacity coefficient at the saturation pressure. The term fugacity coefficient is explained in Appendix A. Replacing P_i^{sat} in Equation 2.8 with $P_i^{sat,modified}$ gives

$$ln\left(K_i \times P\right) = Poynt_i \times ln\left(\varphi_i^{sat}\right) + Poynt_i \times ln\left(P_i^{sat}\right) \tag{2.11}$$

Assume

$$Con\_A = Poynt_i \times ln\left(\varphi_i^{sat}\right) \tag{2.12}$$

$$Con\_B = Poynt_i \tag{2.13}$$

where *Con_A* and *Con_B* are constants independent of the component index. Combined with Equation 2.5, that gives

$$ln\left(K_i \times P\right) = Con\_A + Con\_B \times \left(b_i\left(\frac{1}{T_{Bi}} - \frac{1}{T}\right)\right) \tag{2.14}$$

which reproduces Equation 2.1.

Several questionable approximations have been made, which makes the Hoffmann relation uncertain:

- The pure component vapor pressure does not in general follow the simplified Antoine equation.
- A separator gas is not an ideal gas.
- A separator oil is not an ideal mixture.
- The Poynting correction is an approximation and component dependent.

At first glance, the Hoffmann plot in Figure 2.12 might give the impression that the assumed linear relationship is close to being fulfilled, but this is not the case. Table 2.2 shows the composition of the sampled separator gas and of the gas composition according to the Hoffman relation; that is, Table 2.2 shows what the plot would look like if all the dots in Figure 2.12 had been on the dashed line. The concentrations of the gas components according to the Hoffmann relation deviate significantly from the measured ones.

TABLE 2.2

Measured Separator Gas Composition at 69 bar and 65°C and the Separator Gas Composition (Hoffmann Gas) That Obeys the Hoffmann Relation

Component	Separator Gas Mol%	Hoffmann Gas Mol%	%Dev
N_2	2.32	1.735	−25.2
CO_2	3.39	3.015	−11.1
H_2S	10.669	10.670	0.0
C_1	76.443	78.383	2.5
C_2	2.873	2.642	−8.0
C_3	1.786	1.497	−16.2
iC_4	0.498	0.387	−22.3
nC_4	0.851	0.698	−18.0
iC_5	0.332	0.256	−22.8
nC_5	0.289	0.221	−23.5
C_6	0.268	0.212	−20.9

The Hoffmann method was good practice at a time when computers and cubic equations of state had not yet made their way into oil industry. Cubic equations were known at the time (1953) Hoffmann et al. presented their method, but phase equilibrium calculations using cubic equations were out of reach until computers in the 1970s became a widely used engineering tool in the oil industry.

With a cubic equation, it is straightforward to perform a PT flash calculation of the recombined reservoir fluid composition at separator conditions and compare the simulated K-factors with those derived from the sampled separator compositions. A plot of experimental K-factors versus simulated K-factors should give a straight line with $y = x$, as illustrated in Figure 2.11. As illustrated in Figure 2.8, the phase envelopes of the separator gas and separator liquid are to intersect at the separator P and T, which can be used as a further quality check.

2.4 COMPOSITIONAL ANALYSES

Oil and gas condensate mixtures consist of thousands of different components, which makes it almost impossible to carry out a complete component analysis. Instead, components heavier than nC_5 are grouped into boiling point fractions, often referred to as *carbon number fractions*. Table 2.3 shows the boiling point temperatures separating the carbon number fractions (Katz and Firoozabadi 1978). The cut points are determined from the boiling points (T_B) of *n*-paraffins. The C_7 fraction, for example, consists of the hydrocarbons with a boiling point from 0.5°C above T_B of nC_6 to 0.5°C above T_B of nC_7. The C_8 fraction consists of the hydrocarbons with a boiling point from 0.5°C above T_B of nC_7 to 0.5°C above T_B of nC_8, and so on. Table 2.3 also shows generalized densities and molecular weights of each carbon number fraction up to C_{45}. These are based on a US study about gas condensate mixtures by Bergman et al. (1975).

Two standard analytical techniques are used in compositional analyses:

Gas chromatography (GC)
True boiling point (TBP) analysis or carbon number distillation

2.4.1 Gas Chromatography

2.4.1.1 Preparation of Oil Mixtures

The gas and liquid phases to be analyzed using GC are prepared using a spike flash technique. A spike flash apparatus is shown schematically in Figure 2.13 and consists of an oven with a flask inside where the reservoir fluid is flashed to atmospheric pressure and approximately 49°C (120°F). This will separate the single-phase fluid into a gas and an oil phase. The spike flash can also be performed at ambient temperature, but the oil phase is easier to handle if the separation takes place at an elevated temperature. The evolved gas resulting from the flash is collected in a coiled metal tube, which is placed in an oven at approximately 65°C (~150°F). A gas meter is attached to this coil, and the volume of gas collected is noted. The residual liquid from the flask and the evolved gas are analyzed separately.

2.4.1.2 Preparation of Gas Condensate Mixtures

Cryogenic distillation is an appropriate technique for preparing gas condensate samples for chromatographic analyses. It has an advantage over the spike flash technique because it requires less sample volume. The operating principle is sketched in Figure 2.14. A sub-sample of single phase reservoir fluid is pumped into a glass receiver, which is submerged in liquid nitrogen. Because of the low temperature, the entire sample will condense. This liquid phase is then allowed to warm to ambient temperature, and the evolved gas is collected into a cylinder also held in liquid nitrogen. This process is continued until the temperature is just above ambient temperature. The gas and liquid separated in this manner are analyzed separately.

TABLE 2.3
Generalized Properties of Petroleum Hexane Plus Groups

Carbon Number	Boiling Range (°C)	"Average" Boiling Point (°C)	Density (g/cm³)	Molecular Weight
C_6	36.5–69.2	63.9	0.685	84
C_7	69.2–98.9	91.9	0.722	96
C_8	98.9–126.1	116.7	0.745	107
C_9	126.1–151.3	142.2	0.764	121
C_{10}	151.3–174.6	165.8	0.778	134
C_{11}	174.6–196.4	187.2	0.789	147
C_{12}	196.4–216.8	208.3	0.800	161
C_{13}	216.8–235.9	227.2	0.811	175
C_{14}	235.9–253.9	246.4	0.822	190
C_{15}	253.9–271.1	266	0.832	206
C_{16}	271.1–287.3	283	0.839	222
C_{17}	287–303	300	0.847	237
C_{18}	303–317	313	0.852	251
C_{19}	317–331	325	0.857	263
C_{20}	331–344	338	0.862	275
C_{21}	344–357	351	0.867	291
C_{22}	357–369	363	0.872	305
C_{23}	369–381	375	0.877	318
C_{24}	381–392	386	0.881	331
C_{25}	392–402	397	0.885	345
C_{26}	402–413	408	0.889	359
C_{27}	413–423	419	0.893	374
C_{28}	423–432	429	0.896	388
C_{29}	432–441	438	0.899	402
C_{30}	441–450	446	0.902	416
C_{31}	450–459	455	0.906	430
C_{32}	459–468	463	0.909	444
C_{33}	468–476	471	0.912	458
C_{34}	476–483	478	0.914	472
C_{35}	483–491	486	0.917	486
C_{36}	—	493	0.919	500
C_{37}	—	500	0.922	514
C_{38}	—	508	0.924	528
C_{39}	—	515	0.926	542
C_{40}	—	522	0.928	556
C_{41}	—	528	0.930	570
C_{42}	—	534	0.931	584
C_{43}	—	540	0.933	598
C_{44}	—	547	0.935	612
C_{45}	—	553	0.937	626

Source: Data from Katz, D.L. and Firoozabadi, A., Predicting phase behavior of condensate/crude-oil systems using methane interaction coefficients, *J. Petroleum Technol.* 20, 1649–1655, 1978.

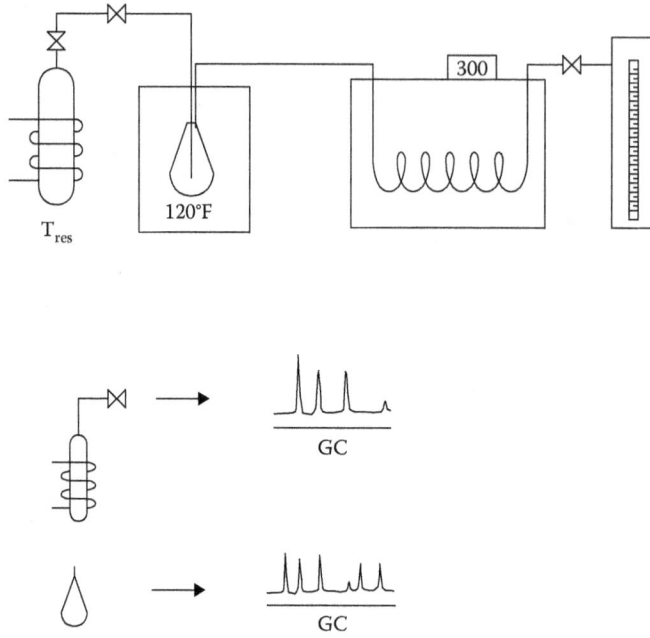

FIGURE 2.13 Spike flash apparatus.

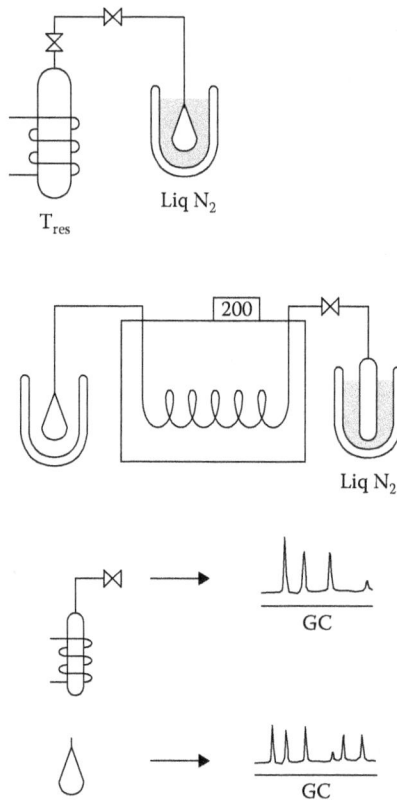

FIGURE 2.14 Cryogenic distillation apparatus.

2.4.1.3   Gas Chromatograph

A gas chromatograph (GC) can be used to analyze gas and liquid compositions. A GC consists of an injector, a column, and a detector. A typical GC is sketched in Figure 2.15.

2.4.1.3.1   Injector

A subsample of either a gas or a liquid is injected into the column of the GC. A valve system is used for gases and a syringe for liquids.

2.4.1.3.2   Column

The column of a GC contains a stationary phase over which a carrier gas is continuously moving to effect separation. If the fluid is liquid, a capillary column is used, which is made of fused silica bonded with a liquid phase. If the fluid is a gas, three different sequential columns are used with different packing—one made of a porous polymer, the next one with a molecular sieve, and finally one with a liquid phase (capillary). The GC utilizes three columns to clearly identify all eluted components.

2.4.1.3.3   Detector

Once the separation of components is done in the column, the carrier gas flows into the detector, which produces an electrical signal proportional to the weight concentration of each component in the sample. For liquid chromatograms, a flame ionization detector is used, whereas for gas samples, a thermal conductivity detector is used along with the flame ionization detector.

Because each component in the sample elutes at a particular temperature, the sample injector, column, and detectors are enclosed in a temperature-controlled environment (oven).

2.4.1.3.4   GC Analyses

A GC analysis provides a weight percent composition. Table 2.4 shows an example of the compositions of a gas and a liquid from a flash of a reservoir fluid to standard conditions and of the recombined reservoir fluid composition (Sah et al. 2011). The components between C_6 and C_7 are called C_7 isomers, those between C_7 and C_8 are C_8 isomers, and those between C_8 and C_9 are C_9 isomers. Some

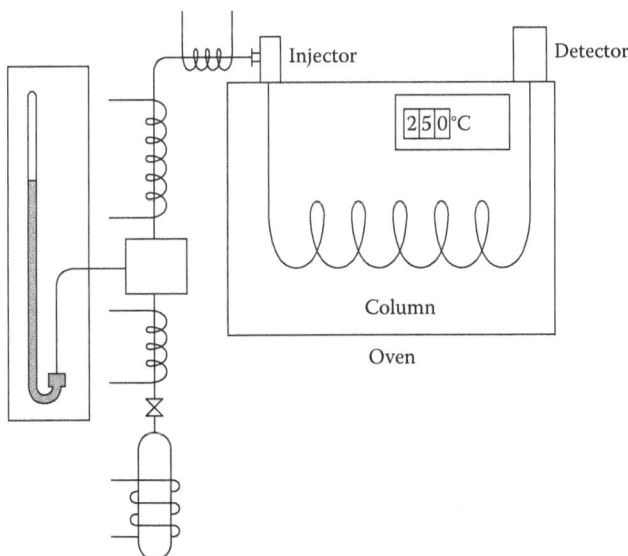

FIGURE 2.15   Sketch of gas chromatograph.

TABLE 2.4

Gas and Liquid Compositions after Flash of Bottom Hole Fluid Sample to Standard Conditions and Recombined Reservoir Fluid Composition. The Reservoir Fluid Composition Has a C_{7+} Molecular Weight of 208.4 and a C_{7+} Density of 0.846 g/cm³

Component	Flashed Gas		Flashed Liquid		Reservoir Fluid		Reservoir Fluid	
	Weight%	Mol%	Weight%	Mol%	Weight%	Mol%	Weight%	Mol%
N_2	0.252	0.244	0.000	0.000	0.075	0.183	0.075	0.183
CO_2	9.800	6.027	0.000	0.000	2.918	4.524	2.918	4.524
H_2S	5.792	4.600	0.000	0.000	1.725	3.453	1.725	3.453
C_1	36.275	61.211	0.000	0.000	10.801	45.950	10.801	45.950
C_2	12.045	10.842	0.048	0.307	3.620	8.216	3.620	8.216
C_3	12.417	7.621	0.214	0.934	3.848	5.954	3.848	5.954
iC_4	3.328	1.550	0.143	0.473	1.091	1.281	1.091	1.281
nC_4	7.778	3.622	0.578	1.912	2.722	3.196	2.722	3.196
iC_5	3.428	1.286	0.668	1.781	1.490	1.409	1.490	1.409
nC_5	3.705	1.390	1.026	2.733	1.824	1.725	1.824	1.725
C_6	3.138	1.011	2.274	5.204	2.531	2.056	2.531	2.056
m-c-C5	0.376	0.121	0.476	1.087	0.446	0.362		
Benzene	0.127	0.044	0.224	0.550	0.195	0.170		
c-C_6	0.165	0.053	0.341	0.779	0.289	0.234		
C_7	0.841	0.237	3.083	6.173	2.416	1.717	3.346	2.483
m-c-C_6	0.134	0.037	0.720	1.409	0.546	0.379		
Toluene	0.112	0.033	1.011	2.108	0.743	0.550		
C8	0.225	0.057	3.682	6.613	2.653	1.692	3.942	2.621
et-benzene	0.004	0.001	0.365	0.660	0.257	0.165		
m and p-xylene	0.012	0.003	0.756	1.368	0.534	0.343		
o-xylene	0.000	0.000	0.229	0.414	0.161	0.103		
C_9	0.045	0.010	3.908	6.208	2.758	1.555	3.710	2.166
C_{10}	0.000	0.000	5.391	7.733	3.786	1.928	3.786	1.928
C_{11}	0.000	0.000	4.529	5.922	3.181	1.476	3.181	1.476
C_{12}	0.000	0.000	4.003	4.779	2.811	1.191	2.811	1.191
C_{13}	0.000	0.000	4.015	4.410	2.820	1.099	2.820	1.099
C_{14}	0.000	0.000	3.842	3.886	2.698	0.969	2.698	0.969
C_{15}	0.000	0.000	3.675	3.429	2.581	0.855	2.581	0.855
C_{16}	0.000	0.000	3.246	2.810	2.279	0.701	2.279	0.701
C_{17}	0.000	0.000	3.016	2.446	2.118	0.610	2.118	0.610
C_{18}	0.000	0.000	2.895	2.217	2.033	0.553	2.033	0.553
C_{19}	0.000	0.000	2.884	2.108	2.025	0.525	2.025	0.525
C_{20}	0.000	0.000	2.691	1.881	1.890	0.469	1.890	0.469
C_{21}	0.000	0.000	2.383	1.574	1.674	0.392	1.674	0.392
C_{22}	0.000	0.000	2.149	1.377	1.509	0.343	1.509	0.343
C_{23}	0.000	0.000	2.036	1.254	1.430	0.313	1.430	0.313
C_{24}	0.000	0.000	1.884	1.118	1.323	0.279	1.323	0.279
C_{25}	0.000	0.000	1.781	1.016	1.251	0.253	1.251	0.253
C_{26}	0.000	0.000	1.727	0.951	1.212	0.237	1.212	0.237
C_{27}	0.000	0.000	1.656	0.884	1.163	0.220	1.163	0.220
C_{28}	0.000	0.000	1.610	0.832	1.130	0.207	1.130	0.207
C_{29}	0.000	0.000	1.545	0.777	1.085	0.194	1.085	0.194
C_{30}	0.000	0.000	1.375	0.671	0.966	0.167	0.966	0.167

Component	Flashed Gas		Flashed Liquid		Reservoir Fluid		Reservoir Fluid	
	Weight%	Mol%	Weight%	Mol%	Weight%	Mol%	Weight%	Mol%
C_{31}	0.000	0.000	1.306	0.621	0.917	0.155	0.917	0.155
C_{32}	0.000	0.000	1.198	0.555	0.841	0.138	0.841	0.138
C_{33}	0.000	0.000	1.099	0.496	0.772	0.124	0.772	0.124
C_{34}	0.000	0.000	1.009	0.444	0.708	0.111	0.708	0.111
C_{35}	0.000	0.000	0.922	0.398	0.647	0.099	0.647	0.099
C_{36+}	0.000	0.000	16.389	4.701	11.509	1.172	11.509	1.172

Source: Sah, P., Pedersen, K. S., Shaikh, J. A., EOS modeling, the first step in detailed EOR potential evaluation—A case study, SPE 144025, presented at the *SPE Enhanced Oil Recovery Conference* in Kuala Lumpur, Malaysia, July 19–21, 2011.

---

components, for example, benzene, are not classified according to their carbon number. A benzene molecule contains six carbon atoms, but because the boiling point of benzene is in the C_7 cut (see Tables 1.1 and 2.3), benzene is classified as a C_7 component. From C_{10} to C_{35}, all components are lumped into carbon number fractions according to the boiling ranges in Table 2.3. The last fraction is C_{36+} and contains C_{36} and heavier components.

The C_{7+} isomers are also called BTEX components. While it can be desirable to have these components quantified in process simulations, they provide an unnecessary overhead in compositional reservoir simulations. Prior to assigning equation of state model parameters to a reservoir fluid to be used in compositional reservoir simulation studies, the C_{7+} isomers will be combined with the respective carbon number fraction, as shown in the last two columns of Table 2.4. Table 2.5 shows a typical GC analysis as reported from a commercial PVT laboratory. It has most, but not all, components identified to C_9 and a single defined C_{10} component.

The mole fraction, z_i, of component i in an N-component mixture is calculated from

$$z_i = \frac{\dfrac{w_i}{M_i}}{\sum_{j=1}^{N} \dfrac{w_j}{M_j}} \tag{2.15}$$

where w stands for weight fraction, M for molecular weight, and i and j are component indices.

GC is a non-preparative technique, which does not allow the molecular weights of individual carbon number fractions to be quantitatively determined. Because in general no component identification takes place for the fractions above C_9, the GC technique provides no information about molecular weights of the heavier fractions. For C_{10}–C_{35}, the molecular weights in Table 2.3 are generally accepted. The PVT laboratory will measure the average molecular weight, M_{oil}, of the stable oil from the flash to atmospheric pressure (Figures 2.13 and 2.14). The molecular weight, M_+, of a plus fraction can be calculated as

$$M_+ = \frac{M_{oil} w_+}{1 - M_{oil} \sum_{i=1}^{N-1} \dfrac{w_i}{M_i}} \tag{2.16}$$

where w_i is the weight fraction of carbon number fraction i, and w_+ is the weight fraction of the plus fraction. The C_{7+} molecular weight in Table 2.5 has been calculated using Equation 2.16 with N equal to the carbon number of the plus fraction.

TABLE 2.5

Typical Gas Chromatographic Analysis to C_{36+}. The C_{7+} Molecular Weight is 276, and the C_{7+} Density 0.8651 g/cm^3

Formula	Component	Weight%	Mol%
N_2	Nitrogen	0.080	0.338
CO_2	Carbon dioxide	0.210	0.565
C_1	Methane	4.715	34.788
C_2	Ethane	2.042	8.039
C_3	Propane	2.453	6.583
iC_4	i-butane	0.601	1.223
nC_4	n-butane	1.852	3.771
C_5	Neo-pentane	0.001	0.002
iC_5	i-pentane	0.991	1.626
nC_5	n-pentane	1.341	2.200
C_6	Hexanes	2.433	3.341
	M-C-pentane	0.310	0.436
	Benzene	0.070	0.106
	Cyclohexane	0.240	0.338
C_7	Heptane	2.142	2.641
	M-C-hexane	0.380	0.459
	Toluene	0.280	0.360
C_8	Octanes	2.433	2.691
	E-benzene	0.190	0.212
	m/p-xylene	0.390	0.435
	o-xylene	0.190	0.212
C_9	Nonane	2.353	2.301
	1,2,4-TMB	0.230	0.227
C_{10}	Decane	2.843	2.511
C_{11}	Undecane	2.793	2.249
C_{12}	Dodecane	2.573	1.891
C_{13}	Tridecane	2.513	1.700
C_{14}	Tetradecane	2.312	1.441
C_{15}	Pentadecane	2.322	1.334
C_{16}	Hexadecane	2.212	1.180
C_{17}	Heptadecane	2.032	1.015
C_{18}	Octadecane	1.962	0.925
C_{19}	Nonadecane	1.992	0.897
C_{20}	Eicosane	1.812	0.780
C_{21}	Heneicosane	1.722	0.700
C_{22}	Docosane	1.662	0.645
C_{23}	Tricosane	1.552	0.578
C_{24}	Tetracosane	1.472	0.526
C_{25}	Pentacosane	1.401	0.481
C_{26}	Hexacosane	1.321	0.436
C_{27}	Heptacosane	1.281	0.406
C_{28}	Octacosane	1.261	0.385
C_{29}	Nonacosane	1.241	0.365
C_{30}	Triacontane	1.221	0.348
C_{31}	Hentriacontane	1.221	0.336
C_{32}	Dotriacontane	1.131	0.302
C_{33}	Tritriacontane	1.091	0.282
C_{34}	Tetratriacontane	1.031	0.259
C_{35}	Pentatriacontane	1.021	0.249
C_{36+}	Hexatriacontane plus	29.074	4.885

The density, ρ_+, of a plus fraction is calculated using the formula

$$\rho_+ = \frac{\rho_{oil} w_+}{1 - \rho_{oil} \sum_{i=1}^{N-1} \frac{w_i}{\rho_i}} \tag{2.17}$$

where ρ_{oil} is the density of the stable oil, and ρ_i is the liquid density of component (or carbon number fraction) i at standard conditions. Standard liquid densities are used for the components, which are gaseous at standard conditions (see Section 2.5). When a C_{36+} density is reported for a GC composition, the densities assumed for C_7–C_{35} are in most cases those in Table 2.3. This is unfortunate, as the actual densities of the C_7–C_{35} fractions in most reservoir fluids are higher than those in Table 2.3. Assuming C_7–C_{35} densities that are too low will result in a too high C_{36+} density being reported.

Table 2.6 shows the composition to C_{12+} of a heavy gas condensate with relatively high concentrations of CO_2 and H_2S (Takeshi et al. 2020). A fluid with high concentrations of CO_2 and H_2S is called an acid fluid.

High-temperature capillary GC is a technique permitting compositional analyses of oil samples to around C_{80+} (Curvers and van den Engel 1989). Table 2.7 shows a composition to C_{80+} analyzed by this technique.

2.4.1.3.5 Pitfalls in GC Analyses

Often the same GC column is used to perform consecutive GC analyses for black oils and gas condensates. Black oils have a large concentration of C_{30+} components, and part of the C_{30+} fraction may

TABLE 2.6

Heavy Acid Gas Condensate Reservoir Fluid. The Fluid Has a C_{7+} Molecular Weight of 171.0 and a C_{7+} Density of 0.812 g/cm³

Component	Reservoir Fluid Weight %	Reservoir Fluid Mol% %
CO_2	5.86	5.292
H_2S	16.82	19.614
C_1	21.33	52.837
C_2	3.78	4.996
C_3	3.29	2.965
iC_4	0.88	0.602
nC_4	2.14	1.463
iC_5	1.18	0.650
nC_5	1.53	0.843
C_6	3.07	1.416
C_7	3.02	1.250
C_8	3.86	1.434
C_9	3.86	1.268
C_{10}	3.53	1.047
C_{11}	2.79	0.745
C_{12+}	23.06	3.571

Source: Takeshi, A., Tetsuro, F., Leekumjorn, S., Shaikh, J. A., Pedersen, K.S., Alobeidli, A., Mogensen, K., A new technique for common EoS model development for multiple reservoir fluids with gas injection, SPE-202923-MS, presented at the *Abu Dhabi International Petroleum Exhibition & Conference*, November 9–12, 2020.

TABLE 2.7

Composition of Stable Oil to C_{80+} Analyzed by Use of High-Temperature Capillary Gas Chromatography

Component	Mol%	Molecular Weight	Density (g/cm³)
C_1	0.13	16.0	—
C_2	0.50	30.1	—
C_3	0.47	44.1	—
iC_4	0.55	58.1	—
nC_4	0.62	58.1	—
iC_5	1.08	72.1	—
nC_5	0.50	72.1	—
C_6	1.89	86.2	—
C_7	5.34	90.9	0.749
C_8	8.54	105.0	0.768
C_9	7.04	117.7	0.793
C_{10}	6.80	132	0.808
C_{11}	5.51	148	0.815
C_{12}	5.00	159	0.836
C_{13}	5.58	172	0.850
C_{14}	5.08	185	0.861
C_{15}	4.66	197	0.873
C_{16}	3.80	209	0.882
C_{17}	2.67	227	0.873
C_{18}	2.49	243	0.875
C_{19}	2.14	254	0.885
C_{20}	2.23	262	0.903
C_{21}	1.71	281	0.898
C_{22}	1.42	293	0.898
C_{23}	1.63	307	0.899
C_{24}	1.50	320	0.900
C_{25}	1.25	333	0.905
C_{26}	1.45	346	0.907
C_{27}	1.33	361	0.911
C_{28}	1.23	374	0.915
C_{29}	1.15	381	0.920
C_{30}	1.09	(+) 624	(+) 0.953
C_{31}	0.90	—	—
C_{32}	0.92	—	—
C_{33}	0.79	—	—
C_{34}	0.67	—	—
C_{35}	0.70	—	—
C_{36}	0.59	—	—
C_{37}	0.49	—	—
C_{38}	0.52	—	—
C_{39}	0.46	—	—
C_{40}	0.37	—	—
C_{41}–C_{45}	1.59	—	—
C_{46}–C_{50}	1.06	—	—
C_{51}–C_{55}	0.74	—	—

Component	Mol%	Molecular Weight	Density (g/cm³)
C_{56}–C_{60}	0.56	—	—
C_{61}–C_{65}	0.41	—	—
C_{66}–C_{70}	0.33	—	—
C_{71}–C_{75}	0.27	—	—
C_{76}–C_{80}	0.25	—	—
C_{80+}	0.29	—	—

Source: Data from Pedersen, K.S., Blilie, A.L., and Meisingset, K.K., PVT calculations on petroleum reservoir fluids using measured and estimated compositional data for the plus fraction, *Ind. Eng. Chem. Res.* 31, 1379–1384, 1992.

Note: The density is at 1.01 bar and 15°C.

"get delayed" in the column and not reported for the oil. Mawlod et al. (2021) have reported such case in a Round Robin study, and Stephen et al. (2008) have reported similar issues. If the same column is subsequently used to analyze a gas condensate, the delayed C_{30+} fraction from the previous black oil analysis might come out and be reported as belonging to the gas condensate. Too high a C_{30+} concentration will in such case be reported for the gas condensate. To the extent possible, different GC columns should be used for analyses of black oils and gas condensates, or the GC column should be thoroughly cleaned between a black oil and a gas condensate analysis.

2.4.2 TBP ANALYSIS

A TBP analysis separates the components of a stable oil into boiling point cuts. A TBP distillation column is sketched in Figure 2.16. The cut points are given in Table 2.3. Each distillation cut contains sufficient material to measure the density and molecular weight. Because there is a limited span in molecular weight within a carbon number fraction, its molecular weight can be measured with a higher accuracy than the average molecular weight of the oil sample as a whole. An example of data from a TBP analysis is given in Table 2.8. Up to C_{10+}, the distillation has been carried out at atmospheric pressure. From C_{10} to C_{19}, the pressure was 26.6 mbar and, finally, the fractions C_{20}–C_{29} were separated at a pressure of 2.66 mbar. The pressure is reduced to avoid decomposition (cracking). Even though the distillation of C_{10} and heavier fractions takes place at a reduced pressure, it is still customary to report the boiling point temperature at atmospheric pressure (second column of Table 2.8). The densities in Table 2.8 are at atmospheric pressure (1.01 bar) and 15°C. Figure 2.17 shows the percentage of weight distilled off as a function of temperature.

2.4.2.1 Molecular Weight from Freezing Point Depression

Appendix A introduces the phase equilibrium criterion between gas and liquid. At equilibrium, component i will have the same fugacity in the vapor phase and in the liquid phase. Similarly, at equilibrium between a liquid phase and a solid phase, component i will have the same fugacity in the liquid phase (l) as in the solid phase (s):

$$f_i^l = f_i^s \tag{2.18}$$

Assume that a given weight amount of a stable oil is dissolved in a liquid solvent, which in the following is assumed to be benzene even though there may be environmental reasons to not use benzene. The oil dissolved in benzene will decrease (depress) the freezing point of benzene, which in pure form is 5.5°C. At the new freezing point, benzene in the benzene–oil solution will

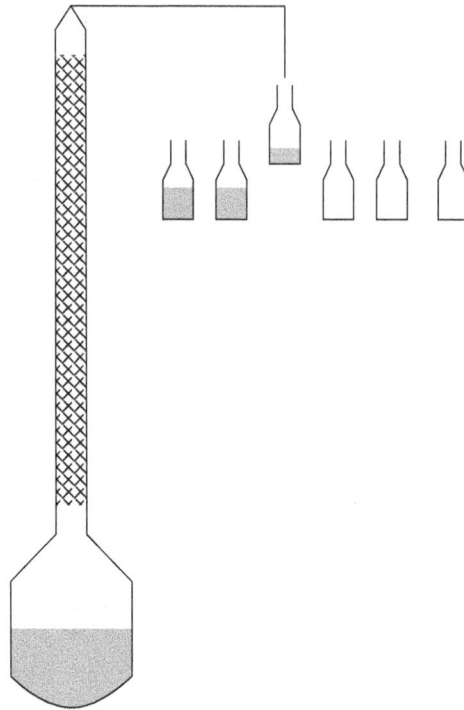

FIGURE 2.16   Sketch of true boiling point distillation column.

TABLE 2.8
Example of True Boiling Point Analysis

Fraction	Cut Point (°C)	Actual Temperature (°C)	Density (g/cm³)	M	Weight%	Cumulative Weight%
P = 1.01 bar						
Gas	—	—	—	33.5	0.064	0.064
<C_6	36.5	36.5	0.598	62.5	3.956	4.020
C_6	69.2	69.2	0.685	82.0	2.016	6.036
C_7	98.9	98.9	0.737	98.7	6.125	12.161
C_8	126.1	126.1	0.754	109.6	4.606	16.767
C_9	151.3	151.3	0.774	121.9	5.046	21.813
P = 26.6 mbar						
C_{10}	174.6	70.9	0.789	134.7	4.020	25.833
C_{11}	196.4	88.7	0.794	150.3	3.953	29.786
C_{12}	216.8	105.7	0.806	166.4	4.061	33.847
C_{13}	235.9	121.8	0.819	181.4	3.800	37.647
C_{14}	253.9	136.9	0.832	194.0	4.421	42.068
C_{15}	271.1	151.2	0.834	209.4	3.765	45.833
C_{16}	287.3	164.3	0.844	222.4	2.969	48.802
C_{17}	303	178	0.841	240.9	3.800	52.602
C_{18}	309	191	0.847	256.0	2.813	55.415
C_{19}	331	203	0.860	268.2	3.364	58.779

Fraction	Cut Point (°C)	Actual Temperature (°C)	Density (g/cm³)	M	Weight%	Cumulative Weight%
P = 2.66 mbar						
C_{20}	344	161	0.874	269.4	1.115	59.894
C_{21}	357	172	0.870	282.5	2.953	62.847
C_{22}	369	181	0.872	297.7	2.061	64.908
C_{23}	381	191	0.875	310.1	1.797	66.705
C_{24}	392	199	0.877	321.8	1.421	68.126
C_{25}	402	208	0.881	332.4	2.083	70.209
C_{26}	413	217	0.886	351.1	1.781	71.990
C_{27}	423	226	0.888	370.8	1.494	73.484
C_{28}	432	234	0.895	381.6	1.625	75.109
C_{29}	441	241	0.898	393.7	1.233	76.342
C_{30+}	>441	—	0.935	612.0	23.658	100.000

Note: M = Molecular weight; the density of the oil is 0.828 g/cm³ and the average molecular weight 191.1.

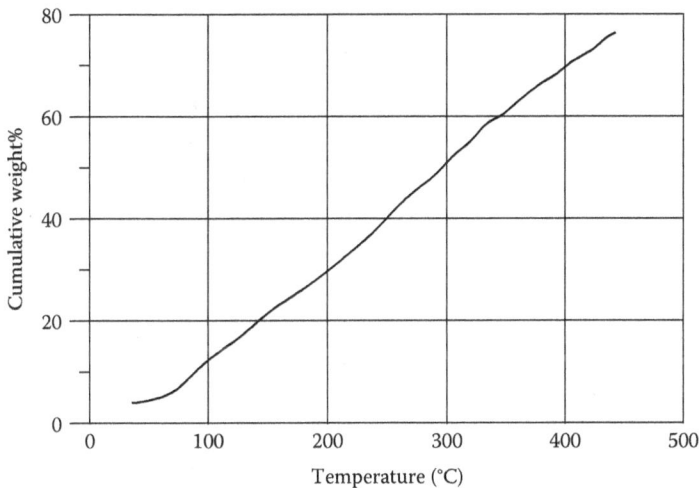

FIGURE 2.17 Cumulative weight percentage distilled off as a function of temperature in true boiling point distillation tabulated in Table 2.8.

be in equilibrium with pure solid benzene. Assuming an ideal liquid solution, the equilibrium criterion will be

$$x_i f_i^{0l} = f_i^{0s} \tag{2.19}$$

where x_i is the mole fraction of benzene in the benzene–oil solution and f_i^{0l} and f_i^{0s} are, respectively, the fugacity of pure benzene in liquid and solid forms.

Using the fundamental thermodynamic relations in Appendix A, Equation 2.18 can be rearranged as

$$ln\, x_i = ln\left(f_i^{0s}\right) - ln\left(f_i^{0l}\right) = \frac{\Delta G_i^f}{RT} = \frac{\Delta H_i^f - T\Delta S_i^f}{RT} \tag{2.20}$$

where ΔG_i^f is the change in Gibbs free energy by solidification. Similarly, ΔH_i^f and ΔS_i^f are the changes in enthalpy and entropy by solidification. The change in entropy can be approximated as

$$\Delta S_i^f \approx \frac{\Delta H_i^f}{T_i^f} \tag{2.21}$$

which allows Equation 2.19 to be rewritten as

$$ln\ x_i = \frac{\Delta H_i^f}{RT}\left(1-\frac{T}{T_i^f}\right) \tag{2.22}$$

The mole fraction of benzene in the benzene–oil solution can be found by using the depressed freezing point of benzene (T) in Equation 2.22. The molar fraction of oil in solution equals $(1-x_i)$. With the molar amount of benzene known, the molar amount of oil is also known. The average molecular weight of the oil can therefore be found as the ratio of the weight and molar amounts of oil in solution.

The uncertainty on the molecular weight of the individual carbon number fractions is around 2%, whereas it is around 5% for a C_{20+} fraction. The uncertainty on the molecular weight of a stable oil from a flash of a bottom hole sample to standard conditions will be at least 10%.

2.5   RESERVOIR FLUID COMPOSITION FROM BOTTOM HOLE SAMPLE

This section details the steps required to generate a compositional analysis of a bottom hole sample. The composition considered ends at C_{10+}, but the same principle will apply for a composition to a higher carbon number.

Consider a bottom hole sample taken at 104°C and 340 bar, at which conditions the sample volume is 1 l (= 1000 cm³). In the laboratory, the sample is flashed to standard conditions (1.01 bar and 15°C), giving a liquid of the volume 0.462 l and a gas of the volume 175.2 l. The density of the oil at standard conditions is 0.836 g/cm³, and it has an average molecular weight of 187.1.

A GC analysis to C_{10+} is carried out for the oil sample giving the composition in Table 2.9. The composition is measured as a weight composition. The weight fraction (w) of each component contained in the fractions from C_7 to C_9 is determined, and the C_{10+} molecular weight is calculated from Equation 2.16. M_{oil} is the average molecular weight of the oil sample, which is 187.1. Similarly, the density of the C_{10+} fraction is calculated from Equation 2.17, where the overall density of the oil, ρ_{oil}, at standard conditions equals 0.836 g/cm³. The densities of the components N_2, CO_2, and C_1–C_6 in Table 2.9 are the pure component densities recommended by the American Petroleum Institute (1982) for use in calculations of densities of oil mixtures at standard conditions. The densities of C_7–C_{10} are at 1.01 bar and 15°C. The molar composition is calculated using Equation 2.15.

The number of moles contained in the oil at standard conditions is calculated:

$$\text{Moles oil} = \frac{\text{Oil volume}\,(\text{cm}^3)\times\text{Oil density}\,(\text{g/cm}^3)}{\text{Average molecular weight}\,(\text{g/mol})} = \frac{462.0\times0.836}{187.1} = 2.064\ \text{mol}$$

A GC analysis of the gas from the flash to standard conditions gives the composition in Table 2.10. All components present in significant amounts are identified, permitting calculation of the average molecular weights of each of the fractions C_7, C_8, C_9, and C_{10+}. The average molecular weight of the gas, 24.25, is calculated from:

$$M_{gas} = \sum_{i=1}^{N} z_i M_i \tag{2.22}$$

TABLE 2.9
Composition to C_{10+} of Oil Sample from Flash of Bottom Hole Sample to Standard Conditions Measured by Gas Chromatography

Component	Weight%	Molecular Weight	Density (g/cm³)	Mol%
N_2	0.0001	28.014	0.804	0.001
CO_2	0.0136	44.010	0.809	0.058
H_2S	0.0000	34.080	0.943	0.000
C_1	0.0298	16.043	0.300	0.348
C_2	0.0608	30.070	0.356	0.378
C_3	0.2317	44.097	0.508	0.983
iC_4	0.1295	58.124	0.563	0.417
nC_4	0.4573	58.124	0.584	1.472
iC_5	0.4639	72.151	0.625	1.203
nC_5	0.8010	72.151	0.631	2.077
C_6	2.2413	86.178	0.664	4.866
C_7	5.0940	91.5	0.738	10.416
C_8	6.4978	101.2	0.765	12.013
C_9	4.9302	119.1	0.781	7.745
C_{10+}	79.0489	254.9	0.871	58.022

Source: The densities for N_2, CO_2 and C_1–C_6 are from American Petroleum Institute, *Technical Data Book—Petroleum Refining*, API, New York, 1982.

TABLE 2.10
Composition of Gas from Flash to Standard Conditions Measured by Gas Chromatography

Fraction	Weight%	Molecular Weight	Mol%
N_2	0.805	28.014	0.697
CO_2	6.518	44.01	3.591
C_1	46.858	16.043	70.817
C_2	13.473	30.07	10.864
C_3	12.840	44.097	7.060
iC_4	2.812	58.124	1.173
nC_4	6.475	58.124	2.701
iC_5	2.410	72.151	0.810
nC_5	3.003	72.151	1.009
C_6	2.396	86.178	0.674
C_7	1.434	91.5	0.380
C_8	0.785	101.2	0.188
C_9	0.142	119.1	0.029
C_{10+}	0.049	147.8	0.008

Assuming the gas at standard conditions behaves as an ideal gas, its volume is given by

$$\text{Molar gas volume (cm}^3) = \frac{R\left(\frac{\text{cm}^3\,\text{bar}}{\text{mol K}}\right) \times T(K)}{P(\text{bar})} = \frac{83.14 \times 288.15}{1.01325} = 23644 \text{ cm}^3/\text{mol},$$

TABLE 2.11

Composition of Reservoir Fluid (Bottom Hole Sample) Found by Recombination of Gas Composition in Table 2.10 and Oil Composition in Table 2.9

Fraction	Mol%	Molecular Weight	Density (g/cm³)
N_2	0.545	28.014	—
CO_2	2.821	44.010	—
C_1	55.465	16.043	—
C_2	8.580	30.070	—
C_3	5.736	44.097	—
iC_4	1.008	58.124	—
nC_4	2.433	58.124	—
iC_5	0.896	72.151	—
nC_5	1.242	72.151	—
C_6	1.587	86.178	—
C_7	2.566	91.5	0.738
C_8	2.764	101.2	0.765
C_9	1.710	119.1	0.781
C_{10+}	12.647	254.9	0.871

Note: The density is at 1.01 bar and 15°C.

which allows the number of moles of gas at standard conditions to be calculated:

$$\text{Mole gas} = \frac{\text{Gas volume (cm}^3)}{\text{Molar gas volume (cm}^3/\text{mole})} = \frac{175200}{23644} = 7.410 \text{ mol.}$$

The total number of moles in the bottom hole sample is then:

Mole reservoir fluid = Mole oil + Mole gas = 2.064 + 7.410 = 9.474 mol.
The reservoir fluid composition (component mole fractions) is now calculated:

$$z_i = \frac{y_i \times \text{Mole gas} + x_i \times \text{Mole oil}}{\text{Mole reservoir fluid}} = \frac{y_i \times 7.410 + x_i \times 2.064}{9.474}$$

where y_i is the mole fraction of component i in the gas at standard conditions (Table 2.10) and x_i the mole fraction of component i in the oil at standard conditions (Table 2.9). The reservoir fluid composition calculated using this formula and converted to mol% can be seen in Table 2.11.

2.6 RESERVOIR FLUID COMPOSITION FROM SEPARATOR SAMPLES

Consider a gas and a liquid sample taken from a separator operating at 70 bar and 50°C. The GOR equals 215.2 Sm^3/m^3. It is defined as the volume of gas from the separator after a flash to standard conditions divided by volume of oil at separator conditions. In the laboratory, the separator gas and the separator oil are flashed to standard conditions. Usually, a small amount of liquid is formed when flashing the gas to standard conditions, but this is neglected in the following calculation. This gives three compositions to analyze:

Gas from the flash of separator gas to standard conditions
Gas from the flash of separator oil to standard conditions
Oil from the flash of separator oil to standard conditions

Each composition is analyzed separately. Table 2.12 shows the various gas and oil volumes, Table 2.13 the composition of separator gas, and Table 2.14 the composition of gas from the flash of separator oil to standard conditions.

The density of oil at standard conditions is 0.825 g/cm³. A GC analysis to C_{10+} is carried out for the oil sample from the flash of separator oil to standard conditions, resulting in the composition shown in Table 2.15. The densities of the components N_2, CO_2, and C_1–C_6 are the pure component densities from Table 2.9. The densities of C_7–C_{10} are at 1.01 bar and 15°C. The molecular weight and density of the C_{10+} fraction are calculated from Equations 2.16 and 2.17, respectively.

Furthermore, a true boiling point analysis is carried out for the oil from the flash of separator oil to standard conditions. The resulting composition is seen in Table 2.16. The compositions in Tables 2.15 and 2.16 are combined (<C_5 from Table 2.15 and C_{5+} from Table 2.16) to give the oil composition shown in Table 2.17. The mole percentages in the last column are calculated from Equation 2.15. The average molecular weight of the oil is 176.6.

The number of moles contained in oil from the flash of separator oil to standard conditions can be calculated as

$$\text{Mole oil} = \frac{\text{Oil volume (cm}^3) \times \text{Oil density (g/cm}^3)}{\text{Average molecular weight (g/mole)}} = \frac{2.55 \times 10^3 \times 0.825}{176.6} = 11.91 \text{ mol}$$

TABLE 2.12
Sampled Volumes and Average Molecular Weights of Sampled Compositions

Sample	Volume (l)	Molecular Weight
Gas from flash of separator gas to standard conditions	694.5	—
Liquid from flash of separator gas to standard conditions	4.2×10^{-5}	—
Separator oil at separator conditions	3.22	—
Gas from flash of separator oil to standard conditions	204.3	31.05
Oil from flash of separator oil to standard conditions	2.55	176.6

TABLE 2.13
Composition of Separator Gas Sampled at 70 bar and 50°C

Component	Weight%	Molecular Weight	Mol%
N_2	1.183	28.014	0.870
CO_2	7.479	44.01	3.502
C_1	62.639	16.043	80.459
C_2	13.202	30.07	9.047
C_3	8.439	44.097	3.944
IC_4	1.332	58.124	0.472
nC_4	2.592	58.124	0.919
iC_5	0.710	72.151	0.203
nC_5	0.822	72.151	0.235
C_6	0.618	86.178	0.148
C_7	0.460	91.5	0.104
C_8	0.327	101.2	0.067
C_9	0.102	119.1	0.018
C_{10}	0.045	133	0.007
C_{11}	0.026	145	0.004
C_{12}	0.012	158	0.002
C_{13}	0.008	171	0.001
C_{14}	0.003	185	0.000
C_{15}	0.001	198	0.000

TABLE 2.14

Composition of Gas from Flash of Separator Oil to Standard Conditions

Component	Weight%	Molecular Weight	Mol%
N_2	0.174	28.014	0.193
CO_2	6.132	44.010	4.336
C_1	24.262	16.043	47.066
C_2	17.484	30.070	18.096
C_3	23.620	44.097	16.670
iC_4	5.313	58.124	2.845
nC_4	11.716	58.124	6.273
iC_5	3.422	72.151	1.476
nC_5	3.860	72.151	1.665
C_6	2.190	86.178	0.791
C_7	1.100	91.5	0.374
C_8	0.576	101.2	0.177
C_9	0.103	119.1	0.027
C_{10}	0.030	133	0.007
C_{11}	0.014	145	0.003
C_{12}	0.005	158	0.001

TABLE 2.15

Gas Chromatographic Composition to C_{10+} of Oil from Flash of Separator Oil to Standard Conditions

Component	Weight%	Molecular Weight	Density (g/cm³)	Mol%
N_2	0.000	28.014	0.804	0.000
CO_2	0.017	44.010	0.809	0.068
C_1	0.021	16.043	0.300	0.231
C_2	0.107	30.070	0.356	0.628
C_3	0.580	44.097	0.508	2.323
iC_4	0.333	58.124	0.563	1.012
nC_4	1.123	58.124	0.584	3.412
iC_5	0.893	72.151	0.625	2.186
nC_5	1.396	72.151	0.631	3.417
C_6	2.776	86.178	0.664	5.689
C_7	5.267	91.5	0.738	10.166
C_8	6.427	101.2	0.765	11.216
C_9	4.758	119.1	0.781	7.055
C_{10+}	76.302	256.2	0.873	52.597

TABLE 2.16

Composition to C_{20+} of Oil from Flash of Separator Oil to Standard Conditions Generated Based on True Boiling Point Distillation

Fraction	Weight%	Molecular Weight	Density (g/cm³)	Mol%
$<C_5$	2.182	50.18	—	7.678
iC_5	0.893	72.151	—	2.185
nC_5	1.400	72.151	—	3.426

Fraction	Weight%	Molecular Weight	Density (g/cm³)	Mol%
C_6	2.776	86.178	—	5.688
C_7	5.267	91.5	0.738	10.164
C_8	6.428	101.2	0.765	11.215
C_9	4.758	119.1	0.781	7.054
C_{10}	3.940	133	0.792	5.231
C_{11}	3.830	145	0.796	4.664
C_{12}	3.478	158	0.810	3.887
C_{13}	4.277	171	0.825	4.416
C_{14}	3.918	185	0.836	3.739
C_{15}	3.691	198	0.842	3.292
C_{16}	2.955	209	0.849	2.497
C_{17}	3.656	226	0.845	2.856
C_{18}	3.220	242	0.848	2.349
C_{19}	3.180	251	0.858	2.237
C_{20+}	40.156	407	0.905	17.421

Note: The density is at 1.01 bar and 15°C.

TABLE 2.17
Composition of Oil from Flash of Separator Oil to Standard Conditions

Fraction	Weight%	Molecular Weight	Density (g/cm³)	Mol%
N_2	0.000	28.014	—	0.000
CO_2	0.017	44.01	—	0.070
C_1	0.021	16.043	—	0.232
C_2	0.107	30.07	—	0.630
C_3	0.580	44.097	—	2.321
iC_4	0.333	58.124	—	1.011
nC_4	1.123	58.124	—	3.413
iC_5	0.893	72.151	—	2.185
nC_5	1.396	72.151	—	3.416
C_6	2.776	86.178	—	5.688
C_7	5.267	91.5	0.738	10.166
C_8	6.428	101.2	0.765	11.216
C_9	4.758	119.1	0.781	7.055
C_{10}	3.940	133	0.792	5.231
C_{11}	3.830	145	0.796	4.664
C_{12}	3.478	158	0.810	3.887
C_{13}	4.277	171	0.825	4.417
C_{14}	3.918	185	0.836	3.740
C_{15}	3.691	198	0.842	3.292
C_{16}	2.955	209	0.849	2.497
C_{17}	3.656	226	0.845	2.857
C_{18}	3.220	242	0.848	2.350
C_{19}	3.180	251	0.858	2.237
C_{20+}	40.156	407	0.905	17.423

Note: This composition has been obtained by combining the gas chromatographic composition in Table 2.15 and the true boiling point composition in Table 2.16. The density is at 1.01 bar and 15°C.

Similarly, the number of moles contained in gas from the flash of separator oil to standard conditions can, assuming ideal gas behavior, be calculated as

$$\text{Mole gas} = \frac{\text{Gas volume (cm}^3)}{\text{Molar gas volume (cm}^3/\text{mole})} = \frac{204300}{23644} = 8.64 \text{ mol}$$

The total number of moles in the separator oil sample is then

Mole separator oil = Mole oil + Mole gas = 11.91 + 8.64 = 20.55 mol

The composition of separator oil (component mole fractions) may now be calculated from

$$z_i = \frac{y_i \times \text{Mole gas} + x_i \times \text{Mole oil}}{\text{Mole separator fluid}} = \frac{y_i \times 8.64 + x_i \times 11.91}{20.55}$$

where y_i is the mole fraction of component i in the gas from the flash of separator oil to standard conditions (Table 2.14) and x_i is the mole fraction of component i in the oil from the flash of separator oil to standard conditions (Table 2.17). The separator oil composition calculated from this formula can be seen in Table 2.18.

TABLE 2.18
Composition of Separator Oil

Fraction	Mol%	Molecular Weight	Density (g/cm³)
N_2	0.081	28.014	—
CO_2	1.862	44.01	—
C_1	19.922	16.043	—
C_2	7.972	30.07	—
C_3	8.356	44.097	—
iC_4	1.783	58.124	—
nC_4	4.616	58.124	—
iC_5	1.887	72.151	—
nC_5	2.686	72.151	—
C_6	3.629	86.178	—
C_7	6.048	91.5	0.738
C_8	6.574	101.2	0.765
C_9	4.100	119.1	0.781
C_{10}	3.035	133	0.792
C_{11}	2.704	145	0.796
C_{12}	2.253	158	0.810
C_{13}	2.559	171	0.825
C_{14}	2.167	185	0.836
C_{15}	1.908	198	0.842
C_{16}	1.447	209	0.849
C_{17}	1.655	226	0.845
C_{18}	1.361	242	0.848
C_{19}	1.296	251	0.858
C_{20+}	10.097	407	0.905

Note: This composition has been found by recombination of gas composition in Table 2.14 and oil composition in Table 2.17. The density is at 1.01 bar and 15°C.

The number of moles in the separator gas can, assuming ideal gas behavior, be calculated as:

$$\text{Mole gas} = \frac{\text{Gas volume at standard conditions (cm}^3)}{\text{Molar gas volume (cm}^3/\text{mole)}} = \frac{694.5 \times 10^3}{23644} = 29.37 \text{ mol}$$

The total number of moles sampled from the separator is then:

Mole separator fluid = Mole separator oil + Mole separator gas = 20.55 + 29.37 = 49.92 mol
The reservoir fluid composition (component mole fractions) may now be calculated from

$$z_i = \frac{y_i \times \text{Mole separator gas} + x_i \times \text{Mole separator oil}}{\text{Mole separator fluid}} = \frac{y_i \times 29.37 + x_i \times 20.55}{49.92}$$

where y_i is the mole fraction of component i in the separator gas (Table 2.13) and x_i is the mole percentage of component i in the separator oil (Table 2.18). The reservoir fluid composition is seen in Table 2.19.

TABLE 2.19
Composition of Reservoir Fluid

Fraction	Mol%	Molecular Weight	Density (g/cm³)
N_2	0.545	28.014	—
CO_2	2.827	44.01	—
C_1	55.538	16.043	—
C_2	8.606	30.07	—
C_3	5.760	44.097	—
iC_4	1.012	58.124	—
nC_4	2.441	58.124	—
iC_5	0.896	72.151	—
nC_5	1.244	72.151	—
C_6	1.581	86.178	—
C_7	2.551	91.5	0.738
C_8	2.746	101.2	0.765
C_9	1.698	119.1	0.781
C_{10}	1.254	133	0.792
C_{11}	1.115	145	0.796
C_{12}	0.929	158	0.810
C_{13}	1.054	171	0.825
C_{14}	0.892	185	0.836
C_{15}	0.785	198	0.842
C_{16}	0.596	209	0.849
C_{17}	0.681	226	0.845
C_{18}	0.560	242	0.848
C_{19}	0.534	251	0.858
C_{20+}	4.157	407	0.905

Note: This composition is obtained by recombining separator compositions in Tables 2.13 and 2.18. The density is at 1.01 bar and 15°C

2.7 MUD-CONTAMINATED SAMPLES

Bottom hole samples collected in wellbore systems using oil-based muds (OBMs) are likely to be contaminated by medium to heavy hydrocarbon fractions present in the OBM (Gozalpour et al. 2002). The OBM generally consists of components in the range C_8–C_{34} and is dominated by the paraffinic C_{11}–C_{18} components.

A plot of ln(mol%) versus carbon number for the fractions C_7–C_{35} will for a clean reservoir fluid show an approximately straight line, as is further dealt with in Chapter 5. For a contaminated fluid, such a plot will, as sketched in Figure 2.18, show two or three peaks originating from the dominant OBM components.

PVT laboratories will usually report the OBM concentration as the weight percent contained in the oil from a flash of the reservoir fluid to standard conditions. Otherwise, the content of OBM may either be estimated by removing the material above the straight line in a plot of ln(mol%) versus carbon number. For the fluid in Figure 2.18, that would mean removing the material above the straight line.

It is of course preferable to have clean samples. However, when only contaminated samples exist, the operator must get the best out of them. PVT measurements, presented in Chapter 3, are used to get data for the behavior of a reservoir fluid. If the fluid sampled is contaminated, the PVT data measured will be for the contaminated sample and the measured PVT properties may deviate substantially from those of the clean reservoir fluid. Figure 2.19 shows how OBM will affect the phase envelope of reservoir fluids. The OBM will lower the saturation pressure of an oil mixture, whereas it will increase the saturation pressure of a gas condensate fluid.

To get a true picture of the reservoir fluid in a field from which only a mud-contaminated sample exists, the contaminated reservoir fluid composition must be numerically cleaned. In general, it is impossible to validate whether the cleaned fluid matches the uncontaminated reservoir fluid, but Sah et al. (2012) have presented a paper allowing numerical cleaning methods to be tested. A clean reservoir sample was intentionally contaminated by OBM. Table 2.20 shows the composition of the clean reservoir fluid and of the reservoir fluid with 21 weight% contaminate of the STO oil. Also

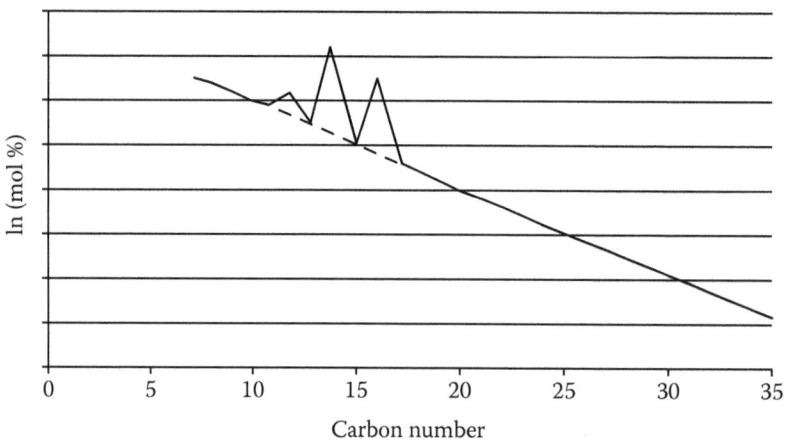

FIGURE 2.18 Schematic illustration of the ln(mol%) versus carbon number trend for a reservoir fluid contaminated by oil-based drilling mud. The dashed line shows the trend seen for a clean reservoir fluid.

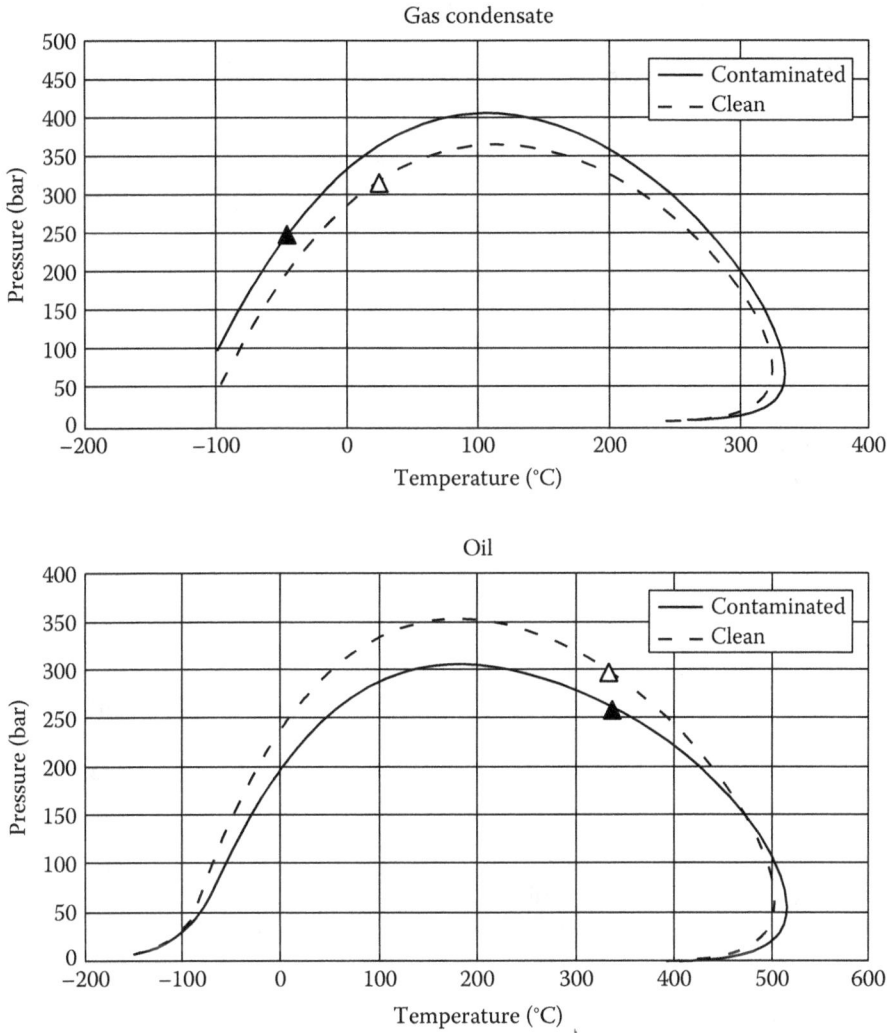

FIGURE 2.19 Phase envelopes for oil and gas condensate mixture showing impact of oil-based mud contamination.

shown in Table 2.20 is the composition of the OBM. It contains components from C_8 to C_{34} and is dominated by C_{11}–C_{14}. It can be seen that the densities of all OBM carbon number fractions are lower than of the corresponding carbon number fractions in the clean reservoir fluid. The OBM is dominated by n-paraffins, which have lower densities than the corresponding naphthenic and aromatic components present in the reservoir fluid. The numerical cleaning must therefore not only correct the component mol% for the presence of OBM but also correct the densities of the contaminated fractions for the influence of OBM.

Table 2.21 outlines the procedure used by Sah et al. to numerically clean the contaminated reservoir fluid composition in Table 2.20 for mud contaminate. An almost perfect correspondence was seen with the clean reservoir fluid composition also shown in Table 2.20.

TABLE 2.20

Clean and Contaminated Reservoir Fluids. The Last Column Shows the Composition of the Mud Contaminate

Component	Clean Reservoir Fluid			Reservoir Fluid with 21 wt% OBM of Stock Tank Oil		OBM	
	Mol%	Mol Weight	Density (g/cc)	Mol%	Density (g/cc)	Mol%	Density (g/cc)
N_2	0.816	—	—	0.718	—	—	—
CO_2	1.271	—	—	1.119	—	—	—
C_1	42.706	—	—	37.582	—	—	—
C_2	2.477	—	—	2.179	—	—	—
C_3	2.323	—	—	2.044	—	—	—
iC_4	0.720	—	—	0.634	—	—	—
nC_4	0.966	—	—	0.850	—	—	—
iC_5	0.683	—	—	0.601	—	—	—
nC_5	0.623	—	—	0.548	—	—	—
C_6	1.369	—	—	1.205	—	—	—
C_7	3.708	96	0.738	3.263	0.738	—	—
C_8	6.212	107	0.765	5.467	0.765	0.006	0.737
C_9	4.492	121	0.781	3.955	0.781	0.015	0.753
C_{10}	3.762	134	0.792	3.325	0.792	0.119	0.763
C_{11}	2.904	147	0.796	4.032	0.787	12.304	0.767
C_{12}	2.675	161	0.810	6.446	0.793	34.086	0.781
C_{13}	3.010	175	0.825	6.521	0.809	32.253	0.795
C_{14}	2.707	190	0.836	4.342	0.823	16.320	0.806
C_{15}	3.263	206	0.842	3.309	0.838	3.648	0.812
C_{16}	2.061	222	0.849	1.877	0.848	0.528	0.818
C_{17}	1.699	237	0.845	1.516	0.845	0.171	0.815
C_{18}	1.898	251	0.848	1.687	0.848	0.144	0.817
C_{19}	1.231	263	0.858	1.095	0.858	0.100	0.827
C_{20}	0.911	275	0.863	0.810	0.863	0.069	0.832
C_{21}	0.782	291	‚0.868	0.694	0.868	0.048	0.837
C_{22}	0.663	305	0.873	0.588	0.873	0.036	0.842
C_{23}	0.566	318	0.877	0.501	0.877	0.027	0.845
C_{24}	0.498	331	0.881	0.441	0.881	0.021	0.849
C_{25}	0.427	345	0.885	0.378	0.885	0.014	0.853
C_{26}	0.358	359	0.889	0.316	0.889	0.010	0.857
C_{27}	0.322	374	0.893	0.285	0.893	0.007	0.861
C_{28}	0.286	388	0.897	0.253	0.897	0.006	0.865
C_{29}	0.286	402	0.900	0.256	0.900	0.004	0.868
C_{30}	0.238	416	0.903	0.209	0.903	0.005	0.870
C_{31}	0.199	430	0.907	0.175	0.907	0.005	0.874
C_{32}	0.134	444	0.910	0.118	0.910	0.005	0.877
C_{33}	0.100	458	0.913	0.088	0.913	0.005	0.880
C_{34}	0.064	472	0.916	0.057	0.916	0.006	0.883
C_{35}	0.046	486	0.919	0.040	0.919	—	—
C_{36+}	0.543	572	0.930	0.478	0.930	—	—

Source: Reproduced from Sah, P. et al., Equation-of-state modeling for reservoir fluid samples contaminated by oil-based drilling mud using contaminated fluid PVT data, *SPE Reservoir Eval. Eng.* 15, 139–149, 2012.

Note: M stands for molecular weight and OBM for oil-based mud.

TABLE 2.21

Numerical Cleaning Procedure Used for Contaminated Reservoir Fluid in Table 2.20

Mol% OBM in reservoir fluids sample from weight% contaminate in stock tank oil (STO)

1. Volume STO = 23,646/GOR
2. Grams STO = Volume STO × Density of STO
3. Grams contaminate = Grams STO × (Weight% OBM of STO)/100
4. Grams clean STO = Grams STO × (100 −Weight% OBM of STO)/100
5. Mole STO = Grams STO/Molecular weight of contaminated STO
6. Mole OBM = Grams contaminate/Molecular weight of OBM
7. Mole clean STO = Mole STO − Mole OBM
8. Mol% OBM of reservoir fluid = 100 × Mole OBM/(1 + Mole STO)

Volume balance equation to get density of cleaned C_{7+} carbon number fraction

$$\frac{z_i^{contam} M_i}{\rho_i^{contam}} = \frac{z_i^{res} M_i}{\rho_i^{res}} + \frac{z_i^{OBM} M_i}{\rho_i^{OBM}} \quad i = 1,...,N$$

where,

z_i^{contam} is the mole fraction of carbon number fraction i in the contaminated reservoir fluid (containing clean i plus mud component i)

ρ_i^{contam} is the density of carbon number fraction i in the contaminated reservoir fluid (containing clean i plus mud component i)

M_i is the molecular weight of component i

z_i^{res} is the mole fraction of clean carbon number fraction i in the contaminated reservoir fluid

ρ_i^{res} is the density of clean carbon number fraction i in the contaminated reservoir fluid

z_i^{OBM} is the mole fraction of mud component i in the contaminated reservoir fluid

ρ_i^{OBM} is the density of mud component i in the contaminated reservoir fluid

N is number of components

Note: OBM stands for oil-based mud.

REFERENCES

American Petroleum Institute, *Technical Data Book—Petroleum Refining*, API, New York, 1982.

Antoine, C., Tensions des vapeurs; nouvelle relation entre les tensions et les températures, *Comptes Rendus des Séances de l'Académie des Sciences* (in French), 107, 681–684, 1888.

Bergman, D.F., Tek, M.R., and Katz, D.L., *Retrograde Condensation in Natural Gas Pipelines*, Monograph Series, American Gas Association, New York, 1975.

Curvers, J. and van den Engel, P., Gas chromatographic method for simulated distillation up to a boiling point of 750°C using temperature programmed injection and high temperature fused silica wide-bore columns, *J. High Resol. Chromatogr.* 20, 16–22, 1989.

Gozalpour, F., Danesh, A., Tehrani, D.H., Todd, A.C., and Tohidi, B., Predicting reservoir fluid phase and volumetric behavior from samples contaminated with oil-based mud, SPE 78130, *SPE Reserv. Evaluation Eng.* 197–205, June 2002.

Hoffmann, A.E., Crump, J.S., and Hocott, C.R., Equilibrium constants for a gas condensate system, *Pet. Trans.*, *AIME* 198, 1–10, 1953.

Katz, D.L. and Firoozabadi, A., Predicting phase behavior of condensate/crude-oil systems using methane interaction coefficients, *J. Pet. Technol.* 20, 1649–1655, 1978.

Mawlod, A., Memon, A., and Nighswander, J., Accuracy and precision of reservoir fluid characterization tests through blind Round-Robin testing, SPE-207749-MS, presented at the *Abu Dhabi International Petroleum Exhibition & Conference*, Abu Dhabi, November 15–18, 2021.

Pedersen, K.S., Blilie, A.L., and Meisingset, K.K., PVT calculations on petroleum reservoir fluids using measured and estimated compositional data for the plus fraction, *Ind. Eng. Chem. Res.* 31, 1379–1384, 1992.

Poynting, J.H., Radiation-pressure, *Philos. Mag.* 9, 393–406, 1905.

Raoult, F.-M., Loi générale des tensions de vapeur des dissolvants, *Comptes Rendus* (in French). 104, 1430–1433, 1886.

Sah, P., Gurdial, G., Pedersen, K.S., Izwan, H., and Ramli, F., Equation-of-state modeling for reservoir fluid samples contaminated by oil-based drilling mud using contaminated fluid PVT data, *SPE Reserv. Evaluation Eng.* 15, 139–149, 2012.

Sah, P., Pedersen, K.S., and Shaikh, J.A., EOS modeling, the first step in detailed EOR potential evaluation— A case study, SPE 144025, presented at the *SPE Enhanced Oil Recovery Conference in Kuala Lumpur*, Malaysia, July 19–21, 2011.

Stephen, A.G., Bergman, D., Todd, T., and Kriel, W., PVT data quality: Round robin results, SPE 116162 presented at the *SPE ATCE in Denver*, Colorado, September 21–24, 2008.

Takeshi, A., Tetsuro, F., Leekumjorn, S., Shaikh, J.A., Pedersen, K.S., Alobeidli, A., Mogensen, K., A new technique for common EoS model development for multiple reservoir fluids with gas injection, SPE-202923-MS, presented at the *Abu Dhabi International Petroleum Exhibition & Conference*, Abu Dhabi, November 9–12, 2020.

3 PVT Experiments

To optimize production from oil and gas fields, it is essential to have extensive knowledge of the volumetric and phase changes the reservoir fluid will undergo on its way from petroleum reservoir to oil refinery. Reservoir pressures typically range from 100 to 2000 bar and reservoir temperatures from 25°C to 200°C. The well connecting the reservoir to the topside facilities can have a length of several kilometers. The pressure and temperature will gradually decrease in the production well, in flow lines connecting the well to the process plant, and in the process plant itself. Figure 3.1 illustrates schematically the production path of a reservoir fluid.

The conditions in the reservoir will also change during production. A reservoir fluid which in the exploration phase was either single-phase gas or single-phase oil may after some time of production split into two phases. The phase split is the result of material being removed from the reservoir. With more space available for the remaining reservoir fluid, the pressure will decrease and may after some time reach the saturation pressure, at which a second phase (gas or oil) will start to form.

PVT properties is the general term used to express the volumetric (V) behavior of a reservoir fluid as a function of pressure (P) and temperature (T). An essential PVT property is the saturation pressure at the reservoir temperature. From the time the reservoir pressure reaches the saturation pressure, and a second phase starts forming, the composition of the produced well stream will change because the production comes primarily from either the gas or the liquid zone.

It is customary to use the volumes of oil and gas at atmospheric pressure and 15°C as reference values. Atmospheric pressure (1 atm or 1.01325 bar) and 15°C are referred to as standard conditions. At standard conditions, a gas will behave approximately like an ideal gas, for which the ideal gas law applies:

$$\frac{PV}{RT} = 1 \tag{3.1}$$

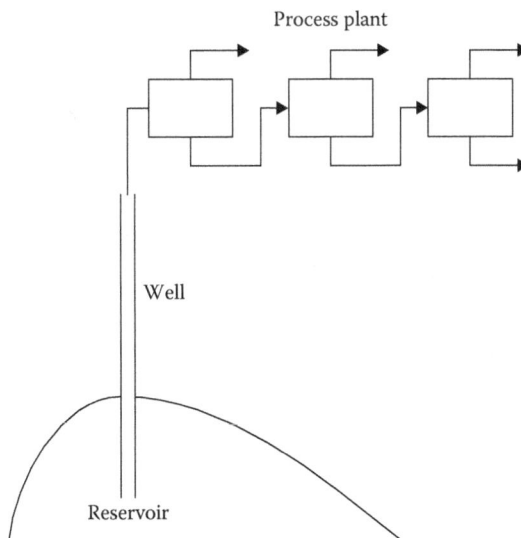

FIGURE 3.1 Production path of a reservoir fluids.

DOI: 10.1201/9780429457418-3

P stands for pressure, V for molar volume, T for temperature, and R is the gas constant. Any deviation from ideal gas behavior may be expressed through the compressibility factor, Z:

$$Z = \frac{PV}{RT} \qquad (3.2)$$

For an ideal gas, Z equals 1.0. For a nonideal gas or a liquid, Z may attain values below or above 1.

The gas dissolved in an oil at reservoir conditions will start evaporating once the saturation pressure is reached. The oil at standard conditions has little content of gas. The loss of gas components makes the oil volume decrease. It seems that the oil shrinks during production.

The volumetric changes taking place in the reservoir, during passage of the well and in the process plant, can be studied by performing PVT experiments on the reservoir fluid. This chapter describes the most commonly performed PVT experiments. Table 3.1 gives an overview of some important PVT properties measured in PVT experiments, all of which are described in more detail in the following. Pedersen et al. (1989) and Shaikh and Sah (2011) have given further descriptions of PVT experiments.

It is customary to distinguish between routine and enhanced oil recovery (EOR) PVT experiments. The routine PVT experiments emulate the processes taking place in a reservoir that is produced through natural depletion, also called primary recovery. As long as the reservoir pressure is higher than the saturation pressure, the fluid is produced in single phase. Pressure will drop as a result of the produced fluid being removed from the reservoir. When the pressure goes

---

TABLE 3.1
Definitions of PVT Properties Measured in a PVT Experiment

Relative volume: $V^{rel} = \dfrac{V^{tot}}{V^{sat}}$

Percentage liquid dropout = $100 \times \dfrac{V^{liq}}{V^{sat}}$

Isothermal compressibility: $c_o = -\dfrac{1}{V}\left(\dfrac{\partial V}{\partial P}\right)$

Y-factor: $\dfrac{\dfrac{P^{sat} - P}{P}}{\dfrac{V^{tot} - V^{sat}}{V^{sat}}}$

Gas gravity = $\dfrac{\text{Molecular weight of gas}}{\text{Molecular weight of atmospheric air}}$

Gas formation volume factor: $B_g = \dfrac{\text{Gas volume at cell condition}}{\text{Gas volume at standard conditions}}$

Oil formation volume factor of oil from stage N: $B_o = \dfrac{V_N^{oil}}{V_{std}^{oil}}$

Differential liberation gas/oil ratio of oil from stage N: $R_S = \dfrac{\sum\limits_{n=N+1}^{NST} V_{std,n}^{gas}}{V_{std}^{oil}}$

Separator gas/oil ratio: $\dfrac{V_{N,std}^{gas}}{V_{std}^{oil}}$

Note: The terms are further explained in the text.

TABLE 3.2
Key Data (Approximate Numbers) for PVT Cells Used for Different Types of Reservoir Fluids

Fluid Type	Cell Volume (ml)	Max Working Pressure (bar)	Max Working Temperature (°C)	Cell View (ml)
Oil	650	585	150	10
Gas condensate	500	585	150	140
HP/HT	4000	1030	177	Full cell

Note: HP/HT Stands for high pressure/high temperature.

TABLE 3.3
Routine PVT Experiments Carried Out on Various Fluid Types

Fluid	Constant-Mass Expansion	Differential Liberation	Separator Test	Constant Volume Depletion	Viscosity
Black oil	X	x	x		x
Volatile oil	X	x	x	x	x
Gas condensate	X		x	x	x[b]
Dry gas	x[a]				

[a] Only Z-factors.
[b] Not performed by standard.

below the saturation pressure, the fluid will split into two phases, a gas and a liquid. Due to its lower viscosity, the lighter gas phase will in most cases be produced faster than the liquid. If the natural depletion process is continued, the total recovery is likely to be low as a significant fraction of the heavier components is left behind in the reservoir. Various techniques exist for EOR, one being gas injection. Gas is injected to maintain the reservoir pressure above the saturation pressure. The aim is to continue producing a single-phase fluid with a high concentration of heavier components.

PVT experiments are heavily dependent on the availability of PVT cells that can provide accurate volumetric information at the relevant pressure and temperature conditions. It is also important to be aware how much sample volume is required to perform a PVT study. Table 3.2 provides key data for PVT cells in common use by PVT laboratories.

A PVT experiment is started by filling the cell with the pressurizing medium (often water) at the temperature and initial pressure of the experiment. The cell is connected to a calibrated pump. The sample is introduced in the cell displacing the pressurizing medium. The volume of displaced fluid is recorded, whose volume equals the initial cell fluid volume.

3.1 ROUTINE PVT EXPERIMENTS

As can be seen in Table 3.3, the routine PVT experiments to be performed depend on the fluid type. Around 600 cc of sample volume at reservoir conditions is required to conduct a routine PVT study. Figure 3.2 shows the workflow of a routine PVT study for an oil and a gas condensate mixture.

PVT STUDY OIL

| Collect samples |
| Validate samples |
| Constant Composition Expansion |
| Differential Liberation |
| Separator test |
| Viscosity measurements |

PVT STUDY CONDENSATE

| Collect samples |
| Validate samples |
| Constant Composition Expansion |
| Constant Volume Depletion |
| Separator test |

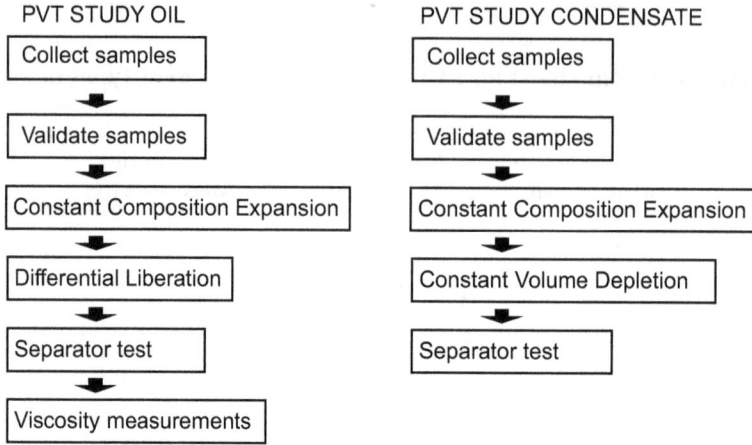

FIGURE 3.2 Workflows for routine PVT study on oil and gas condensate mixtures.

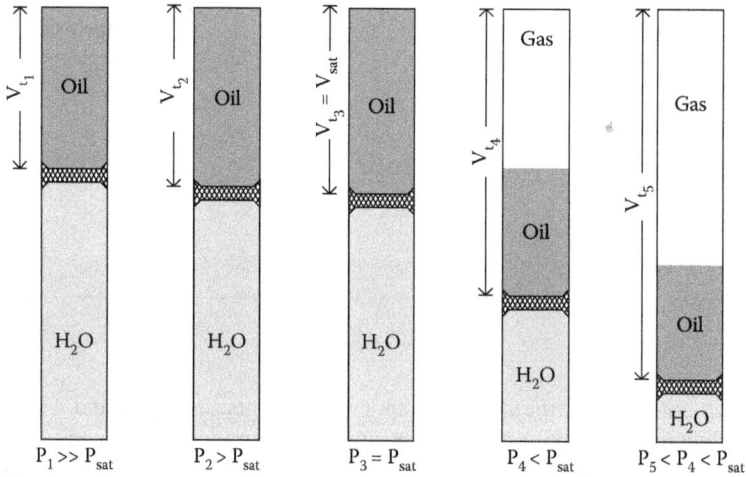

FIGURE 3.3 Schematic representation of a constant mass expansion experiment for an oil mixture.

3.1.1 CONSTANT MASS EXPANSION EXPERIMENT

The constant mass expansion (CME) experiment is also referred to as a constant composition expansion (CCE) or simply as a pressure-volume (PV) test. The CME experiment is performed to investigate the PV relationship of the reservoir fluid. Irrespective of fluid type, it is common practice to at least carry out a CME test at the reservoir temperature. One or two additional CME experiments may be carried out at a lower temperature.

3.1.1.1 Oil Mixtures

The CME experiment for an oil mixture is sketched in Figure 3.3. A known volume of single-phase sample is charged to a PVT cell and heated to the experimental temperature. At this temperature, the fluid is stabilized at single phase conditions at a pressure above the reservoir pressure. Once the sample is stable, the volume is noted. The sample is expanded to increase the volume of the fluid, which will make the pressure decrease. At a predefined pressure, the sample is stabilized and the volume noted. This process of establishing the PV relation is repeated in a number of steps from above the reservoir pressure down to an abandonment pressure. The saturation pressure is visually observed. For an oil mixture the saturation point is a bubble point. The term V^{sat} is used for the saturation point volume. At each stage of the experiment, the relative volume is recorded, defined as the ratio between the actual volume (V^{tot}) and the volume at the saturation pressure:

$$V^{rel} = \frac{V^{tot}}{V^{sat}}$$ (3.3)

The isothermal compressibility, c_o, is recorded above the saturation point:

$$c_o = -\frac{1}{V}\left(\frac{\partial V}{\partial P}\right)_T$$ (3.4)

In this expression, V is the oil volume. Below the saturation point, the Y-factor is recorded:

$$Y\text{-factor} = \frac{\frac{P^{sat}-P}{P}}{\frac{V^{tot}-V^{sat}}{V^{sat}}}$$ (3.5)

The Y-factor is a measure of the ratio between the relative changes in pressure and total volume in the two-phase region. As gas takes up more volume than liquid, the volumetric changes with decreasing pressure will be larger in the two-phase region than in the single-phase region. An oil that releases much gas with decreasing pressure will have a small Y-factor, whereas an oil that only releases small amounts of gas with decreasing pressure will have a large Y-factor.

A CME experiment is usually stopped at a pressure somewhere in the interval from 50 to 100 bar.

Table 3.4 shows the primary results from a CME experiment performed on an oil mixture. Table 3.5 shows CME results for the oil composition in Table 3.6. The Y-factor and relative volume results are plotted in Figure 3.4.

The reservoir fluid density is measured at a single-phase pressure. It can be at the reservoir pressure or any other pressure above the saturation pressure. Fluid is pumped into a pre-weighed vessel of a known volume and the weight increase is measured at the relevant pressure and temperature. The single-phase density is the ratio of mass and volume. The densities at other pressures down to the saturation pressure are calculated using this (reference) density and the relative oil volume data.

3.1.1.2   Gas Condensate Mixtures

The CME experiment is sketched for a gas condensate mixture in Figure 3.5. A known volume of single-phase sample is charged to a windowed PVT cell and heated to the experimental temperature.

TABLE 3.4
Primary Results from a Constant-Mass Expansion Experiment Performed on an Oil Mixture

Relative volume	V^{tot}/V^{sat}, where V^{tot} is the total fluid volume and V^{sat} is the bubble point (or saturation point) volume
Isothermal compressibility	Defined in Equation 3.4. Only reported above saturation point
Oil density	Only above saturation point
Y-factor	Defined in Equation 3.5. Only reported below saturation point

TABLE 3.5

Results of Constant Mass Expansion Experiment at 97.5°C for the Oil Mixture with the Composition Shown in Table 3.6

Pressure (bar)	Relative Volume (V/V^{sat})	Compressibility (1/bar)	Y-Factor
351.4	0.9765	0.000185	—
323.2	0.9721	0.000200	—
301.5	0.9762	0.000211	—
275.9	0.9818	0.000225	—
250.1	0.9874	0.000238	—
226.1	0.9933	0.000249	—
205.9	0.9986	0.000260	—
200.0[a]	1.0000	0.000263	—
197.3	1.0043	—	3.07
189.3	1.0189	—	3.01
183.3	1.0313	—	2.95
165.0	1.0776	—	2.80
131.2	1.2136	—	2.51
108.3	1.3715	—	2.31
85.3	1.6343	—	2.11
55.6	2.3562	—	1.86

Note: Y-factor and relative volume results are plotted in Figure 3.4.

[a] Saturation point.

TABLE 3.6

Molar Composition of the Oil Mixture for Which Constant Mass Expansion Data Is Shown in Table 3.5 and Differential Depletion Data in Table 3.13

Component	Mol%	Molecular Weight (g/mol)	Density at 1.01 bar, 15°C (g/cm³)
N_2	0.39	—	—
CO_2	0.30	—	—
C_1	40.20	—	—
C_2	7.61	—	—
C_3	7.95	—	—
iC_4	1.19	—	—
nC_4	4.08	—	—
iC_5	1.39	—	—
nC_5	2.15	—	—
C_6	2.79	—	—
C_7	4.28	95	0.729
C_8	4.31	106	0.749
C_9	3.08	121	0.770
C_{10}	2.47	135	0.786
C_{11}	1.91	148	0.792
C_{12}	1.69	161	0.804
C_{13}	1.59	175	0.819
C_{14}	1.22	196	0.833
C_{15}	1.25	206	0.836
C_{16}	1.00	224	0.843
C_{17}	0.99	236	0.840
C_{18}	0.92	245	0.846
C_{19}	0.60	265	0.857
C_{20+}	6.64	453	0.918

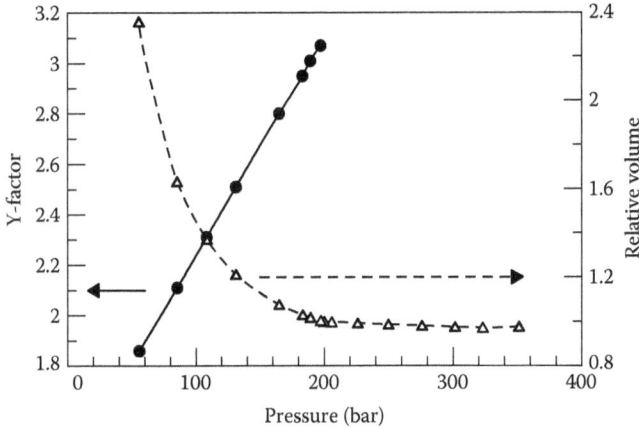

FIGURE 3.4 Y-factor (circles, full-drawn line, and left y-axis) and relative volume (triangles, dashed line, and right y-axis) for constant-mass expansion experiment at 97.5°C on the oil mixture in Table 3.6. The results are tabulated in Table 3.5.

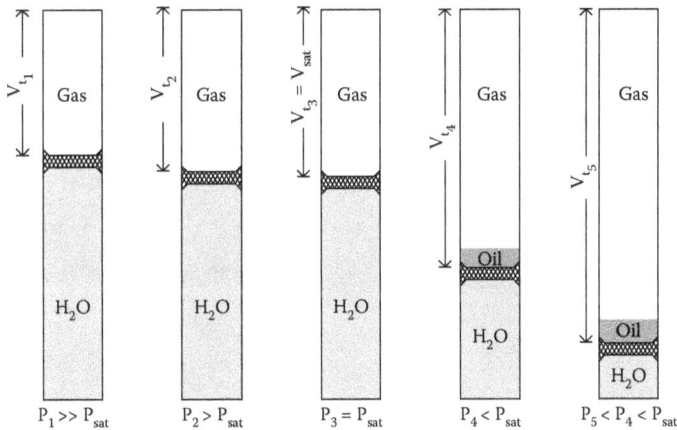

FIGURE 3.5 Schematic representation of a constant mass expansion experiment for a gas condensate.

At this temperature, the fluid is stabilized at a pressure above the reservoir pressure and saturation pressure. A CCE is performed as explained earlier for oil mixtures to a final pressure of around 50 bar. The dew point pressure is visually measured. The relative volume as defined in Equation 3.3 is reported at all pressures. The gas phase compressibility factor as defined in Equation 3.2 is recorded for pressures above the saturation pressure. Below the saturation pressure, the liquid volume as a percentage of the saturation point volume is reported:

$$\% \text{ Liquid dropout} = 100 \times \frac{V^{liq}}{V^{sat}} \tag{3.6}$$

This liquid volume is often referred to as the *liquid dropout*.

The single-phase fluid density is measured in the same way as explained earlier for oil mixtures. This measured (reference) density is then used to calculate the density and compressibility factor at other pressures, using the measured relative volume data.

Table 3.7 lists the primary results from a CME experiment performed on a gas condensate mixture.

Table 3.8 shows the composition of a gas condensate mixture and Table 3.9 the results of a CME experiment performed on this fluid. The relative volume and the liquid dropout versus pressure are plotted in Figure 3.6.

TABLE 3.7

Primary Results from a Constant-Mass Expansion Experiment Performed on a Gas Condensate Mixture

Relative volume	$V^{rel} = \dfrac{V^{tot}}{V^{sat}}$, where V^{tot} is the total fluid volume and V^{sat} is the dew point (or saturation point) volume
Liquid volume	Liquid volume percentage of V^{sat}
Z-factor	Defined in Equation 3.2. Only reported above saturation point

TABLE 3.8

Molar Composition of the Gas Condensate Mixture for Which Constant Mass Expansion Data Is Shown in Table 3.9

Component	Mol%	Molecular Weight (g/mol)	Density at 1.01 bar, 15°C (g/cm³)
N_2	0.60	—	—
CO_2	3.34	—	—
C_1	74.16	—	—
C_2	7.90	—	—
C_3	4.15	—	—
iC_4	0.71	—	—
nC_4	1.44	—	—
iC_5	0.53	—	—
nC_5	0.66	—	—
C_6	0.81	—	—
C_7	1.20	91	0.746
C_8	1.15	104	0.770
C_9	0.63	119	0.788
C_{10}	0.50	133	0.795
C_{11}	0.29	144	0.790
C_{12}	0.27	155	0.802
C_{13}	0.28	168	0.814
C_{14}	0.22	181	0.824
C_{15}	0.17	195	0.833
C_{16}	0.15	204	0.836
C_{17}	0.14	224	0.837
C_{18}	0.09	234	0.839
C_{19}	0.13	248	0.844
C_{20+}	0.47	362	0.877

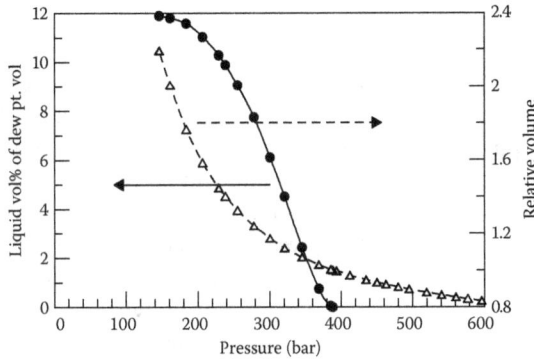

FIGURE 3.6 Liquid dropout curve (circles, full-drawn line, and left y-axis) and relative volume (triangles, dashed line, and right y-axis) for constant-mass expansion experiment at 155°C on the gas condensate mixture in Table 3.8. The results are tabulated in Table 3.9.

TABLE 3.9
Results of Constant Mass Expansion Experiment at 155°C for a Gas Condensate Mixture of the Composition Shown in Table 3.8

Pressure (bar)	Relative Volume (V/Vsat)	Liquid Volume (Percentage of Vsat)	Z-Factor
597.1	0.8338	—	1.3729
577.8	0.8441	—	1.3450
560.9	0.8539	—	1.3208
540.5	0.8656	—	1.2902
519.5	0.8793	—	1.2596
495.1	0.8968	—	1.2244
479.8	0.9090	—	1.2027
462.7	0.9232	—	1.1779
449.9	0.9341	—	1.1589
434.8	0.9481	—	1.1367
412.0	0.9720	—	1.1043
393.0	0.9959	—	1.0793
388.0[a]	1.0000	0.00	1.0740
385.1	1.0035	0.05	—
368.6	1.0299	0.75	—
345.1	1.0707	2.43	—
320.7	1.1200	4.52	—
300.5	1.1727	6.11	—
278.7	1.2411	7.75	—
255.6	1.3249	9.06	—
238.6	1.4021	9.89	—
229.3	1.4476	10.29	—
206.7	1.5843	11.03	—
183.7	1.7651	11.58	—
161.3	2.0047	11.80	—
146.2	2.1923	11.89	—

Note: Relative volume and liquid volume results are plotted in Figure 3.6.

[a] Saturation point.

3.1.1.3 Dry Gases

Dry gases will not have a saturation point at the typical reservoir temperature. It is therefore not possible to conduct a full CME study on a dry gas. Instead, a PVTZ study is performed, which reports gas Z-factors at the relevant temperature and decreasing pressures. For a gas to be used as injection gas for EOR purposes, the temperature will most often be equal to the reservoir temperature. A pressure interval around the reservoir pressure will be covered. The compressibility factors of the gas are determined by initially measuring the volume (V^{ref}) of a given amount of gas at the temperature in question (T^{ref}) and the highest pressure (P^{ref}) to be considered (reference conditions). The gas is flashed to atmospheric (standard) conditions and the gas volume recorded. Assuming that the Z-factor is 1.0 at atmospheric conditions, the number of moles of gas, n, contained in the total volume at standard conditions (V^{std}) can be determined using a slightly rewritten Equation 3.1:

$$n = \frac{P^{std}\, V^{std}}{RT^{std}} \tag{3.7}$$

The same number of moles will be contained in the total volume (V^{ref}) at the reference pressure and temperature. That allows the Z-factor (Z^{ref}) at reference conditions to be determined from:

$$Z^{ref} = \frac{P^{ref}\, V^{ref}\, T^{std}}{P^{std}\, V^{std}\, T^{ref}} \tag{3.8}$$

The same equation may be used to calculate the Z-factor at other pressures and temperatures for which the total volume of the same molar amount of gas has been measured.

Any deviation from 1.0 of the Z-factor at atmospheric conditions will transfer into the Z-factors at elevated pressure and temperature. If, for example, the gas Z-factor at standard conditions is not 1.0 but only 0.99, the reported gas Z-factor at reservoir conditions will be 1% too low.

Table 3.10 shows the composition of a dry gas for which gas Z-factors at 155°C are shown in Table 3.11.

Measurements of gas Z-factors are further dealt with in Section 3.1.7.

3.1.2 Differential Liberation Experiment

The differential liberation (DL) experiment is sketched in Figure 3.7. It is also known as a differential vaporization or differential depletion experiment and is performed on black oil reservoir fluids

---

TABLE 3.10

Molar Composition of the Dry Gas Composition for Which Z-Factor Data Is Shown in Table 3.11

Component	Mol%
N_2	0.60
CO_2	3.34
C_1	74.16
C_2	7.90
C_3	4.15
iC_4	0.71
nC_4	1.44
iC_5	0.53
nC_5	0.66
C_6	0.81

TABLE 3.11
Z-Factor Data for the Dry Gas in Table 3.10 at 155°C

Pressure (bar)	Z-Factor
388.0[a]	1.1109
385.1	1.1089
368.6	1.0974
345.1	1.0816
320.7	1.0660
300.5	1.0538
278.7	1.0414
255.6	1.0294
238.6	1.0212
229.3	1.0170
206.7	1.0078
183.7	0.9998
161.3	0.9936
146.2	0.9903

[a] Reservoir pressure.

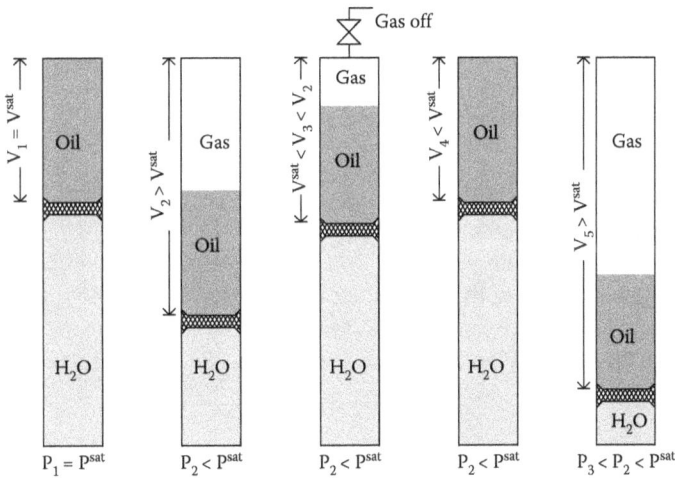

FIGURE 3.7 Schematic representation of a differential depletion experiment.

and volatile oils. The DL experiment emulates the compositional and volumetric changes that occur in oil reservoirs during production.

The experiment is started by transferring reservoir fluid to a cell kept at a fixed temperature, most often the reservoir temperature, and a pressure above the reservoir pressure. The DL cell is equipped with a valve on top allowing gas to be depleted (removed) during the experiment. The experiment is started at the saturation pressure.

The sample is equilibrated at the first selected pressure below the saturation pressure. The evolved gas is pumped out of the cell at constant pressure and its volume and composition measured at standard conditions. The pressure is further reduced, the evolved gas taken off, and so on in typically six pressure stages between saturation and atmospheric pressure.

Figure 3.8 illustrates a DL experiment in a pressure–temperature diagram.

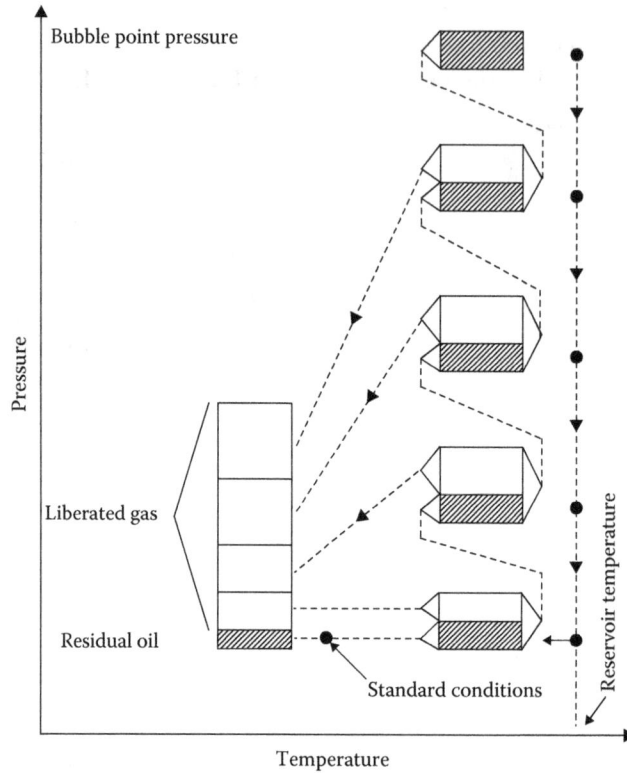

FIGURE 3.8   Differential liberation experiment in a pressure–temperature diagram.

A differential depletion experiment is usually continued down to atmospheric pressure before cool-ing off the cell to the standard temperature of 15°C. The volume of the cell content at atmospheric (standard) conditions is reported as the residual (or standard) oil volume, V_{std}^{oil}. The liquid volumes at the remaining pressure stages are reported relative to the residual oil volume through the oil formation volume or shrinkage factor, B_o. If the oil volume at stage N is V_N^{oil}, B_o is for stage N defined as:

$$B_o(N) = \frac{V_N^{oil}}{V_{std}^{oil}} \qquad (3.9)$$

The oil at standard conditions is often referred to as *stable oil* to indicate that it can be trans-ported at standard conditions without further release of gas. The B_o-factor defined in Equation 3.9 is a measure of how much the oil would shrink during production. Assume an oil has a B_o-factor of $B_{o,x}$ at a reservoir pressure of P_x and that it takes up a volume of VOL_x at P_x. The oil volume will at atmospheric conditions have shrunk to $VOL_x/B_{o,x}$. The oil shrinks because it releases gas when the pressure decreases and because of thermal contraction with decreasing temperature. The SI unit for B_o is m^3/Sm^3, where Sm^3 indicates that the residual oil volume is in m^3 and at standard conditions.

The solution gas/oil ratio (GOR), R_S, is another important quantity measured in a DL experi-ment. The gas/oil ratio of the oil at a given stage in a DL experiment is calculated by adding the standard volumes of the gas liberated in each of the subsequent stages and then dividing this sum of gas volumes by the residual oil volume. For the oil at stage N in a DL experiment with a total of NST pressure stages, R_S is given by

$$R_S(N) = \frac{\sum_{n=N+1}^{NST} V_{std,n}^{gas}}{V_{std}^{oil}} \qquad (3.10)$$

The standard volumes of gas are determined by flashing the gas liberated at each stage to standard conditions. That will cause a small volume of liquid to drop out. The molar volume of the liquid can be determined as the ratio of density and molecular weight. The molar amount, n, is the ratio between the actual liquid volume and the molar volume. This volume is converted to an equivalent gas volume by solving Equation 3.7 for V^{std}. That volume is added to the gas volume entering into Equation 3.10.

The volume of the liberated gas is measured at both cell conditions and at standard conditions. This enables calculation of the gas formation volume factor, B_g:

$$B_g = \frac{\text{Gas volume at cell conditions}}{\text{Gas volume at standard conditions}} \qquad (3.11)$$

Cell conditions refer to the pressure and temperature in the cell at the pressure stage at which the gas was depleted. The SI unit for B_g is m^3/Sm^3. S stands for standard and indicates that the volume is reported at standard conditions.

The gas gravity is defined as the average molecular weight of the gas divided by the average molecular weight of atmospheric air:

$$\text{Gas gravity} = \frac{\text{Molecular weight of gas}}{\text{Molecular weight of atmospheric air}} \qquad (3.12)$$

The molecular weight of atmospheric air is usually taken to be equal to 28.964 g/mol. By expressing the molecular weight relative to that of atmospheric air, the gas gravity becomes a measure of the low-pressure density of the liberated gas relative to that of atmospheric air.

Table 3.12 lists the primary results reported from a DL experiment.

The reservoir pressure will decrease during production. From the time the pressure reaches the saturation pressure, two phases will be present, an oil phase and a gas phase. Owing to continuous liberation of gas, the amount of gas dissolved in the oil will decrease with decreasing pressure. This will result in decreasing B_o-factors and solution gas/oil ratios with decreasing pressure. Figure 3.9 shows a plot of the B_o-factor of the oil in Table 3.6 against pressure. The results are tabulated in Table 3.13. It is seen that the B_o-factor increases with decreasing pressure above the

---

TABLE 3.12

Primary Results from a Differential Liberation Experiment Performed on an Oil Mixture

B_o	Oil formation volume factor, that is, oil volume at actual pressure, divided by volume of residual oil at standard conditions
R_s	Solution gas/oil ratio, that is, total standard volume of gas liberated at lower pressure stages than the actual one, divided by the volume of the residual oil at standard conditions
Oil density	Density of oil phase at cell conditions
B_g	Gas formation volume factor defined as gas volume at the actual pressure divided by the volume of the same gas at standard conditions
Z-factor gas	Defined in Equation 3.2. Refers to depleted gas at cell conditions
Gas gravity	Molecular weight of the gas liberated divided by the molecular weight of atmospheric air (= 28.964)

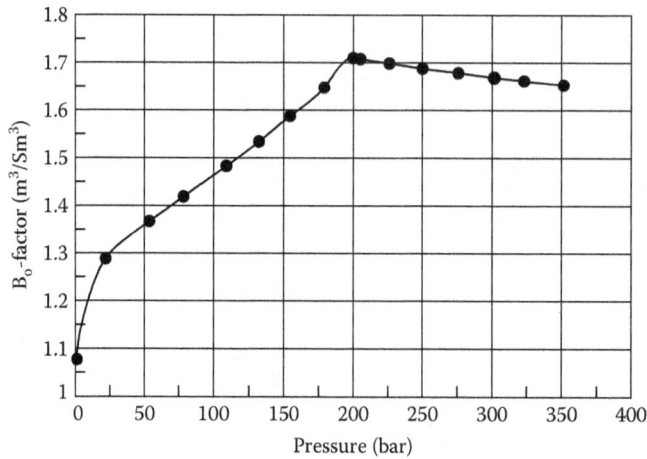

FIGURE 3.9 B_o-factor as a function of pressure in the differential liberation experiment at 97.5°C on the oil composition in Table 3.6. The results are tabulated in Table 3.13.

TABLE 3.13
Results of Differential Liberation Experiment at 97.5°C on the Oil with the Composition Shown in Table 3.6

Pressure (bar)	B_o (m³/Sm³)	R_S (m³/Sm³)	Oil Density (g/cm³)	B_g (m³/Sm³)	Z-Factor Gas	Gas Gravity (air = 1)
351.4	1.653	198.3	0.670	—	—	—
323.2	1.662	198.3	0.667	—	—	—
301.5	1.669	198.3	0.664	—	—	—
275.9	1.679	198.3	0.660	—	—	—
250.1	1.688	198.3	0.656	—	—	—
226.1	1.699	198.3	0.652	—	—	—
205.9	1.708	198.3	0.649	—	—	—
200.0[a]	1.710	198.3	0.645	—	—	—
179.1	1.648	176.2	0.656	0.00610	0.844	0.791
154.6	1.588	154.3	0.668	0.00713	0.851	0.779
132.1	1.534	134.5	0.679	0.00839	0.857	0.764
109.0	1.483	115.5	0.691	0.01030	0.868	0.758
78.6	1.413	91.7	0.706	0.01440	0.882	0.772
53.6	1.367	72.8	0.719	0.02150	0.901	0.805
22.0	1.288	46.1	0.739	0.05280	0.933	0.953
1.0	1.077	0.0	0.778	—	—	2.022
1.01, 15°C	1.000	0.0	0.838	—	—	—

[a] Saturation point.

saturation point. This is because the oil expands with decreasing pressure until it starts releasing gas. Figure 3.10 shows a plot of R_S towards pressure for the same oil. Above the saturation point, R_S is constant because the composition of the produced reservoir fluid is constant until the saturation point is reached. Below the saturation point, R_S decreases with decreasing pressure. The gas liberated from the oil just below the saturation point primarily consists of lighter gas components.

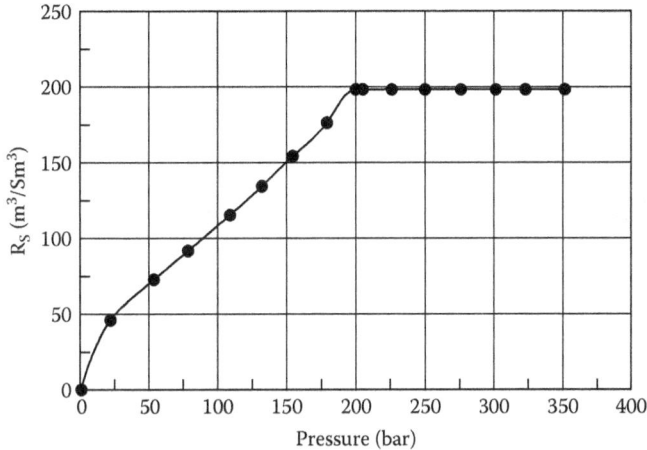

FIGURE 3.10 R_S as a function of pressure in the differential liberation experiment at 97.5°C on oil composition in Table 3.6. The results are tabulated in Table 3.13.

As the pressure is further decreased, the content of heavier compounds in the gas will increase. This is reflected in an increasing gas gravity with decreasing pressure, as can be seen in Table 3.13.

3.1.3 Constant-Volume Depletion Experiment

The constant-volume depletion (CVD) experiment is performed on gas condensate and volatile oil mixtures. It is sketched for a gas condensate in Figure 3.11 and for a volatile oil in Figure 3.12.

The test consists of a series of volume expansions and constant pressure displacements of excess volume of gas to return the cell content to a constant volume. The constant volume is equal to the volume at the saturation pressure. This procedure is repeated for typically six stages down to the abandonment pressure of approximately 50 bar.

The cell is constructed in the same manner as for a CME experiment but is equipped with a valve on top allowing depletion of gas during the experiment. The experiment is started at the saturation point. The saturation point pressure, P^{sat}, and the saturation point volume, V^{sat}, are recorded. The volume is increased, which will make the pressure decrease, and two separate phases form in the cell. The mixture volume is subsequently decreased to V^{sat} by letting out the excess gas through the valve on top, maintaining a constant pressure. The molar amount of gas depleted as a percentage of the gas initially in the cell and the liquid volume in the cell as a percentage of the saturation point volume are recorded. The volume is increased again, the excess volume is depleted, and so on until the pressure is somewhere between 100 and 40 bar.

The composition of each displaced gas can be determined using the techniques explained in Chapter 2. The number of moles of depleted gas is found as the ratio of mass depleted to molecular weight. The molar volume of the depleted gas at cell conditions equals the volume depleted divided by the number of moles. That allows the compressibility factor to be determined from Equation 3.2.

The liquid volume at each depletion pressure is measured and expressed as a percentage of the sample volume at the saturation pressure.

It is customary to also report the gas viscosity at cell conditions from a CVD experiment. Usually, the reported gas viscosities are not measured values but calculated from a correlation, often the one of Lee et al. (1966), which is further described in Chapter 10.

The primary results from a constant-volume depletion experiment are summarized in Table 3.14.

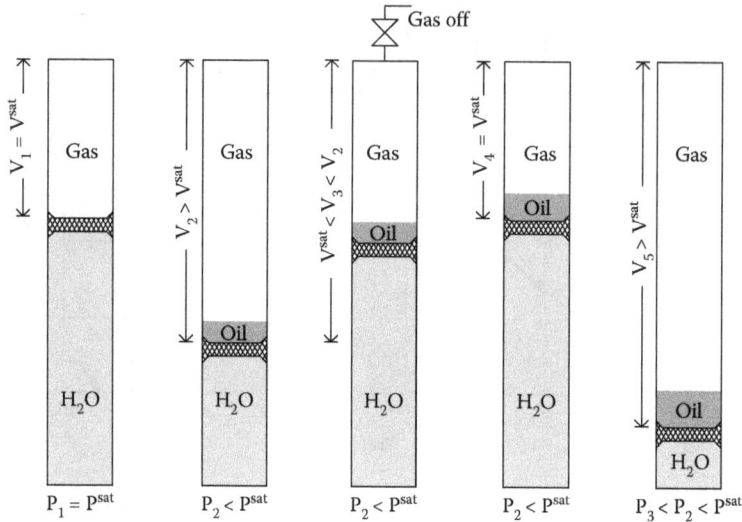

FIGURE 3.11   Schematic representation of a constant volume depletion experiment for a gas condensate fluid.

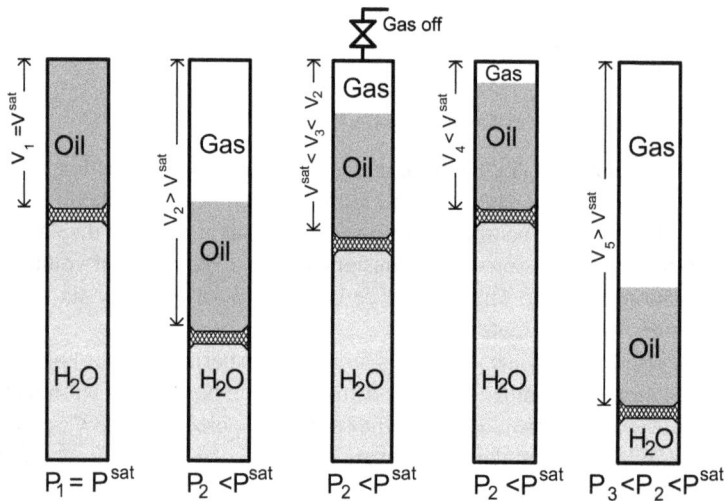

FIGURE 3.12   Schematic representation of a constant volume depletion experiment for a volatile oil.

TABLE 3.14

Primary Results from a Constant-Volume Depletion Experiment Performed on a Gas Condensate or Volatile Mixture

Liquid volume	Liquid volume percentage of saturation point volume
Percentage produced	Cumulative molar percentage of initial mixture removed (depleted) from cell
Z-factor gas	Defined in Equation 3.2. Refers to depleted gas at cell conditions
Two-phase Z-factor	Z-factor defined in Equation 3.2. Average of gas and liquid in cell after removal of excess gas
Viscosity of gas	Viscosity of the gas in cell (usually not measured but calculated)
Gas compositions	Molar compositions of gas liberated from each pressure stage $< P^{sat}$

The constant-volume depletion experiment has been designed to gain knowledge about the changes with time in PVT properties and composition of the produced well streams from gas condensate and volatile oil reservoirs. The reservoir is seen as a tank of fixed volume and at a fixed temperature. During production the pressure decreases because material is removed from the field, whereas the volume and temperature remain (almost) constant. When the pressure reaches the saturation point, the mixture splits into a gas and a liquid phase. If all the production comes from the gas zone, the mixture produced will have the same composition as the gas removed from the cell in a constant-volume depletion experiment. This gas will gradually become less enriched in heavy hydrocarbons, and less liquid will be produced from the topside separation plant.

The amount of reservoir fluid removed from the reservoir from the time the pressure is P_1 until it has decreased to P_2 corresponds to the amount of gas removed through the valve on top of the PVT cell in the depletion stage at pressure P_2.

Table 3.15 shows the molar composition of a gas condensate mixture that has undergone a constant volume depletion experiment at a temperature of 150.5 °C. The primary data from the CVD experiment is shown in Table 3.16. Table 3.17 shows the compositions of the depleted gas at the various pressure stages. The C_{7+} concentration is seen to decrease with decreasing pressure until the pressure is below 50 bar. The molecular weight of the C_{7+} fraction of the depleted gas decreases with decreasing pressure until the pressure is below 150 bar.

TABLE 3.15
Molar Composition of the Gas Condensate Mixture for Which Constant-Volume Depletion Results Are Shown in Tables 3.16 and 3.17

Component	Mol%	Wt%	Molecular Weight (g/mol)	Density at 1.01 bar, 15°C (g/cm³)
N_2	0.64	0.57	—	—
CO_2	3.53	4.92	—	—
C_1	70.78	35.94	—	—
C_2	8.94	8.51	—	—
C_3	5.05	7.05	—	—
iC_4	0.85	1.56	—	—
nC_4	1.68	3.09	—	—
iC_5	0.62	1.42	—	—
nC_5	0.79	1.80	—	—
C_6	0.83	2.26	—	—
C_7	1.06	3.09	92.2	0.7324
C_8	1.06	3.51	104.6	0.7602
C_9	0.79	2.98	119.1	0.7677
C_{10}	0.57	2.40	133	0.790
C_{11}	0.38	1.86	155	0.795
C_{12}	0.37	1.90	162	0.806
C_{13}	0.32	1.79	177	0.824
C_{14}	0.27	1.69	198	0.835
C_{15}	0.23	1.47	202	0.840
C_{16}	0.19	1.29	215	0.846
C_{17}	0.17	1.26	234	0.840
C_{18}	0.13	1.03	251	0.844
C_{19}	0.13	1.11	270	0.854
C_{20+}	0.62	7.48	381	0.880

TABLE 3.16

Results of Constant-Volume Depletion Experiment at 150.5°C for Gas Condensate Mixture with the Composition Shown in Table 3.15

Pressure (bar)	Liquid Volume (Percentage of V^{sat})	Gas-Phase Z-Factor	Two-Phase Z-Factor	Cumulative Mole Percentage Depleted
381.5[a]	0.0	1.084	1.084	0.00
338.9	3.1	1.031	1.019	6.57
290.6	6.9	0.981	0.971	15.84
242.3	9.9	0.941	0.933	26.86
194.1	11.3	0.911	0.900	39.58
145.8	11.1	0.896	0.876	53.95
97.5	10.5	0.910	0.852	69.89
49.3	9.6	0.940	0.798	84.43

[a] Saturation point.

TABLE 3.17

Molar Compositions (Mol%) of the Gas Depleted in a Constant-Volume Depletion Experiment at 150.5°C for the Mixture of Table 3.15

Pressure (bar)	381.5[a]	338.9	290.6	242.3	194.1	145.8	97.5	49.3
N_2	0.64	0.65	0.66	0.67	0.67	0.67	0.66	0.63
CO_2	3.53	3.50	3.52	3.55	3.59	3.61	3.63	3.68
C_1	70.78	72.29	73.27	73.92	74.31	74.44	74.24	73.29
C_2	8.94	8.83	8.89	9.01	9.02	9.04	9.20	9.30
C_3	5.04	4.99	4.96	4.92	4.93	4.97	5.01	5.19
iC_4	0.85	0.82	0.81	0.80	0.80	0.81	0.84	0.89
nC_4	1.67	1.65	1.64	1.62	1.63	1.66	1.68	1.72
iC_5	0.61	0.59	0.57	0.56	0.56	0.57	0.58	0.61
nC_5	0.78	0.76	0.74	0.72	0.72	0.72	0.74	0.76
C_6	0.81	0.77	0.73	0.70	0.68	0.68	0.71	0.77
C_{7+}	6.35	5.15	4.21	3.53	3.09	2.83	2.71	3.16
C_{7+} molecular weight	161	151	141	132	125	121	121	123

[a] Saturation point.

Table 3.18 shows the molar composition of a volatile oil mixture that has undergone a constant volume depletion experiment at a temperature of 119.4°C. The primary data from the CVD experiment is shown in Table 3.19. Table 3.20 shows the compositions of the depleted gas at the various pressure stages.

3.1.4 Separator Test

Separator experiments are carried out for both oil and gas condensate mixtures. A three-stage separator test is sketched in Figure 3.13. The separator test is performed to measure the compositional and volumetric changes the fluid will undergo during production. The separator test conditions are often selected to correspond with those of the field separators.

TABLE 3.18

Molar Composition of the Volatile Oil Mixture for Which Constant Volume Depletion Results Are Shown in Tables 3.19 and 3.20

Component	Mol %
N_2	0.11
CO_2	2.87
H_2S	1.83
C_1	54.33
C_2	8.17
C_3	4.46
iC_4	0.99
nC_4	2.75
iC_5	1.14
nC_5	1.57
C_6	2.10
C_7	2.39
C_8	2.69
C_9	2.34
C_{10+}	12.26

TABLE 3.19

Results of Constant Volume Depletion Experiment at 119.4 °C for Volatile Oil Composition Shown in Table 3.18

Pressure bar	Liquid Volume (Percentage of V^{sat})	Cum Mole Percentage Depleted	Gas Phase Z-Factor	Two Phase Z-Factor
312.7	100.00	0.00		
266.5	84.21	6.55	0.876	0.982
225.1	76.09	14.39	0.861	0.907
180.3	70.88	24.43	0.877	0.825
138.9	67.45	35.07	0.907	0.743
83.8	63.45	47.90	0.942	0.631
49.3	56.63	62.97	0.973	0.472

TABLE 3.20

Molar Compositions (Mole%) of the Fluid Depleted in a Constant Volume Depletion Experiment at 119.4 °C for the Volatile Oil in Table 3.18

Pressure (bar)	312.7	266.5	225.1	180.3	138.9	83.8	49.3
N_2	0.11	0.13	0.21	0.23	0.22	0.21	0.18
CO_2	2.84	2.78	3.18	3.26	3.38	3.53	3.71
H_2S	1.81	2.07	1.79	1.85	1.94	2.10	2.42
C_1	53.74	54.65	73.26	75.24	75.93	76.09	74.08
C_2	8.08	8.21	8.44	8.43	8.62	9.02	9.87

(*Continued*)

TABLE 3.20   (Continued)

Molar Compositions (Mole%) of the Fluid Depleted in a Constant Volume Depletion Experiment at 119.4 °C for the Volatile Oil in Table 3.18

Pressure (bar)	312.7	266.5	225.1	180.3	138.9	83.8	49.3
C_3	4.41	4.40	3.77	3.75	3.75	3.79	4.24
iC_4	0.98	0.94	0.72	0.69	0.69	0.70	0.76
nC_4	2.72	2.62	1.82	1.71	1.70	1.71	1.84
iC_5	1.12	1.11	0.68	0.60	0.55	0.53	0.57
nC_5	1.56	1.52	0.92	0.80	0.72	0.65	0.69
C_6	2.08	2.00	1.04	0.86	0.73	0.59	0.60
C_7	2.36	2.28	1.01	0.76	0.55	0.44	0.43
C_8	2.66	2.51	0.91	0.64	0.50	0.32	0.31
C_9	2.28	2.16	0.65	0.44	0.30	0.17	0.15
C_{10+}	13.25	12.62	1.60	0.75	0.42	0.15	0.15
C_{10+} Mol Wgt	215	182	172	163	161	146	147
C_{10+} Density (g/cm^3)	0.855	0.818	0.813	0.806	0.799	0.794	0.795

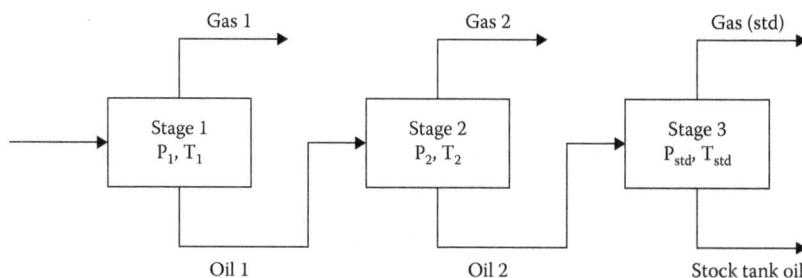

FIGURE 3.13   Schematic representation of a three-stage separator experiment.

The reservoir fluid is placed in a closed cell (a *separator*) at pressure and temperature conditions, at which the fluid mixture separates in a gas and a liquid phase. Once the two phases are in equilibrium, the gas is pumped out of the separator at the test conditions through the top and is transferred to standard conditions, where its volume, gas gravity, and composition are measured. As for the DL experiment, liquid dropping out from the gas is converted to an equivalent gas volume at standard conditions. The liquid from the first stage separator is subjected to a second stage separation at a lower pressure and temperature than the first one. More gas will be liberated. As for the first separator this gas is transferred to standard conditions and its properties measured. The oil from the last separator at standard conditions is called *stock tank oil*, and the volume of this oil is called *stock tank oil volume*. The term stock tank signals that the oil can be stored at atmospheric conditions without liberating gas. The stock tank oil is weighed and the density measured at 15°C to calculate the volume of stock tank liquid. The composition of the stock tank liquid is determined.

The purpose of a separator experiment is to get a rough idea about the relative volumetric amounts of gas and oil produced from a particular petroleum reservoir. Table 3.21 gives a summary of the results reported from a separator experiment.

Table 3.22 shows the results of a four-stage separator test on the oil composition in Table 3.23. The compositions of the gas liberated in each stage are shown in Table 3.24. The separator results in Table 3.22 also include a B_o-factor for the reservoir fluid at 199.7 bar (saturation point) and 97.8°C

TABLE 3.21
Primary Results from a Separator Experiment Performed on an Oil or a Gas Condensate Mixture

Separator gas/oil ratio	Volume of gas from actual separator stage at standard conditions divided by the volume of the oil from the last stage (at atmospheric conditions)
Gas gravity	Molecular weight of the gas from actual separator stage divided by the molecular weight of atmospheric air (=28.964)
Separator B_o	Oil formation volume factor, that is, volume of oil at actual separator stage, divided by volume of oil from last stage (at atmospheric conditions). For oil mixtures it is customary also to report B_o of the saturated reservoir oil
Gas compositions	Molar compositions of separator gas in each stage and of the oil at standard conditions

TABLE 3.22
Results of a Separator Experiment for the Oil of the Composition in Table 3.23

	Pressure (bar)	Temperature (°C)	Gas/Oil Ratio (Sm^3/Sm^3)	B_o-Factor (m^3/Sm^3)
Saturation point	199.7	97.8	—	1.605
Stage 1	68.9	89.4	109.0	1.279
Stage 2	22.7	87.2	33.7	1.182
Stage 3	6.9	83.9	17.1	1.126
Stage 4	2.0	77.2	12.3	1.053
Standard	1.0	15.0	0.0	1.000

TABLE 3.23
Molar Composition of the Oil for Which Separator Test Results Are Given in Tables 3.22 and 3.24

Component	Mol%	Molecular Weight (g/mol)	Density at 1.01 bar, 15°C (g/cm³)
N_2	0.59	—	—
CO_2	0.36	—	—
C_1	40.81	—	—
C_2	7.38	—	—
C_3	7.88	—	—
iC_4	1.20	—	—
nC_4	3.96	—	—
iC_5	1.33	—	—
nC_5	2.09	—	—
C_6	2.84	—	—
C_7	4.15	97	0.711
C_8	4.37	113	0.740
C_9	3.40	129	0.763
C_{10}	2.52	144	0.780
C_{11}	1.87	158	0.794

(*Continued*)

TABLE 3.23 (*Continued*)
Molar Composition of the Oil for Which Separator Test Results Are Given in Tables 3.22 and 3.24

Component	Mol%	Molecular Weight (g/mol)	Density at 1.01 bar, 15°C (g/cm³)
C_{12}	1.66	171	0.806
C_{13}	1.28	184	0.814
C_{14}	1.40	196	0.826
C_{15}	1.24	210	0.834
C_{16}	0.90	223	0.841
C_{17}	0.88	234	0.848
C_{18}	0.82	246	0.853
C_{19}	0.82	257	0.858
C_{20+}	6.25	458	0.926

TABLE 3.24
Molar Compositions (Mol%) of the Separator Gases from the Experiment in Table 3.22

Component	Stage 1	Stage 2	Stage 3	Stage 4
N_2	1.07	0.42	0.01	0.00
CO_2	0.49	0.62	0.60	0.28
C_1	77.43	64.63	36.04	8.89
C_2	9.56	14.42	20.23	16.19
C_3	6.70	12.01	24.89	35.52
iC_4	0.71	1.30	3.13	5.74
nC_4	2.01	3.65	9.02	17.22
iC_5	0.44	0.74	1.86	3.77
nC_5	0.59	0.98	2.31	4.61
C_6	0.47	0.66	1.28	2.92
C_{7+}	0.53	0.57	0.63	4.86

(reservoir temperature). This B_o-factor expresses the shrinkage from saturated oil at the reservoir temperature to stable oil from a four-stage separation with separator temperatures and pressures as in Table 3.22. The remaining B_o-factors in Table 3.22 express the shrinkage of the oil from the current stage of the separation down to stock tank conditions. The separator gas/oil ratio equals the ratio between the volume of the gas liberated from the current stage taken to standard conditions and the volume of the oil from the last separator stage, which is at standard conditions. The separator gas/oil ratio for separator number N becomes:

$$\text{Stage N separator gas/oil ratio} = \frac{V_{N,std}^{gas}}{V_{std}^{oil}} \quad (3.13$$

3.1.5 VISCOSITY EXPERIMENT

In a viscosity experiment, the oil viscosity is measured at decreasing pressure at a constant temperature, typically the reservoir temperature. Much existing viscosity data has been measured using a rolling-ball viscometer as sketched in Figure 3.14, where the viscosity is related to the time it takes

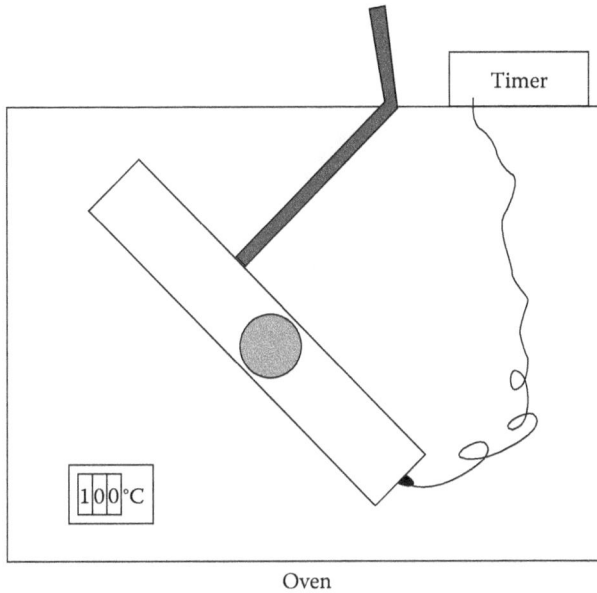

FIGURE 3.14 Rolling ball viscometer.

for a ball of a given weight and diameter to fall from the top to the bottom of a cell filled with the oil under investigation.

A newer and more accurate technique for determining viscosity is by electromagnetic viscometer (EMV) as sketched in Figure 3.15. This instrument is based on the technique in which the magnetic force is applied to a piston immersed in the fluid under the test. The motion of the piston is resisted by a viscous drag of the fluid and provides the measure of viscosity. The EMV technique is applicable to both oil and gas viscosities. To measure gas viscosities, a piston is used with a hollow in the center and the chamber is placed horizontally. Most often gas viscosities reported along with oil viscosities are not measured, but found from a gas viscosity correlation, often the one of Lee et al. (1966) described in Chapter 10.

The approximate development in oil viscosity with pressure is sketched in Figure 3.16. Above the saturation pressure, the viscosity decreases with decreasing pressure due to the compressibility effect, which increases the distance between the molecules. The minimum viscosity is seen at the saturation pressure. Below the saturation point the viscosity increases with decreasing pressure. Lighter (gaseous) components are released and the viscosity becomes increasingly influenced by heavier components.

Experimental viscosity data is presented in Chapter 10.

3.1.6 MATERIAL BALANCE TESTS

For the PVT experiments in which material is being removed, it is customary for PVT labs to perform a mass balance test to check whether the mass of the fluid charged to the cell equals the mass of the fluid removed from the cell plus the material left in the cell when the test is over. A such material balance test is outlined in the following for a differential liberation test as described in section 3.1.2.

A differential liberation test is started at the saturation point (stage NSAT) and continued for N stages at a constant temperature. The last stage is at atmospheric pressure after which the oil is cooled to standard temperature (15°C). In the following mass balance calculation, the reference is

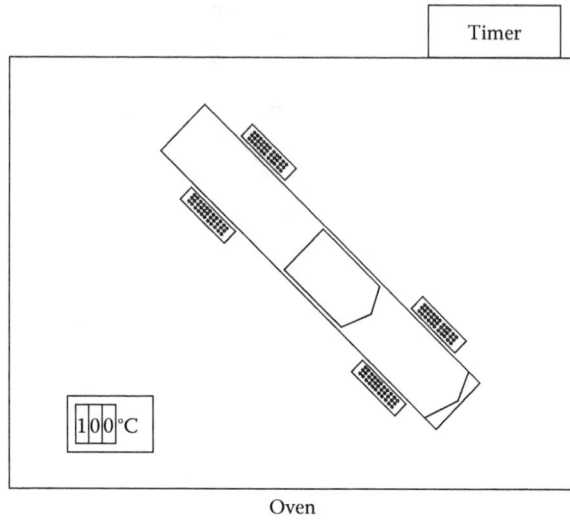

FIGURE 3.15 Electromagnetic viscometer.

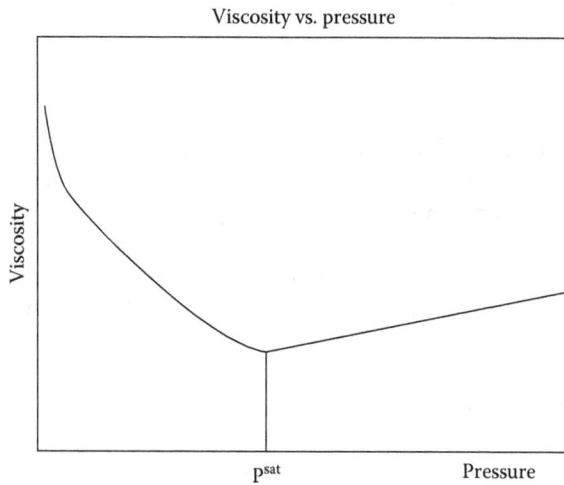

FIGURE 3.16 Qualitative development in oil viscosity with pressure.

one volume unit of oil at standard conditions (STO oil). With this reference, the mass of oil charged to the cell becomes

$$Mass\, In = B_o^{NSAT} \rho_{oil}^{NSAT} \qquad (3.14)$$

B_o is the oil formation volume factor defined in Table 3.12 and ρ_{oil} is the density of the reservoir oil and NSAT the saturation stage. At each differential liberation stage, all liberated gas is removed. The total mass of removed gas is

$$Mass\, Out\, with\, Gas = \sum_{i=NSAT+1}^{N} \left(R_s^{(i-1)} - R_s^i \right) \times Gas\, Gravity^i \times \rho_{air} \qquad (3.15)$$

where ρ_{air} is the density of atmospheric air, which is 0.0012250 g/cm³. The mass of the stock tank (residual) oil equals

$$Mass\,STO\,Oil = \rho_{oil}^{STO} \qquad (3.16)$$

To fulfil the mass balance, the following equation should be met

$$Mass\,In = Mass\,Out\,with\,Gas + Mass\,STO\,Oil \qquad (3.17)$$

Table 3.25 shows the molar composition of a reservoir oil for which a differential liberation experiment has been carried out at a temperature of 126.7°C and of the gas liberated at each pressure stage. The differential liberation data is given in Table 3.26. Following is a mass balance calculation for the differential liberation data in Table 3.26.

$$Mass\,In = 1.411 \times 0.6906 = 0.9745$$
$$Mass\,out\,with\,Gas = 0.00989 + 0.00949 + 0.00998 + 0.01222 + 0.02449 + 0.07434 = 0.14041$$

The mass of the stock tank (residual) oil is

$$Mass\,STO\,Oil = 0.8325$$
$$Mass\,Out = 0.14041 + 0.8325 = 0.9729$$

which deviates from *Mass In* by 0.16%. A deviation of up to ±2% is usually considered acceptable.

TABLE 3.25

Reservoir Oil and Liberated Gas and Residual Oil Compositions (Mole%) for Differential Liberation Experiment Carried Out at a Temperature of 126.7°C. The C_{7+} Molecular Weight Is 212.0, and the Average Molecular Weight of the Reservoir Fluid Is 135.7. The C_{7+} Density Is 0.835 g/cm³. The Molecular Weight of the STO Oil Is 209.2. The Volumetric Differential Liberation Data Is Given in Table 3.26

Component	Res Fluid 79.5 bar	Liberated Gas 63.1 bar	49.3 bar	35.5 bar	21.7 bar	7.9 bar	1.01 bar	STO Oil 1.01 bar
N_2	0.341	2.385	1.735	1.114	0.572	0.161	0.010	0.000
CO_2	1.183	3.327	3.530	3.782	3.995	3.205	0.542	0.000
H_2S	0.441	0.554	0.801	0.921	1.015	1.987	0.964	0.000
C_1	18.297	74.013	71.542	66.775	58.102	32.431	3.281	0.000
C_2	4.201	7.636	8.594	10.084	12.477	14.633	4.796	0.000
C_3	4.880	5.061	5.867	7.295	10.048	16.648	12.360	0.070
iC_4	1.534	1.082	1.244	1.585	2.238	4.363	5.337	0.110
nC_4	3.815	2.194	2.547	3.231	4.628	9.564	14.427	0.490
iC_5	2.254	0.832	0.963	1.224	1.766	4.132	9.551	0.800
nC_5	2.914	0.942	1.083	1.374	1.947	4.676	11.969	1.270
C_6	4.584	0.862	1.033	1.304	1.776	4.515	13.203	3.720
C_7	5.511	0.661	0.702	0.873	1.094	2.741	10.404	6.520
C_8	5.897	0.351	0.351	0.421	0.422	1.058	6.772	8.680
C_9	5.337	0.140	0.120	0.130	0.090	0.201	3.351	8.590
C_{10+}	38.810	0.060	0.030	0.050	0.010	0.030	3.201	69.750

TABLE 3.26

Differential Liberation Results for Reservoir Oil in Table 3.25 at a Temperature of 126.7°C. The Oil at Atmospheric Conditions (STO oil) Has a Molecular Weight of 209.3

Pressure bar	Oil FVF (B$_o$) m³/Sm³	GOR (R$_s$) Sm³/Sm³	Oil Density g/cm³	B$_g$ m³/Sm³	Z-Factor Gas	Gas Gravity (Air = 1)
79.48	1.411	76.054	0.6906			
63.06	1.385	66.258	0.6965	0.02047	0.9180	0.824
49.27	1.360	57.174	0.7023	0.02683	0.9400	0.853
35.48	1.336	48.269	0.7074	0.03807	0.9606	0.915
21.69	1.305	38.472	0.7150	0.06347	0.9793	1.018
7.90	1.267	24.401	0.7170	0.17686	0.9947	1.421
1.01	1.111	0.000	0.7504		1.0000	2.487
1.01*[1)]	1.000		0.8325			

*[1)] at 15°C

A material balance can also be carried out as a molar balance

$$Moles\,In = \frac{B_o^{NSAT}\,\rho_{oil}^{NSAT}}{M_{oil}^{NSAT}} \qquad (3.18)$$

where M_{oil}^{NSAT} is the molecular weight of the reservoir oil. The number of moles removed with the gas at stage i equals

$$\left(Moles\,Out\,with\,Gas\right)^i = \frac{\left(R_s^{(i-1)} - R_s^i\right) \times \rho_{air}}{M_{air}} \qquad (3.19)$$

where the molecular weight of air is 28.964. The total number of moles removed with the gas is

$$Moles\,Out\,with\,Gas = \sum_{i=NSAT+1}^{N}\left(Moles\,\,Out\,with\,Gas\right)^i \qquad (3.20)$$

The number of moles in the stock tank (residual) oil is

$$Moles\,\,STO\,Oil = \frac{\rho_{oil}^{STO}}{M_{oil}^{STO}} \qquad (3.21)$$

where M_{oil}^{STO} is the molecular weight of the STO oil.

The following is a molar balance for the differential liberation data in Table 3.26

$$Moles\,In = \frac{1.411 \times 0.6906}{135.7} = 0.007181$$

From Equations 3.19 and 3.20:

$$Moles\,\,Out\,with\,Gas = 0.000414 + 0.000384 + 0.000377 + 0.000414 + 0.000595 + 0.001032 = 0.003217$$

From Equation 3.21:

$$Moles\ STO\ Oil = \frac{0.8325}{209.3} = 0.003978$$

This gives the total number of moles out

$$Moles\ Out = 0.003217 + 0.003978 = 0.0.007195$$

which agrees with the number of *Moles In* within 0.19%. As for a mass balance test, a deviation of up to ±2% is usually considered acceptable.

The molar balance can be extended to a component level

$$Moles\ In\ of\ Component\ j = z_j \times Moles\ In \qquad (3.22)$$

where z_j is the mole fraction of component j in the reservoir fluid.

$$Moles\ of\ Component\ j\ Out\ with\ Gas = \sum_{i=NSAT+1}^{N} y_j^i \times (Moles\ Out\ with\ Gas)^i \qquad (3.23)$$

where y_j^i is the mole fraction of component j in the gas released at stage i.

$$Moles\ of\ Component\ j\ Out\ with\ STO\ Oil = x_j^{STO} Moles\ STO\ Oil \qquad (3.24)$$

where x_j^{STO} is the mole fraction of component j in the stock tank oil. Table 3.27 shows a component molar balance for the differential liberation experiment in Table 3.26 and the liberated gas compositions in Table 3.25. Table 3.28 shows the *Moles Out* normalized to 100% (= total composition out) together with the reservoir composition and the deviation between the two compositions. Table 3.29 shows the maximum deviations that are usually considered acceptable for a component molar balance. All deviations are within those limits.

3.1.7 MEASUREMENT OF GAS Z-FACTORS

As mentioned in Section 3.1.1.3., PVT laboratories have the practice to assume that the Z-factor equals 1.0 for a gas at standard conditions. Table 3.30 shows Z-factors for some pure component gases at standard conditions (1.01325 bar and 15°C). For N_2 and C_1, a Z-factor of 1.0 at standard conditions is a good assumption. For the heavier gases, the Z-factors at standard conditions are more than 0.5% lower than 1.0. For gas mixtures with a considerable concentration of C_2 and heavier, the Z-factors reported by PVT labs could therefore be slightly too high.

Accurate Z-factor measurements are conducted in a Burnett cell as sketched in Figure 3.17 (Burnett 1936 and Fredenslund et al. 1973). The cell consists of two thick-walled vessels of a volume of 125–275 cm³, connected to each other by means of an expansion valve. The Burnett cell is located in a thermostat and equipped with a pressure measuring system. Prior to the start of a Z-factor measurement, both vessels are evacuated. The valve between the two vessels is closed, and one is filled with the gas mixture to be studied. After isothermal conditions are obtained, the temperature and pressure are measured. The valve between the two vessels is opened, and the pressure and temperature are measured after stabilization. The valve between the two vessels is closed and one vessel evacuated. These expansions are continued until a low pressure is reached.

TABLE 3.27

Component Molar Balance for Differential Liberation Experiment in Tables 3.25 and 3.26

Component	Moles In	Liberated Gas						STO Oil	Moles Out	Deviation
		63.1 bar	49.3 bar	35.5 bar	21.7 bar	7.9 bar	1.01 bar			
Moles	0.00718100	0.00041400	0.00038400	0.00037700	0.00041400	0.00059500	0.00103200	0.00397800	0.00719400	−0.00001300
N_2	0.00002449	0.00000985	0.00000664	0.00000418	0.00000236	0.00000095	0.00000010	0.00000000	0.00002410	0.00000039
CO_2	0.00008495	0.00001374	0.00001352	0.00001421	0.00001648	0.00001894	0.00000557	0.00000000	0.00008247	0.00000248
H_2S	0.00003167	0.00000269	0.00000361	0.00000407	0.00000493	0.00001382	0.00001166	0.00000000	0.00004078	−0.00000911
C_1	0.00131391	0.00030581	0.00027395	0.00025092	0.00023967	0.00019163	0.00003375	0.00000000	0.00129572	0.00001819
C_2	0.00030167	0.00003155	0.00003291	0.00003789	0.00005147	0.00008647	0.00004933	0.00000000	0.00028962	0.00001206
C_3	0.00035043	0.00002091	0.00002246	0.00002741	0.00004145	0.00009837	0.00012713	0.00000278	0.00034052	0.00000991
iC_4	0.00011016	0.00000447	0.00000476	0.00000596	0.00000923	0.00002578	0.00005489	0.00000438	0.00010947	0.00000069
nC_4	0.00027396	0.00000907	0.00000975	0.00001214	0.00001909	0.00005651	0.00014839	0.00001949	0.00027445	−0.00000049
iC_5	0.00016186	0.00000344	0.00000369	0.00000460	0.00000729	0.00002441	0.00009824	0.00003182	0.00017348	−0.00001162
nC_5	0.00020925	0.00000389	0.00000415	0.00000516	0.00000803	0.00002763	0.00012311	0.00005052	0.00022249	−0.00001324
C_6	0.00032918	0.00000356	0.00000396	0.00000490	0.00000733	0.00002668	0.00013580	0.00014798	0.00033021	−0.00000103
C_7	0.00039574	0.00000273	0.00000269	0.00000328	0.00000451	0.00001620	0.00010701	0.00025937	0.00039578	−0.00000004
C_8	0.00042346	0.00000145	0.00000134	0.00000158	0.00000174	0.00000625	0.00006965	0.00034529	0.00042731	−0.00000385
C_9	0.00038325	0.00000058	0.00000046	0.00000049	0.00000037	0.00000119	0.00003447	0.00034171	0.00037927	0.00000398
C_{10+}	0.00278695	0.00000025	0.00000012	0.00000019	0.00000004	0.00000018	0.00003292	0.00277466	0.00280835	−0.00002140

TABLE 3.28
Total Component Molar Balance for Differential Liberation Data in Tables 3.25 and 3.26. The Maximum Acceptable Deviations Can Be Seen from Table 3.29

	Inlet Composition Mol	Outlet Composition Mol	Deviation
N_2	0.341	0.335	−0.006
CO_2	1.183	1.149	−0.034
H_2S	0.441	0.483	0.042
C_1	18.297	18.042	−0.255
C_2	4.201	4.034	−0.167
C_3	4.880	4.744	−0.136
iC_4	1.534	1.525	−0.009
nC_4	3.815	3.822	0.007
iC_5	2.254	2.415	0.161
nC_5	2.914	3.097	0.183
C_6	4.584	4.595	0.011
C_7	5.511	5.505	−0.006
C_8	5.897	5.942	0.045
C_9	5.337	5.273	−0.064
C_{10+}	38.810	39.038	0.228

TABLE 3.29
Acceptance Criteria for Component Molar Balance

Concentration Range (Mol%)	Acceptance Criteria (Mol%)
0.00–0.10	± 0.03
0.10–1.00	± 0.08
1.00–10.00	± 0.30
10.0–30.00	± 0.75
> 30%	± 1.50

TABLE 3.30
Z-Factors for Pure Component Gases at Standard Conditions

Component	Z-Factor at 1.01325 bar and 15°C	Reference
N_2	0.9997	Hilsenrath (1955)
CO_2	0.9943	Hilsenrath (1955)
C_1	0.9981	American Petroleum Institute (1982)
C_2	0.9916	American Petroleum Institute (1982)
C_3	0.9820	American Petroleum Institute (1982)
iC_4	0.9703	American Petroleum Institute (1982)

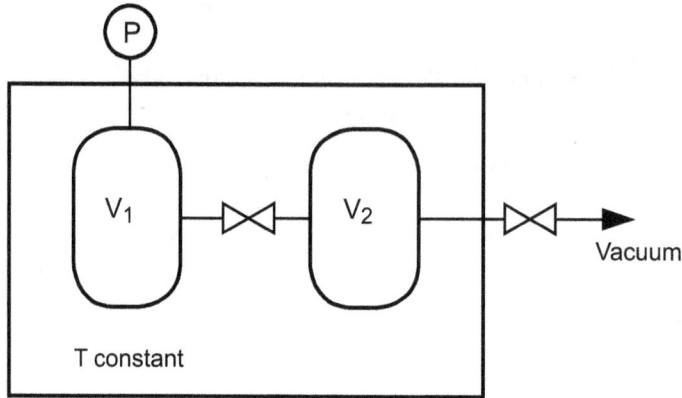

FIGURE 3.17　Burnett cell for accurate gas Z-factor measurements.

In the following, it is assumed for simplicity that both cells have the same volume, that initially n_0 moles are charged to the system at a pressure of P_0, and that N expansions are carried out. The initial Z-factor is given by

$$Z_0 = \frac{P_0 V^{vessel}}{n_0 RT} \tag{3.25}$$

The Z-factor after the ith expansion is given by

$$Z_i = \frac{P_i V^{vessel}}{\frac{n_0}{2^i} RT} \tag{3.26}$$

The experiment is continued until it (for expansion N) applies that

$$\frac{P_{(N-1)}}{P_N} = 2 \Rightarrow Z_N = Z_{(N-1)} = 1 \tag{3.27}$$

The Z-factor after the ith expansion equals

$$Z_i = \frac{Z_{(i+1)} P_i}{2 P_{(i+1)}} \tag{3.28}$$

Table 3.31 shows an example of how the Z-factors of a natural gas mixture of the composition in Table 3.32 are calculated. At the lowest pressures, the ratio between two subsequent pressures is 2.0 and the Z-factors equal to 1.0. For the preceding steps, the Z-factors are calculated from Equation 3.28.

TABLE 3.31

Example of How Pressure Data from a Burnett Cell Could Look and Corresponding Z-factors. The Data Is Simulated for the Natural Gas Mixture in Table 3.32 for a Temperature of 50°C. The Simulations Are Performed Using the GERG-2008 Equation of State Presented in Chapter 4

Pressure bar	$P_i/P_{(i+1)}$	Z-factor
750	3.501	1.5135
214.2412	1.978	0.8647
108.3380	1.893	0.8745
57.2366	1.931	0.9240
29.6359	1.956	0.9569
15.1547	1.979	0.9786
7.6588	1.989	0.9892
3.8503	1.994	0.9946
1.9305	1.997	0.9973
0.9666	1.999	0.9987
0.4836	1.999	0.9993
0.2419	2.000	0.9998
0.12098	2.000	1.0000
0.06049	2.000	1.0000
0.03025		

TABLE 3.32

Natural Gas Composition for Which Simulated Z-factors Are Shown in Table 3.31

Component	Mol %
N_2	0.340
CO_2	0.840
C_1	90.400
C_2	5.199
C_3	2.060
iC_4	0.360
nC_4	0.550
iC_5	0.140
nC_5	0.097
nC_6	0.014

3.2 EOR PVT EXPERIMENTS

Figure 3.18 shows the workflow for a typical EOR or gas injection PVT study. Such a study requires three to four times as much sample volume (around 2,000 cc at reservoir conditions) than a routine PVT study.

EOR PVT STUDY

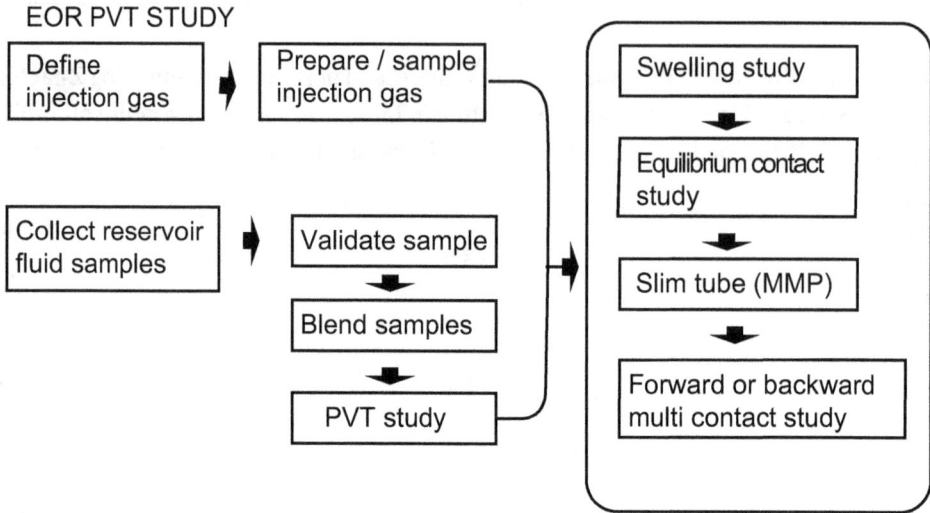

FIGURE 3.18   Workflow for gas injection EOR PVT study.

3.2.1 SOLUBILITY SWELLING TEST

Solubility swelling studies (or swelling experiments) are carried out on reservoir oil mixtures to study how the reservoir fluid will react to gas injection. In rare cases, swelling experiments are also performed on gas condensate mixtures. The saturation pressure of the reservoir fluid is determined by carrying out a CME experiment as described in Section 3.1.1. It is likely that a different fluid sample will be used to carry out the EOR PVT experiments than was used in the routine experiments. An initial CME test on the sample to be used in the EOR experiments in such cases also serves as a quality check of consistency between fluid samples.

A PVTZ study as described in Section 3.1.1.3 is carried out on the injection gas. Using the experimentally determined Z-factor and Equation 3.2, it is straightforward to convert injection gas volumes to injection gas moles. A known molar amount of injection gas is transferred to the PVT cell. The pressure is increased to a pressure high enough to ensure all gas is dissolved in the oil (could be around 550 bar), and a CME experiment is caried out for the mixture of oil and gas. Some PVT labs will report the CME data, while others may only report the saturation pressure and the swelling volume at the saturation pressure. More gas is injected, and the new saturation pressure and swelling volume are recorded. This process is repeated for a number of stages, as sketched in Figure 3.19. At the last stage(s), the saturation point may have changed from a bubble point to a dew point, signaling that the fluid would be critical somewhere between the last bubble point and the first dew point. At the critical point, two identical fluid phases are in equilibrium. The critical composition and critical pressure are key parameters in gas injection EOR evaluations. This will be further dealt with in Chapter 15.

As can be seen in Table 3.33, the swelling experiment gives information on the volume increase (swelling behavior) as a result of a particular amount of gas dissolving in the oil. It further tells how much extra pressure is needed to dissolve the injected gas. The swelling gas/oil ratio is defined as the cumulative volume of the injection gas at standard conditions per initial oil volume. This is different from how the GOR is defined in other PVT experiments.

A separate study of the viscosity of swelled mixtures may be carried out.

Table 3.34 presents swelling data for the reservoir fluid composition in Table 3.35, which also shows the composition of the injection gas that contains 60.32 mol% CO_2. The saturation point shifts from a bubble point to a dew point between 100 and 150 mol% gas added per initial mole oil.

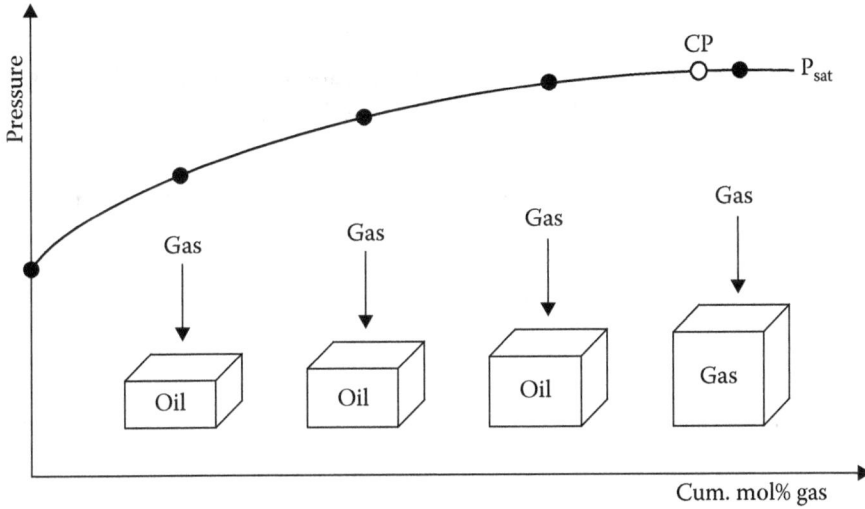

FIGURE 3.19 Schematic representation of swelling experiment. CP stands for critical point.

TABLE 3.33
Results from a Swelling Experiment Performed on an Oil Mixture

Mole percentage gas	Cumulative mole percentage gas added per initial mole oil
Gas/oil ratio	Standard volume of gas added per initial volume of oil
Saturation pressure	Saturation pressure after each addition of gas
Swelling volume	Volume of oil-injection gas mixture at saturation point per initial volume of oil
Density	Density of each swelled mixture at saturation point
Saturation point	Bubble or dew point pressure after each addition of gas
Liquid shrinkage or build-up	Liquid volume percent of saturation point volume as a function of pressure for each mixture of oil and gas

TABLE 3.34
Swelling Data at 90°C for Reservoir Oil and Injection Gas in Table 3.35

Stage	Mol% Gas/Initial Mol Oil	Saturation Pressure (bar)	Saturation Point	Swelling Volume/ Initial Oil Volume
1	0	186.0	Bubble	1.00
2	33.3	195.7	Bubble	1.03
3	100	231.2	Bubble	1.32
4	150	272.3	Dew	1.72
5	233	327.5	Dew	2.05
6	400	380.9	Dew	2.33

Source: Memon, A., et al., Miscible Gas Injection and Asphaltene Flow Assurance Fluid Characterization: A Laboratory Case Study for Black Oil Reservoir, SPE 1509238, presented at the *SPE EOR Conference Muscat*, Oman, 16–18 April, 2012

TABLE 3.35

Molar Composition of the Reservoir Oil and the Injection Gas for Which Swelling Data Is Shown in Table 3.34. The C_{7+} Molecular Weight of the Reservoir Fluid Is 237 and the C_{7+} Density 0.878 g/cm^3

Component	Reservoir Fluid (Mol%)	Injection Gas (Mol%)
N_2	0.39	
CO_2	0.84	60.32
C_1	36.63	10.73
C_2	8.63	7.55
C_3	6.66	9.09
iC_4	1.21	
nC_4	3.69	6.47
iC_5	1.55	0.03
nC_5	2.25	5.82
C_6	3.36	
C_7	3.34	
C_8	3.44	
C_9	3.04	
C_{10}	2.77	
C_{11}	2.23	
C_{12}	1.82	
C_{13}	1.66	
C_{14}	1.45	
C_{15}	1.31	
C_{16}	1.07	
C_{17}	0.98	
C_{18}	0.87	
C_{19}	0.83	
C_{20+}	9.98	

Table 3.36 shows swelling data for the reservoir fluid composition and injection gas in Table 3.37. For this fluid, the shift from bubble point to dew point is observed in the interval between 225 and 275 mol% gas added per initial mole oil.

Table 3.38 shows swelling data for the oil composition in Table 3.39 using the injection gas also shown in Table 3.39. Table 3.40 shows the liquid volume percentages below the saturation point in each of the four swelling stages. The liquid dropout curves are plotted in Figure 3.20.

3.2.2 EQUILIBRIUM CONTACT EXPERIMENT

Although the swelling experiment provides information about the reservoir fluid when saturated with injection gas in various quantities, it does not provide any information about the phases in equilibrium below the saturation pressure. As is illustrated in Figure 3.21, the equilibrium contact experiment is designed to provide two-phase information on mixtures of reservoir oil and injection gas. It is customary that one of the gas-oil mixing ratios is close to the one for which the saturation point shifts from a bubble to a dew point in the swelling experiment. The mixed fluid composition will, in other words, be near critical at its saturation point.

TABLE 3.36
Swelling Data at 77°C for Reservoir Oil and Injection Gas in Table 3.37

Stage	Mol% Gas/ Initial Mol Oil	GOR Sm³/Sm³	Saturation Pressure (bar)	Saturation Point	Swelling Volume/ Initial Oil Volume	Density g/cm³
1	0.0	0.0	87.5	Bubble	1.0000	0.7961
2	25.0	28.7	110.5	Bubble	1.0761	0.7880
3	75.0	86.2	143.7	Bubble	1.2314	0.7729
4	125.0	143.6	165.9	Bubble	1.3916	0.7586
5	200.0	229.8	198.1	Bubble	1.6296	0.7434
6	225.0	258.4	209.6	Bubble	1.7082	0.7396
7	275.0	316.0	230.3	Dew	1.8656	0.7328

Source: Al-Ajmi, M., et al., EoS modeling for two major Kuwaiti oil reservoirs, SPE 141241-PP, presented at the *SPE Middle East Oil and Gas Show,* Manama, Bahrain, 20–23 March, 2011.

TABLE 3.37
Molar Composition of the Reservoir Oil and the Injection Gas for Which Swelling Data Is Shown in Table 3.36 and Equilibrium Contact Data in Table 3.42. The C_{7+} Molecular Weight of the Reservoir Fluid Is 291 and the C_{7+} Density 0.8945 g/cm³

Component	Reservoir Fluid (Mol%)	Injection Gas (Mol%)
N_2	0.293	
CO_2	0.223	59.660
C_1	21.657	10.300
C_2	6.758	7.690
C_3	7.024	9.500
iC_4	1.325	
nC_4	4.229	6.790
iC_5	1.817	
nC_5	2.680	6.060
C_6	4.146	
C_7	4.114	
C_8	4.181	
C_9	3.698	
C_{10}	3.412	
C_{11}	2.954	
C_{12}	2.535	
C_{13}	2.332	
C_{14}	1.958	
C_{15}	1.854	
C_{16}	1.661	
C_{17}	1.431	
C_{18}	1.306	
C_{19}	1.259	

(Continued)

TABLE 3.37   (Continued)

Molar Composition of the Reservoir Oil and the Injection Gas for Which Swelling Data Is Shown in Table 3.36 and Equilibrium Contact Data in Table 3.42. The C_{7+} Molecular Weight of the Reservoir Fluid Is 291 and the C_{7+} Density 0.8945 g/cm³

Component	Reservoir Fluid (Mol%)	Injection Gas (Mol%)
C_{20}	1.126	
C_{21}	0.997	
C_{22}	0.940	
C_{23}	0.845	
C_{24}	0.762	
C_{25}	0.727	
C_{26}	0.625	
C_{27}	0.587	
C_{28}	0.583	
C_{29}	0.542	
C_{30}	0.528	
C_{31}	0.541	
C_{32}	0.458	
C_{33}	0.440	
C_{34}	0.399	
C_{35}	0.391	
C_{36+}	6.659	

TABLE 3.38

Swelling Data at 136.7°C for Reservoir Oil and Injection Gas in Table 3.39

Stage	Mol% Gas/ Initial Mol Oil	Saturation Pressure bar	Swelling Volume	Saturation Point
1	0.0	176.1	1.0000	Bubble
2	25.4	245.8	1.2239	Bubble
3	101.8	353.7	1.7235	Bubble
4	305.3	476.8	2.7699	Dew

Source: Takeshi, A., Tetsuro, F., Leekumjorn, S., Shaikh, J. A., Pedersen, K.S., Alobeidli, A., Mogensen, K., A new technique for common EoS model development for multiple reservoir fluids with gas injection, SPE-202923-MS, presented at the *Abu Dhabi International Petroleum Exhibition & Conference*, November 9–12, 2020.

TABLE 3.39

Reservoir Fluid and Injection Gas for Which Swelling Data Is Shown in Table 3.38 and CME Liquid Dropout Data in Table 3.40. The C_{7+} Molecular Weight Is 192.5 and the C_{7+} Density Is 0.8231 g/cm³

Component	Reservoir Fluid Mol%	Injection Gas Mol%
N_2	0.137	0.90
CO_2	0.557	

Component	Reservoir Fluid Mol%	Injection Gas Mol%
H_2S	8.743	
C_1	33.707	83.22
C_2	9.323	10.61
C_3	7.573	3.72
iC_4	1.436	0.40
nC_4	4.032	0.79
iC_5	1.773	0.20
nC_5	2.198	0.16
C_6	2.835	
C_7	3.160	
C_8	3.508	
C_9	3.007	
C_{10}	2.698	
C_{11+}	15.314	

TABLE 3.40
CME Data for Oil + Gas Mixtures in Swelling Experiment in Table 3.38

0.0 Mole% Gas Added		25.4 Mole% Gas Added		101.8 Mole% Gas Added		305.3 Mole% Gas Added	
Pressure bar	Liquid Volume%	Pressure bar	Liquid Volume%	Pressure bar	Liquid Volume%	Pressure bar	Liquid Volume%
176.1	100.0	245.8	100.0	353.7	100.0	476.8	0.0
173.4	97.9	242.3	98.2	352.6	98.1	473.3	0.4
169.9	96.2	235.4	95.4	349.2	94.4	469.9	0.9
166.5	94.8	232.0	94.2	345.8	90.5	463.0	1.8
163.0	93.5	214.8	89.3	338.9	84.8	452.6	3.1
159.6	92.4	197.5	85.0	328.5	77.9	438.8	5.0
152.7	90.3	180.3	81.1	314.7	71.4	421.6	7.1
145.8	88.5	163.0	77.5	297.5	66.3	404.4	9.1
138.9	86.7	145.8	73.9	280.3	62.9	387.1	11.1
128.6	84.2	128.6	70.6	263.0	59.9	369.9	13.1
114.8	81.1	111.3	67.3	245.8	57.2	352.6	15.0
97.5	77.4	94.1	64.2	228.5	54.6	335.4	16.8
80.3	73.9	76.9	60.9	211.3	52.1	318.2	18.4
63.1	70.5			194.1	50.0	300.9	20.0
				176.8	48.3	283.7	21.4
				159.6	46.6	266.5	22.6
				142.4	45.0	249.2	23.5
						232.0	24.0

Source: Takeshi, A., Tetsuro, F., Leekumjorn, S., Shaikh, J. A., Pedersen, K.S., Alobeidli, A., Mogensen, K., A new technique for common EoS model development for multiple reservoir fluids with gas injection, SPE-202923-MS, presented at the *Abu Dhabi International Petroleum Exhibition & Conference*, November 9–12, 2020.

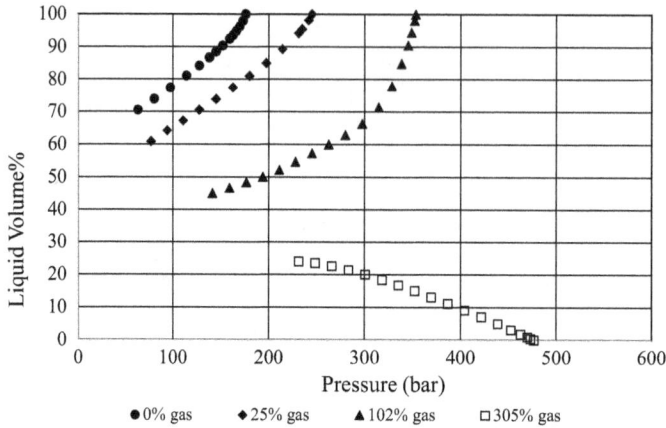

FIGURE 3.20 Experimental liquid dropout data in Table 3.40 for mixtures of the reservoir fluid and the injection gas in Table 3.39 at a temperature of 136.7°C.

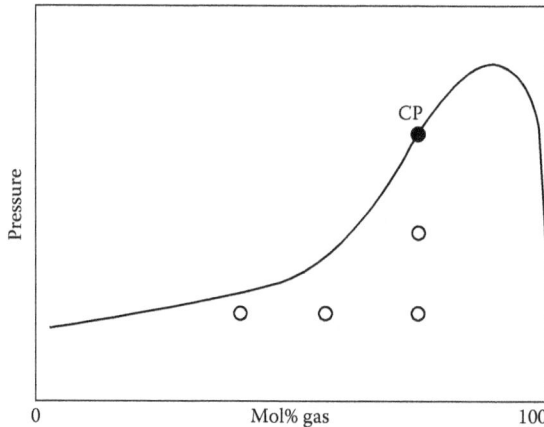

FIGURE 3.21 Gas mole percentages and pressure conditions used in equilibrium contact experiment.

A portion of the recombined reservoir fluid sample is charged to a high-pressure visual cell at a pressure of around 550 bar and thermally expanded to the reservoir temperature. The Z-factor of the injection gas is known from a PVTZ study, as described in Section 3.1.1.3. Using this experimentally determined Z-factor, the volume of gas corresponding to a specified molar amount of gas is determined. After adding the gas, the cell content is pressurized to around 550 bar to dissolve all gas in the oil. The pressure is lowered to the first contact pressure and kept there for 24 hours to be sure phase equilibrium has established.

The gas phase is displaced (as in a differential liberation experiment) and the gas phase composition determined. The molecular weight of the gas phase can be derived from the composition.

A partial CCE experiment is performed on the liquid phase to confirm the saturation pressure used for the equilibrium contact phase separation. The density of the liquid phase is determined by pumping portions of the fluid under pressure from the cell at around 550 bar into a pre-weighed vessel. The density at the saturation pressure is then calculated using the measured relative volumes from the preceding partial CCE experiment.

The viscosity on the liquid phase is measured using one of the techniques described in Section 3.1.5.

The data reported from an equilibrium contact experiment is listed in Table 3.41.

Table 3.42 shows the results of an equilibrium contact experiment on the reservoir fluid and the injection gas in Table 3.37.

TABLE 3.41
Results from an Equilibrium Contact Experiment

Mixing ratio	Molar amount of gas per amount of oil
Phase volume percent	Volume percent gas and oil at contact pressure
Phase densities	Densities of gas and oil at contact pressure
Phase molecular weights	Molecular weights of gas and oil at contact pressure
Gas/oil weight ratio	Relative weight amounts of gas and oil at each contact
Phase viscosities	Viscosities of gas and oil at contact pressure

---

TABLE 3.42
Equilibrium Contact Data at 77°C for Reservoir Fluid and Injection Gas in Table 3.37

Pressure (bar)	Gas/Liquid Molar Mixing Ratio	Gas Volume (%)	Liquid Volume (%)	Gas Density (g/cm³)	Liquid Density (g/cm³)
138.9	2.25	27.48	72.52	0.3656	0.7622
157.9	2.25	23.03	76.97	0.4651	0.7551

Source: Al-Ajmi, et al., "EoS modeling for two major Kuwaiti oil reservoirs", SPE 141241-PP, presented at the *SPE Middle East Oil and Gas Show*, Manama, Bahrain, 20–23 March, 2011.

3.2.3 MULTI-CONTACT EXPERIMENT

When gas is injected into a reservoir containing undersaturated oil, some of the gas will dissolve in the oil. If the oil phase cannot accommodate all the injected gas, the fluid will split into two equilibrium phases, a gas and an oil. The composition of the gas phase will be different from that of the injection gas. The gas has taken up some components from the reservoir oil, and it has lost other components to the oil phase. The equilibrium gas will be pushed away from the injection well by new injection gas and will contact fresh oil in some distance from the injection well. The oil near the injection well will be contacted by fresh injection gas.

A multi-contact experiment as sketched in Figure 3.22 is designed to simulate those processes. The cell initially contains a known amount of reservoir oil. The experiment can be conducted either forward or backward. Forward means that the gas phase is moved to the subsequent stage and mixed with fresh reservoir oil. In a backward experiment the oil phase from the first contact is mixed with fresh injection gas. This process is continued for a number of stages.

A multi-contact study is performed in the same type of cell as is used for the swelling study (Section 3.2.1), and the first contact is experimentally carried out in the same way as for an equilibrium contact experiment (Section 3.2.2).

In a forward multi-contact experiment, a known volume of reservoir fluid is in a second cell mixed with the equilibrium gas from the first cell. The mixing and equilibration take place the same way as in the first cell. This procedure is repeated for three to four stages using the gas recovered from the previous contact. After four stages, the amount of equilibrium gas is generally insufficient to form a gas phase large enough for analyses if again mixed with reservoir oil.

In a backward multi-contact experiment, the liquid phase remaining after the equilibrium gas has been removed is contacted by a known volume and known molar amount of fresh injection gas. The mixing and equilibration take place the same way as for the first contact. This procedure is repeated for a number of stages (usually maximum four) until the volume of gas formed at the selected test pressure is insufficient for any analyses.

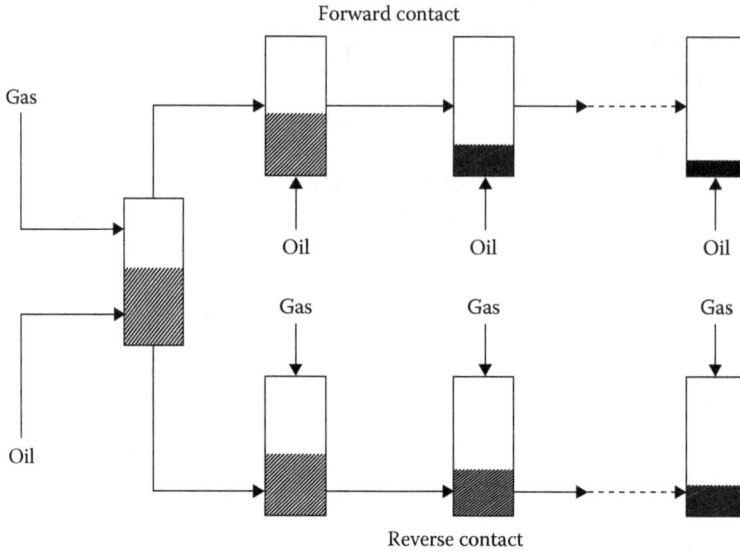

FIGURE 3.22   Schematic representations of forward and reverse (backward) multi-contact experiments.

As a quality check, mass and molar balances are performed for each contact. The mass and moles of fluid charged to the cells are calculated using the densities, molecular weights, and volumes. The densities, molecular weights, and volumes of the gas and oil equilibrium phases (at the selected test pressure) are then used to calculate the mass and moles at each stage. The mass/mole in the last cell plus the mass/mole removed should add up to the mass/moles charged to the cell. As a rule of thumb, the deviation should not be more than 2%.

The quantities reported from multi-contact experiments are listed in Table 3.43.

A multi-contact experiment requires careful design. The purpose of the study is to get experimental data for the phase behavior at near-miscible conditions. On the other hand, miscibility must not develop, as there would be no gas or oil to move forward to the next contract from a miscible (single phase) fluid. It is an idea to choose the mixing ratios in such a way that the two first contacts are for oil systems and the last two contacts for gas systems. That means the saturation point of the total fluid in the first two contacts will be bubble points, whereas in the last two contacts they will be dew points. At the optimum mixing ratio, the volumes of gas and oil at the contact pressure will be approximately equal. A PVT simulator can be helpful to design a multi-contact experiment.

In a reservoir context, a forward multi-contact experiment will follow the fluid at the gas-oil front and a backward experiment will follow the fluid near the injection well. The experiment is therefore mainly designed for 100% vaporizing or 100% condensing and not for combined vaporizing/condensing drives. The terms vaporizing and condensing are explained in Chapter 15.

Table 3.44 shows molar compositions of a reservoir fluid for which multi-contact data is shown in Tables 3.45 and 3.46. Two multi-contact experiments were carried out, one with the hydrocarbon injection gas in Table 3.44 and another with CO_2 as injection gas.

3.2.4   SLIM TUBE EXPERIMENT

The oil recovery it is possible to achieve with gas injection can be measured in a slim tube apparatus as sketched in Figure 3.23. Gas is displacing oil in a ¼-inch OD diameter, approximately 60-foot-long stainless-steel tube (dimensions may vary depending on the laboratory), which is filled with sand or

TABLE 3.43
Results from a Multi-Contact Experiment

Mixing ratio	Molar amount of gas per amount of oil. Each stage may have its own mixing ratio
Phase volume percent	Volume percent of gas and oil at each contact pressure
Phase densities	Densities of gas and oil at each contact
Phase molecular weights	Molecular weights of gas and oil at each contact
Gas/oil weight ratio	Relative weight amounts of gas and oil at each contact pressure
Phase viscosities	Viscosities of gas and oil at each contact pressure

TABLE 3.44
Molar Compositions of Reservoir Fluid and Hydrocarbon Injection Gas for Which Multi-Contact Data Is Shown in Tables 3.45 and 3.46

Component	Reservoir Fluid			Hydrocarbon Injection Gas		
	Mol%	Mol Weight	Density (g/cm^3)	Mol%	Mol (Weight)	Density (g/cm^3)
N_2	0.360			0.56		
CO_2	2.969			8.52		
H_2S	0.000			1.40		
C_1	28.847			53.58		
C_2	7.109			11.25		
C_3	6.528			10.97		
iC_4	1.831			2.90		
nC_4	4.199			5.16		
iC_5	2.257			1.88		
nC_5	2.842			1.79		
C_6	4.073			1.23		
C_7	4.503	98.3	0.7143	0.54	96	0.738
C_8	4.418	112.9	0.7358	0.17	107	0.765
C_9	3.801	122.8	0.7568	0.03	121	0.781
C_{10}	2.807	135.7	0.7736	0.01	134	0.792
C_{11}	4.337	150.7	0.7865	0.01	147	0.796
C_{12}	3.015	167.6	0.8025			
C_{13}	1.912	187.3	0.8142			
C_{14}	2.036	206.0	0.8249			
C_{15}	1.436	215.4	0.8323			
C_{16}	1.278	223.3	0.8431			
C_{17}	1.443	244.6	0.8475			
C_{18}	1.011	259.7	0.8600			
C_{19}	0.730	272.0	0.8703			
C_{20+}	6.258	456.6	0.9396			

TABLE 3.45

Results of Forward Multi-Contact Experiments for the Reservoir Fluid in Table 3.44 Using, Respectively, the Hydrocarbon Gas in Table 3.44 and CO_2 as Injection Gas

Stage	Gas/Liquid Molar Mixing Ratio	Gas Volume (%)	Liquid Volume (%)	Gas Density (g/cc)	Liquid Density (g/cc)	Gas Mol Weight	Liquid Mol Weight	Gas/Liquid Weight Ratio	Gas Viscosity (cP)	Liquid Viscosity (cP)
CO_2 Injection Gas at 221 bar and 121°C										
1	1.10	49.0	51.0	0.476	0.691	47.3	78.4	0.663	0.0838	0.198
2	1.19	52.5	47.5	0.465	0.645	47.5	70.7	0.798	0.0788	0.158
3	1.78	63.0	37.0	0.454	0.638	46.7	68.6	1.211	0.0744	0.150
4	1.51	60.1	39.9	0.428	0.621	45.7	66.6	1.038	0.0647	—
Hydrocarbon Injection Gas at 277 bar and 121°C										
1	1.7	42.3	57.7	0.294	0.569	31.8	67.5	0.380	0.0350	0.210
2	3.0	49.5	50.5	0.323	0.538	34.4	62.0	0.588	0.0396	0.236
3	4.0	50.5	49.5	0.326	0.523	35.2	58.5	0.635	0.0399	0.103
4	4.0	42.3	57.7	0.333	0.521	35.7	57.4	0.469	0.0413	—

Source: Negahban, S., et al., An EoS Model for a Middle East Reservoir fluid with an extensive EOR PVT data material, SPE-136530-PP presented at the *Abu Dhabi International Petroleum Exhibition & Conference*, Abu Dhabi, UAE November 1–4, 2010.

TABLE 3.46

Phase Equilibrium Compositions Measured in Multi-Contact Experiments on Reservoir Fluid in Table 3.44 Using, Respectively, the Hydrocarbon Gas in Table 3.44 and CO_2 as Injection Gas

Component	Stage 1 Gas Mol%	Stage 1 Liquid Mol%	Stage 2 Gas Mol%	Stage 2 Liquid Mol%	Stage 3 Gas Mol%	Stage 3 Liquid Mol%	Stage 4 Gas Mol%	Stage 4 Liquid Mol%
CO_2 Injection Gas at 221 bar and 121°C								
N_2	0.17	0.07	0.38	0.25	0.52	0.33	0.74	0.45
CO_2	72.56	57.79	61.38	50.87	55.81	46.92	50.48	43.11
C_1	12.02	8.47	17.91	13.33	21.46	16.13	24.94	19.16
C_2	2.60	2.45	3.70	3.49	4.30	4.08	4.82	4.62
C_3	2.31	2.59	3.21	3.50	3.64	3.94	4.00	4.38
iC_4	0.62	0.78	0.85	1.02	0.95	1.13	1.05	1.24
nC_4	1.40	1.94	1.90	2.48	2.12	2.68	2.35	2.90
iC_5	0.70	1.08	0.90	1.32	1.00	1.41	1.15	1.50
nC_5	0.85	1.37	1.10	1.67	1.21	1.76	1.39	1.86
C_6	1.08	2.01	1.40	2.42	1.52	2.46	1.71	2.52
C_7+	5.69	21.45	7.27	19.65	7.47	19.16	7.37	18.26
Hydrocarbon Injection Gas at 277 bar and 121°C								
N_2	0.58	0.30	0.66	0.40	0.77	0.49	0.91	0.60
CO_2	4.12	3.40	4.08	3.47	4.10	3.56	4.10	3.55
H_2S	0.42	0.42	0.32	0.31	0.21	0.23	0.09	0.04
C_1	66.97	48.41	65.15	49.39	64.43	50.24	64.21	50.82

Component	Stage 1 Gas Mol%	Stage 1 Liquid Mol%	Stage 2 Gas Mol%	Stage 2 Liquid Mol%	Stage 3 Gas Mol%	Stage 3 Liquid Mol%	Stage 4 Gas Mol%	Stage 4 Liquid Mol%
C_2	8.26	7.84	8.12	7.70	8.06	7.72	8.02	7.58
C_3	5.88	6.53	5.77	6.31	5.75	6.25	5.62	6.05
iC_4	1.40	1.73	1.38	1.70	1.40	1.69	1.36	1.64
nC_4	2.79	3.71	2.79	3.74	2.88	3.76	2.82	3.62
iC_5	1.09	1.70	1.17	1.88	1.23	1.83	1.23	1.81
nC_5	1.23	2.00	1.38	2.25	1.44	2.22	1.47	2.20
C_6	1.33	2.55	1.58	2.84	1.63	2.90	1.81	2.94
C_{7+}	7.26	23.96	9.18	22.85	9.73	22.01	7.26	23.96

Source: Negahban, S. et al., An EoS Model for a Middle East Reservoir fluid with an extensive EOR PVT data material, SPE-136530-PP presented at the *Abu Dhabi International Petroleum Exhibition & Conference*, Abu Dhabi, UAE November 1–4, 2010.

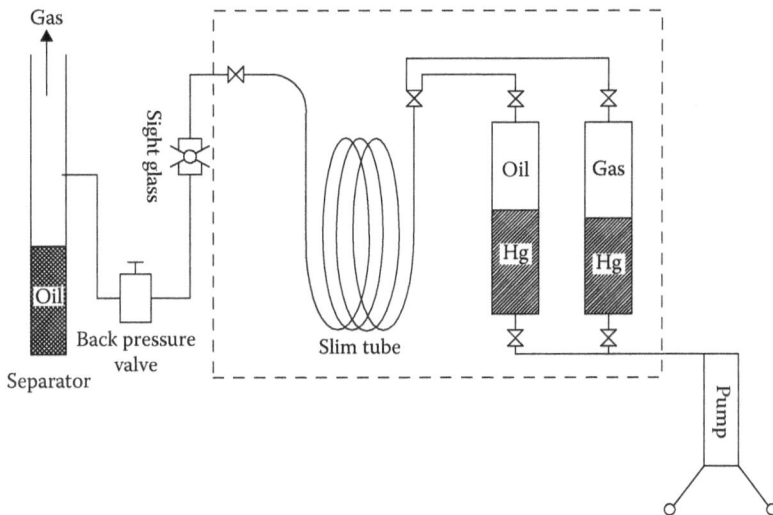

FIGURE 3.23   Schematic representation of slim tube apparatus.

glass beads. The pore volume is around 65 ml. The experiment is carried out at the reservoir temperature at a series of pressures between the saturation pressure and the highest operating pressure, which is typically around 700 bar. This coil is cleaned, evacuated, and weighed before the test commences.

The coil is precharged with toluene at the first test pressure and the free (pore) volume determined. The coil is then heated in an oven to the reservoir temperature and stabilized at the run pressure. Reservoir fluid is injected into the coil, and toluene is displaced at a constant pressure. Once all the toluene is displaced, a quick atmospheric test is performed on the produced reservoir fluid. The atmospheric test determines the GOR, stock tank density, and formation volume factor of the reservoir fluid. This is to check that the data corresponds with the reservoir fluid and all toluene has been displaced.

The test gas is injected at a fixed rate (~6 ml per hour), and the evolved products (displaced fluid) are continuously collected. The evolved gas volume, weight, and liquid density are measured every hour. This injection rate is continued for the first 6 hours and then increased to approximately 8 ml per hour for the rest of the test. At a certain point, gas breakthrough occurs. It will be observed by a significant increase in GOR, decrease in liquid density, and change in gas density. A windowed PVT cell and camera can also be used to visually observe the time at which gas breakthrough occurs. A high pressure Paar densitometer can be used to measure the density of the produced liquid.

The recovery is reported for 1.2 pore volume of test gas injected, but the test is allowed to continue until slightly more pore volume (say, 1.4) has been injected into the coil. The remainder of the gas in the coil is allowed to flow to atmospheric pressure ("blow down"). Any residual oil produced during the blow down is collected and weighed. The coil is then disconnected and weighed to determine the weight of the residual oil remaining at the end of the test. This is used for the material balance. This test is normally run at four to six different test pressures. A plot of recovery at 1.2 pore volume versus the test pressure as sketched in Figure 3.24 is used to determine the minimum miscibility pressure (MMP). It is taken to be the intersection of the line joining the high recoveries (typically above 90%) and the line joining the low recoveries.

Table 3.47 gives the composition of a reservoir oil, which has been used in a slim tube experiment with pure CO_2 as injection gas. The results (Zuo et al. 1993) of the slim tube experiment are presented in Table 3.48 and plotted in Figure 3.25. The minimum miscibility pressure is seen to be at approximately 207 bar.

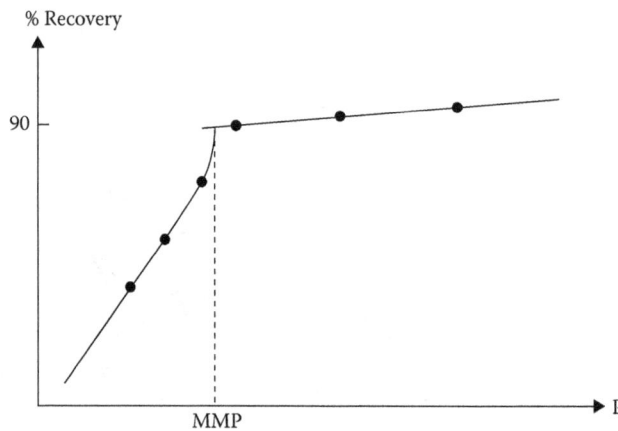

FIGURE 3.24 Stylistic slim tube recovery curve.

TABLE 3.47

Molar Composition of the Oil for Which Experimental Slim Tube Data Is Shown in Table 3.48

Component	Mol%	Molecular Weight (g/mol)	Density 1.01 bar, 15°C (g/cm³)
N_2	1.025	—	—
CO_2	0.251	—	—
C_1	17.243	—	—
C_2	5.295	—	—
C_3	4.804	—	—
iC_4	0.948	—	—
nC_4	1.644	—	—
iC_5	0.542	—	—
nC_5	0.348	—	—
C_6	0.134	—	—
C_{7+}	67.769	254	0.8367

TABLE 3.48

Slim Tube Data for a Pure-CO_2 Displacement at 85.7°C of the Oil in Table 3.47

Displacement Pressure (bar)	Recovery (%)
170.2	76.1
186.2	81.6
200.0	87.6
206.9	90.1
213.7	90.4
227.5	91.4

Source: Zuo, Y et al., A study on the minimum miscibility pressure for miscible flooding systems, *J. Petroleum Sci. Eng.* 8, 315–328, 1993.

Note: Results are plotted in Figure 3.25.

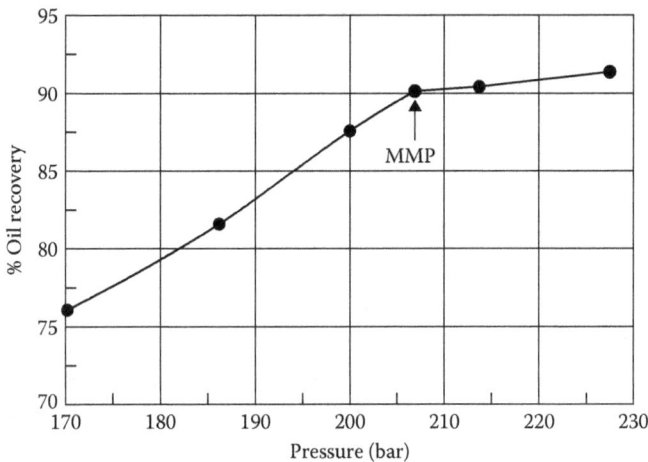

FIGURE 3.25 Percentage of oil recovery in slim tube experiment on the oil composition of Table 3.47 displaced with pure CO_2 at 85.7°C. The results are tabulated in Table 3.48.

3.2.5 Gas Revaporization Experiment

A gas revaporization experiment is sketched in Figure 3.26 and is designed for gas condensates undergoing gas injection. A portion of reservoir fluid (around 50 ml) is charged to a gas condensate PVT cell and stabilized at reservoir temperature and working pressure. A CCE test is performed on this reservoir fluid to determine the dew point pressure and the volume at the dew point. The test is continued until the fluid is expanded to the gas revaporization pressure. At this pressure, a known portion of gas is pumped off from the cell until the volume returns to the volume at dew point pressure. The composition of the pumped-off gas phase is measured.

A known volume of injection gas is charged to the cell, and the content in the cell is then mixed to get a new mixture at the test conditions. At these conditions the sample is allowed to stabilize to have the gas and liquid phases in equilibrium. Once equilibrium is attained, the total sample volume and the volumes of the gas and liquid phases are determined. The excess volume of the gas phase is pumped off to return the cell volume to that at the dew point pressure. The composition of the pumped-off phase is measured. The test is repeated for (typically) four more steps and

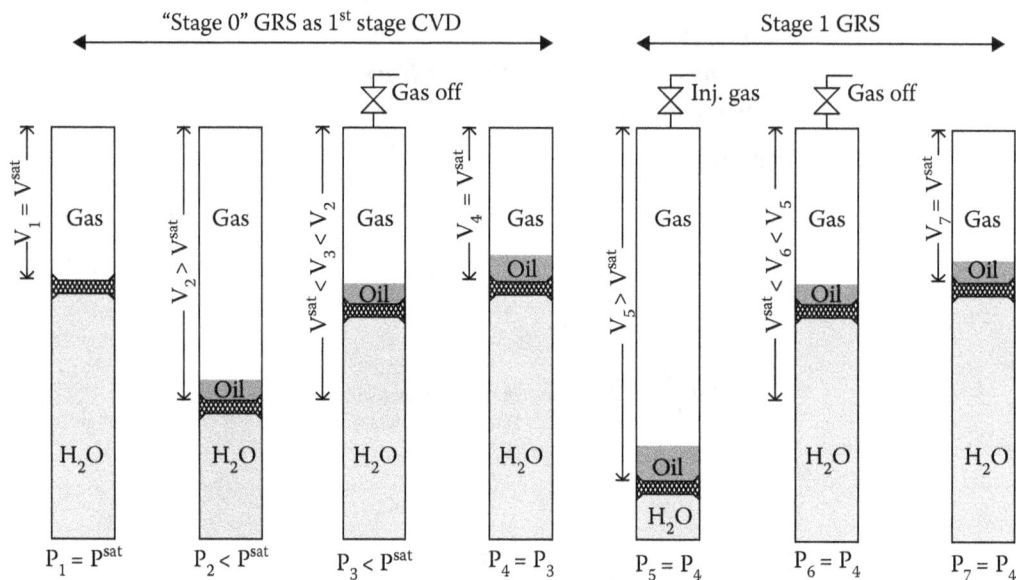

FIGURE 3.26 Schematic representation of gas revaporization experiment. GRS stands for gas revaporization study.

TABLE 3.49

Primary Results from a Gas Revaporization Experiment

Liquid volume	Liquid volume percentage of dew point volume
Percentage produced	Cumulative molar percentage of initial mixture removed (depleted) from cell
Percentage charged	Molar percentage of initial mixture charged to cell in each stage
Z-factor gas	Defined in Equation 3.2. Refers to depleted gas at cell conditions
Viscosity of gas	Viscosity of the gas in cell (usually not measured but calculated)
Gas compositions	Molar compositions of gas liberated from each pressure stage $< P^{sat}$

---

the change in fluid properties, composition, and liquid volume as a function of gas revaporization process is determined. The primary results from a gas revaporization experiment are listed in Table 3.49.

Three gas revaporization experiments have been carried out on the reservoir fluid in Table 3.50 (Kumar et al. 2015) using three different injection gases, also shown in Table 3.50. One is pure nitrogen (N_2), the second one is pure carbon dioxide (CO_2), and the third one is a gas mixture with a methane (C_1) mole percentage of 91.64. The results of the gas revaporization experiments are shown in Table 3.51. Of the three injection gases, CO_2 has the highest ability to revaporize already condensed liquid. After injection of 0.4 mole CO_2 per initial mole reservoir fluid, the liquid volume percentage is reduced by around 56%, whereas the reduction of the liquid volume after injection of the same molar amount of the gas is only 11% with N_2 as injection gas and 28% when the gas mixture is used as injection gas.

TABLE 3.50

Compositions of Reservoir Fluid and Injection Gases Used in Gas Revaporization Experiment in Table 3.51. The C_{7+} Molecular Weight of the Reservoir Fluid Is 138 and the C_{7+} Density 0.780 g/cm³

Component	Reservoir Fluid Mole%	Injection Gas 1 Mole%	Injection Gas 2 Mole%	Injection Gas 3 Mole %
N_2	0.13	100		0.15
CO_2	2.99		100	2.30
H_2S	0.81			
C_1	74.37			91.64
C_2	7.95			5.40
C_3	3.74			0.49
iC_4	0.82			
nC_4	1.69			
iC_5	0.71			
nC_5	0.82			
C_6	1.00			
C_7	1.00			
C_8	0.97			
C_9	0.76			
C_{10}	0.55			
C_{11}	0.38			
C_{12}	0.28			
C_{13}	0.22			
C_{14}	0.17			
C_{15}	0.14			
C_{16}	0.11			
C_{17}	0.08			
C_{18}	0.06			
C_{19}	0.05			
C_{20+}	0.18			

TABLE 3.51

Gas Revaporization Results for Fluid Composition in Table 3.50

N_2 as Injection Gas at 133°C and 208 bar			
	Stage 1	Stage 2	Stage 3
Cum. mol% added gas	0.0	40	140
Liq vol% of V^{sat} at contact P	7.35	6.54	4.60
Saturation pressure (bar)	140	460	-

CO_2 as Injection Gas at 133°C and 222 bar					
	Stage 1	Stage 2	Stage 3	Stage 4	Stage 5
Cum. mol% added gas	0.0	9.91	24.76	44.55	74.26
Liq vol% of V^{sat} at contact P	7.18	5.91	4.20	2.90	1.48
Saturation pressure (bar)	310	307	303	297	287

(Continued)

TABLE 3.51   (Continued)
Gas Revaporization Results for Fluid Composition in Table 3.50

	Injection Gas 3 in Table 3.50 as Injection Gas at 133°C and 222 bar				
	Stage 1	Stage 2	Stage 3	Stage 4	Stage 5
Cum. mol% added gas	0.0	9.89	19.78	29.66	39.97
Liq vol% of V^{sat} at contact P	7.19	6.60	6.04	5.62	5.16
Saturation pressure (bar)	312	337	358	377	397

Source: Kumar, A., Gohary, M.E., Pedersen, K.S., and Azeem, J., Gas injection as an enhanced recovery technique for gas condensates. A comparison of three injection gases, SPE 177778-MS, presented at the *Abu Dhabi International Petroleum Exhibition and Conference* in Abu Dhabi, UAE, November 9–12, 2015.

REFERENCES

Al-Ajmi, M., Tybjerg, P., Rasmussen, C.P., and Azeem, J.S., EoS modeling for two major Kuwaiti oil reservoirs, SPE 141241-PP, presented at the *SPE Middle East Oil and Gas Show*, Manama, Bahrain, March 20–23, 2011.

American Petroleum Institute, *Technical Data Book—Petroleum Refining*, API, New York, 1982.

Burnett, E.S., Compressibility determinations without volume measurements, *J. Appl. Mech. Trans. ASME* 3, A136–A146, 1936.

Fredenslund, Aa., Mollerup, J., and Christiansen, L.J., An apparatus for accurate determinations of vapour-liquid properties and gas PVT properties, *Cryogenics* 13, 414–419, 1973.

Hilsenrath, J., Tables of thermal properties of gases, *US Department of Commerce*, National Bureau of Standards, 1955.

Kumar, A., Gohary, M.E., Pedersen, K.S., and Azeem, J., Gas injection as an enhanced recovery tech-nique for gas condensates. A comparison of three injection gases, SPE 177778-MS, presented at the *Abu Dhabi International Petroleum Exhibition and Conference*, Abu Dhabi, November 9–12, 2015.

Lee, A., Gonzalez, M., and Eakin, B., The viscosity of natural gases, SPE Paper 1340-PA, *J. Pet. Technol.* 18, 997–1000, 1966.

Memon, A., Qassim, B., Al-Ajmi, M., Kumar, A., Gao, J., Ratulowski, J., and Al-Otaibi, B., Miscible gas injection and asphaltene flow assurance fluid characterization: A laboratory case study for Black Oil Reservoir, SPE 150938, presented at the *SPE EOR Conference Muscat*, Oman, April 16–18, 2012.

Negahban, S., Pedersen, K.S., Baisoni, M.A., Sah, P., and Azeem, J.S., An EoS model for a middle east reservoir fluid with an extensive EOR PVT data material, SPE-136530-PP, presented at the *Abu Dhabi International Petroleum Exhibition & Conference*, Abu Dhabi, November 1–4, 2010.

Pedersen, K.S., Fredenslund, Aa., and Thomassen, P., Properties of oils and gases, *Contributions in Petroleum Geology and Engineering*, Vol. 5, Gulf Publishing Company, Houston, TX, 1989.

Shaikh, J.A. and Sah, P., Experimental PVT data needed to develop EOS model for EOR projects, SPE-144023, presented at the *EORC*, Kuala Lumpur, Malaysia, July 19–21, 2011.

Zuo, Y., Chu, J., Ke, S., and Guo, T., A study on the minimum miscibility pressure for miscible flooding systems, *J. Pet. Sci. Eng.* 8, 315–328, 1993.

4 Equations of State

The majority of calculations of the pressure-volume-temperature (PVT) relation carried out for oil and gas mixtures are based on a cubic equation of state. Cubic equations date back more than 100 years to the famous van der Waals equation (van der Waals 1873). The most commonly used equations in the petroleum industry today are similar to the van der Waals equation, but it took almost a century for the industry to accept this type of equation as a valuable engineering tool. The first cubic equation of state to obtain widespread use was the one presented by Redlich and Kwong (1949). Soave (1972) and Peng and Robinson (1976 and 1978) further developed this equation in the 1970s. In 1982, Peneloux et al. (1982) presented a volume-shift concept for improving liquid density predictions of the two former equations. Computer technology has made it possible, within seconds, to perform millions of multicomponent phase equilibrium and physical property calculations with an equation of state as the thermodynamic basis. This chapter presents some of the most popular cubic equations of state as well as the noncubic PC-SAFT and GERG-2008 equations of state. Chapter 6 describes the application of cubic equations of state in phase equilibrium (flash) calculations, Chapter 8 the derivation of physical properties from cubic equations of state, and Chapter 16 the application of cubic equations of state to mixtures with water and other aqueous components as well as an extension of cubic equations with an association term.

4.1 VAN DER WAALS EQUATION

When deriving the first cubic equation of state, van der Waals used the phase behavior of a pure component as the starting point. Figure 4.1 shows schematically pressure (P) versus molar volume (V) curves for a pure component at various temperatures. At temperatures far above the critical (T_1 in Figure 4.1), the pressure–volume (PV) curves exhibit a hyperbolic shape suggesting that the pressure is inversely proportional to the molar volume. This behavior is known from the ideal gas law:

$$P = \frac{RT}{V} \qquad (4.1)$$

where R equals the gas constant and T the absolute temperature. The molar volume of a component behaving like an ideal gas also at high pressures would asymptotically approach zero for the pressure going toward infinity. As seen from Figure 4.1, this is not the case in reality. With increasing pressure, the molar volume approaches a limiting value, which van der Waals named b. Rearrangement of Equation 4.1 to

$$V = \frac{RT}{P} \qquad (4.2)$$

suggests that the b parameter should enter the equation as follows:

$$V = \frac{RT}{P} + b \qquad (4.3)$$

which would give the following expression for P:

$$P = \frac{RT}{V - b} \qquad (4.4)$$

DOI: 10.1201/9780429457418-4

At temperatures below the critical (T_3 in Figure 4.1), a vapor-to-liquid phase transition may take place. Consider a component at temperature T_3 initially at a low pressure and in vapor form. By decreasing the volume while maintaining a constant temperature, T_3, the pressure will increase and at some stage a liquid phase may start to form showing that the dew point pressure has been reached. A further lowering of the volume will take place at a constant pressure until all the vapor has been transformed into liquid. As a liquid is almost incompressible, a further reduction of the volume will be associated with a steep increase in pressure, as is also illustrated in Figure 4.1. The fact that the substance may undergo a transition from a gaseous form with the molecules far apart to a liquid form with the molecules much closer together shows that also attractive forces act between the molecules. These attractive forces are not accounted for in Equation 4.4, which is therefore incapable of describing a vapor-to-liquid phase transition.

Figure 4.2 shows a container filled with gas. The two small volume elements, v_1 and v_2, initially contain one molecule each. Suppose the force between the two volume elements is f. If another molecule is added to v_2 keeping one molecule in v_1, the force acting between the two elements will

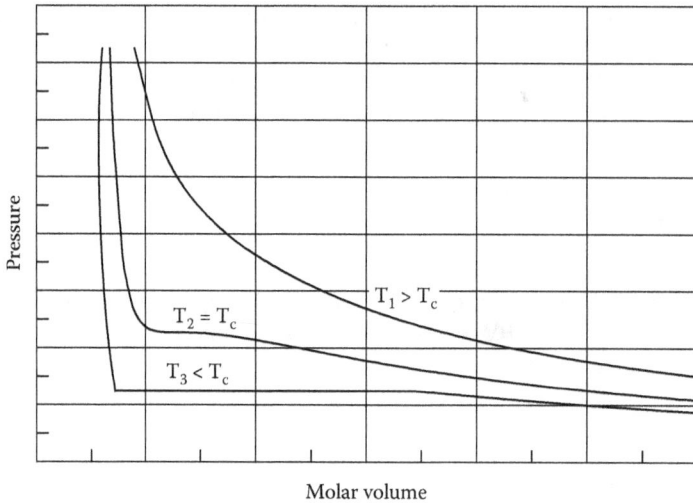

FIGURE 4.1 Pressure-volume curves for pure component.

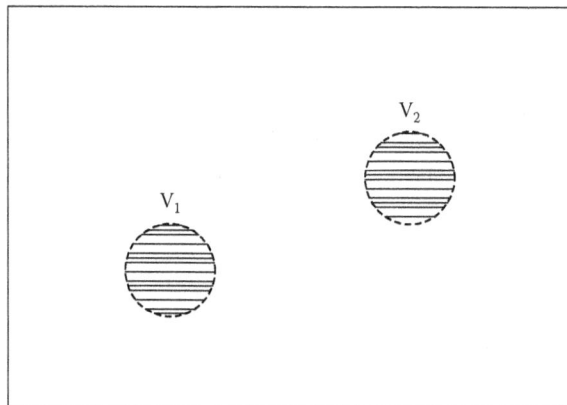

FIGURE 4.2 Interaction between two volume elements in a container filled with gas.

be 2f. Addition of a third molecule to v_2 will increase the force to 3f, and so on. The force of attraction between the two volume elements is therefore proportional to c_2, the concentration of molecules in v_2. If a second, third, and so on, molecule is added to v_1, keeping the number of molecules in v_2 constant, the force will double, triple, and so on. The force is therefore also proportional to c_1, the concentration of molecules in v_1. Thus, the force acting between the two volume elements is proportional to $c_1 \times c_2$. In reality, the concentration in the gas is the same everywhere, that is, $c = c_1 = c_2$, where c is the molecular concentration in the container. The concentration c is inversely proportional to the molar volume, V, implying that the attractive force is proportional to the $1/V^2$. Based on that type of consideration, van der Waals found that the attractive term should be a constant a times $1/V^2$ leading to:

$$P = \frac{RT}{V-b} - \frac{a}{V^2}$$ (4.5)

which is the final form of the van der Waals equation. The constants a and b are equation of state parameters, the values of which are found by evaluating the PV curve for the critical temperature. This curve is also called the critical isotherm. As is illustrated in Figure 4.1, this curve has what looks like an inflexion point right at the critical point (T_c, P_c), implying

$$\left(\frac{\partial P}{\partial V}\right)_{T \text{ at } T=T_c, P=P_c} = \left(\frac{\partial^2 P}{\partial V^2}\right)_{T \text{ at } T=T_c, P=P_c} = 0$$ (4.6)

At the critical point, V equals the molar critical volume V_c, which is related to T_c and P_c through Equations 4.5 and 4.6. A total of five constants $(T_c, P_c, V_c, a, \text{ and } b)$ enter into Equations 4.5 and 4.6 (a total of three equations). One equation is used to eliminate the molar volume V, which at the critical point is V_c. The two remaining equations are rearranged to give the following expressions for a and b:

$$a = \frac{27 R^2 T_c^2}{64 P_c}$$ (4.7)

$$b = \frac{R T_c}{8 P_c}$$ (4.8)

Equation 4.5 may be used for any pure substance for which T_c and P_c are known. By rearranging the equation to

$$V^3 - \left(b + \frac{RT}{P}\right)V^2 + \frac{a}{P}V - \frac{ab}{P} = 0$$ (4.9)

it is seen that the van der Waals equation is cubic in V, which explains why the van der Waals and related equations are called cubic. Figure 4.3 shows PV curves for methane calculated using the van der Waals equation. At temperatures $T_1 > T_c$ and $T_2 = T_c$, the curves qualitatively agree with experimental observations as sketched in Figure 4.1. At $T_3 < T_c$, the vapor-to-liquid phase transition is not as in Figure 4.1 represented by a constant line in P. Starting from the high-volume (vapor) side, the PV curve crosses the experimental vapor pressure curve at point A after which it goes through a maximum, crosses the vapor pressure line a second time at point B, passes through a minimum, and finally crosses the vapor pressure line a third time at point C. The molar volume at point A equals the vapor-phase molar volume at the saturation point. The molar volume at point C equals

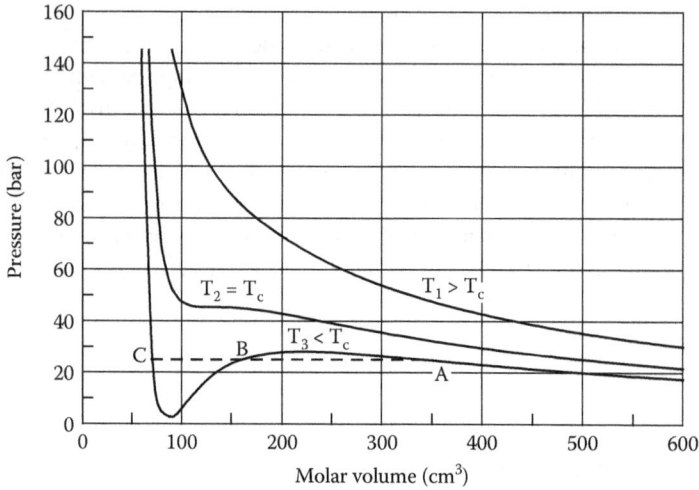

FIGURE 4.3   Pressure–volume curves for methane at $T_1 = 248$ K ($>T_c$), $T_2 = 190.6$ K ($= T_c$), and $T_3 = 162$ K ($< T_c$) calculated from the van der Waals equation. The dashed line indicates the location of the vapor pressure at $T = T_3$.

the liquid-phase molar volume at the saturation point. The molar volume at point B has no physical importance, and the whole section of the PV curve between A and C may be disregarded. Hence, the van der Waals equation is seen to qualitatively describe the pure-component phase behavior at temperatures above, equal to, and below the critical temperature.

Later developments of cubic equations of state have primarily served to improve the quantitative predictions of either vapor pressure or liquid density. In addition, much effort has been used to extend the application area of cubic equations of state from pure components to mixtures.

4.2   REDLICH–KWONG EQUATION

The equation of Redlich and Kwong (1949) is, by many, considered the first modern equation of state and takes the form

$$P = \frac{RT}{V - b} - \frac{a}{\sqrt{T}\, V(V + b)} \tag{4.10}$$

By comparing this equation with the van der Waals equation (Equation 4.5), it is seen that the attractive term has a more complicated temperature dependence. This temperature modification serves to improve the vapor pressure predictions. To also improve the predictions of the liquid-phase molar volumes, the term V^2 in the denominator of the attractive term in the van der Waals equation has, in the Redlich–Kwong equation, been replaced by $V(V + b)$. The parameters a and b are found by imposing the critical point criteria expressed in Equation 4.6. This leads to the following expressions:

$$a = \frac{0.42748\, R^2\, T_c^{2.5}}{P_c} \tag{4.11}$$

$$b = \frac{0.08664\, R\, T_c}{P_c} \tag{4.12}$$

For an N-component mixture, the parameters a and b in Equation 4.10 are found using the following mixing rules:

$$a = \sum_{i=1}^{N} \sum_{j=1}^{N} z_i z_j a_{ij} \tag{4.13}$$

$$b = \sum_{i=1}^{N} z_i b_i \tag{4.14}$$

where z_i and z_j are the mole fractions of component i and j, respectively. The term b_i is the b parameter of component i found from Equation 4.12. The term a_{ij} is determined from:

$$a_{ij} = \frac{0.42748 \, R^2 \, T_{cij}^{2.5}}{P_{cij}} \tag{4.15}$$

which corresponds to Equation 4.11, but T_c and P_c of the pure component are, in Equation 4.15, replaced by the cross terms T_{cij} and P_{cij}. T_{cij} is related to the pure component critical temperatures T_{ci} and T_{cj} of components i and j as follows:

$$T_{cij} = \sqrt{T_{ci} T_{cj}} (1 - k_{ij}) \tag{4.16}$$

In this expression, k_{ij} is a binary interaction parameter for components i and j. For two identical components, k_{ij} is zero by definition. For two different nonpolar compounds, k_{ij} is equal to or close to zero. For a binary pair of at least one polar component, nonzero k_{ij}s are often appropriate. With a $k_{ij} > 0$ the simulated attraction between molecules of components i and j will be reduced as compared to a k_{ij} of zero. The mixing rule in Equation 4.16 is based on considerations regarding the attractive energy acting between two molecules or two bodies. P_{cij} is found from:

$$P_{cij} = \frac{Z_{cij} \, R \, T_{cij}}{V_{cij}} \tag{4.17}$$

where

$$Z_{cij} = \frac{Z_{ci} + Z_{cj}}{2} \tag{4.18}$$

and

$$V_{cij} = \left(\frac{V_{ci}^{1/3} + V_{cj}^{1/3}}{2} \right)^3 \tag{4.19}$$

Z_{ci} and Z_{cj} are the compressibility factors of component i and j at the pure-component critical points. The mixing rule used for the critical volume is based on the idea that the molecules of components i and j have linear dimensions proportional to the cubic roots of V_{ci} and V_{cj}, respectively. The term within the parenthesis in Equation 4.19 is then proportional to linear length of an average molecule of i and j.

4.3 SOAVE–REDLICH–KWONG EQUATION

Soave (1972) found that the pure-component vapor pressures calculated from the Redlich–Kwong (RK) equation could be improved. He suggested replacing the term $\frac{a}{\sqrt{T}}$ in the RK equation by a more general temperature-dependent term, a(T), giving an equation of state of the form:

$$P = \frac{RT}{V-b} - \frac{a(T)}{V(V+b)} \tag{4.20}$$

This equation is usually referred to as the Soave–Redlich–Kwong or just SRK equation. Soave plotted $\sqrt{\dfrac{a}{a_c}}$ versus $\sqrt{\dfrac{T}{T_c}}$ for a number of pure hydrocarbons. The term $\sqrt{\dfrac{a}{a_c}}$ was determined from vapor pressure data. The type of plots Soave made is shown schematically in Figure 4.4. He observed an almost linear relationship indicating that a linear dependence should be chosen for the square root of the preceding a parameter ratio versus the square root of the reduced temperature $T_r = T/T_c$. Soave proposed the following temperature dependence:

$$a(T) = a_c \, \alpha(T) \tag{4.21}$$

$$a_c = \frac{0.42747R^2T_c^2}{P_c} \tag{4.22}$$

$$b = \frac{0.08664R \, T_c}{P_c} \tag{4.23}$$

$$\alpha(T) = \left(1 + m\left(1 - \sqrt{\frac{T}{T_c}}\right)\right)^2 \tag{4.24}$$

$$m = 0.480 + 1.574\,\omega - 0.176\,\omega^2 \tag{4.25}$$

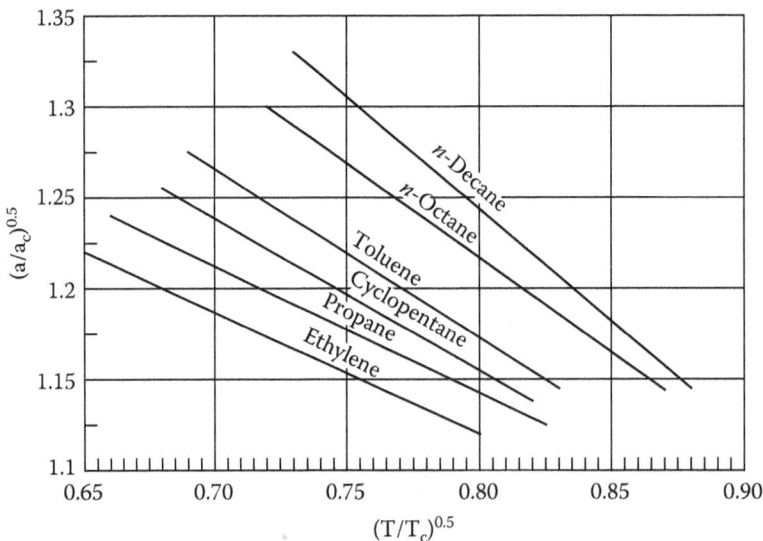

FIGURE 4.4   Relationship between $(a/a_c)^{0.5}$ and $(T/T_c)^{0.5}$ observed by Soave (1972).

In Equation 4.25, ω is the acentric factor as defined in Equation 1.1. Equations 4.21 and 4.24 may be combined to give:

$$\sqrt{\frac{a(T)}{a_c}} = (1+m) - m\sqrt{\frac{T}{T_c}} \qquad (4.26)$$

which is in accordance with Soave's observations and expresses a linear relationship between

$$\sqrt{\frac{a}{a_c}} \text{ and } \sqrt{\frac{T}{T_c}}.$$

The coefficients in the expression for m (Equation 4.25) were determined by a data fit to experimental vapor pressure data for nine pure hydrocarbons.

It is explained in Chapter 6 how to make vapor pressure calculations for pure components using a cubic equation of state.

With the Soave temperature dependence, $\alpha(T) = 1$ at the critical temperature, where a(T) therefore becomes equal to a_c. The terms a_c and b in Equations 4.22 and 4.23 are found making use of the critical point criteria expressed in Equation 4.6. The constants 0.42747 and 0.08664 are often referred to as Ω_a and Ω_b.

Graboski and Daubert (1978) refitted the three coefficients in Equation 4.25 to 0.48508, 1.55171, and −0.15613 based on vapor pressure data for more components including aromatics and iso-paraffins, but these coefficients have not obtained as widespread use as those proposed by Soave.

Mathias and Copeman (1983) have presented a more flexible temperature dependence for α:

$$\alpha(T) = \left(1 + C_1\left(1 - \sqrt{T_r}\right) + C_2\left(1 - \sqrt{T_r}\right)^2 + C_3\left(1 - \sqrt{T_r}\right)^3\right)^2 \quad T_r < 1 \qquad (4.27)$$

$$\alpha(T) = \left(1 + C_1\left(1 - \sqrt{T_r}\right)\right)^2 \quad T_r \geq 1 \qquad (4.28)$$

It is seen that the Mathias–Copeman expression reduces to Equation 4.24 for $C_1 = m$ and $C_2 = C_3 = 0$. Khashayar and Moshfeghian (1998) have presented Mathias–Copeman coefficients for C_1–C_4 hydrocarbons for use with the SRK equation as shown in Table 4.1. Table 4.1 also shows Mathias–Copeman coefficients for water and methanol (Dahl and Michelsen 1990). In general, the Mathias and Copeman temperature dependence is more widely used for polar compounds, for example, water and methanol, than for hydrocarbons.

Recalling that the compressibility factor Z is defined as follows:

$$Z = \frac{PV}{RT} \qquad (4.29)$$

Equation 4.20 may be rewritten in terms of Z:

$$Z^3 - Z^2 + \left(A - B - B^2\right)Z - AB = 0 \qquad (4.30)$$

where A and B are given by the following expressions:

$$A = \frac{a(T)P}{R^2 T^2} \qquad (4.31)$$

$$B = \frac{bP}{RT} \qquad (4.32)$$

TABLE 4.1

Mathias and Copeman Coefficients (Equations 4.27 and 4.28) for Use with the Soave–Redlich–Kwong Equation

Component	C_1	C_2	C_3	Reference
Methane	0.5857	−0.7206	1.2899	Khashayar and Moshfeghian (1998)
Ethane	0.7178	−0.7644	1.6396	Khashayar and Moshfeghian (1998)
Propane	0.7863	−0.7459	1.8454	Khashayar and Moshfeghian (1998)
Iso-butane	0.8284	−0.8285	2.3201	Khashayar and Moshfeghian (1998)
n-butane	0.8787	−0.9399	2.2666	Khashayar and Moshfeghian (1998)
Water	1.0873	−0.6377	0.6345	Dahl and Michelsen (1990)
Methanol	1.4450	−0.8150	0.2486	Dahl and Michelsen (1990)

With the SRK equation, the compressibility factor of a pure component at its critical point will always be equal to 0.333. This can be shown by replacing T and P in Equations 4.31 and 4.32 by T_c and P_c, a(T) in Equation 4.31 by a_c from Equation 4.22, and b in Equation 4.32 by the expression for b in Equation 4.23. The derived expressions for A and B are inserted into Equation 4.30, which is solved for Z.

For an N-component mixture, Soave suggested to find a and b from

$$a = \sum_{i=1}^{N}\sum_{j=1}^{N} z_i\, z_j\, a_{ij} \tag{4.33}$$

$$b = \sum_{i=1}^{N} z_i\, b_i \tag{4.34}$$

where z stands for mole fraction, i and j are component indices, and

$$a_{ij} = \sqrt{a_i a_j}\,(1 - k_{ij}) \tag{4.35}$$

With these mixing rules, Soave, as outlined in Figure 4.5, defines a hypothetical component with one a and one b parameter representing the entire fluid mixture. In phase equilibrium calculations with two or more phases, each phase will have its own a and b parameters.

The parameter k_{ij} is a binary interaction coefficient similar to the one entering into the RK mixing rule of Equation 4.16. Examples of binary interaction parameters recommended for the SRK equation are given in Table 4.2. It is seen that the interaction parameter between N_2 and CO_2 is negative, which suggests that the attraction between an N_2 and a CO_2 molecule is higher than would

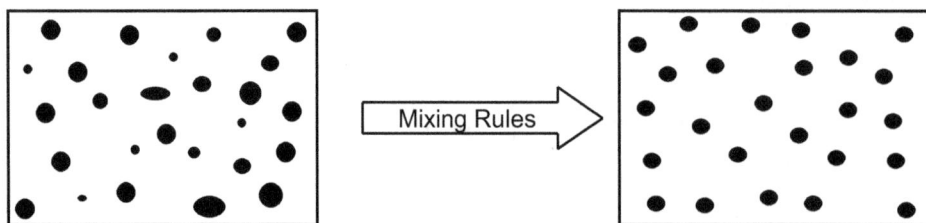

FIGURE 4.5 Schematic illustration of mixing rules applied for a cubic equation of state. The a and b parameters for the individual components are averaged to one a and one b representing the whole mixture.

TABLE 4.2

Nonzero Binary Interaction Coefficients for Petroleum Reservoir Fluid Constituents for Use with the Soave–Redlich–Kwong Equation of State

Component Pair	N_2	CO_2	H_2S
Soave–Redlich–Kwong			
N_2	0.0000	−0.0315	0.1696
CO_2	−0.0315	0.0000	0.0989
H_2S	0.1696	0.0989	0.0000
C_1	0.0278	0.1200	0.0800
C_2	0.0407	0.1200	0.0852
C_3	0.0763	0.1200	0.0885
iC_4	0.0944	0.1200	0.0511
nC_4	0.0700	0.1200	0.0600
iC_5	0.0867	0.1200	0.0600
nC_5	0.0878	0.1200	0.0689
C_6	0.0800	0.1200	0.0500
C_{7+}	0.0800	0.0100	0.0000
Peng–Robinson			
N_2	0.0000	0.0170	0.1767
CO_2	0.0170	0.0000	0.0974
H_2S	0.1767	0.0974	0.0000
C_1	0.0311	0.1200	0.0800
C_2	0.0515	0.1200	0.0833
C_3	0.0852	0.1200	0.0878
iC_4	0.1033	0.1200	0.0474
nC_4	0.0800	0.1200	0.0600
iC_5	0.0922	0.1200	0.0600
nC_5	0.1000	0.1200	0.0630
C_6	0.0800	0.1200	0.0500
C_{7+}	0.0800	0.0100	0.0000

Source: Data from Knapp, H.R., et al., Vapor-liquid equilibria for mixtures of low boiling substances, *Chem. Data Ser.* Vol. VI, DECHEMA, Frankfurt am Main, Germany, 1982.

be simulated with an interaction parameter of zero. The mixing rule used for b says that pure-component molar volumes at high pressures are additive.

4.4 PENG–ROBINSON EQUATION

The liquid-phase densities predicted using the SRK equation are in general too low. Peng and Robinson (1976) traced this deficiency to the fact that the SRK equation predicts the pure-component critical compressibility factor to be 0.333. Table 4.3 presents experimental critical compressibility factors for the C_1–C_{10} *n*-paraffins. The critical compressibility factors are generally of the order 0.25–0.29, that is, somewhat lower than simulated using the SRK equation. Peng and Robinson suggested an equation of the form:

$$P = \frac{RT}{V-b} - \frac{a(T)}{V(V+b)+b(V-b)}$$ (4.36)

TABLE 4.3

Critical Compressibility Factors of C_1–C_{10} n-Paraffins

Compound	Z_c
C_1	0.288
C_2	0.285
C_3	0.281
nC_4	0.274
nC_5	0.251
nC_6	0.260
nC_7	0.263
nC_8	0.259
nC_9	0.260
nC_{10}	0.247

Source: Data from Poling, B.E, Prausnitz, J.M., and O'Connell, J.P., *The Properties of Gases and Liquids,* McGraw-Hill, New York, 2000.

---

where

$$a(T) = a_c \, \alpha(T) \tag{4.37}$$

$$a_c = 0.45724 \frac{R^2 \, T_c^2}{P_c} \tag{4.38}$$

$$\alpha(T) = \left(1 + m \left(1 - \sqrt{\frac{T}{T_c}} \right) \right)^2 \tag{4.39}$$

$$m = 0.37464 + 1.54226 \, \omega - 0.26992 \, \omega^2 \tag{4.40}$$

$$b = \frac{0.07780 \, R \, T_c}{P_c} \tag{4.41}$$

Equation 4.36 gives a universal critical compressibility factor of 0.307 for pure substances, which is lower than 0.333 as found with the SRK equation but still high compared to the experimental critical compressibility factors presented in Table 4.3. For mixtures, Peng and Robinson recommend to calculate a and b using the same mixing rules as for SRK and shown in Equations 4.33 and 4.34.

Two years later, Peng and Robinson (1978) presented a modification of Equation 4.40 to be used for $\omega > 0.49$:

$$m = 0.379642 + 1.48503 \, \omega - 0.164423 \, \omega^2 + 0.016666 \, \omega^3 \tag{4.42}$$

4.5   PENELOUX VOLUME CORRECTION

Until 1982, the application of the SRK equation was essentially limited to phase equilibrium and gas-phase density calculations. Owing to poor liquid density predictions, the SRK equation was often applied with external liquid density correlations. This caused problems, for example, for near-critical systems for which it is difficult to distinguish between a gas and a liquid phase. In 1982,

Peneloux et al. (1982) presented a SRK modification with a volume translation parameter. The Peneloux equation (SRK–Peneloux) takes the form:

$$P = \frac{RT}{V-b} - \frac{a(T)}{(V+c)(V+b+2c)} \tag{4.43}$$

The parameter c is called a volume translation or volume-shift parameter. It is possible to relate the molar volumes and the b parameters entering into the SRK and SRK–Peneloux equations as follows:

$$V_{Pen} = V_{SRK} - c \tag{4.44}$$

$$b_{Pen} = b_{SRK} - c \tag{4.45}$$

where the subindex SRK stands for SRK equation and Pen for SRK–Peneloux equation.

The parameter c is assigned a value that at a particular pressure and temperature makes the calculated liquid molar volume agree with experimental observations. This can be done without affecting phase equilibrium results. The SRK–Peneloux equation will, for example, give the same pure-component vapor pressures and the same mixture dew and bubble point pressures as the classical SRK equation, as presented in Equation 4.20. The term volume translation or volume-shift parameter is used for c to signal that it affects volume without affecting equilibrium phase compositions. Peneloux et al. (1982) recommended the following expression to be used for c of nonhydrocarbons and hydrocarbons lighter than C_7

$$c = \frac{0.40768 \, RT_c \, (0.29441 - Z_{RA})}{P_c} \tag{4.46}$$

where Z_{RA} is the Rackett compressibility factor (Rackett 1970; Spencer and Danner 1973):

$$Z_{RA} = 0.29056 - 0.08775\,\omega \tag{4.47}$$

The constants in Equation 4.46 were found by a fit to densities of saturated liquid C_1–C_6 hydrocarbons at atmospheric pressure. It is explained in Chapter 5 how to determine c for hydrocarbons heavier than C_6.

The Peneloux volume translation concept is not limited to the SRK equation but is equally applicable to the Peng–Robinson (PR) equation (Jhaveri and Youngren 1988). With the Peneloux volume correction, the PR equation becomes (PR–Peneloux):

$$P = \frac{RT}{V-b} - \frac{a(T)}{(V+c)(V+2c+b)+(b+c)(V-b)} \tag{4.48}$$

For nonhydrocarbons and hydrocarbons lighter than C_7, the volume-shift parameter may be found from:

$$c = \frac{0.50033\,RT_c}{P_c}(0.25969 - Z_{RA}) \tag{4.49}$$

where Z_{RA} is defined in Equation 4.47.

As for the SRK–Peneloux equation, the constants in this equation were found by a fit to densities of saturated liquid of C_1–C_6 hydrocarbons at atmospheric pressure.

Although it is generally acknowledged that it is necessary to volume-correct SRK liquid densities, it is less obvious whether such correction is needed for the PR equation, as the PR equation was developed with more focus on liquid density predictions. Figure 4.6 shows experimental and calculated liquid densities of three n-paraffins at their saturation points at different temperatures. The densities have been calculated using the SRK, the PR, and the SRK–Peneloux equations. The highest temperature for which results are shown is the critical temperature. The best overall agreement with experimental data is obtained with the SRK–Peneloux equation. The SRK equation with no volume correction generally predicts too-low liquid densities. This is more pronounced for propane and n-hexane than for methane. For methane and propane, the PR equation predicts too-high liquid densities at the lowest temperatures. Good results are obtained for n-hexane with the PR-equation but still not as good as those obtained with the SRK–Peneloux equation.

The phase equilibrium results obtained with the Peneloux-volume-corrected SRK and PR equations are identical to those obtained with the original equations with no volume correction. This comes from the fact that (taking the SRK equation as an example) the SRK and SRK–Peneloux fugacity coefficients of component i are interrelated through:

$$\ln\varphi_{i,SRK} = \ln\varphi_{i,Pen} + \frac{c_i P}{RT} \qquad (4.50)$$

The term *fugacity coefficient* is explained in Appendix A, where it is also explained that at equilibrium between a vapor phase (V) and a liquid phase (L) the following relation will apply for component i:

$$\frac{y_i}{x_i} = \frac{\varphi_{i,SRK}^L}{\varphi_{i,SRK}^V} \qquad (4.51)$$

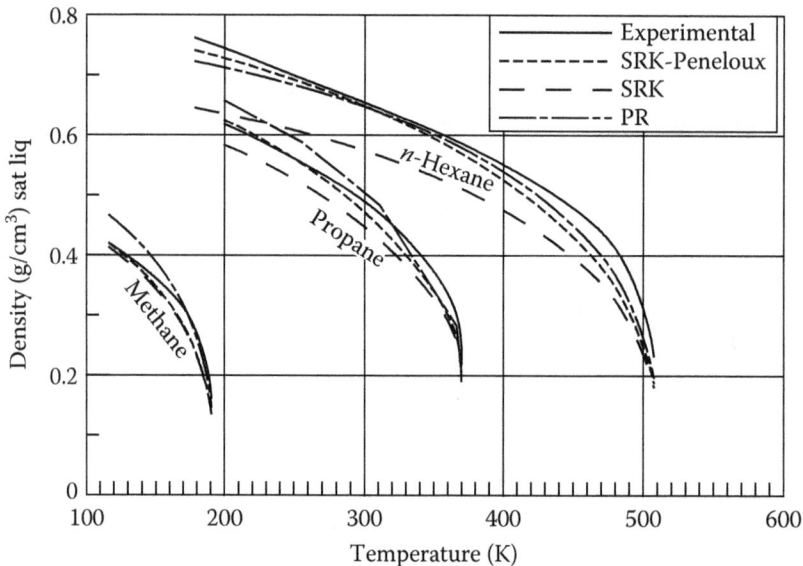

FIGURE 4.6 Experimental and calculated densities of saturated liquids. (Adapted from American Petroleum Institute, *Technical Data Book—Petroleum Refining*, API, New York, 1982).

y_i is the mole fraction of component i in the vapor phase and x_i the mole fraction of component i in the liquid phase.

Using Equation 4.50, this equilibrium relation may be rewritten as:

$$\frac{y_i}{x_i} = \frac{\varphi_{i,Pen}^{L} \exp\left(\frac{c_i P}{RT}\right)}{\varphi_{i,Pen}^{V} \exp\left(\frac{c_i P}{RT}\right)} = \frac{\varphi_{i,Pen}^{L}}{\varphi_{i,Pen}^{V}} \tag{4.52}$$

showing that SRK and SRK–Peneloux equations will provide exactly the same results for phase compositions and therefore also for phase amounts. It is only the molar volume (phase densities) and, as outlined in Chapter 8, some other physical properties that differ. The same applies for the PR and PR–Peneloux equations (Equations 4.36 and 4.48).

The Peneloux equation not only corrects liquid-phase densities but also the vapor-phase densities. This is illustrated in Figure 4.7, which shows a PV curve for n-hexane at 15°C calculated using the SRK equation (Equation 4.20) and using the SRK–Peneloux equation (Equation 4.43). At 1 bar, the SRK molar volume is found to be 148 cm^3, while the real molar volume of n-hexane at these conditions is 130 cm^3. By assigning the Peneloux volume-shift parameter (c in Equation 4.43) a value of 148−130 = 18 cm^3/mol, it is possible to match the liquid volume of n-hexane at 15°C and 1 bar. The PV curve calculated using SRK–Peneloux is also shown in Figure 4.7. The SRK–Peneloux molar volumes are consistently a constant c (18 cm^3/mol in Figure 4.7) lower than SRK volumes. Because the gas-phase molar volumes are high, the volume correction, however, has only a minor influence on the gas volumes but a significant influence on the liquid volumes. This is the whole idea of the correction.

Chapter 5 will introduce temperature dependent volume corrections.

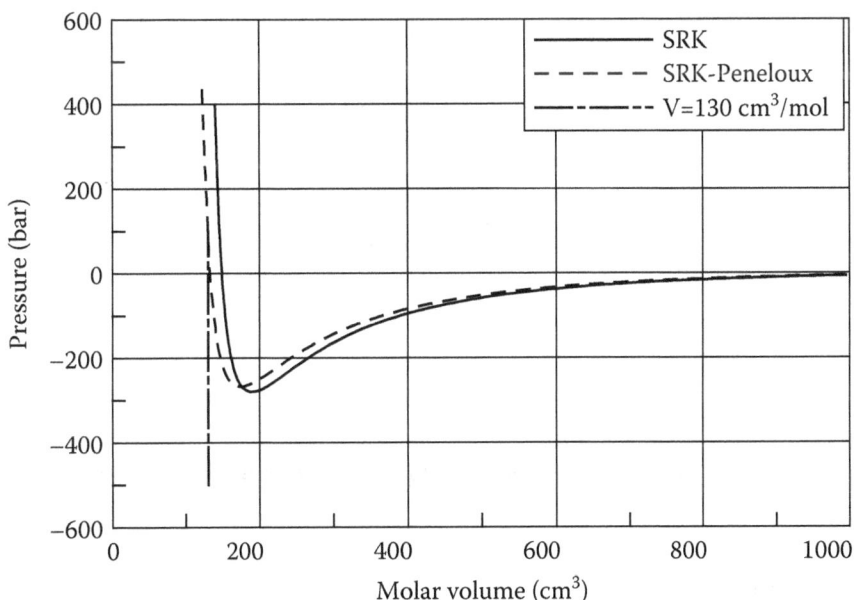

FIGURE 4.7 Molar volume of n-hexane at 15°C calculated using the Soave–Redlich–Kwong (SRK) equation and the SRK–Peneloux equation. The volume-shift parameter has been adjusted to a molar volume of 130 cm^3 at 1 bar.

4.6 OTHER CUBIC EQUATIONS OF STATE

The increasing popularity of cubic equations of state in the 1970s and 1980s inspired thermodynamics research groups to propose alternatives to the SRK and PR equations. Many of these equations have the general form:

$$P = \frac{RT}{V + \delta_1} - \frac{a(T)}{(V + \delta_2)(V + \delta_3)} \tag{4.53}$$

The SRK, PR, SRK–Peneloux, and PR–Peneloux equations (Equations 4.20, 4.36, 4.43, and 4.48) all belong to the general class of equations expressed through Equation 4.53. The values of the parameters δ_1–δ_3 for each equation are shown in Table 4.4. Equation 4.53 offers the opportunity to include three different volumetric correction parameters, δ_1, δ_2, and δ_3. The SRK and PR equations only make use of one parameter, b in Equations (4.20) and (4.36). The Peneloux corrected SRK and PR equations make use of two volume correction parameters, b and c in Equations (4.43) and (4.48)

The Adachi–Lu–Sugie (ALS) equation (1983) is an example of an equation making full use of the extra flexibility presented by the 3 δ-parameters in Equation 4.53. It takes the form:

$$P = \frac{RT}{V - b_1} - \frac{a(T)}{(V - b_2)(V + b_3)} \tag{4.54}$$

ALS uses the following temperature dependence for the a parameter:

$$a = a_c\, \alpha(T) \tag{4.55}$$

$$a_c = \frac{\Omega_a\, R^2\, T_c^2}{P_c} \tag{4.56}$$

$$\alpha(T) = \left(1 + m\left(1 - \sqrt{\frac{T}{T_c}}\right)\right)^2 \tag{4.57}$$

$$\Omega_a = 0.44869 + 0.04024\,\omega + 0.01111\,\omega^2 - 0.00576\,\omega^3 \tag{4.58}$$

$$m = 0.4070 + 1.3787\,\omega - 0.2933\,\omega^2 \tag{4.59}$$

TABLE 4.4

Values of Equation of State Parameters in the Generalized Equation of State Expressed in Equation 4.53

Equation	δ_1	δ_2	δ_3
Soave–Redlich–Kwong (Equation 4.20)	$-b$	0	b
Peng–Robinson (Equation 4.36)	$-b$	$(1+\sqrt{2})\,b$	$(1-\sqrt{2})\,b$
Soave–Redlich–Kwong–Peneloux (Equation 4.43)	$-b$	c	$b + 2c$
Peng–Robinson–Peneloux (Equation 4.48)	$-b$	$c + (1+\sqrt{2})(b+c)$	$c + (1-\sqrt{2})(b+c)$
Adachi–Lu–Sugie (Equation 4.54)	$-b_1$	$-b_2$	b_3

The volume correction parameters b_1–b_3 are found from:

$$b_k = \frac{B_k T_c R}{P_c}, \quad k = 1, 2, 3 \qquad (4.60)$$

where the constants B_1, B_2, and B_3 are functions of the acentric factor:

$$B_1 = 0.08974 - 0.03452\,\omega + 0.00330\,\omega^2 \qquad (4.61)$$

$$B_2 = 0.03686 + 0.00405\,\omega - 0.01073\,\omega^2 + 0.00157\,\omega^3 \qquad (4.62)$$

$$B_3 = 0.15400 + 0.14122\,\omega - 0.00272\,\omega^2 - 0.00484\,\omega^3 \qquad (4.63)$$

The parameters entering into Equation 4.61 through 4.63 were determined by regression to volumetric phase equilibrium and enthalpy data for pure components and mixtures. The classical mixing rules in Equations 4.33 and 4.34 are used for the parameters a and b_1–b_3.

Making use of all three volume parameters in Equation 4.54, the ALS equation is more flexible than the SRK and PR equations even when the two latter are volume corrected. The ALS equation has, however, not achieved the same popularity as SRK and PR. In the petroleum industry, it is important to have some kind of industrial standards to enable different companies working on the same project to produce consistent calculation results. PR seems to be the preferred choice in North America. Europe generally prefers SRK, while the rest of the world is more divided between the two equations of state. Since SRK and PR are by far the most commonly applied cubic equations of state in the petroleum industry, the example calculations for cubic equations presented in this book will all be based on either the SRK or PR equation, most often with Peneloux volume correction.

4.7 FUGACITY COEFFICIENTS

Phase equilibrium calculations using an equation of state are carried out as described in Chapter 6. Fugacity coefficients are required in these calculations. The term fugacity coefficient is further explained in Appendix A. The SRK expression for the fugacity coefficient of component i in a mixture is

$$\ln \varphi_i = -\ln(Z - B) + (Z - 1)\frac{b_i}{b} - \frac{A}{B}\left[\frac{1}{a}\left(2\sqrt{a_i}\sum_{j=1}^{N} z_j \sqrt{a_j}(1 - k_{ij})\right) - \frac{b_i}{b}\right]\ln\left(1 + \frac{B}{Z}\right) \qquad (4.64)$$

A and B are defined in Equations 4.31 and 4.32. For the PR equation, the expression for the fugacity coefficient takes the form:

$$\ln \varphi_i = -\ln(Z - B) + (Z - 1)\frac{b_i}{b} - \frac{A}{2^{1.5}B}\left[\frac{1}{a}\left(2\sqrt{a_i}\sum_{j=1}^{N} z_j \sqrt{a_j}(1 - k_{ij})\right) - \frac{b_i}{b}\right]$$

$$\times \ln\left(\frac{Z + (2^{0.5} + 1)B}{Z - (2^{0.5} - 1)B}\right) \qquad (4.65)$$

The fugacity coefficients for the Peneloux volume-corrected SRK and PR equations (Equations 4.43 and 4.48) can be found from Equations 4.64 and 4.65 making use of Equation 4.50. When using Equation 4.50 for the PR equation, the subindex SRK must be replaced by PR.

4.8   NONCLASSICAL MIXING RULES

Cubic equations of state were originally intended for hydrocarbons and other essentially nonpolar systems, but the widespread use of cubic equations for oil and gas mixtures has inspired work on extending its application area to mixtures containing polar compounds. This will be further dealt with in Chapter 16.

4.9   PC-SAFT EQUATION

PC-SAFT (Gross and Sadowski 2001) is an example of a non-cubic equation of state that has found application in the oil industry. It is based in statistical mechanics, which is a molecular approach to describe macroscopic systems. PC-SAFT stands for *perturbed chain statistical association fluid theory*. The model concept was developed by Chapman et al. (1988, 1990).

The compressibility factor Z is defined as

$$Z = \frac{PV}{RT} \tag{4.66}$$

where P is pressure, V molar volume, R the gas constant, and T the absolute temperature. $Z = 1$ for an ideal gas, which is one of a high molar volume thanks to a low pressure. For lower molar volumes (higher pressures), Z deviates from 1, which can be expressed through a Taylor series expansion:

$$Z = 1 + \frac{A}{V} + \frac{B}{V^2} + \frac{C}{V^3} + \ldots \tag{4.67}$$

This Z-factor expression is called the *virial equation*, and A, B, and C are called *virial coefficients*. The viral equation may be truncated after the first, second, or third term, depending on the molar volume. The lower the molar volume (the higher the pressure), the more terms are needed.

Similar to the virial equation, the PC-SAFT model expresses the compressibility factor as a deviation from the ideal gas compressibility factor of 1.0:

$$Z = 1 + Z^{hc} + Z^{disp} \tag{4.68}$$

Z^{hc} is the hard-chain contribution to the compressibility factor accounting for repulsive molecular interactions and Z^{disp} is an attractive (dispersive) term.

PC-SAFT represents each molecule through three parameters:

- Number of segments: m
- Segment diameter: σ
- Segment energy: ε

The number of segments is 1 for methane. For heavier hydrocarbons, it is a little lower than the number of hydrocarbon segments.

The PC-SAFT approach is schematically illustrated in Figure 4.8. PC-SAFT sees a pure fluid as consisting of equal-sized hard spheres or segments (a in Figure 4.8). These hard spheres are then combined to hard chain molecules (b in Figure 4.8). These hard-chain molecules interact with each other (c in Figure 4.8).

Figure 4.9 shows a schematic view of a PC-SAFT mixture containing two molecules with m_1 and m_2 number of segments, diameters of σ_1 and σ_2, and an intersegment radial distance $r_{1,2}$. ε is the segment energy, which can be understood as the maximum attraction between two molecules.

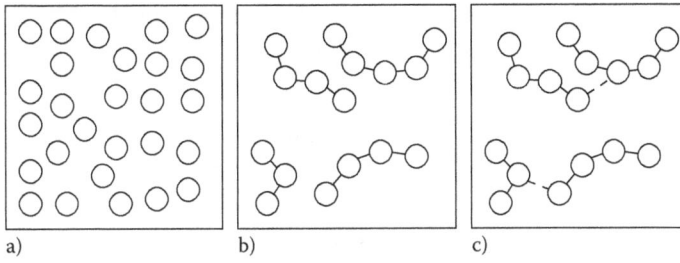

a)   b)   c)

FIGURE 4.8   Graphical representation of the PC-SAFT concept.

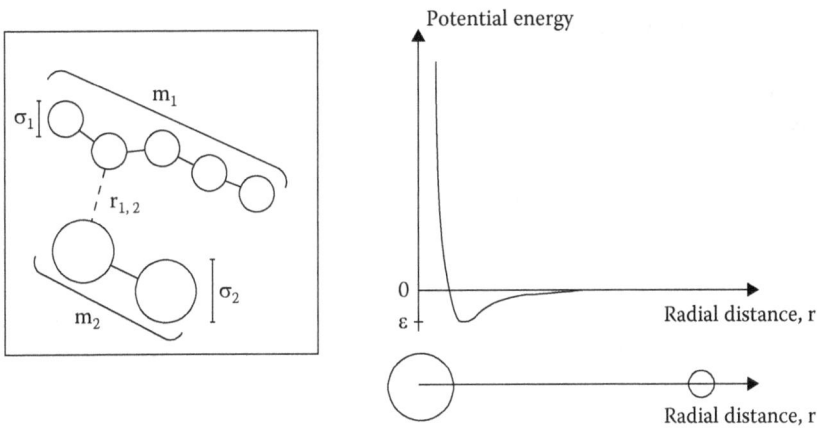

FIGURE 4.9   Schematic view of a PC-SAFT mixture consisting of two components.

The hard chain term to the PC-SAFT compressibility factor is expressed as

$$Z^{hc} = \bar{m}Z^{hs} - \sum_{i=1}^{N} x_i(m_i - 1)\frac{\rho}{g_{ii}^{hs}}\frac{\partial \ln g_{ii}^{hs}}{\partial \rho} \tag{4.69}$$

where N is the number of components, x_i the mole fraction of component i, and

$$\bar{m} = \sum_{i=1}^{N} x_i m_i \tag{4.70}$$

Z^{hs} is the hard sphere contribution to Z^{hc}, which term is expressed as

$$Z^{hs} = \frac{\zeta_3}{1-\zeta_3} + \frac{3\zeta_1\zeta_2}{\zeta_0(1-\zeta_3)^2} + \frac{3\zeta_2^3 - 3\zeta_3\zeta_2^3}{\zeta_0(1-\zeta_3)^3} \tag{4.71}$$

where

$$\zeta_n = \frac{\pi}{6}\rho\sum_{i=1}^{N} x_i m_i d_i^n \tag{4.72}$$

The exponent n may take the values 0, 1, 2, and 3. The term *packing fraction* is used for ζ_3. The temperature-dependent diameter, d, is expressed through

$$d_i = \sigma_i \left[1 - 0.12 \exp\left(-\frac{3\varepsilon_i}{kT} \right) \right]$$ (4.73)

where k is the Boltzmann constant.

In Equation 4.72, ρ is the total number density of molecules

$$\rho = \frac{6\zeta_3}{\pi \sum\limits_{i=1}^{N} x_i m_i d_i^3}$$ (4.74)

where k is the Boltzmann constant.

The term g_{ii}^{hs} in Equation 4.69 is the molar radial pair distribution function for two segments of component i in the hard sphere system. The radial pair distribution function takes the general form for segments of component i and j:

$$g_{ij}^{hs} = \frac{1}{1-\zeta_3} + \frac{d_i d_j}{d_i + d_j} \frac{3\zeta_2}{(1-\zeta_3)} + \left(\frac{d_i d_j}{d_i + d_j} \right)^2 \frac{2\zeta_2^2}{(1-\zeta_3)^3}$$ (4.75)

The radial pair distribution function is a measure of the probability of finding a particle of type i in a given distance from a fixed particle of type j in the fluid. The density derivative of the radial distribution function may be found from

$$\rho \frac{\partial g_{ij}^{hs}}{\partial \rho} = \frac{\zeta_3}{(1-\zeta_3)^2} + \frac{d_i d_j}{d_i + d_j} \left(\frac{3\zeta_2}{(1-\zeta_3)^2} + \frac{6\zeta_2\zeta_3}{(1-\zeta_3)^3} \right) + \left(\frac{d_i d_j}{d_i + d_j} \right)^2 \left(\frac{4\zeta_2^2}{(1-\zeta_3)^3} + \frac{6\zeta_2^2\zeta_3}{(1-\zeta_3)^4} \right)$$ (4.76)

PC-SAFT uses the following expression for the dispersion contribution to the compressibility factor, Z^{disp}

$$Z^{disp} = -2\pi\rho \frac{\partial(\zeta_3 I_1)}{\partial \zeta_3} \overline{m^2 \varepsilon \sigma^3} - \pi \rho \overline{m} \left[C_1 \frac{\partial(\zeta_3 I_2)}{\partial \zeta_3} + C_2 \zeta_3 I_2 \right] \overline{m^2 \varepsilon^2 \sigma^3}$$ (4.77)

where

$$C_1 = 1 + \overline{m} \frac{8\zeta_3 - 2\zeta_3^2}{(1-\zeta_3)^4} + (1-\overline{m}) \frac{20\zeta_3 - 27\zeta_3^2 + 12\zeta_3^3 - 2\zeta_3^4}{[(1-\zeta_3)(2-\zeta_3)]^2}$$ (4.78)

$$C_2 = -C_1^2 \left(\overline{m} \frac{-4\zeta_3^2 + 20\zeta_3 + 8}{(1-\zeta_3)^5} + (1-\overline{m}) \frac{2\zeta_3^3 + 12\zeta_3^2 - 48\zeta_3 + 40}{[(1-\zeta_3)(2-\zeta_3)]^3} \right)$$ (4.79)

$$\overline{m^2 \varepsilon \sigma^3} = \sum_{i=1}^{N} \sum_{j=1}^{N} x_i x_j m_i m_j \left(\frac{\varepsilon_{ij}}{kT} \right) \sigma_{ij}^3$$ (4.80)

$$\overline{m^2 \varepsilon^2 \sigma^3} = \sum_{i=1}^{N} \sum_{j=1}^{N} x_i x_j m_i m_j \left(\frac{\varepsilon_{ij}}{kT} \right)^2 \sigma_{ij}^3$$ (4.81)

$$I_1 = \sum_{j=0}^{6} a_j(\overline{m})\zeta_3^j \qquad (4.82)$$

$$I_2 = \sum_{j=0}^{6} b_j(\overline{m})\zeta_3^j \qquad (4.83)$$

In Equations 4.80 and 4.81

$$\varepsilon_{ij} = \sqrt{\varepsilon_i \varepsilon_j}(1 - k_{ij}) \qquad (4.84)$$

and

$$\sigma_{ij} = \frac{1}{2}(\sigma_i + \sigma_j) \qquad (4.85)$$

where k_{ij} is a binary interaction parameter similar to that in the mixing rule for the a parameter of a cubic equation of state (Equations 4.33 and 4.35). In Equation 4.82

$$a_j(\overline{m}) = a_{0j} + \frac{\overline{m}-1}{\overline{m}}a_{1j} + \frac{\overline{m}-1}{\overline{m}}\frac{\overline{m}-2}{\overline{m}}a_{2j} \qquad (4.86)$$

and in Equation 4.83

$$b_j(\overline{m}) = b_{0j} + \frac{\overline{m}-1}{\overline{m}}b_{1j} + \frac{\overline{m}-1}{\overline{m}}\frac{\overline{m}-2}{\overline{m}}b_{2j} \qquad (4.87)$$

The universal constants for a_{0j}, a_{1j}, a_{2j}, b_{0j}, b_{1j}, and b_{2j} are given in Table 4.5.

Gross and Sadowski (2001) have given values for m, σ and ε of N_2, CO_2, and hydrocarbons from C_1 to C_{20}.

Figure 4.10 shows a plot of the Z-factor of n-heptane for a temperature of 500 K calculated using the PC-SAFT equation. Also shown are the contributions from the hard chain (hc) term, the ideal gas (id) term, and the dispersion (disp) term. Recalling Equation 4.68, it can be seen from Figure 4.10 that the Z-factor is a fine balance between the hard chain contribution and the dispersion term.

TABLE 4.5
Universal Constants for Constants in Equations 4.86 and 4.87

j	a_{0j}	a_{1j}	a_{2j}	b_{0j}	b_{1j}	b_{2j}
0	0.9105631445	−0.3084016918	−0.0906148351	0.7240946941	0.5755498075	0.0976883116
1	0.6361281449	0.1860531159	0.4527842806	2.2382791861	0.6995095521	−0.2557574982
2	2.6861347891	−2.5030047259	0.5962700728	−4.0025849485	3.8925673390	−9.1558561530
3	−26.547362491	21.419793629	−1.7241829131	−21.003576815	−17.215471648	20.642075974
4	97.759208784	−65.255885330	−4.1302112531	26.855641363	192.67226447	−38.804430052
5	−159.59154087	83.318680481	13.776631870	206.55133841	−161.82646165	93.626774077
6	91.297774084	−33.746922930	−8.6728470368	−355.60235612	−165.20769346	−29.666905585

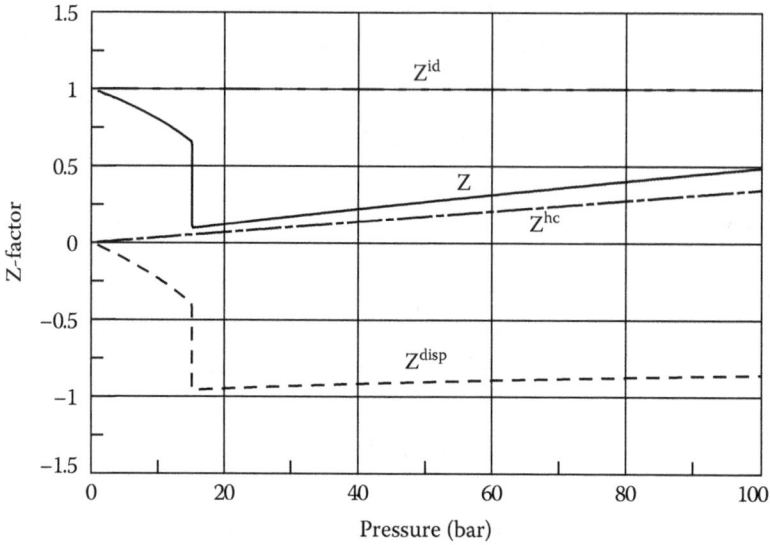

FIGURE 4.10 Contributions to Z-factor of *n*-heptane at 500 K according to PC-SAFT equation. The vapor pressure of ~15 bar is seen as a shift in the Z-factor curves except for the ideal gas contribution.

To better understand the qualitative differences between the PC-SAFT equation and a cubic equation of state, it may be useful to consider the latter the equivalent of Equation 4.68. A cubic equation of state consists of a repulsive term and an attractive term

$$Z = 1 + Z^{\text{repulsive}} + Z^{\text{attractive}} \tag{4.88}$$

Neglecting the attractive term, $\left(-\dfrac{a(T)}{V(V+b)} \right)$, the SRK equation of state becomes

$$P = \frac{RT}{V - b} \tag{4.89}$$

which equation may be rewritten to

$$Z = 1 + \frac{Pb}{RT} \tag{4.90}$$

and the Z factor repulsive term becomes

$$Z^{\text{repulsive}} = \frac{Pb}{RT} \tag{4.91}$$

This leads to the following expression for the attractive term of the Z-factor

$$Z^{\text{attractive}} = Z^{\text{cubic EoS}} - 1 - \frac{Pb}{RT} \tag{4.92}$$

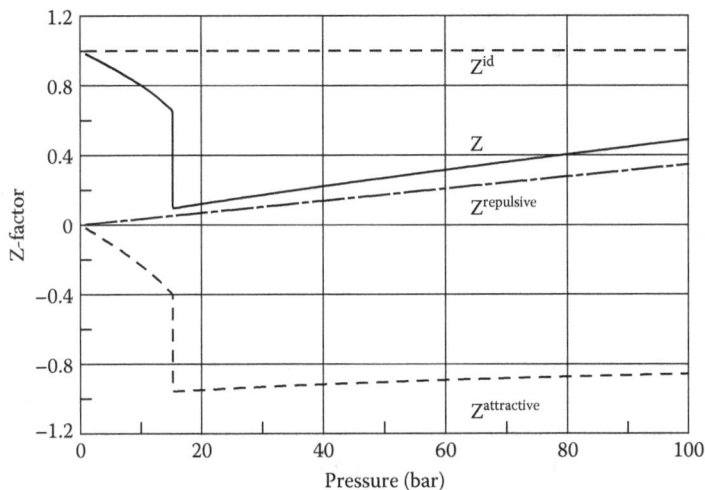

FIGURE 4.11 Contributions to Z-factor of *n*-heptane at 500 K according to Soave–Redlich–Kwong equation. The vapor pressure of ~15 bar is seen as a shift in the Z-factor curves except for the ideal gas contribution.

Figure 4.11 shows a plot of the Z-factor of *n*-heptane calculated using the SRK equation with Peneloux volume correction. Also shown are the contributions from the

- Ideal gas term (Z^{id})
- Cubic equation repulsive term ($Z^{repulsive}$)
- Cubic equation attractive term ($Z^{attractive}$)

As can be seen from Equation 4.91, the repulsive term of a cubic equation of state for a constant temperature is bound to increase proportionally with the pressure. For higher pressures, the attractive term will approach −1, for which value the ideal gas term and the attractive term will cancel out. This means that the Z-factor for high pressure asymptotically will approach the repulsive term (Equation 4.91). There is no experimental evidence that the liquid volume should decrease (exactly) linearly with 1/P for pressures above a certain level, and cubic equations of state do have some problems matching experimental data for the isothermal compressibility at high pressures (the compressibility is defined in Equation 3.4). With the PC-SAFT equation, both the hard chain term and the dispersion term (as can be seen from Figure 4.10) influence the liquid phase Z-factor at high pressures and none of them approaches a constant value. Furthermore, these two terms are found from more flexible expressions, enabling a more accurate description of the molecular interactions and the volumetric response to pressure changes.

4.10 GERG-2008 EQUATION OF STATE

When very accurate simulation results are required for a natural gas mixture, the GERG-2008 equation (Kunz and Wagner 2012) can be a better choice than a cubic equation. GERG stands for Groupe Européen de Recherches Gazières (European Gas Research Group), and the equation was developed to provide accurate physical properties of natural gas mixtures. The GERG-2008 equation has a different basis than cubic equations, which are based on considerations about the repulsive and attractive forces acting between two molecules. Once these are well described, it is implicitly assumed that all other properties will fall in place. A volume-corrected cubic equation has three equation of state parameters (a, b, and c), while the GERG-2008 many more parameters, which for each component have been fitted to comply with experimental data for

- Relation between pressure (P), molar volume (V), and absolute temperature (T)
- Heat capacity, C_V, at constant volume
- Heat capacity, C_P, at constant pressure
- Sound velocity (u)
- Enthalpy (H)
- Saturated liquid density
- Equilibrium phase compositions

The mentioned physical properties are further dealt with in Chapter 8. GERG-2008 handles mixtures through a correction (departure) term added to molar average terms for the pure components.

While cubic equations of state express the relation between pressure (P), molar volume (V), and temperature (T) directly, the GERG-2008 equation expresses the thermodynamic behavior in terms of the Helmholtz energy, A, which is defined as

$$A = U - TS \tag{4.93}$$

U is internal energy, T absolute temperature, and S is entropy. These thermodynamics quantities are explained in Appendix A, and entropy is further dealt with in Chapter 8. The Helmholtz energy is a convenient choice as thermodynamic basis for an equation of state as it, as outlined by, for example, Bell and Jäger (2016), is relatively easy to derive other thermodynamic quantities from the Helmholtz energy. The pressure is, for example, related to the Helmholtz energy as follows:

$$P = -\frac{1}{RT}\left(\frac{\partial A}{\partial V}\right)_T \tag{4.94}$$

where V is molar volume.

A dimensionless term, α, is defined

$$\alpha = \frac{A}{RT} = \alpha^0 + \alpha^r \tag{4.95}$$

where R is the gas constant and T the absolute temperature. The term α^0 is the ideal gas part and α^r the residual part of α. The residual part expresses how much the dimensionless Helmholtz energy at the actual temperature and density deviates from that of an ideal gas of the same composition. The ideal gas part of α is for component i at temperature, T, and density, ρ, found from:

$$\alpha_{oi}^o(\rho,T) = \ln\left(\frac{\rho}{\rho_{c,i}}\right)$$

$$+ \frac{R^*}{R}\left[\begin{array}{l} n_{oi,1}^\circ + n_{oi,2}^\circ \dfrac{T_{c,i}}{T} + n_{oi,3}^\circ \ln\left(\dfrac{T_{c,i}}{T}\right) + \sum_{k=4,6} n_{oi,k}^\circ ln\left(\left|\sinh(\vartheta_{oi,k}^\circ \dfrac{T_{c,i}}{T})\right|\right) \\[3mm] \\ -\sum_{k=5,7} n_{oi,k}^\circ ln\left(\cosh\left(\vartheta_{oi,k}^{o\circ} \dfrac{T_{c,i}}{T}\right)\right) \end{array}\right] \tag{4.96}$$

where ρ is density, T absolute temperature, and the sub-index c stands for property at the critical point. The parameters $n_{oi,k}^\circ$ and $\vartheta_{oi,k}^\circ$ are component specific. The gas constant R is assigned a value of 8.314472 J mol^{-1} K^{-1} and R* a value of 8.314510 J mol^{-1} K^{-1}.

The residual part of α is for component i found from

$$\alpha_{oi}^r\left(\delta,\tau\right) = \sum_{k=1}^{K_{Pol,i}} n_{oi,k}\,\delta^{d_{oi,k}}\,\tau^{t_{oi,k}} + \sum_{k=K_{Pol,i+1}}^{K_{Pol,i}+K_{Exp,i}} n_{oi,k}\,\delta^{d_{oi,k}}\,\tau^{t_{oi,k}}\,e^{-\delta^{c_{oi,k}}}$$ (4.97)

$K_{Pol,i}$, $n_{oi,k}$, $d_{oi,k}$, $T_{oi,k}$, and $c_{oi,k}$ are component-specific parameters. For a pure component, δ equals the ratio of the actual density and the critical density and τ the ratio of the critical temperature and the actual temperature.

The idea behind the GERG-2008 equation is that with enough component-specific parameters, it is possible to represent all properties of a pure component very well.

For a mixture, the GERG-2008 equation takes the form in Figure 4.12, with

$$\alpha^o\left(\delta,\tau,\bar{x}\right) = \sum_{i=1}^{N} x_i\left[\alpha_{oi}^o\left(\rho,T\right) + \ln x_i\right]$$ (4.98)

$$\Delta\alpha^r\left(\delta,\tau,\bar{x}\right) = \sum_{i=1}^{N-1}\sum_{j=i+1}^{N} x_i x_j F_{ij}\alpha_{ij}^r\left(\delta,\tau\right)$$ (4.99)

In Figure 4.12, the following convention is used for the molar composition

$$\bar{x} = x_i;\quad i = 1,2,\ldots,N$$ (4.100)

where x_i is the mole fraction of component i and N the number of components. F_{ij} is specific for the component pair, i and j. The reducing functions in Figure 4.10 are given by

$$\frac{1}{\rho_r\left(\bar{x}\right)} = \sum_{i=1}^{N}\frac{x_i^2}{\rho_{c,i}} + \sum_{i=1}^{N-1}\sum_{j=i+1}^{N} 2x_i x_j \beta_{v,ij}\gamma_{v,ij} \times \frac{x_i+x_j}{\beta_{v,ij}^2 x_i + x_j} \times \frac{1}{8}\left(\frac{1}{\rho_{c,i}^{\frac{1}{3}}} + \frac{1}{\rho_{c,j}^{\frac{1}{3}}}\right)^3$$ (4.101)

$$T_r\left(\bar{x}\right) = \sum_{i=1}^{N} x_i^2 T_{c,i} + \sum_{i=1}^{N-1}\sum_{j=i+1}^{N} 2x_i x_j \beta_{T,ij}\gamma_{T,ij} \times \frac{x_i+x_j}{\beta_{T,ij}^2 x_i + x_j} \times \left(T_{c,i}T_{c,j}\right)^{0.5}$$ (4.102)

The parameters $\beta v_{,ij}$, $\gamma v_{,ij}$, and $\beta_{T,ij}$ are specific for the component pair, i and j.

$$\alpha(\delta,\tau,\bar{x}) = \alpha^0(\rho,T,\bar{x}) + \sum_{i=1}^{N} x_i\ \alpha_{0i}^r\left(\delta,\tau\right) + \Delta\alpha^r(\delta,\tau,\bar{x})$$

Ideal gas part   Pure substance part   Departure part

Reduced density and temperature of the mixture

$$\delta = \frac{\rho}{\rho_r(\bar{x})} \qquad\qquad \tau = \frac{T_r(\bar{x})}{T}$$

T_r and ρ_r are reducing functions
(only dependent of the composition of the mixture)

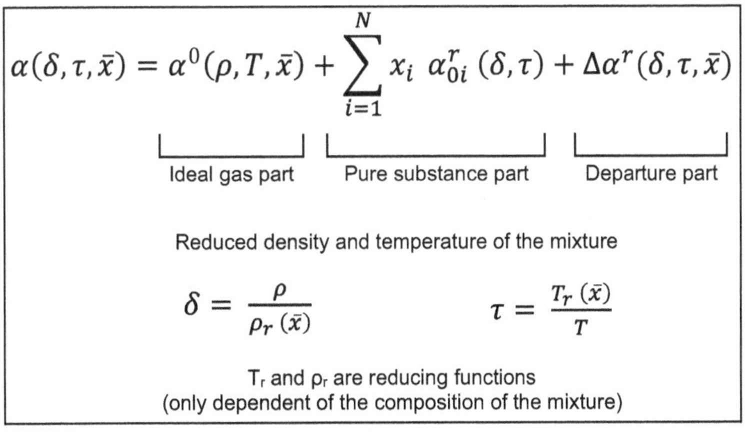

FIGURE 4.12 Terms of GERG-2008 equation for a mixture.

TABLE 4.6

Components Handled by the GERG-2008 Equation

Component Type	Component	Short Name
Inorganic	Water	H_2O
	Hydrogen	H_2
	Helium	He
	Argon	Ar
	Oxygen	O_2
	Carbon monoxide	CO
	Nitrogen	N_2
	Carbon dioxide	CO_2
	Hydrogen sulfide	H_2S
Organic	Methane	C_1
	Ethane	C_2
	Propane	C_3
	Iso-butane	iC_4
	n-butane	nC_4
	Iso-pentane	iC_5
	n-pentane	nC_5
	n-hexane	nC_6
	n-heptane	nC_7
	n-octane	nC_8
	n-nonane	nC_9
	n-decane	nC_{10}

Source: Kunz, O. and Wagner, W., The GERG-2008 wide-range equation of state for natural gases and other mixtures: An expansion of GERG-2004, *J. Chem. Eng. Data* 57, 3032–3091, 2012.

The GERG-2008 equation can be used for the components in Table 4.6 and mixtures of those components. Because several parameters are needed that are either specific for a particular component or for a particular component pair, it would be quite cumbersome to extend the application of the GERG-2008 equation to new components.

GERG-2008 is developed to provide a high accuracy in the temperature and pressure range:

90 K ≤ T ≤ 450 K and P ≤ 350 atm (355 bar)

Good accuracy can also be expected outside these ranges when within the following ranges:

60 K ≤ T ≤ 700 K and P ≤ 700 atm (709 bar)

The EOS-CG equation (Gernert and Span 2016) is a further development of the GERG-2008 equation that uses the mathematical approach of the GERG-2008 equation and presents new mixing parameters for mixtures of carbon dioxide, water, nitrogen, oxygen, argon, and carbon monoxide. The aim was to improve the calculation results for phase boundaries of binary and multi-component mixtures, where water and/or carbon dioxide is involved.

4.11 SPAN–WAGNER EQUATION

Span and Wagner (1996) have presented an equation for pure CO_2 which is very accurate and is seen as a standard for simulations on pure CO_2. Like the GERG-2008 equation, it expresses the

Helmholtz energy through Equation 4.90 as the sum of an ideal gas term and a residual term. The ideal gas term is found from

$$\alpha^{o}(\delta,\tau) = \ln(\delta) + a_1^{\circ} + a_2^{\circ}\tau + a_3^{\circ}\ln(\tau) + \sum_{i=4}^{8} \ln\left[1 - e^{-\tau\theta_i^{\circ}}\right] \tag{4.103}$$

and the residual term from

$$\alpha^{r}(\delta,\tau) = \sum_{i=1}^{7} n_i \delta^{d_i} \tau^{t_i} + \sum_{i=8}^{34} n_i \delta^{d_i} \tau^{t_i} e^{-\delta^{c_i}} + \sum_{i=35}^{39} n_i \delta^{d_i} \tau^{t_i} e^{-\alpha_i(\delta-\epsilon_i)^2 - \beta_i(\tau-\gamma_i)^2}$$
$$+ \sum_{i=40}^{42} n_i \Delta^{b_i} \delta e^{-C_i(\delta-1)^2 - D_i(\tau-1)^2} \tag{4.104}$$

with

$$\Delta = \left[(1-\tau) + A_i\left[(\delta-1)^2\right]^{\frac{1}{2\beta_i}}\right]^2 + B_i\left[(\delta-1)^2\right]^{a_i} \tag{4.105}$$

The terms δ and τ are defined in Figure 4.12. The parameters A_i, a_i°, B_i, b_i, C_i, c_i D_i, d_i, n_i, t_i, a_i, β_i, ϵ_i, γ_i, and θ_i° are fitted constants determined from experimental data in a similar way as explained for the GERG-2008 equation in Section 4.10.

4.12   OTHER EQUATIONS OF STATE

Much exploration activity is directed toward deep reservoirs at high temperature and high pressure. The ability of the classical cubic equations of state to represent the molecular interactions at such conditions has often been questioned. More sophisticated equations of state have been proposed, some of which include terms to account for the strong repulsive forces acting at high pressures, such as Benedict et al. (1940), Lee and Kesler (1975), Donohue and Vimalchand (1988), and Lin et al. (1983). None of these equations have obtained widespread use in the oil industry.

Wei and Sadus (2000) have given an extensive review of equation of states, cubic as well as noncubic.

4.13   POSTSCRIPT ON THE ROLE OF CUBIC
EQUATIONS IN THE OIL INDUSTRY

This chapter has presented the volume-corrected SRK and PR equations that are today's standard in the oil industry. Strictly speaking, these equations are only minor modifications of the van der Waals equation presented in 1873. It was not until the late 1970s that real effort was invested in making cubic equations available as an engineering tool for the oil industry. One may ask why it did not happen earlier and what made the use of cubic equations as widespread in the oil industry as it is today. The year 1973 was a turning point. That year not only saw a sharp increase in the price of oil but also a shortage of oil that was so great that some countries in Europe imposed car-free Sundays. This initiated extensive oil exploration in, among other places, the North Sea.

The increased costs of producing oil and gas condensates led to a desire in the oil industry for better tools to simulate the properties of reservoir fluids. With Soave's and Peng and Robinson's improvements to the van der Waals equation, the basic models were available. However, it was not until the mid-1980s that sufficiently comprehensive compositional analyses became available which allowed the determination of representative critical properties (T_c, P_c, and acentric factor) of the components contained in the C_{7+} fractions. Characterization of heavy hydrocarbons is the subject of the next chapter. Even a correct characterization of the reservoir fluid is not sufficient to carry out

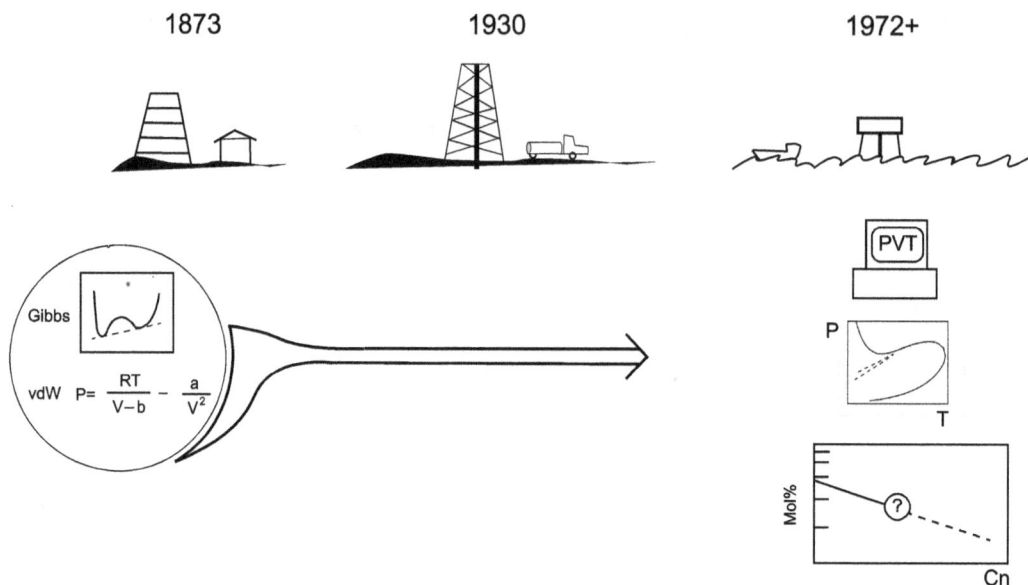

FIGURE 4.13 Historical development of PVT simulation methods.

efficient and fast flash calculations. As described in Chapter 6, it requires methods to determine the number of phases and numerical methods to converge the flash equations quickly and efficiently.

The development of computers is what binds these technologies together. Only very simple calculations with a cubic equation of state such as determination of single-phase Z-factors can be performed without the use of a computer. Therefore, for almost 100 years, the van der Waals equation was more of a theoretical invention than a practical tool. The same applies to a method proposed by Gibbs around the year 1900 for determining the number of phases. The latter method is described in Chapter 6.

Figure 4.13 schematically shows the historical development of PVT simulation methods.

REFERENCES

Adachi, Y., Lu, B.C.-Y., and Sugie, H., A four-parameter equation of state, *Fluid Phase Equilib.* 11, 29–48, 1983.

American Petroleum Institute, *Technical Data Book—Petroleum Refining*, API, New York, 1982.

Bell, J.H. and Jäger, A., Helmholtz energy transformations of common cubic equations of state for use with pure fluids and mixtures, *J. Res. Natl. Inst. Stand. Technol.* 121, 238–263, 2016.

Benedict, M., Webb, G.R., and Rubin, L.C., An empirical equation for thermodynamic properties of light hydrocarbons and their mixtures. I. Methane, ethane, propane and butane, *J. Chem. Phys.* 8, 334–345, 1940.

Chapman, W.G., Gubbins, K.E., Jackson, G., and Radosz, M., New reference equation of state for associating liquids, *Ind. Eng. Chem. Res.* 29, 1709–1721, 1990.

Chapman, W.G., Jackson, G., and Gubbins, K.E., Phase equilibria of associating fluids: Chain molecules with multiple bonding sites, *Mol. Phys.* 65, 1057–1079, 1988.

Dahl, S. and Michelsen, M.L., High-pressure vapor-liquid equilibrium with a UNIFAC-based equation of state, *AIChE J.* 36, 1829–1836, 1990.

Donohue, M.D. and Vimalchand, P., The perturbed-hard-chain theory. Extensions and applications, *Fluid Phase Equilib.* 40, 185–211, 1988.

Gernert, J. and Span, R., EOS–CG: A Helmholtz energy mixture model for humid gases and CCS mixtures, *J. Chem. Thermodyn.* 93, 274–293, 2016.

Graboski, M.S. and Daubert, T.E., A modified Soave equation of state for phase equilibrium calculations. 1. Hydrocarbon systems, *Ind. Eng. Chem. Process Des. Dev.* 17, 443–448, 1978.

Gross, J. and Sadowski, G., Perturbed-chain SAFT: An equation of state based on perturbation theory for chain molecules, *Ind. Eng. Chem. Res.* 40, 1244–1260, 2001.

Jhaveri, B.S. and Youngren, G.K., Three-parameter modification of the Peng–Robinson equation of state to improve volumetric predictions, *SPE Res. Eng.* 1033–1040, August 1988.

Khashayar, N. and Moshfeghian, M., A saturated density equation in conjunction with the Predictive-Soave–Redlich–Kwong equation of state for pure refrigerants on LNG multicomponent systems, *Fluid Phase Equilib.* 153, 231–242, 1998.

Knapp, H.R., Doring, R., Oellrich, L., Plocker, U., and Prausnitz, J.M., *Vapor-Liquid Equilibria for Mixtures of Low Boiling Substances*, Chemistry Data Series, Vol. VI, DECHEMA, Frankfurt am Main, Germany, 1982.

Kunz, O. and Wagner, W., The GERG-2008 wide-range equation of state for natural gases and other mixtures: An expansion of GERG-2004, *J. Chem. Eng. Data* 57, 3032–3091, 2012.

Lee, B.I. and Kesler, M.G., A generalized thermodynamic correlation based on three-parameter corresponding states, *AICHE J.* 21, 510–527, 1975.

Lin, H.-M., Kim, H., Guo, T.M., and Chao, K.C., Cubic chain-of-rotators equation of state and VLE calculations, *Fluid Phase Equilib.* 13, 143–152, 1983.

Mathias, P.M. and Copeman, T.W., Extension of the Peng–Robinson equation of state to complex mixtures: Evaluation of the various forms of the local composition concept, *Fluid Phase Equilib.* 13, 91–108, 1983.

Peneloux, A., Rauzy, E., and Fréze, R., A consistent correction for Redlich-Kwong-Soave volumes, *Fluid Phase Equilib.* 8, 7–23, 1982,

Peng, D.-Y. and Robinson, D.B., A new two-constant equation of state, *Ind. Eng. Chem. Fundam.* 15, 59–64, 1976.

Peng, D.-Y. and Robinson, D.B., The characterization of the heptanes and heavier fractions for the GPA Peng–Robinson Programs, *GPA Research Report RR-28*, Tulsa, Oklahoma, 1978.

Poling, B.E., Prausnitz, J.M., and O'Connell, J.P., *The Properties of Gases and Liquids*, McGraw-Hill, New York, 2000.

Rackett, H.G., Equation of state for saturated liquids, *J. Chem. Eng. Data* 15, 514–517, 1970.

Redlich, O. and Kwong, J.N.S., The thermodynamics of solutions. V. An equation of state. Fugacities of gaseous solutions, *Chem. Rev.* 44, 233–244, 1949.

Soave, G., Equilibrium constants from a modified Redlich-Kwong equation of state, *Chem. Eng. Sci.* 27, 1197–1203, 1972.

Span, R. and Wagner, W., A new equation of state for carbon dioxide covering the fluid region from the triple-point temperature to 1100 K at pressures up to 800 MPa, *J. Phys. Chem. Ref. Data* 25, 1509–1596, 1996.

Spencer, C.F. and Danner, R.P., Prediction of bubble-point density of mixtures, *J. Chem. Eng. Data* 18, 230–234, 1973.

van der Waals, J.D., *Over de Continuiteit van der Gas-en Vloeistoftoestand Leiden*, Doctoral dissertation, University, The Netherlands, 1873 (In Dutch).

Wei, Y.S. and Sadus, R.J., Equations of state for the calculation of fluid-phase equilibria, *AICHE J.* 46, 169–196, 2000.

5 C_{7+} Characterization

To perform phase equilibrium calculations on a reservoir fluid composition using a cubic equation of state, the critical temperature (T_c), the critical pressure (P_c), and the acentric factor (ω) are required for each component contained in the mixture. In addition, a binary interaction parameter (k_{ij}) is needed for each pair of components. If an equation of state with volume correction is used (Peneloux et al. 1982), a volume shift parameter must further be assigned to each component. Naturally occurring oil or gas condensate mixtures may contain thousands of different components. Such a high number is impractical in flash calculations. Some components must be lumped together and represented as pseudocomponents. C_{7+} characterization consists of representing the hydrocarbons with seven and more carbon atoms (the heptane plus or C_{7+} fraction) as a convenient number of pseudocomponents and finding the needed equation of state parameters (T_c, P_c, and ω) for each of these pseudocomponents. The characterization and lumping problem is illustrated in Figure 5.1.

5.1 CLASSES OF COMPONENTS

The components contained in oil and gas condensate mixtures can be divided into three classes:

- *Defined components to C_6*: These components are N_2, CO_2, H_2S, C_1, C_2, C_3, iC_4, nC_4, iC_5, nC_5, and C_6 (C_6 is usually considered to be pure nC_6, though branched and cyclic C_6 components may also be present in the C_6 fraction).
- *C_{7+} fractions*: It is common to see some defined C_7–C_{10} components quantitatively analyzed for, but a complete component analysis of the C_{7+} fraction will not be doable. The number of components is simply too high. Instead the C_{7+} fraction is split into carbon number fractions, each of which contains hydrocarbons with boiling points within a given temperature interval. The temperature intervals can be seen in Table 2.3 and are determined by the boiling points of the *n*-paraffins. If a true boiling point (TBP) analysis as presented in Chapter 2 has been carried out, measured densities at standard conditions (atmospheric pressure and 15°C) and measured molecular weights will also be available for each C_{7+} fraction. When characterizing a C_{7+} fraction, it is essential to take into consideration the diversity of hydrocarbon components contained in the fraction. Figure 5.2 shows four different components belonging to the C_9 fraction. The importance of structural differences for the phase behavior is illustrated in Figure 5.3. The dew point of a binary mixture of C_1 and C_9 is seen to depend heavily on the chemical structure of the C_9 component. The maximum dew point temperature is around 20°C higher when C_9 is nC_9 than when it is dimethylcyclohexane.
- *Plus fraction*: The plus fraction consists of the components that are too heavy to be separated into individual carbon number fractions. If a TBP analysis has been carried out, the average molecular weight and density of the plus fraction will be measured and reported.

Each of the preceding component classes will be dealt with separately in the following subsections.

5.1.1 DEFINED COMPONENTS TO C_6

T_c, P_c, and ω of the defined components can be determined experimentally and the experimental values looked up on the Internet or in textbooks on applied thermodynamics. Literature values are listed in Table 5.1.

DOI: 10.1201/9780429457418-5

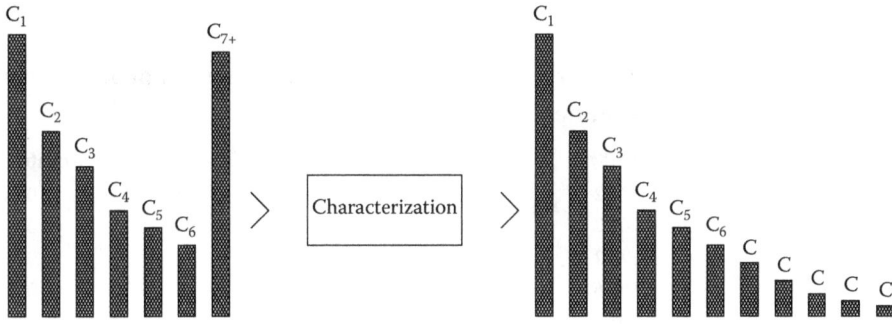

FIGURE 5.1   Characterization and lumping process.

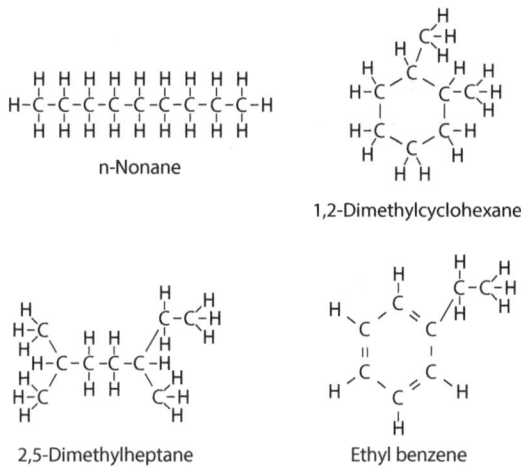

FIGURE 5.2   Four different components belonging to the C_9 fraction. *n*-nonane is an *n*-paraffin (P), 2,5-dimethylheptane an iso-paraffin (P), 1,2-dimethylcyclohexane a naphthene (N), and ethyl benzene an aromatic (A). The P, N, and A component classes are further dealt with in Chapter 1.

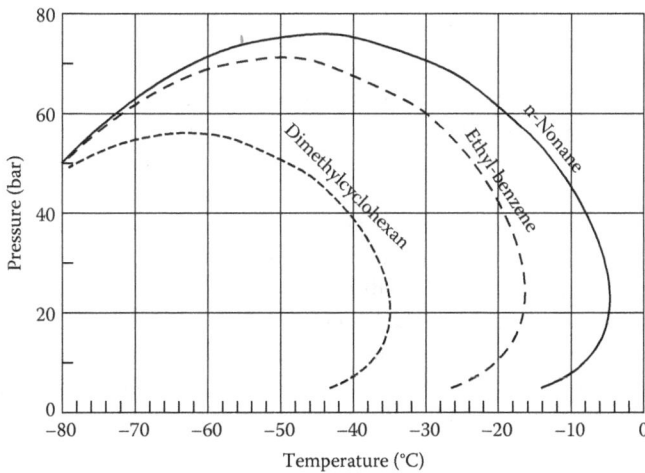

FIGURE 5.3   Simulated phase envelopes for mixtures of 99.99 mole percentage C_1 and 0.01 mole percentage of nC_9, dimethylcyclohexane, and ethyl benzene, respectively. PR equation is used.

TABLE 5.1
Critical Temperature (T_c), Critical Pressure (P_c), and Acentric Factor (w) of Some Common Petroleum Reservoir Fluid Constituents

Component	T_c (K)	P_c (bar)	Acentric Factor
N_2	126.2	33.9	0.040
CO_2	304.2	73.8	0.225
H_2S	373.2	89.4	0.100
C_1	190.6	46.0	0.008
C_2	305.4	48.8	0.098
C_3	369.8	42.5	0.152
iC_4	408.1	36.5	0.176
nC_4	425.2	38.0	0.193
iC_5	460.4	33.8	0.227
nC_5	469.5	33.7	0.251
nC_6	507.4	29.7	0.296

TABLE 5.2
Densities at 15°C and 1.01 bar of Compounds in Figure 5.2

Component	Component Class	Density (g/cm³)
n-nonane	P	0.718
2,5-dimethylheptane	P	0.720
1,2-dimethylcyclohexane	N	0.796
Ethyl benzene	A	0.867

Source: Data from Poling, B.E., Prausnitz, J.M., and O'Connell, J.P., *The Properties of Gases and Liquids*, McGraw-Hill, New York, 2000.

Note: The density is at 1.01 bar and 15°C.

5.1.2 C_{7+} Fractions

Coats (1985) wrote one of the earliest papers on fluid characterization, and his paper exemplifies that in the early 1980s, reservoir fluid compositions would usually stop at C_{7+}. The C_{7+} characterization therefore had to be based on a guessed molar C_{7+} distribution.

In the mid-1980s, fluid compositions to C_{20+} and C_{30+} were published (Pedersen et al. 1985), which meant that the C_{7+} characterization could at least partly be based on measured compositions. A C_{7+} fraction will typically contain paraffinic (P), naphthenic (N), and aromatic (A) compounds. Table 5.2 shows the densities at standard conditions of each of the four C_9 components in Figure 5.2. It is seen that the density increases in the order paraffin (P), naphthene (N), and aromatic (A). The density is therefore a good measure of the PNA distribution. The higher the density, the more aromatic the fraction. This density dependence is reflected in property correlations of Pedersen et al. (1989, 1992). T_c (K), P_c (atm), and ω of a carbon number fraction are expressed in terms of its molecular weight, M (g/mol), and density, ρ (g/cm³), at atmospheric conditions:

$$T_c = c_1\rho + c_2\ln M + c_3M + \frac{c_4}{M} \tag{5.1}$$

$$\ln P_c = d_1 + d_2\rho^{d_5} + \frac{d_3}{M} + \frac{d_4}{M^2} \tag{5.2}$$

$$m = e_1 + e_2 M + e_3\rho + e_4 M^2 \tag{5.3}$$

For the SRK equation (Equation 4.20), m is related to the acentric factor, ω, through

$$m = 0.480 + 1.574\omega - 0.176\omega^2 \tag{5.4}$$

and for the PR equation (Equation 4.36) through

$$m = 0.37464 + 1.54226\omega - 0.26992\omega^2 \tag{5.5}$$

The coefficients c_1–c_4, d_1–d_5, and e_1–e_4 in Equations 5.1 through 5.3 have been determined from experimental PVT data. Because the SRK and PR equations are different, the optimum coefficients differ between the two. Two sets of coefficients are shown in Table 5.3, an SRK and a PR set (Pedersen et al. 1989, 1992, 2004). The coefficients are the same with and without Peneloux volume correction.

The Peneloux parameters (c in Equations 4.43 and 4.48) for defined components can be found from Equations 4.46 and 4.47. The Peneloux volume shift parameter of C_{7+} pseudocomponent i can be found from

$$c_i = \frac{M_i}{\rho_i} - V_i^{EOS} \tag{5.6}$$

M_i is the molecular weight and ρ_i the density of pseudocomponent i at 15°C and atmospheric pressure. V_i^{EOS} is the molar volume of pseudocomponent i at the same conditions found using the

---

TABLE 5.3
Coefficients in the Correlations in Equations 5.1 through 5.3 for Use with the Soave–Redlich–Kwong and the Peng–Robinson Equations (T$_c$ Is in K and P$_c$ in atm)

Subindex/Coefficient	1	2	3	4	5
SRK/SRK-Peneloux[a]					
c	1.6312×10^2	8.6052×10	4.3475×10^{-1}	-1.8774×10^3	—
d	-1.3408×10^{-1}	2.5019	2.0846×10^2	-3.9872×10^3	1.0
e	7.4310×10^{-1}	4.8122×10^{-3}	9.6707×10^{-3}	-3.7184×10^{-6}	—
PR/PR-Peneloux[b]					
c	7.34043×10	9.73562×10	6.18744×10^{-1}	-2.05932×10^3	—
d	7.28462×10^{-2}	2.18811	1.63910×10^2	-4.04323×10^3	1/4
e	3.73765×10^{-1}	5.49269×10^{-3}	1.17934×10^{-2}	-4.93049×10^{-6}	—

[a] Data from Pedersen, K.S., Blilie, A.L., and Meisingset, K.K., PVT calculations on petroleum reservoir fluids using measured and estimated compositional data for the plus fraction, *Ind. Eng. Chem. Res.* 31, 1378–1384, 1992.

[b] Data from Pedersen, K.S., Milter, J., and Sørensen, H., Cubic equations of state applied to HT/HP and highly aromatic fluids, *SPE J.* 9, 186–192, 2004.

appropriate equation of state (SRK or PR) with no volume correction. Equation 5.6 ensures that the Peneloux volume of pseudocomponent i agrees with the experimentally determined density at 15°C and atmospheric pressure. Pedersen et al. (2004) have shown that the thermal expansion of heavy oils as dealt with in Section 5.7 is somewhat underpredicted with a constant Peneloux correction determined from Equation 5.6. At higher temperatures, the simulated liquid densities are higher than seen experimentally. According to the ASTM 1250–80 correlation, the density, ρ, of a stable oil varies with temperature according to the formula (ρ in kg/m^3)

$$\rho_{T_1} = \rho_{T_0} e^{[(-A(T_1-T_0)(1+0.8A(T_1-T_0)))]} \tag{5.7}$$

T_0 is a reference temperature at which the density is known, and T_1 is the temperature for which the density is to be calculated. The constant A is found as

$$A = \frac{613.9723}{\rho_{T_0}^2} \tag{5.8}$$

Pedersen et al. (2004) suggest to make use of Equation 5.7 for C_{7+} components to introduce a temperature-dependent Peneloux parameter in the SRK-P and PR-P equations:

$$c_i = c_{0i} + c_{1i}(T - 288.15) \tag{5.9}$$

T is the absolute temperature in K, c_{0i} the usual Peneloux parameter of component i as determined from Equation 5.6 for a temperature of 288.15 K (15°C), and c_{1i} a temperature-dependent term determined to give a density variation of component i from $T_0 = 288.15$ K to $T_1 = 353.15$ K compliant with Equation 5.7.

Other property correlations have been presented by Cavett (1964), Kesler and Lee (1976), Daubert (1980), Sim and Daubert (1980), Riazi and Daubert (1980), Twu (1983, 1984), Jalowka and Daubert (1986), Watanasiri et al. (1985), Teja et al. (1990), and Riazi (1997). Newman (1981) has evaluated a number of T_c and P_c correlations for use on aromatic fluids, and Whitson (1982) has investigated what difference it makes for equation of state predictions to use different correlations. A correlation that works well with one cubic equation of state may not work equally well with a different equation of state.

5.1.3 PLUS FRACTION

Characterization of the plus fraction involves the following:

- Estimation of the mole fraction of each carbon number contained in the plus fraction
- Estimation of T_c, P_c, and ω of the resulting carbon number fractions
- Lumping of the carbon number fractions into a reasonable number of pseudocomponents

Pedersen et al. (1983, 1984) observed a pattern in the compositions of oil and gas condensate reservoir fluids. For the carbon number fractions above C_6, an approximate linear relationship is seen between carbon number and the logarithm of the corresponding mole fraction, z_N:

$$C_N = A + B \ln z_N \tag{5.10}$$

The dots in Figure 5.4 show a plot of the logarithm of the mole fractions of the C_7–C_{19} fractions against carbon number for the reservoir fluid in Table 5.4. It is seen that the molar distribution for this mixture is in reasonable agreement with Equation 5.10. This suggests that the mole fractions of the

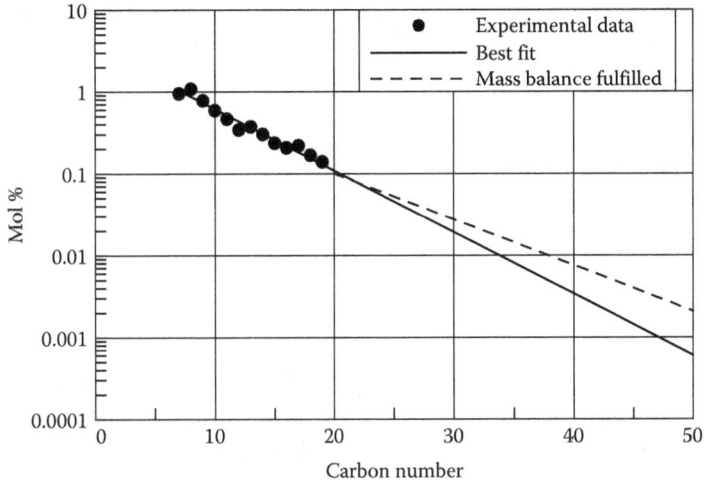

FIGURE 5.4 Mole percentage versus carbon number for gas condensate mixture in Table 5.4.

TABLE 5.4
Molar Composition of North Sea Gas Condensate

Component Group	Component	Mol%	M	ρ (g/cm³) 15°C, 1.01 bar
Defined	N_2	0.12	—	—
	CO_2	2.49	—	—
	C_1	76.43	—	—
	C_2	7.46	—	—
	C_3	3.12	—	—
	iC_4	0.59	—	—
	nC_4	1.21	—	—
	iC_5	0.50	—	—
	nC_5	0.59	—	—
	C_6	0.79	—	—
C_{7+} fractions	C_7	0.95	95	0.726
	C_8	1.08	106	0.747
	C_9	0.78	116	0.769
	C_{10}	0.592	133	0.781
	C_{11}	0.467	152	0.778
	C_{12}	0.345	164	0.785
	C_{13}	0.375	179	0.802
	C_{14}	0.304	193	0.815
	C_{15}	0.237	209	0.817
	C_{16}	0.208	218	0.824
	C_{17}	0.220	239	0.825
	C_{18}	0.169	250	0.831
	C_{19}	0.140	264	0.841
Plus fraction	C_{20+}	0.833	377	0.873

carbon number fractions heavier than C_{19} can be determined by extrapolating the best-fit line for the carbon number fractions C_7–C_{19} (full-drawn line in Figure 5.4). These mole fractions are, however, constrained by the mass balance equations:

$$z_+ = \sum_{i=C_+}^{C_{max}} z_i \tag{5.11}$$

$$M_+ = \frac{\sum_{i=C_+}^{C_{max}} z_i M_i}{\sum_{i=C_+}^{C_{max}} z_i} \tag{5.12}$$

C_+ is the carbon number of the plus fraction (20 for the mixture in Table 5.4) and C_{max} the heaviest carbon number fraction considered. Equations 5.11 and 5.12 can be used to determine the constants A and B in Equation 5.10. For ordinary reservoir fluids, C_{80} is a reasonable choice as the heaviest component (C_{max}) to be considered. In heavy oils, components as heavy as C_{200} may influence the phase behavior (Pedersen et al. 2004). Having determined the constants A and B, the mole fractions of each carbon number fraction contained in the plus fraction may be determined from Equation 5.10. As is illustrated by the dashed line in Figure 5.4, the slope of the line relating mole fractions to carbon numbers for the C_{20+} subfractions may deviate slightly from the line found by extrapolating the best-fit line for the carbon number fractions. The mass balance equations (Equations 5.11 and 5.12) must be fulfilled, and the deviation from a straight line simply means that the logarithmic dependence expressed in Equation 5.10 is only approximate.

The molar distribution function in Equation 5.10 can be explained using the theory of chemical reaction equilibria (Sørensen et al. 2013; Boesen et al. 2023). If the normal paraffins $C_n H_{2n+2}$ and $C_{n+1} H_{2(n+1)+2}$ were to form from the pure elements, the reaction equilibrium would be

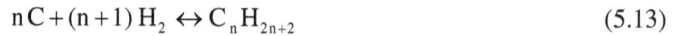

$$n\,C + (n+1)\,H_2 \leftrightarrow C_n H_{2n+2} \tag{5.13}$$

The equilibrium constant for this reaction is defined as

$$K_{C_n} = \frac{[C_n H_{2n+2}]}{[C]^n [H_2]^{n+1}} \tag{5.14}$$

and related to the Gibbs free energy of formation, ΔG_i^0, as

$$-RT\ln K_{C_n} = \sum v_i \Delta G_i^0 \tag{5.15}$$

Gibbs energy is introduced in Appendix A. The v_is are stoichiometric coefficients for the reactants and product in Equation 5.13, and the term C_n is used for $C_n H_{2n+2}$. ΔG_i^0 is zero for pure elements, which allows Equation 5.15 to be reduced to

$$-RT\ln K_{C_n} = \Delta G_{C_n}^0 \tag{5.16}$$

where $\Delta G_{C_n}^0$ is Gibbs free energy of formation for C_n. The ratio between the equilibrium constants for the reactions leading to C_n and C_{n+1} becomes

$$-RT\ln \frac{K_{C_n}}{K_{C_{n+1}}} = \Delta G_{C_n}^0 - \Delta G_{C_{n+1}}^0 \tag{5.17}$$

which, using Equation 5.14, can be rewritten to

$$-RT\ln\frac{[C_n][C][H_2]}{[C_{n+1}]} = \Delta G^0_{C_n} - \Delta G^0_{C_{n+1}} \qquad (5.18)$$

or

$$-RT\left(\ln[C_n] - \ln[C_{n-1}]\right) = \Delta G^0_{C_n} - \Delta G^0_{C_{n+1}} + RT\ln[C] + RT\ln[H_2] \qquad (5.19)$$

Table 5.5 shows the Gibbs free energies of formation for the normal paraffins from nC_7 to nC_{20} in gas and liquid forms (Journal of Physical and Chemical Reference Data 1982). Also shown is how much ΔG of formation increases from one C_n to the next one. For both gas and liquid states the increase is seen to be almost constant independent of carbon number. That means the term $\Delta G^0_{C_n} - \Delta G^0_{C_{n+1}}$ in Equation 5.19 is a constant and so are the terms $RT\ln[C]$ and $RT\ln[H_2]$. With a constant difference between $\ln[C_n]$ and $\ln[C_{n-1}]$ Equation 5.10 is reproduced.

The data in Table 5.5 is for pure substances at 25°C. There will be an additional contribution to ΔG from transferring the hydrocarbons from pure form into a hydrocarbon mixture at a different temperature, but this contribution is an order of magnitude lower than ΔG of formation. Also, the C_{7+} fractions will contain other components than n-paraffins. For heavy oils dominated by aromatics including biodegraded oils, Equation 5.10 may not apply until after ~C15 (Krejbjerg and Pedersen 2006). Most reservoir fluids are, however, dominated by paraffins and paraffinic side branches on aromatic and naphthenic molecules. For those reservoir fluids it can be concluded that the observed dependence of C_{7+} mole fractions versus carbon number expressed in Equation 5.10 has a foundation in the theory of chemical reaction equilibria.

The densities of the C_{7+} fractions usually increase with carbon number. As is illustrated by the circles in Figure 5.5, the density is reasonably represented, assuming the relation

$$\rho_N = C + D\ln C_N \qquad (5.20)$$

TABLE 5.5
Gibbs Energies of Formation of Normal Paraffins at 25°C (Journal of Physical and Chemical Reference Data 1982)

Component	ΔG Formation Gas (J/mol)	ΔG Formation Liquid (J/mol)	Difference in ΔG Gas between C_n and C_{n-1} (J/mol)	Difference in ΔG Liquid between C_n and C_{n-1} (J/mol)
nC_7	8,033	1,004		
nC_8	16,401	6,360	8,368	5,356
nC_9	24,811	11,757	8,410	5,397
nC_{10}	33,221	17,280	8,410	5,523
nC_{11}	41,631	22,719	8,410	5,439
nC_{12}	50,041	28,075	8,410	5,356
nC_{13}	58,450	33,556	8,410	5,481
nC_{14}	66,818	38,869	8,368	5,314
nC_{15}	75,228	44,350	8,410	5,481
nC_{16}	83,764	49,999	8,535	5,648
nC_{17}	92,090	55,187	8,326	5,188
nC_{18}	100,458	60,919	8,368	5,732
nC_{19}	108,951	66,275	8,494	5,356
nC_{20}	117,319	71,630	8,368	5,356

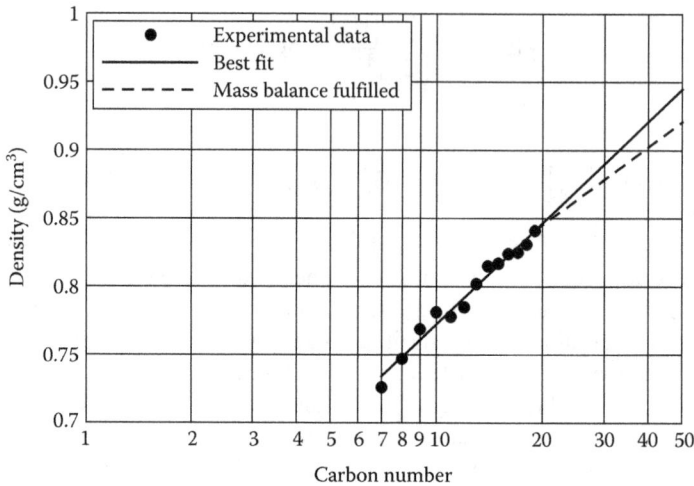

FIGURE 5.5 Density versus carbon number for gas condensate mixture in Table 5.4.

C and D are constants determined from the overall density of the plus fraction, ρ_+

$$\rho_+ = \frac{\sum\limits_{i=C_+}^{C_{max}} z_i\, M_i}{\sum\limits_{i=C_+}^{C_{max}} \frac{z_i\, M_i}{\rho_i}} \tag{5.21}$$

and the density of the last carbon number fraction before the plus fraction (e.g., density of C_{19} for a composition to C_{20+}). It is seen from Figure 5.5 that the slope of the (dashed) line giving the correct overall density of the plus fraction deviates slightly from that found by extrapolating the best-fit (full drawn) line for the densities of the C_{7+} fractions analyzed for (e.g., C_7–C_{19} in Figure 5.5). The density correlation expressed in Equation 5.20 is, in other words, only approximate.

Finally, it is assumed that the molecular weight, M_N, of a given carbon number fraction, C_N, can be determined from the equation

$$M_N = 14\, C_N - 4 \tag{5.22}$$

The constant 14 expresses that approximately two hydrogen atoms accompany each extra carbon atom. The atomic weight of carbon is 12 and that of hydrogen is 1, giving a total molecular weight increment of 14 per extra carbon atom. The term (-4) in Equation 5.22 accounts for the presence of aromatic structures in the reservoir fluids. An aromatic contains fewer hydrogen atoms per carbon atom than a paraffin.

From the densities and molecular weights estimated in this manner, Equations 5.1 through 5.5 may readily be used to determine T_c, P_c, and ω of the subfractions of the plus fraction. The Peneloux parameter of the subfractions of the plus fraction may be determined in the same manner as for the fractions C_7–C_{19} (Equation 5.6).

A compositional analysis to C_{20+} like the one in Table 5.4 is established by combining compositional data from a gas chromatographic (GC) analysis and a TBP analysis. These two techniques are further dealt with in Chapter 2. A TBP analysis will seldom be made, and the fluid characterization must in most cases be performed solely based on a GC analysis. This type of analysis is not preparative; that is, it does not produce enough sample to enable molecular weight and density measurements on individual carbon number fractions. The default densities of Katz and Firoozabadi given in Table 2.3 are often used to fill out the gap in molecular weight and density data. This is, however, not to be recommended as the Katz and Firoozabadi densities are for paraffinic oils and therefore very low. Table 5.6 presents generalized C$_{7+}$ densities for paraffinic reservoir fluids and for naphthenic and aromatic fluids from the North Sea (Rønningsen et al. 1989). These values are based on compositional data for 77 different reservoir fluids. The Katz and Firoozabadi densities in Table 2.3 are close to those for the paraffinic fluids but significantly lower than those for the naphthenic and aromatic fluids.

When no TBP analysis exists, Equation 5.20 may be used to split up the C$_{7+}$ density on carbon number fractions. The constants C and D can be determined subject to the mass balance constraint in Equation 5.21 and forcing Equation 5.20 to comply with a C$_6$ density 0.86 times the density of the total C$_{7+}$ fraction.

Whitson (1983) expresses the molar distribution with molecular weight through an asymmetric mathematical function (Pearson 1895)

$$p(M) = \frac{(M-\eta)^{\alpha-1}\exp\left(-\frac{M-\eta}{\beta}\right)}{\beta^\alpha\Gamma(\alpha)} \tag{5.23}$$

TABLE 5.6
Generalized Densities (g/cm^3) of Carbon Number Fractions

Carbon Number	Paraffinic	Aromatic and Naphthenic
C$_6$	0.675	0.669
C$_7$	0.739	0.746
C$_8$	0.762	0.762
C$_9$	0.780	0.787
C$_{10}$	0.790	0.809
C$_{11}$	0.793	0.820
C$_{12}$	0.806	0.837
C$_{13}$	0.821	0.848
C$_{14}$	0.833	0.857
C$_{15}$	0.838	0.866
C$_{16}$	0.844	0.874
C$_{17}$	0.839	0.875
C$_{18}$	0.842	0.878
C$_{19}$	0.852	0.888
C$_{20}$	0.869	0.899
C$_{21}$	0.870	0.897
C$_{22}$	0.871	0.899
C$_{23}$	0.872	0.900
C$_{24}$	0.874	0.901
C$_{25}$	0.876	0.905
C$_{26}$	0.879	0.908
C$_{27}$	0.883	0.910
C$_{28}$	0.888	0.917
C$_{29}$	0.892	0.921

Source: Data from Rønningsen, H.P., Skjevrak, I., and Osjord. E., Characterization of North Sea petroleum fractions: Hydrocarbon group types, density and molecular weight, *Energy Fuels* 3, 744–755, 1989.

where η is the minimum molecular weight found in the C_{7+} fraction (typically M of C_7), and β is defined as

$$\beta = \frac{M_{C_{7+}} - \eta}{\alpha} \tag{5.24}$$

$M_{C_{7+}}$ is the average molecular weight of the C_{7+} fraction, and Γ is the gamma function, which for $0 \leq x \leq 1$ can be estimated by the relation (Abramowitz and Stegun 1972)

$$\Gamma(x+1) = 1 + \sum_{i=1}^{8} a_i x^i \tag{5.25}$$

The recurrence formula $\Gamma(x + 1) = x\Gamma(x)$ is used for $x > 1$. The coefficients a_1 to a_8 can be seen in Table 5.7.

To find the total mole fraction of the components with a molecular weight in the interval from M_1 to M_2, the probability function in Equation 5.23 must be integrated from M_1 to M_2 and multiplied by the total mole fraction of components with a molecular weight $> \eta$.

The distribution function used by Whitson may at first hand appear to be quite different from that of Pedersen et al. (Equation 5.10). In fact, the two distribution functions are closely related, which can be seen by assuming $\alpha = 1$ in Equation 5.23. With this assumption, the equation reduces to

$$p(M) = \frac{\exp\left(-\frac{M-\eta}{M_{C_{7+}} - \eta}\right)}{M_{C_{7+}} - \eta} \tag{5.26}$$

or

$$\ln(p(M)) = -\frac{M-\eta}{M_{C_{7+}} - \eta} \ln(M_{C_{7+}} - \eta) \tag{5.27}$$

If the molecular weight is assumed to increase linearly with carbon number as expressed in Equation 5.22, the probability density function can be rewritten to

$$C_N = Con1 + Con2 \ln\left(p(M)\right) \tag{5.28}$$

where Con1 and Con2 are constants. This relation is equivalent to Equation 5.10. Whitson uses α as a regression parameter when matching experimental PVT data. Figure 5.6 shows a comparison of a

TABLE 5.7
Coefficients in Equation 5.25

Coefficient	Value
a_1	−0.577191652
a_2	0.988205891
a_3	−0.897056937
a_4	0.918206857
a_5	−0.756704078
a_6	0.482199394
a_6	−0.193527818
a_8	0.035868343

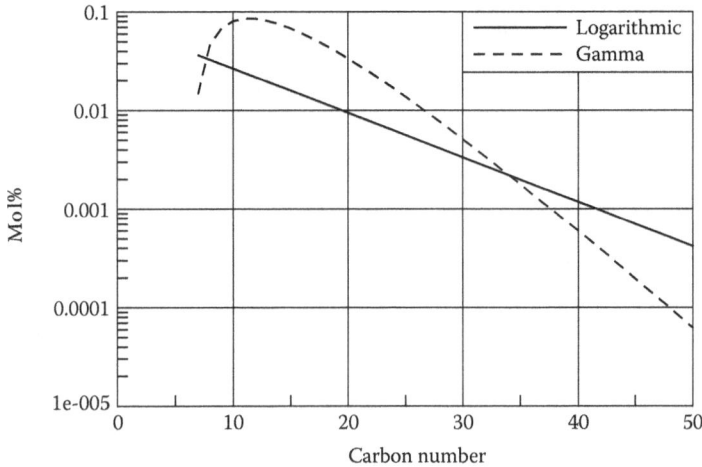

FIGURE 5.6 Comparison of logarithmic distribution with gamma distribution with $\alpha = 2.27$.

logarithmic molar distribution with that of a gamma distribution with $\alpha = 2.27$. Extended compositional analyses (Pedersen et al. 1992) and the theory of chemical reaction equilibria (Sørensen et al. 2013) support the logarithmic distribution expressed in Equation 5.10 and gives no justification of Equation 5.23 with $\alpha \neq 1$. This is consistent with work of Zuo and Zhang (2000), who have reviewed the two characterization procedures.

Which molar distribution function to use was a hot topic in the 1980s, when many compositional analyses stopped at C_{7+} or C_{10+} and compositional analyzes to higher carbon numbers than C_{20+} were rare. Most compositional analyses newer than the year 2010 extends to C_{36+}, and this has made the discussion of molar distribution functions less relevant, as a measured composition is always preferable to an estimated one.

The most frequently used C_{7+} characterization methods describe a reservoir fluid as a continuous distribution. In the 1980s, this inspired researchers to propose that phase equilibrium calculations on petroleum reservoir fluids should use a continuous distribution instead of splitting the C_{7+} fraction into pseudocomponents (e.g. Cotterman and Prausnitz 1985). This idea has not gained traction in the oil industry because a continuous distribution is not preserved after a flash calculation if the fluid splits into two or more phases. After the flash calculation each phase composition must be recharacterized, which means that the recombined phase compositions will differ from the feed composition.

5.2 BINARY INTERACTION COEFFICIENTS

To determine parameter a in a cubic equation of state, for example, the SRK or PR equation, it is necessary to know a binary interaction parameter, k_{ij}, for each binary component pair, that is, for any components i and j. The mixing rule used for parameter a can be seen from Equations 4.32 and 4.34. k_{ij} is zero by definition for $i = j$. Figure 5.7 illustrates the consideration behind the assignment of a binary interaction parameters to pairs of two different components. For a hydrocarbon, for example, C_1 it makes little difference for the intermolecular attraction whether the neighboring molecule is another C_1 molecule or a hydrocarbon (HC) with more carbon atoms, for example, nC_4. $k_{ij} = 0$ is therefore an appropriate value not only for a C_1–C_1 pair but also for C_1–nC_4 and for other HC-HC pairs. The attraction between two CO_2 molecules is high. If $k_{ij} = 0$ were also used for a CO_2-HC pair, part of the strong attraction between two CO_2 molecules would incorrectly also be attributed to the CO_2-HC interaction. To dampen the attraction between CO_2 and a HC molecule,

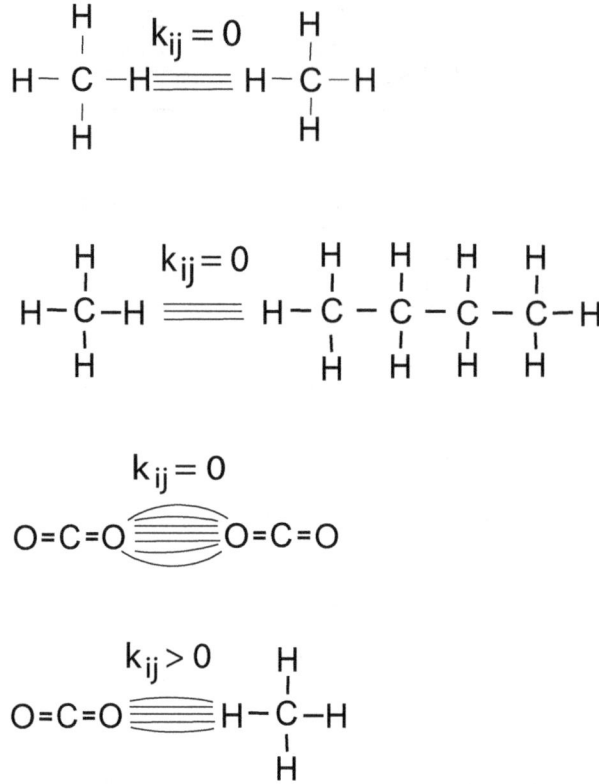

FIGURE 5.7 Considerations behind binary interaction parameters of zero and nonzero values.

$k_{ij} > 0$ is used. Also for pairs of other non-hydrocarbons and HC, it may be appropriate to use $k_{ij} > 0$. The nonhydrocarbons contained in petroleum reservoir fluids are usually limited to N_2, CO_2, and H_2S. It can further be of interest to consider H_2O. Only for binaries comprising at least one of these components is it necessary in general to work with nonzero binary interaction coefficients. Nonzero binary interaction coefficients between pairs of hydrocarbons are, however, often used for regression (parameter fitting) purposes. The nonzero binary interaction coefficients recommended for use with the SRK and PR equations can be seen in Table 4.2. Interaction parameters for mixtures with water are dealt with in Chapter 16.

5.3 LUMPING

Table 5.8 shows the composition in Table 5.4 after characterization. The characterized mixture consists of more than 80 components and pseudocomponents. It is desirable to reduce this number before performing phase equilibrium calculations. Lumping consists of the following:

- Deciding what carbon number fractions to lump (group) into the same pseudocomponent
- Averaging T_c, P_c, and ω of the individual carbon number fractions to one T_c, P_c, and ω representative for the whole lumped pseudocomponent

Pedersen et al. (1984) recommend a weight-based lumping where each lumped pseudocomponent contains approximately the same weight amount and where T_c, P_c, and ω of the pseudocomponents are found as weight mean averages of T_c, P_c, and ω of the individual carbon number fractions.

TABLE 5.8
Mixture in Table 5.4 after Characterization but before Lumping

Component	Mol%	M	Density (g/cm³) at 15°C and 1.01 bar	T$_c$ (°C)	P$_c$ (bar)	Acentric Factor
N$_2$	0.12	28.014	—	−146.95	33.94	0.04
CO$_2$	2.49	44.01	—	31.05	73.76	0.225
C$_1$	76.43	16.043	—	−82.55	46	0.008
C$_2$	7.46	30.07	—	32.25	48.84	0.098
C$_3$	3.12	44.097	—	96.65	42.46	0.152
iC$_4$	0.590	58.124	—	134.95	36.48	0.176
nC$_4$	1.21	58.124	—	152.05	38	0.193
iC$_5$	0.50	72.151	—	187.25	33.84	0.227
nC$_5$	0.59	72.151	—	196.45	33.74	0.251
C$_6$	0.79	86.178	0.664	234.25	29.69	0.296
C$_7$	0.95	95	0.726	258.7	31.44	0.465
C$_8$	1.08	106	0.747	278.4	28.78	0.497
C$_9$	0.78	116	0.769	295.6	27.22	0.526
C$_{10}$	0.592	133	0.781	318.8	23.93	0.574
C$_{11}$	0.467	152	0.778	339.8	20.58	0.626
C$_{12}$	0.345	164	0.785	353.6	19.41	0.658
C$_{13}$	0.375	179	0.802	371.4	18.65	0.698
C$_{14}$	0.304	193	0.815	386.8	18.01	0.735
C$_{15}$	0.237	209	0.817	401.7	16.93	0.775
C$_{16}$	0.208	218	0.824	410.8	16.66	0.798
C$_{17}$	0.220	239	0.825	428.7	15.57	0.849
C$_{18}$	0.169	250	0.831	438.7	15.31	0.874
C$_{19}$	0.140	264	0.841	451.5	15.11	0.907
C$_{20}$	0.1010	275	0.845	460.8	14.87	0.932
C$_{21}$	0.0888	291	0.849	473.6	14.48	0.966
C$_{22}$	0.0780	305	0.853	484.7	14.21	0.996
C$_{23}$	0.0686	318	0.857	494.8	13.99	1.023
C$_{24}$	0.0603	331	0.860	504.7	13.8	1.049
C$_{25}$	0.0530	345	0.864	515.1	13.61	1.075
C$_{26}$	0.0465	359	0.867	525.4	13.43	1.101
C$_{27}$	0.0409	374	0.870	536.1	13.26	1.128
C$_{28}$	0.0359	388	0.873	546.0	13.12	1.151
C$_{29}$	0.0316	402	0.876	555.8	12.99	1.174
C$_{30}$	0.0277	416	0.879	565.5	12.88	1.195
C$_{31}$	0.0244	430	0.881	575.0	12.77	1.216
C$_{32}$	0.0214	444	0.884	584.4	12.68	1.235
C$_{33}$	0.0188	458	0.887	593.7	12.59	1.253
C$_{34}$	0.0165	472	0.889	602.9	12.52	1.270
C$_{35}$	0.0145	486	0.891	612.0	12.44	1.285
C$_{36}$	0.0128	500	0.894	621.0	12.38	1.300
C$_{37}$	0.0112	514	0.896	630.0	12.32	1.313
C$_{28}$	0.00986	528	0.898	638.8	12.26	1.325
C$_{39}$	0.00866	542	0.900	647.6	12.21	1.335

(Continued)

TABLE 5.8   *(Continued)*

Mixture in Table 5.4 after Characterization but before Lumping

Component	Mol%	M	Density (g/cm³) at 15°C and 1.01 bar	T_c (°C)	P_c (bar)	Acentric Factor
C_{40}	0.00761	556	0.902	656.3	12.17	1.344
C_{41}	0.00609	570	0.904	664.9	12.12	1.352
C_{42}	0.00588	584	0.906	673.5	12.09	1.359
C_{43}	0.00517	598	0.908	682.0	12.05	1.364
C_{44}	0.00454	612	0.910	690.5	12.02	1.368
C_{45}	0.00399	626	0.912	698.9	11.99	1.371
C_{46}	0.00351	640	0.914	707.3	11.96	1.372
C_{47}	0.00308	654	0.916	715.6	11.93	1.372
C_{48}	0.00271	668	0.917	723.8	11.91	1.371
C_{49}	0.00238	682	0.919	732.0	11.89	1.369
C_{50}	0.00209	696	0.921	740.2	11.87	1.365
C_{51}	0.00183	710	0.922	748.3	11.85	1.359
C_{52}	0.00161	724	0.924	756.4	11.84	1.353
C_{53}	0.00142	738	0.926	764.4	11.82	1.345
C_{54}	0.00128	752	0.927	772.4	11.81	1.335
C_{55}	0.00109	766	0.929	780.4	11.80	1.325
C_{56}	0.000962	780	0.930	788.3	11.78	1.313
C_{57}	0.000845	794	0.932	796.2	11.77	1.300
C_{58}	0.000743	808	0.933	804.1	11.77	1.286
C_{59}	0.000653	822	0.934	811.9	11.76	1.270
C_{60}	0.000574	836	0.936	819.7	11.75	1.253
C_{61}	0.000504	850	0.937	827.5	11.75	1.236
C_{62}	0.000443	864	0.939	835.2	11.74	1.216
C_{63}	0.000389	878	0.940	843.0	11.74	1.196
C_{64}	0.000342	892	0.941	850.6	11.73	1.175
C_{65}	0.000300	906	0.942	858.3	11.73	1.152
C_{66}	0.000264	920	0.944	866.0	11.73	1.129
C_{67}	0.000232	934	0.945	873.6	11.72	1.104
C_{68}	0.000204	948	0.946	881.2	11.72	1.078
C_{69}	0.000179	962	0.947	888.7	11.72	1.052
C_{70}	0.000157	976	0.949	896.3	11.72	1.024
C_{71}	0.000138	990	0.950	903.8	11.72	0.995
C_{72}	0.000122	1004	0.951	911.3	11.72	0.965
C_{72}	0.000107	1018	0.952	918.8	11.72	0.935
C_{74}	0.0000939	1032	0.953	926.3	11.73	0.903
C_{75}	0.0000825	1046	0.954	933.7	11.73	0.871
C_{76}	0.0000725	1060	0.955	941.2	11.73	0.838
C_{77}	0.0000637	1074	0.956	948.6	11.73	0.804
C_{78}	0.0000560	1088	0.957	956.0	11.74	0.769
C_{79}	0.0000492	1102	0.959	963.4	11.74	0.734
C_{80}	0.0000432	1116	0.960	970.7	11.74	0.697

If the k-th pseudocomponent contains the carbon number fractions m to n, T_c, P_c, and ω are found from the relations

$$T_{ck} = \frac{\sum_{i=m}^{n} z_i M_i T_{ci}}{\sum_{i=m}^{n} z_i M_i} \tag{5.29}$$

$$P_{ck} = \frac{\sum_{i=m}^{n} z_i M_i P_{ci}}{\sum_{i=m}^{n} z_i M_i} \tag{5.30}$$

$$\omega_k = \frac{\sum_{i=m}^{n} z_i M_i \omega_i}{\sum_{i=m}^{n} z_i M_i} \tag{5.31}$$

where z_i is the mole fraction and M_i the molecular weight of carbon number fraction i. The weight-based procedure ensures that all hydrocarbon segments of the C_{7+} fraction are given equal importance. An example of this grouping is given in Table 5.9. The C_{7+} fraction is divided into three groups, which on a weight basis are of approximately equal sizes. The weight percentages of the three C_{7+} pseudocomponents differ slightly because the cut points are made between the carbon number fractions. In Table 5.9, the entire C_{11} fraction is thus found in the first of the three C_{7+} pseudocomponents, although it would have given a more equal weight distribution had C_{11} been divided between the first and the second pseudocomponent.

Several other lumping schemes have been proposed in the literature. Instead of working with pseudocomponents of equal weight amounts, Danesh et al. (1992) suggested that the sum of the mole fractions times the logarithm of the molecular weight ($£z_i \ln M_i$) should be the same for each pseudocomponent. Whitson et al. (1989) have proposed picking the pseudocomponents using a

TABLE 5.9
Mixture in Table 5.4 after Characterization and Lumping

Component	Mol%	Weight%	T_c (K)	P_c (bar)	Acentric Factor
N_2	0.12	0.11	126.2	33.9	0.040
CO_2	2.49	3.51	304.2	73.8	0.225
C_1	76.43	39.30	190.6	46.0	0.008
C_2	7.46	7.19	305.4	48.8	0.098
C_3	3.12	4.41	369.8	42.5	0.152
iC_4	0.59	1.10	408.1	36.5	0.176
nC_4	1.21	2.25	425.2	38.0	0.193
iC_5	0.50	1.16	4604	33.8	0.227
nC_5	0.59	1.36	469.6	33.7	0.251
C_6	0.79	2.18	507.4	29.7	0.296
C_7–C_{11}	3.87	14.26	568.0	26.8	0.530
C_{12}–C_{18}	1.86	11.92	668.9	17.4	0.762
C_{19}–C_{80}	0.97	11.25	817.3	13.5	1.108

quadrature method. This essentially means that each pseudocomponent contains a wider range of molecular weights and that components with the same molecular weight may be distributed between more pseudocomponents.

Leibovici (1993) have suggested that the equation of state parameters, a and b, of the total mixture should be unaffected by lumping. The conventional mixing rule for parameter a is expressed in Equation 4.32. By introducing the expression for a_{ij} (Equation 4.34), parameter a for a mixture may be expressed as

$$a(T) = C_1 \sum_{i=1}^{N} \sum_{j=1}^{N} z_i z_j \frac{T_{ci} T_{cj}}{\sqrt{P_{ci} P_{cj}}} \sqrt{\alpha_i(T) \alpha_j(T)} \, (1 - k_{ij}) \tag{5.32}$$

Similarly, for parameter b (Equations 4.23 and 4.34):

$$b = C_2 \sum_{i=1}^{N} z_i \frac{T_{ci}}{P_{ci}} \tag{5.33}$$

where C_1 and C_2 are constants, T_c critical temperature, P_c critical pressure, and z mole fraction. If some of the N components are lumped into pseudocomponents, parameters a and b of the total mixture will not in general be unaffected. The lumping procedure proposed by Leibovici et al. retains approximately the same mixture parameters after lumping. Say pseudocomponent k consists of the components m to n. To retain the same a and b of the total mixture, a and b of the pseudocomponent must be given by

$$a_k(T) = C_1 \sum_{i=m}^{n} \sum_{j=m}^{n} z_i z_j \frac{T_{ci} T_{cj}}{\sqrt{P_{ci} P_{cj}}} \sqrt{\alpha_i(T) \alpha_j(T)} \, (1 - k_{ij}) \tag{5.34}$$

$$b_k = C_2 \sum_{i=m}^{n} z_i \frac{T_{ci}}{P_{ci}} \tag{5.35}$$

where $\alpha(T)$ is defined in Equation 4.24. Furthermore, the following relation must be fulfilled for pseudocomponent k (from Equation 4.21)

$$a_k(T) = C_1 \frac{T_{ck}^2}{P_{ck}} \alpha_k(T) \tag{5.36}$$

and the following relation can be derived from Equations 5.34 and 5.36

$$\frac{T_{ck}^2}{P_{ck}} \alpha(T) = \sum_{i=m}^{n} \sum_{j=m}^{n} z_i z_j \frac{T_{ci} T_{cj}}{\sqrt{P_{ci} P_{cj}}} \sqrt{\alpha_i(T) \alpha_j(T)} \, (1 - k_{ij}) \tag{5.37}$$

This relation must also be fulfilled for a temperature of T_{ck} for which $\alpha_k(T) = 1$, giving

$$\frac{T_{ck}^2}{P_{ck}} = \sum_{i=m}^{n} \sum_{j=m}^{n} z_i z_j \frac{T_{ci} T_{cj}}{\sqrt{P_{ci} P_{cj}}} \sqrt{\alpha_i(T_{ck}) \alpha_j(T_{ck})} \, (1 - k_{ij}) \tag{5.38}$$

The definition of parameter b is used to derive the relation

$$\frac{T_{ck}}{P_{ck}} = \sum_{i=m}^{N} z_i \frac{T_{ci}}{P_{ci}} \tag{5.39}$$

By elimination of P$_{ck}$, Equations 5.38 and 5.39 are easily reduced to an equation with T$_{ck}$ as the only unknown, allowing T$_{ck}$ and (subsequently) P$_{ck}$ to be determined. Next, the temperature dependence of the parameter α is to be determined. All the parameters on the right-hand side of Equation 5.37 are known, making it possible to calculate numerical values of $\alpha_k(T)$ for a series of temperature values. The numerical values are fitted to a fourth-degree polynomial in T, and this polynomial is used as the temperature dependence of α in phase equilibrium calculations for pseudocomponent k.

Using this lumping procedure, the mixture a parameter for a single-phase system will be independent of the number of pseudocomponents. In a calculation involving two or more phases, the calculation results will be independent of the number of pseudocomponents only if the individual components making up a pseudocomponent split equally (with the same K-factors) between the two phases. This will seldom be the case.

Lomeland and Harstad (1995) have presented a lumping scheme which minimizes the variation in the equation of state parameters a and b within a pseudocomponent. Newley and Merrill (1989) have suggested a somewhat similar lumping scheme which minimizes the variation in K-factors defined in Equation 5.40 rather than the variation in the equation of state parameters.

5.4 DELUMPING

Compositional reservoir simulation studies are often quite time consuming, and the simulation time increases with the number of components. Compositions used in compositional reservoir simulation studies are therefore often heavily lumped. Also, some of the defined components are usually lumped in a compositional reservoir simulation. Table 5.10 shows the composition in Table 5.4 after characterization and lumping into a total of six pseudocomponents. N$_2$ and C$_1$ are lumped together, CO$_2$ is lumped with C$_2$–C$_3$, and all the C$_4$–C$_6$ components are lumped into one fraction. The C$_{7+}$ fractions are lumped in the same manner as in Table 5.9.

In a process plant separating a produced well stream into gas and oil, the pressure is usually much lower than in the reservoir. A lumping that was justified for reservoir conditions is not necessarily justified for process conditions. It would therefore be interesting with a procedure which in a meaningful manner could split a lumped composition from a compositional reservoir simulation into its original constituents. Such a split is called *delumping*.

In a pressure-temperature flash for a hydrocarbon mixture, the relative molar amounts of a component i ending up in the gas and liquid phases are determined by the K-factor of each component:

$$K_i = \frac{y_i}{x_i} \tag{5.40}$$

TABLE 5.10
Mixture in Table 5.4 after Characterization and Lumping into a Total of Six Pseudocomponents

Component	Mol%	Weight%	T$_c$ (K)	P$_c$ (bar)	Acentric Factor
N$_2$ + C$_1$	76.55	39.40	190.4	46.0	0.008
CO$_2$ + C$_2$ + C$_3$	13.07	15.11	323.9	52.8	0.143
C$_4$–C$_6$	3.68	8.06	457.7	34.2	0.233
C$_7$–C$_{11}$	3.87	14.26	568.0	26.8	0.530
C$_{12}$–C$_{18}$	1.86	11.92	668.9	17.4	0.762
C$_{19}$–C$_{80}$	0.97	11.25	817.3	13.5	1.108

where y_i is the mole fraction of component i in the gas phase and x_i the mole fraction of component i in the liquid phase. If two components i and j have approximately the same K-factor, it is justified to lump them together to one pseudocomponent before performing the flash. The K-factor of the lumped component will be approximately the same as the K-factors of the two components treated individually.

A number of papers deal with delumping alternating with flash calculations as a way of speeding up flash calculations (Drohm and Schlijper 1985; Danesh et al. 1992; Leibovici et al. 1996). These flash calculations are carried out for a heavily lumped fluid and the resulting phase compositions delumped after each flash calculation using an appropriate K-factor correlation. It is questionable whether that will save any computation time as compared to carrying out the flash calculation with the full component number using an efficient flash algorithm (dealt with in Chapter 6). Furthermore, the delumping will inevitably introduce inaccuracies.

5.5 MIXING OF MULTIPLE FLUIDS

There is often a need to mix a number of reservoir fluid compositions into one. This is, for example, the case when multiple fluids are let to the same process plant. When representing the mixed stream, one may either work with a weaved composition where the pseudocomponents of each stream are retained or with a truly mixed composition. The difference between a weaved and a mixed composition is here exemplified for the two reservoir fluids in Tables 3.6 and 5.11. These are to be weaved or mixed in equal molar amounts. The weaved composition is shown in Table 5.12. The two compositions have initially been characterized individually. For both fluids, the C_{7+} fraction is represented using three pseudocomponents. As is seen in Table 5.12, the pseudocomponent properties differ between the two fluids. In the weaved composition, the molar amounts of the defined components have been obtained as a simple average of the molar concentrations of these compounds in each individual composition. The weaved composition contains all the pseudocomponents found in each of the two compositions. Weaving of fluids is advantageous when it is essential to keep track of components from individual feed streams in a mixed composition but is impractical when the number of

TABLE 5.11
Molar Composition of Gas Condensate Fluid

Component	Mol%	Molecular Weight	Density at 1.01 bar, 15°C (g/cm³)
N_2	0.96	—	—
CO_2	0.77	—	—
C_1	83.57	—	—
C_2	6.16	—	—
C_3	3.07	—	—
iC_4	0.44	—	—
nC_4	1.12	—	—
iC_5	0.35	—	—
nC_5	0.50	—	—
C_6	0.48	—	—
C_7	0.67	95	0.724
C_8	0.60	103	0.748
C_9	0.38	116	0.765
C_{10+}	0.93	165	0.811

TABLE 5.12

Gas Composition in Table 5.11 Weaved with Oil Composition in Table 3.6 in Equal Molar Amounts

Component	Gas Composition Characterized				Oil Composition Characterized				Weaved Composition			
	Mol%	T_c (K)	P_c (bar)	ω	Mol%	T_c (K)	P_c (bar)	ω	Mol%	T_c (K)	P_c (bar)	ω
N_2	0.96	126.2	33.94	0.040	0.39	126.2	33.94	0.040	0.68	126.2	33.94	0.040
CO_2	0.77	304.2	73.76	0.225	0.30	304.2	73.76	0.225	0.54	304.2	73.76	0.225
C_1	83.56	190.6	46.00	0.008	40.20	190.6	46.00	0.008	61.88	190.6	46.00	0.008
C_2	6.16	305.4	48.84	0.098	7.61	305.4	48.84	0.098	6.89	305.4	48.84	0.098
C_3	3.07	369.8	42.46	0.152	7.95	369.8	42.46	0.152	5.51	369.8	42.46	0.152
iC_4	0.44	408.1	36.48	0.176	1.19	408.1	36.48	0.176	0.82	408.1	36.48	0.176
nC_4	1.12	425.2	38.00	0.193	4.08	425.2	38.00	0.193	2.60	425.2	38.00	0.193
iC_5	0.35	460.4	33.84	0.227	1.39	460.4	33.84	0.227	0.87	460.4	33.84	0.227
nC_5	0.50	469.6	33.74	0.251	2.15	469.6	33.74	0.251	1.33	469.6	33.74	0.251
C_6	0.48	507.4	29.69	0.296	2.79	507.4	29.69	0.296	1.64	507.4	29.69	0.296
C_7–C_8 (gas)	1.27	539.3	30.61	0.476	—	—	—	0.563	0.64	539.3	30.61	0.476
C_9–C_{11} (gas)	0.87	587.7	24.57	0.566	—	—	—	0.894	0.43	587.7	24.57	0.566
C_{12}–C_{42} (gas)	0.44	668.3	18.82	0.752	—	—	—	1.256	0.22	668.4	18.82	0.752
C_7–C_{13} (oil)	—	—	—	—	19.33	584.2	25.42	0.563	9.67	584.2	25.42	0.563
C_{14}–C_{26} (oil)	—	—	—	—	8.64	722.4	16.20	0.894	4.32	722.4	16.20	0.894
C_{27}–C_{80} (oil)	—	—	—	—	3.98	952.8	13.32	1.256	1.99	952.8	13.32	1.256

fluids to be weaved is high. For each new fluid, the number of components increases by the number of pseudocomponents found in that particular fluid. In such a case, it is more attractive to carry out a mixing of the individual compositions. This means that the mixed fluid is represented through pseudocomponents representative of the mixed fluid instead of having to work with all the pseudo-components contained in each composition.

It is recommended to carry out the mixing before lumping into pseudocomponents. Say NFLUID different fluids are to be mixed; the properties of carbon number fraction i of the mixed fluid are found from

$$T_{ci}^{mix} = \frac{\sum_{j=1}^{NFLUID} Frac(j) z_i^j T_{ci}^j}{\sum_{j=1}^{NFLUID} Frac(j) z_i^j} \qquad (5.41)$$

$$P_{ci}^{mix} = \frac{\sum_{j=1}^{NFLUID} Frac(j) z_i^j P_{ci}^j}{\sum_{j=1}^{NFLUID} Frac(j) z_i^j} \qquad (5.42)$$

$$\omega_i^{mix} = \frac{\sum_{j=1}^{NFLUID} Frac(j) z_i^j \omega_i^j}{\sum_{j=1}^{NFLUID} Frac(j) z_i^j} \qquad (5.43)$$

and the mole fraction and average molecular weight from

$$z_i^{mix} = \sum_{j=1}^{NFLUID} Frac(j) z_i^j \tag{5.44}$$

$$M_i^{mix} = \frac{\sum_{j=1}^{NFLUID} Frac(j) z_i^j M_i^j}{\sum_{j=1}^{NFLUID} Frac(j) z_i^j} \tag{5.45}$$

In these equations, z_i^j is the molar fraction of carbon number fraction i in the j-th composition to be mixed. Similarly, T_{ci}^j, P_{ci}^j, and ω_i^j are the critical temperature, critical pressure, and acentric factor of carbon number fraction i in the j-th composition, respectively. Frac(j) is the mole fraction of the j-th composition of the total mixture. By use of Equations 5.41 to 5.45, a mixed composition is obtained, which may be grouped using one of the procedures outlined for a single composition in Section 5.3. If the k-th pseudocomponent comprises the carbon number fractions from m to n, the k-th pseudocomponent will get the following properties:

$$T_{ck}^{mix} = \frac{\sum_{i=m}^{n} z_i^{mix} M_i^{mix} T_{ci}^{mix}}{\sum_{i=m}^{n} z_i^{mix} M_i^{mix}} \tag{5.46}$$

$$P_{ck}^{mix} = \frac{\sum_{i=m}^{n} z_i^{mix} M_i^{mix} P_{ci}^{mix}}{\sum_{i=m}^{n} z_i^{mix} M_i^{mix}} \tag{5.47}$$

$$\omega_k^{mix} = \frac{\sum_{i=m}^{n} z_i^{mix} M_i^{mix} \omega_i^{mix}}{\sum_{i=m}^{n} z_i^{mix} M_i^{mix}} \tag{5.48}$$

Table 5.13 shows the composition consisting of the gas in Table 5.11 mixed with the oil in Table 3.6 in equal molar amounts.

TABLE 5.13

Gas Composition in Table 5.11 Mixed with Oil Composition in Table 3.6 in Equal Molar Amounts

Component	Mol%	T_c (K)	P_c (bar)	Acentric Factor
N_2	0.675	126.2	33.94	0.040
CO_2	0.535	304.2	73.76	0.225
C_1	61.885	190.6	46.00	0.008
C_2	6.885	305.4	48.84	0.098
C_3	5.510	369.8	42.46	0.152
iC_4	0.815	408.1	36.48	0.176
nC_4	2.600	425.2	38.00	0.193
iC_5	0.870	460.4	33.84	0.227
nC_5	1.325	469.6	33.74	0.251
C_6	1.635	507.4	29.69	0.296
C_7–C_{12}	10.008	575.5	26.26	0.545
C_{13}–C_{25}	5.113	708.6	16.76	0.857
C_{26}–C_{80}	2.144	945.6	13.36	1.248

5.6 CHARACTERIZING OF MULTIPLE COMPOSITIONS TO THE SAME PSEUDOCOMPONENTS

In process simulations and compositional reservoir and flow simulations, it may be advantageous to characterize a number of different reservoir fluids to a unique set of pseudocomponents (also called a *Common EoS*). This is practical, for example, when numerous process streams are let to the same separation plant. There is often a need to perform simulations on each stream separately as well as on the mixed stream. If each composition is represented using the same pseudocomponents, the streams can readily be mixed without having to increase the number of components.

Initially, the plus fractions of the compositions to be characterized to the same pseudocomponents are split into carbon number fractions. For each C$_{7+}$ carbon number fraction, T$_c$, P$_c$, and ω are estimated in the usual manner. The lumping uses the same cut points for all fluid compositions as sketched in Figure 5.8 and a common set of T$_c$s, P$_c$s, and ωs is determined for each fraction:

$$T_{ci}^{unique} = \frac{\sum_{j=1}^{NFLUID} Wgt(j) z_i^j T_{ci}^j}{\sum_{j=1}^{NFLUID} Wgt(j) z_i^j} \tag{5.49}$$

$$P_{ci}^{unique} = \frac{\sum_{j=1}^{NFLUID} Wgt(j) z_i^j P_{ci}^j}{\sum_{j=1}^{NFLUID} Wgt(j) z_i^j} \tag{5.50}$$

$$\omega_i^{unique} = \frac{\sum_{j=1}^{NFLUID} Wgt(j) z_i^j \omega_i^j}{\sum_{j=1}^{NFLUID} Wgt(j) z_i^j} \tag{5.51}$$

NFLUID is the number of compositions to be characterized to the same pseudocomponents, z_i^j is the mole fraction of component (carbon number fraction) i in composition number j, and Wgt(j) is the weight to be assigned to composition number j.

Component	Fluid 1	Fluid 2	....	
...				
...				
C$_7$	xx	yy		Pseudo_1
C$_8$	xx	yy		
C$_9$	xx	yy		
C$_{10}$	...	...		Pseudo_2
C$_{11}$	...	...		
C$_{12}$	...	...		
C$_{13}$	...	...		
C$_{14}$	...	...		
...	...	...		
...	...	...		
...	...	...		

FIGURE 5.8 Principle of same pseudocomponents (common EoS).

To decide what carbon number fractions to include in each pseudocomponent, an imaginary (mixed) molar composition is calculated that is assumed to be representative of all individual compositions. In this imaginary composition, (pseudo)component i enters with a mole fraction of

$$z_i^{unique} = \frac{\sum\limits_{j=1}^{NFLUID} Wgt(j) z_i^j}{\sum\limits_{j=1}^{NFLUID} Wgt(j)} \tag{5.52}$$

and is assigned a molecular weight of

$$M_i^{unique} = \frac{\sum\limits_{j=1}^{NFLUID} Wgt(j) z_i^j M_i^j}{\sum\limits_{j=1}^{NFLUID} Wgt(j) z_i^j} \tag{5.53}$$

This imaginary composition is now treated like an ordinary composition to be lumped into pseudocomponents. The lumping determines the carbon number ranges and T_c, P_c, and ω of each pseudocomponent. The component properties of the lumped composition are assumed to apply for all the individual compositions. If the k-th pseudocomponent contains the carbon number fractions from m to n, the mole fraction of this pseudocomponent in the j-th composition will be

$$z_k^j = \sum\limits_{i=m}^{n} z_i^j \tag{5.54}$$

Table 5.14 shows the result of a characterization of the compositions in Tables 3.6 and 5.11 to the same pseudocomponents. The same weight has been applied to both fluids. This is why T_c, P_c, and ω are the same as when the two compositions are mixed in equal molar amounts. This can be seen by comparing the component properties in Tables 5.13 and 5.14.

TABLE 5.14

Gas Composition in Table 5.11 Characterized to Same Pseudocomponents as Oil Composition in Table 3.6

Component	Mol% Gas	Mol% Oil	T_c (K)	P_c (bar)	Acentric Factor
N_2	0.960	0.390	126.2	33.94	0.040
CO_2	0.770	0.300	304.2	73.76	0.225
C_1	83.570	40.200	190.6	46.00	0.008
C_2	6.160	7.610	305.4	48.84	0.098
C_3	3.070	7.950	369.8	42.46	0.152
iC_4	0.440	1.190	408.1	36.48	0.176
nC_4	1.120	4.080	425.2	38.00	0.193
iC_5	0.350	1.390	460.4	33.84	0.227
nC_5	0.500	2.150	469.6	33.74	0.251
C_6	0.480	2.790	507.4	29.69	0.296
C_7–C_{12}	2.276	17.740	575.5	26.26	0.545
C_{13}–C_{25}	0.301	9.925	708.6	16.76	0.857
C_{26}–C_{80}	0.002	4.285	945.6	13.36	1.248

5.7   HEAVY OIL COMPOSITIONS

A heavy oil is one of a high density at standard conditions, and the term *heavy oil* may be used for oil mixtures of an API gravity below 30. API is the density defined by the American Petroleum Institute as

$$API = \frac{141.5}{SG} - 131.5 \qquad (5.55)$$

where SG is the 60°F/60°F specific gravity. *Specific gravity* is defined as the mass ratio of equal volumes of oil and water at the appropriate temperature. As the density of water at 60°F is close to 1 g/cm^3, the specific gravity of an oil sample at atmospheric conditions will take approximately the same value as the density of the oil in g/cm^3.

Heavy oils will have a high content of aromatic components and a non-negligible content of components heavier than C$_{80}$. The melting temperatures of aromatics are lower than of normal and slightly branched paraffins of approximately the same molecular weight. At low temperature, solid wax consisting of heavy paraffins is likely to form from a fluid rich in paraffins (see Chapter 11). Had those paraffins stayed in solution in the oil, the oil viscosity might have become very high. In heavy oils dominated by aromatics, high-molecular-weight compounds are kept in solution even at low temperatures, and the viscosity of heavy oil mixtures can be very high indeed at production conditions and even at reservoir conditions, as is further dealt with in Chapter 10.

Table 5.15 shows a reservoir oil composition with an API gravity of 28. This oil is at the very light end of what is classified as a heavy oil. Figure 5.9 shows a plot of the C$_{7+}$ mole percentages (logarithmic scale) versus carbon number for the reservoir fluid in Table 5.15. As is indicated by the dashed line, an approximately linear relation is seen consistent with Equation 5.10. As will be shown in the next section, biodegradation may cause heavy oils not to exhibit the linear trend all the way from C$_7$.

Table 5.16 shows a heavy oil composition with an API of 15.7. The C$_{7+}$ molecular weight is 450. If the C$_7$–C$_{35}$ molecular weights in Table 2.3 are representative for this fluid, the C$_{36+}$ molecular weight would be 808, corresponding to a C$_{57}$ compound. By assuming that ln(mol%) versus carbon

---

TABLE 5.15

Molar Composition of Heavy Reservoir Fluid with an API Gravity of 28. The Reservoir Temperature Is 74°C and the Saturation Pressure at This Temperature Is 227 bar

Component	Mol%	Molecular Weight	Density at 1.01 bar, 15°C (g/cm³)
N$_2$	0.49	—	—
CO$_2$	0.31	—	—
C$_1$	44.01	—	—
C$_2$	3.84	—	—
C$_3$	1.12	—	—
iC$_4$	0.61	—	—
nC$_4$	0.72	—	—
iC$_5$	0.69	—	—
nC$_5$	0.35	—	—
C$_6$	1.04	—	—
C$_7$	2.87	96	0.738
C$_8$	4.08	107	0.765
C$_9$	3.51	121	0.781

(Continued)

TABLE 5.15 *(Continued)*
Molar Composition of Heavy Reservoir Fluid with an API Gravity of 28. The Reservoir Temperature Is 74°C and the Saturation Pressure at This Temperature Is 227 bar

Component	Mol%	Molecular Weight	Density at 1.01 bar, 15°C (g/cm³)
C_{10}	3.26	134	0.792
C_{11}	2.51	147	0.796
C_{12}	2.24	161	0.810
C_{13}	2.18	175	0.825
C_{14}	2.07	190	0.836
C_{15}	2.03	206	0.842
C_{16}	1.67	222	0.849
C_{17}	1.38	237	0.845
C_{18}	1.36	251	0.848
C_{19}	1.19	263	0.858
C_{20}	1.02	275	0.863
C_{21}	0.89	291	0.868
C_{22}	0.78	305	0.873
C_{23}	0.72	318	0.877
C_{24}	0.64	331	0.881
C_{25}	0.56	345	0.885
C_{26}	0.53	359	0.889
C_{27}	0.48	374	0.893
C_{28}	0.46	388	0.897
C_{29}	0.45	402	0.900
C_{30+}	9.96	449.1	0.989

Source: Data from Krejbjerg, K. and Pedersen, K.S., Controlling VLLE equilibrium with a cubic EoS in heavy oil modeling, presented at *57th Annual Technical Meeting of the Petroleum Society* (Canadian International Petroleum Conference), Calgary, Canada, June 13–15, 2006.

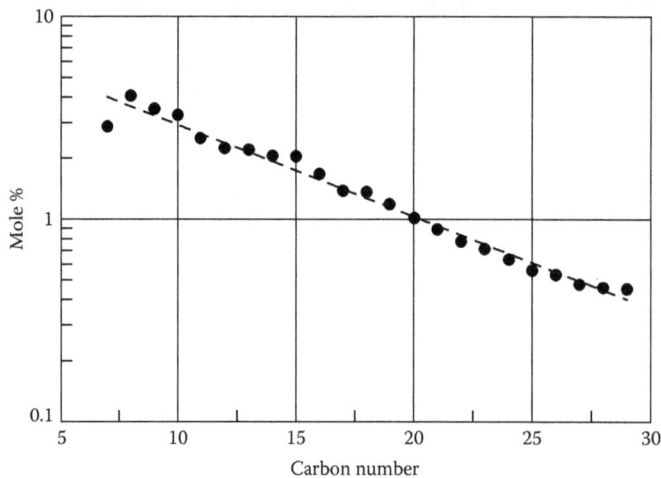

FIGURE 5.9 C_{7+} component mole percentages for fluid in Table 5.15 plotted against carbon number. The mole percentages are shown as dots, and the dashed line is a best-fit line according to Equation 5.10.

TABLE 5.16

Molar Composition of Heavy Oil Reservoir Fluid with an API Gravity of 15.7. The C_{7+} Molecular Weight is 450 and the C_{7+} Density 0.970 g/cm³. At a Temperature of 38°C the Saturation Pressure is 60 bar

Component	Weight%	Mol %
N_2	0.001	0.012
CO_2	0.234	1.615
H_2S	0.365	3.248
C_1	0.972	18.403
C_2	0.283	2.863
C_3	0.354	2.441
iC_4	0.154	0.806
nC_4	0.249	1.299
iC_5	0.251	1.055
nC_5	0.090	0.377
C_6	0.831	2.928
C_7	0.596	1.884
C_8	0.707	2.007
C_9	0.737	1.849
C_{10}	0.940	2.130
C_{11}	1.064	2.198
C_{12}	1.082	2.041
C_{13}	1.251	2.170
C_{14}	1.344	2.148
C_{15}	1.494	2.203
C_{16}	1.469	2.009
C_{17}	1.456	1.866
C_{18}	1.487	1.799
C_{19}	1.454	1.680
C_{20}	1.427	1.576
C_{21}	1.342	1.400
C_{22}	1.294	1.289
C_{23}	1.196	1.142
C_{24}	1.155	1.060
C_{25}	1.156	1.018
C_{26}	1.167	0.987
C_{27}	1.153	0.936
C_{28}	1.161	0.909
C_{29}	1.180	0.891
C_{30}	1.117	0.815
C_{31}	1.105	0.780
C_{32}	1.056	0.722
C_{33}	0.989	0.656
C_{34}	0.956	0.615
C_{35}	0.934	0.584
C_{36+}	62.749	23.586

number exhibits a linear trend from C_{35}, the content of components heavier than C_{80} can be estimated to be 3.2 mole percent, corresponding to 15.1% by weight. The properties of a heavy oil may be influenced by components as heavy as C_{200}. The expression for m in Equation 5.3 is inappropriate components heavier than around C_{80}. The functional forms of Equations 5.1 and 5.2 can be retained for T_c and P_c, and the coefficients in Table 5.3 in the expressions for T_c can be retained for both the Soave–Redlich–Kwong (SRK) and the Peng–Robinson (PR) equations. For P_c, new coefficients are required for SRK, while the coefficients in Table 5.3 can be retained for PR. The following expression may be used for the acentric factor of a heavy oil

$$\omega = e_1 + e_2\rho + \frac{\left(\dfrac{M}{e_3} - e_4\right)^2}{\exp\left(\dfrac{M}{e_3} - e_4\right) - 1} \tag{5.56}$$

where ρ is density in g/cm³ and M is molecular weight. The coefficients c_1–c_4, d_1–d_5, and e_1–e_4 for use with heavy oil mixtures can be seen in Table 5.17. Figure 5.10 shows the phase envelope of the oil mixture in Table 5.16 simulated using the Peng–Robinson equation of state with the fluid characterized using the heavy oil characterization outlined in this section. The phase envelope has a relatively flat section connecting the low- and high-temperature parts of the phase envelope. This is typical for a heavy oil.

5.7.1 Biodegraded Oils

Table 5.18 shows an example of a heavy oil composition that almost lacks hydrocarbons from C_3 to C_6 and further has a relatively low concentration of hydrocarbons from C_7 to C_{17}. This is an example of a biodegraded oil. Biodegradation occurs when bacteria, fungi, or other organisms consume paraffinic components. Up to C_6, most hydrocarbons are n-paraffins. The fractions C_7 and heavier contain naphthenic or aromatic hydrocarbons in addition to paraffins, which are compounds not degraded by biological organisms. Therefore, as is the case for the oil in Table 5.18, the components C_2–C_6 may be almost absent, while components from C_7 to about C_{17}, thanks to aromatics and naphthenes, are still present, although in low concentrations. The mentioned component classes are introduced in Chapter 1. Paraffins heavier than C_{17} are not subject to

TABLE 5.17

Coefficients in the Correlations in Equations 5.1, 5.2, and 5.56 for Use with the Soave–Redlich–Kwong and the Peng–Robinson Equations (T_c Is in K and P_c in atm)

Subindex/Coefficient	1	2	3	4	5
SRK/SRK-Peneloux					
c	1.6312×10^2	8.6052×10	4.3475×10^{-1}	-1.8774×10^3	—
d	-0.42444	2.4792	2.1927×10^2	-5.6413×10^3	0.25
e	8.7888×10^{-1}	1.1293×10^2	1.2902	-3.5120×10^{-2}	—
PR/PR-Peneloux					
c	7.34043×10	9.73562×10	6.18744×10^{-1}	-2.05932×10^3	—
d	7.28462×10^{-2}	2.18811	1.63910×10^2	-4.04323×10^3	0.25
e	7.5968×10^{-1}	1.0930×10^2	1.2363	-3.5120×10^{-2}	—

Source: Data from Calsep, PVTsim Nova Method Documentation, Lyngby, Denmark, 2023.

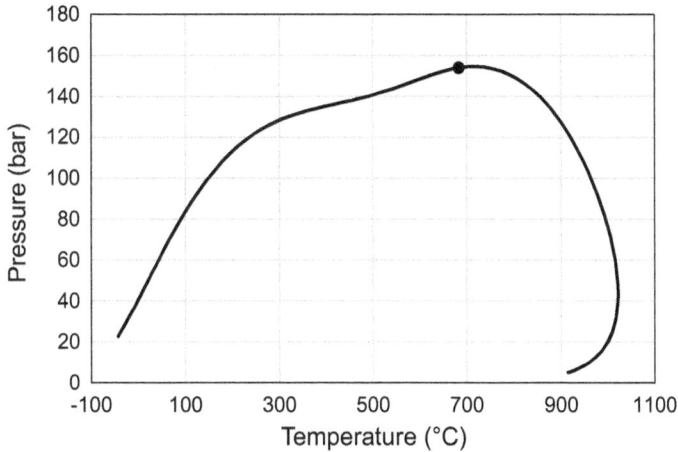

FIGURE 5.10 Simulated phase envelope for heavy oil mixture in Table 5.16. The Peng–Robinson equation of state and the heavy oil characterization outlined in Section 5.7 have been used.

TABLE 5.18
Molar Composition of Biodegraded Reservoir Fluid with an API Gravity of 10. The Reservoir Temperature Is 52°C and the Saturation Pressure at This Temperature Is 71.5 bar

Component	Mol%	Molecular Weight	Density at 1.01 bar, 15°C (g/cm³)
CO_2	1.44	—	—
C_1	18.72	—	—
C_2	0.14	—	—
C_3	0.03	—	—
iC_4	0.01	—	—
nC_4	0.01	—	—
iC_5	0.01	—	—
nC_5	0.27	—	—
C_6	0.41	—	—
C_7	0.13	96	0.722
C_8	0.32	107	0.745
C_9	0.45	121	0.764
C_{10}	0.90	134	0.778
C_{11}	1.45	147	0.789
C_{12}	1.97	161	0.800
C_{13}	2.50	175	0.811
C_{14}	2.57	190	0.822
C_{15}	2.86	206	0.832
C_{16}	2.91	222	0.839
C_{17}	2.96	237	0.870
C_{18}	2.99	251	0.852
C_{19}	3.07	263	0.857
C_{20}	2.72	275	0.862

(Continued)

TABLE 5.18 (*Continued*)
Molar Composition of Biodegraded Reservoir Fluid with an API Gravity of 10. The Reservoir Temperature Is 52°C and the Saturation Pressure at This Temperature Is 71.5 bar

Component	Mol%	Molecular Weight	Density at 1.01 bar, 15°C (g/cm³)
C_{21}	2.90	291	0.867
C_{22}	2.20	305	0.872
C_{23}	2.26	318	0.877
C_{24}	2.14	331	0.881
C_{25}	1.96	345	0.885
C_{26}	1.77	359	0.889
C_{27}	1.68	374	0.893
C_{28}	1.82	388	0.896
C_{29}	1.64	402	0.899
C_{30}	1.63	416	0.902
C_{31}	1.36	430	0.906
C_{32}	1.33	444	0.909
C_{33}	1.12	458	0.912
C_{34}	1.19	472	0.914
C_{35}	1.00	486	0.917
C_{36+}	25.17	1038.1	1.104

Source: Data from Krejbjerg, K. and Pedersen, K.S., Controlling VLLE equilibrium with a cubic EoS in heavy oil modeling, presented at *57th Annual Technical Meeting of the Petroleum Society* (Canadian International Petroleum Conference), Calgary, Canada, June 13–15, 2006.

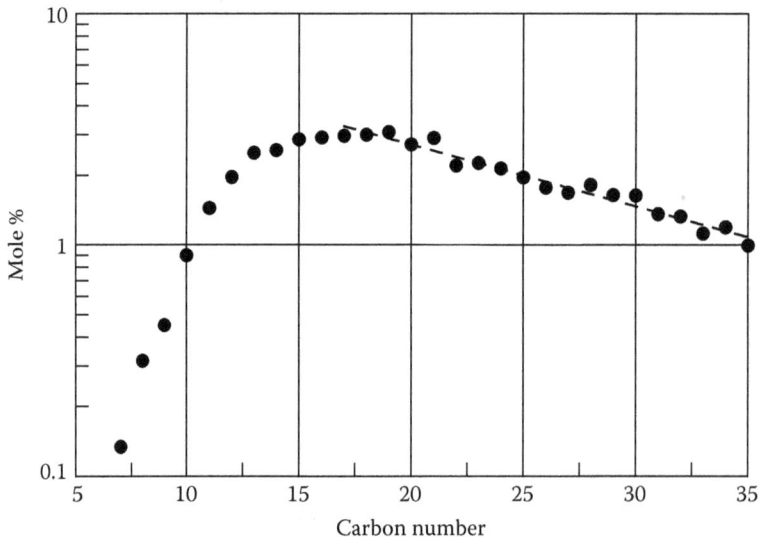

FIGURE 5.11 C_{7+} component mole percentages for fluid in Table 5.18 plotted against carbon number. The mole percentages are shown as dots, and the dashed line is a best-fit line according to Equation 5.10 starting with C_{17}.

biodegradation. Figure 5.11 shows a plot of the C$_{7+}$ mole percentages (logarithmic scale) versus carbon number for the reservoir fluid in Table 5.18. Due to the biodegradation, the usual linear progression is not seen from C$_7$ to C$_{17}$, but the linear relationship is still there for the heavier hydrocarbons. For such a composition, it is essential to have a compositional analysis for at least C$_{20+}$, as extrapolation from C$_{7+}$ or C$_{10+}$ would give an incorrect picture of how the components in the plus fraction are distributed.

5.8 PC-SAFT CHARACTERIZATION PROCEDURE

The PC-SAFT EoS model presented in Section 4.9 uses three component parameters:

- number of segments (m)
- segment diameter (σ)
- segment energy (ε).

The PC-SAFT component parameters are not uniquely tied to the physical behavior of a component as is the case with T$_c$, P$_c$, and ω in a cubic equation of state. Gross and Sadowski (2001) have for defined components determined m, σ, and ε as a best fit to density and saturation pressure data. These values can be seen in Table 5.19 together with the PC-SAFT values for nC7, c-C$_6$, and benzene.

Tybjerg and Pedersen (2017) have presented the PC-SAFT characterization procedure described in the following, which is a further development of the PC-SAFT characterization procedure of Pedersen et al. (2012). For a given component class of the C$_{7+}$ components (Paraffins (P), Naphthenes (N), and Aromatics (A)), the ε-value for the relevant C$_7$ component in Table 5.19 is used independent

---

TABLE 5.19

Values of m, σ, and ε for Defined Components and for a Paraffinic, a Naphthenic and an Aromatic C$_7$ Component

Component	Short Name	m	σ Å	ε/k K
Nitrogen	N$_2$	1.2053	3.3130	90.96
Carbon dioxide	CO$_2$	2.0729	2.7852	169.21
Hydrogen sulphide	H$_2$S	1.6941	3.0214	226.79
Methane	C$_1$	1.0000	3.7039	150.03
Ethane	C$_2$	1.6069	3.5206	191.42
Propane	C$_3$	2.0020	3.6184	208.11
i-butane	iC$_4$	2.2616	3.7574	216.53
n-butane	nC$_4$	2.3316	3.7086	222.88
i-pentane	iC$_5$	2.5620	3.8296	230.75
n-pentane	nC$_5$	2.6896	3.7729	231.20
n-hexane	C$_6$	3.0576	3.7983	236.77
n-heptane	nC$_7$	3.4831	2.8049	238.40
Cyclo-hexane	c-C$_6$	2.5303	3.8499	278.11
Benzene	Benzene	2.4653	3.6478	287.35

Source: Gross, J. and Sadowski, G., Perturbed-chain SAFT: An equation of state based on perturbation theory for chain molecules, *Ind. Eng. Chem. Res.* 40, 1244–1260, 2001.

of molecular weight; that is, the ε-value for n-heptane is used for all C_{7+} paraffins, the ε-value for cyclo-hexane for all naphthenes, and the ε-value for benzene for all aromatics. It is further assumed that m increases linearly with carbon number and is to match the m-values of the C_7 components in Table 5.19 for a molecular weight of 96. The values used for m and ε for the P, N, and A fractions of the C_{7+} components are shown in Table 5.20.

The m and ε parameters of carbon number fraction i are found as average values of the parameters for the P, N and A component classes:

$$m_i = P\text{-}fraction(i) \times m_{Pi} + N\text{-}fraction(i) \times m_{Ni} + A\text{-}fraction(i) \times m_{Ai} \qquad (5.57)$$

$$\varepsilon_i = P\text{-}fraction(i) \times \varepsilon_{Pi} + N\text{-}fraction(i) \times \varepsilon_{Ni} + A\text{-}fraction(i) \times \varepsilon_{Ai} \qquad (5.58)$$

The parameter σ of each lumped pseudocomponent is found to match density at atmospheric pressure and 15°C. The binary interaction parameters in Table 4.2 used with cubic equations are also found to be appropriate for the PC-SAFT equation.

Table 5.21 shows how m and ε are calculated for a C_9 fraction of a molecular weight of 126.885 with the PNA distribution also given in Table 5.21. The PNA distribution is found using the procedure of Nes and Westerns (1951).

The reservoir fluid composition in Table 5.22 has been characterized for the PC-SAFT equation using the method of Tybjerg and Pedersen. The plus fraction is split, as suggested by Pedersen et al. (1992), and the C_{7+} fractions were lumped into 12 pseudocomponents. The characterized fluid is shown in Table 5.23 and the binary interaction parameters in Table 5.24.

TABLE 5.20

Values of PC-SAFT Parameters m and ε for Paraffinic, Naphthenic and Aromatic C_{7+} Fractions. M Is the Molecular Weight of the Fraction

Component type	m	ε/k K
Paraffinic C_{7+}	$0.036282 \times M$	238.40
Naphthenic C_{7+}	$0.026253 \times M$	278.11
Aromatic C_{7+}	$0.025680 \times M$	287.35

TABLE 5.21

P, N and A Contributions to m and ε of a C_9 Fraction with a Molecular Weight of 126.885

	P	N	A	Total
Fraction	0.4613	0.3351	0.2036	1.0000
m	4.604	3.331	3.258	3.903
ε	238.40	278.11	287.35	261.67

TABLE 5.22

Reservoir Fluid Composition for Which PC-SAFT Characterization Is Shown in Table 5.23.
The C$_{7+}$ Molecular Weight Is 196, and the C$_{7+}$ Density Is 0.841 g/cm^3

Component	Mol%
N$_2$	0.186
CO$_2$	3.709
H$_2$S	2.891
C$_1$	46.479
C$_2$	8.307
C$_3$	5.890
iC$_4$	1.245
nC$_4$	3.079
iC$_5$	1.362
nC$_5$	1.697
C$_6$	2.268
C$_{7+}$	22.887

TABLE 5.23

Reservoir Fluid in Table 5.22 Characterized for the PC-SAFT Equation. NonZero Binary
Interaction Parameters Are Shown in Table 5.24

Component	Mol %	m	σ Å	ε/k K
N$_2$	0.186	1.2053	3.3130	90.96
CO$_2$	3.709	2.0729	2.7852	169.21
H$_2$S	2.891	1.6941	3.021	226.79
C$_1$	46.479	1.0000	3.7039	150.03
C$_2$	8.307	1.6069	3.5206	191.42
C$_3$	5.890	2.0020	3.6184	208.11
iC$_4$	1.245	2.2616	3.7574	216.53
nC$_4$	3.079	2.3316	3.7086	222.88
iC$_5$	1.362	2.5620	3.8296	230.75
nC$_5$	1.697	2.6896	3.7729	231.20
C$_6$	2.268	3.0576	3.7983	236.77
C$_7$	2.748	3.0576	3.7983	236.77
C$_8$	2.418	3.0190	3.8373	258.37
C$_9$	2.128	3.3865	3.8159	257.58
C$_{10}$–C$_{11}$	3.520	3.8681	3.7936	256.28
C$_{12}$–C$_{13}$	2.725	4.4956	3.7740	255.82
C$_{14}$–C$_{15}$	2.110	5.3801	3.7561	255.81
C$_{16}$–C$_{17}$	1.634	6.3250	3.7446	256.22
C$_{18}$–C$_{19}$	1.265	7.2974	3.7378	256.96
C$_{20}$–C$_{23}$	1.738	8.2162	3.7177	256.42
C$_{24}$–C$_{27}$	1.042	9.4273	3.7020	256.73
C$_{28}$–C$_{34}$	0.923	11.1405	3.6860	257.35
C$_{35}$–C$_{80}$	0.635	13.3710	3.6718	258.35

TABLE 5.24

NonZero Binary Interaction Parameters for Use with the PC-SAFT Characterization in Table 5.23

Component	N_2	CO_2	H_2S
CO_2	−0.0315		
H_2S	0.1696	0.0989	
C_1	0.0278	0.1200	0.0800
C_2	0.0407	0.1200	0.0852
C_3	0.0763	0.1200	0.0885
iC_4	0.0944	0.1200	0.0511
nC_4	0.0700	0.1200	0.0600
iC_5	0.0867	0.1200	0.0600
iC_5	0.0878	0.1200	0.0689
iC_5	0.0800	0.1200	0.0500
C_{7+}	0.0800	0.1000	

REFERENCES

Abramowitz, M. and Stegun, I.A., eds., *Handbook of Mathematical Functions*, Dover Publications, Inc., New York, 256–257, 1972.

Boesen, R.R., Lekumjorn, S., Agger, C., Sørensen, H., Chemical reaction equilibrium theory—a novel approach to estimate shale reservoir fluid compositions, SPE-213222-MS, presented at the *Middle East Oil, Gas and Geosciences Show*, Manama, Bahrain, February 19–21, 2023.

Cavett, R.H., Physical data for distillation calculation, vapor-liquid equilibria, *27th Midyear Meeting*, API Division of Refining, San Francisco, CA, May 15, 1964.

Chase, Jr., M.W., Curnutt, J.L., Downey, Jr. J.R. McDonald, R.A., Syverud, A. N., Valenzuela, E.A., JANAF Thermochemical Tables, 1982 Supplement, J. Phys. Chem. Ref. Data 11, 695–940, 1982.

Coats, K.H., Simulation of gas condensate reservoir performance, *J. Pet. Technol.* 37, 1870–1866, 1985.

Cotterman, R.L. and Prausnitz, J.M., Flash calculations for continuous or semicontinuous mixtures by use of an equation of state, *Ind. Eng. Chem. Process Des. Dev.* 24, 434–443, 1985.

Danesh, A., Xu, D., and Todd, A.C., A grouping method to optimize oil description for compositional simulation of gas injection processes, SPE 20745, *Res. Eng.* 343–348, August 1992.

Daubert, T.E., State-of-the-art property predictions, *Hydrocarb. Process.* 107–112, March 1980.

Drohm, J.R. and Schlijper, A.G., An inverse lumping method: Estimating compositional data from lumped information, SPE 14267, presented at *SPE ATCE*, Las Vegas, NV, September 22–25, 1985.

Gross, J. and Sadowski, G., Perturbed-chain SAFT: An equation of state based on perturbation theory for chain molecules, *Ind. Eng. Chem. Res.* 40, 1244–1260, 2001.

Jalowka, J.W. and Daubert, T.E., Group contribution method to predict critical temperature and pressure of hydrocarbons, *Ind. Eng. Chem. Process Des. Dev.* 25, 139–142, 1986.

Kesler, M.G. and Lee, B.I., Improve prediction of enthalpy of fractions, *Hydrocarb. Process.* 55, 153–158, 1976.

Krejbjerg, K. and Pedersen, K.S., Controlling VLLE equilibrium with a cubic EoS in heavy oil modeling, presented at the *57th Annual Technical Meeting of the Petroleum Society (Canadian International Petroleum Conference)*, Calgary, Canada, June 13–15, 2006.

Leibovici, C.F., A consistent procedure for the estimation of properties associated to lumped systems, *Fluid Phase Equilib.* 87, 189–197, 1993.

Leibovici, C.F., Stenby, E., and Knudsen, K., A consistent procedure for pseudo-component delumping, *Fluid Phase Equilib.* 117, 225–232, 1996.

Lomeland, F. and Harstad, O., Simplifying the task of grouping fluid components in compositional reservoir simulation, *SPE Comp App* 7, 38–43, 1995.

Nes, K. and Westerns, H.A. van, *Aspects of the Constitution of Mineral Oils*, Elsevier, New York, 1951.

Newley, T.M.J. and Merrill Jr., R.C. Pseudocomponent selection for compositional simulation, SPE 19638, presented at the *SPE ATCE*, San Antonio, TX, October 8–11, 1989.

Newman, S.A., Correlations evaluated for coal tar liquids, *Hydrocarb. Process.* 133–142, December 1981.

Pearson, K., Contributions to the mathematical theory of evolution, II: Shew variation in homogeneous material, Philos. *Trans. Royal Soc.* 186, 343–414, 1895.

Pedersen, K.S., Blilie, A.L., and Meisingset, K.K., PVT calculations on petroleum reservoir fluids using measured and estimated compositional data for the plus fraction, *Ind. Eng. Chem. Res.* 31, 1378–1384, 1992.

Pedersen, K.S., Leekumjorn, S., Krejbjerg, K., and Azeem, J., Modeling of EOR PVT data using PC-SAFT equation, SPE-162346-PP, presented at the *Abu Dhabi International Petroleum Exhibition & Conference*, Abu Dhabi, November 11–14, 2012.

Pedersen, K.S., Milter, J., and Sørensen, H., Cubic equations of state applied to HT/HP and highly aromatic fluids, *SPE J.* 9, 186–192, 2004.

Pedersen, K.S., Thomassen, P., and Fredenslund, Aa., SRK-EOS calculation for crude oils, *Fluid Phase Equilib.* 14, 209–218, 1983.

Pedersen, K.S., Thomassen, P., and Fredenslund, Aa., Thermodynamics of petroleum mixtures containing heavy hydrocarbons. 1. Phase envelope calculations by use of the Soave–Redlich–Kwong equation of state, *Ind. Eng. Chem. Process Des. Dev.* 23, 163–170, 1984.

Pedersen, K.S., Thomassen, P., and Fredenslund, Aa., Thermodynamics of petroleum mixtures containing heavy hydrocarbons. 3. Efficient flash calculation procedures using the SRK equation of state, *Ind. Eng. Chem. Process Des. Dev.* 24, 948–954, 1985.

Pedersen, K.S., Thomassen, P., and Fredenslund, Aa., Characterization of gas condensate mixtures, *Advances in Thermodynamics*, Vol. 1, Taylor & Francis, New York, 137–152, 1989.

Peneloux, A., Rauzy, E., and Fréze, R., A consistent correction for Redlich-Kwong-Soave volumes, *Fluid Phase Equilib.* 8, 7–23, 1982.

Riazi, M.R., A continuous method for C$_{7+}$ characterization of petroleum fluids, *Ind. Eng. Chem. Res.* 36, 4299–4307, 1997.

Riazi, M.R. and Daubert, T.E., Simplify property predictions, *Hydrocarb. Process.* 115–116, March 1980.

Rønningsen, H.P., Skjevrak, I., and Osjord, E., Characterization of North Sea petroleum fractions: Hydrocarbon group types, density and molecular weight, *Energy Fuels.* 3, 744–755, 1989.

Sim, J.S. and Daubert, T.E., Prediction of vapor-liquid equilibria of undefined mixtures, *Ind. Eng. Chem. Process Des. Dev.* 19, 386–393, 1980.

Sørensen, H., Pedersen, K.S., and Christensen, P.L., Method for generating shale gas fluid composition from depleted sample, presented at the *International Gas Injection Symposium*, Calgary, September 24–27, 2013.

Teja, A.S., Lee, R.J., Rosenthal, R.D., and Anselme, M., Correlations of the critical properties of alkanes and alkanols, *Fluid Phase Equilib.* 56, 153–169, 1990.

Twu, C.H., Prediction of thermodynamic properties of normal paraffins using only normal boiling point, *Fluid Phase Equilib.* 11, 65–81, 1983.

Twu, C.H., An internally consistent correlation for predicting the critical properties and molecular weights of petroleum and coal-tar liquids, *Fluid Phase Equilib.* 16, 137–150, 1984.

Tybjerg, P. and Pedersen, K.S., Reservoir fluid characterization procedure for the PC-SAFT equation of state, SPE-187170-MS, presented at *SPE ATCE*, San Antonio, TX, October 9–11, 2017.

Watanasiri, S., Owens, V.H., and Starling, K.E., Correlations for estimating critical constants, acentric factor, and dipole moment for undefined coal-fluid fractions, *Ind. Eng. Chem. Process Des. Dev.* 24, 294–296, 1985.

Whitson, C.H., Effect of physical properties estimation on equation-of-state predictions, SPE 11200, presented at the *SPE ATCE*, New Orleans, LA, September 26–29, 1982.

Whitson, C.H., Characterizing hydrocarbon plus fractions, *SPE J.* 23, 683–694, 1983.

Whitson, C.H., Andersen, T.F., and Søreide, I., C$_{7+}$ characterization of related equilibrium fluids using distribution, *Advances in Thermodynamics*, Vol. 1, Taylor & Francis, New York, 35–56, 1989.

Zuo, J.Y. and Zhang, D., Plus fraction characterization and PVT data regression for reservoir fluids near critical conditions, SPE 64520, presented at the *SPE Asia Pacific Oil and Gas Conference in Brisbane*, Australia, October 16–18, 2000.

6 Flash and Phase Envelope Calculations

Figure 6.1 illustrates a two-phase pressure–temperature (PT)-flash process. A feed stream consisting of a mixture of N components is led to a flash separator kept at a constant temperature and pressure. Two phases are present in the separator. The gas is let out at the top and the oil at the bottom. If P, T, and component mole fractions in the feed $(z_1, z_2, \ldots, z_N)$ are known, a flash calculation will provide the following results:

1. Number of phases.
2. Molar amounts of each phase. Figure 6.1 uses the term β for the vapor mole fraction.
3. Molar compositions of each phase. In Figure 6.1, the component mole fractions in the gas phase are called $(y_1, y_2, \ldots, y_N)$, and the component mole fractions in the liquid phase are called $(x_1, x_2, \ldots, x_N)$.

As is shown in Equation A.36 in Appendix A, the following relations apply for two phases in equilibrium:

$$\frac{y_i}{x_i} = \frac{\varphi_i^L}{\varphi_i^V} \quad i = 1, 2, \ldots, N \tag{6.1}$$

A material balance for each component yields

$$z_i = \beta y_i + (1-\beta)x_i \quad i = 1, 2, \ldots, N \tag{6.2}$$

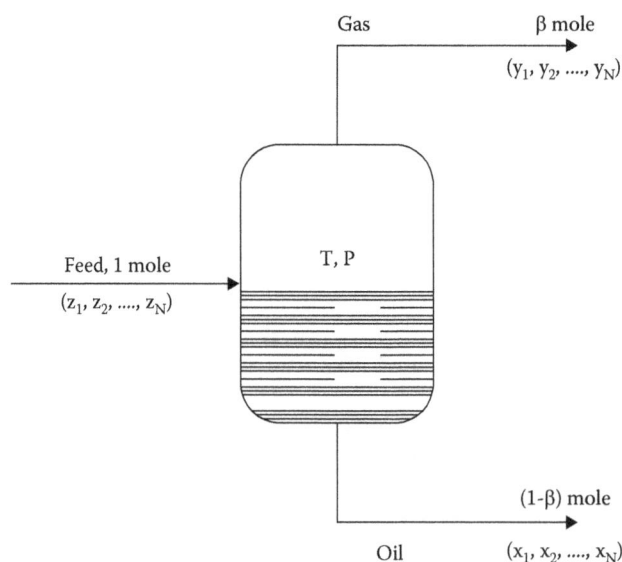

FIGURE 6.1   Principle of pressure–temperature (PT)-flash process for a hydrocarbon reservoir fluid mixture.

          DOI: 10.1201/9780429457418-6

In addition, the component mole fractions must for each phase sum to unity, yielding one additional relation, which is often written in the form suggested by Rachford and Rice (1952):

$$\sum_{i=1}^{N}(y_i - x_i) = 0 \tag{6.3}$$

The preceding equations may be simplified by introducing the equilibrium ratios or K-factors:

$$K_i = \frac{y_i}{x_i} = \frac{\varphi_i^L}{\varphi_i^V} \quad i = 1, 2, \ldots, N \tag{6.4}$$

Using Equation 6.4, Equation 6.2 may be rearranged to give:

$$y_i = \frac{z_i K_i}{1 + \beta(K_i - 1)} \quad i = 1, 2, \ldots, N \tag{6.5}$$

$$x_i = \frac{z_i}{1 + \beta(K_i - 1)} \quad i = 1, 2, \ldots, N \tag{6.6}$$

These 2N equations and Equation 6.3 may be reduced to the following (N + 1) equations:

$$\ln K_i = \ln \varphi_i^L - \ln \varphi_i^V \quad i = 1, 2, \ldots, N \tag{6.7}$$

$$\sum_{i=1}^{N}(y_i - x_i) = \sum_{i=1}^{N} \frac{z_i(K_i - 1)}{1 + \beta(K_i - 1)} = 0 \tag{6.8}$$

With T and P fixed, the number of variables is also (N + 1), these being $(K_1, K_2, \ldots, K_N)$ and β. Before solving Equations 6.7 and 6.8, it is necessary to make sure that there are really two phases present and not just a single gas or a single liquid (oil) phase. The solution of the two equations is further complicated by the fact that the fugacity coefficients entering into Equation 6.7 are functions of the phase compositions resulting from the flash calculation, meaning that the fugacity coefficients have to be determined in an iterative manner. Before dealing with the flash problem in general, some simplified cases will be considered.

6.1 PURE COMPONENT VAPOR PRESSURES FROM CUBIC EQUATIONS OF STATE

Neglecting solid states, a pure component will either form a single-phase gas, a single-phase liquid, or a gas and a liquid phase in equilibrium. For a given temperature, two phases in equilibrium can only exist at the pure component vapor pressure. Vapor pressure curves of methane and benzene can be seen from Figure 1.2. At this pressure, the chemical potential (defined in Appendix A) of the component (called "i" here) is the same in the vapor (V) as in the liquid (L) state:

$$\mu_i^V = \mu_i^L \tag{6.9}$$

For a pure component, equality in chemical potentials also means equality in fugacity coefficients:

$$\varphi_i^V = \varphi_i^L \tag{6.10}$$

Fugacity coefficient expressions for the SRK and PR equations are shown in Equations 4.64 and 4.65.

Pure component vapor pressures may be determined from a cubic equation of state, but in an iterative manner. Figure 6.2 shows gas and liquid Z factor roots for ethane calculated for a temperature of 244 K using the Soave–Redlich–Kwong equation of state. The Z factor roots are calculated from Equation 4.30 solving for the higher (vapor) and lower (liquid) roots. The vapor Z factor curve reaches a maximum at a pressure around 21 bar, indicating that the polynomial in Z has only one real root for higher pressures. The vapor pressure must be found somewhere in the pressure interval for which the compressibility factor polynomial of Equation 4.30 has at least two real roots. It would otherwise be impossible to assign compressibility factors to both phases present at vapor pressure conditions.

Figure 6.3 shows a plot of the fugacity coefficients corresponding to the Z factors in Figure 6.2. Following Equation 6.10, the pure component vapor pressure may in Figure 6.3 be identified as the

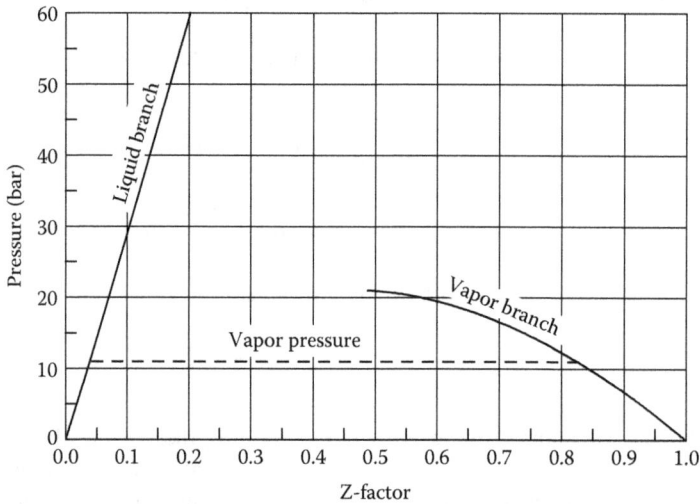

FIGURE 6.2   PZ-curves for ethane at 244 K. The dashed line shows the location of the ethane vapor pressure at the actual temperature. It is found as the point of intersection in Figure 6.3.

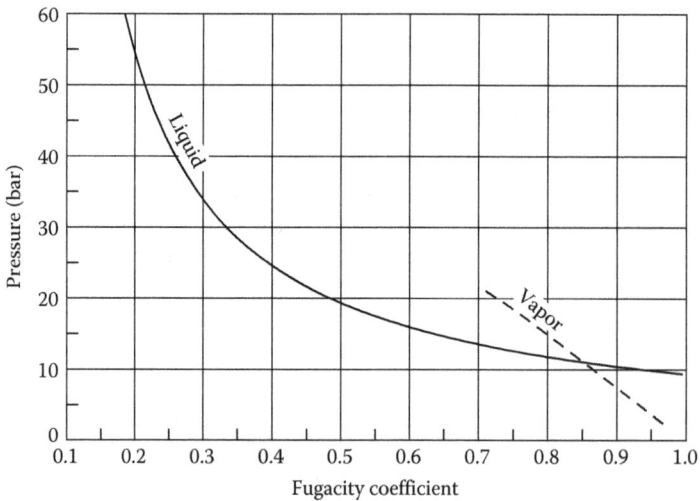

FIGURE 6.3   Vapor and liquid fugacity coefficients of ethane at 244 K. The two curves intersect at the vapor pressure.

pressure for which the liquid and vapor fugacity coefficients are equal, that is, as the point of inter-section between the (full drawn line) liquid and (dashed line) vapor curves. The two curves intersect at a pressure of approximately 11 bar, which is therefore the vapor pressure. This pressure is indicated with a dashed line in Figure 6.2. The part of the liquid Z factor curve in Figure 6.2, which is below the vapor pressure line, does not represent a real physical state. The same applies to the part of the vapor Z factor curve above the vapor pressure line. The part of the vapor Z factor curve below the vapor pressure represents a state of undersaturated gas and the part of the liquid Z factor curve above the vapor pressure a state of undersaturated liquid.

For a saturation point algorithm to be efficient, it is essential to have a reasonable first estimate of the vapor pressure, or at least a pressure estimate for which the polynomial in Z in Equation 4.30 has two real roots. This is needed to be able to compare the fugacity coefficients of the liquid and vapor roots. The correlation of Dong and Lienhard (1986) may be used to provide an initial guess on the reduced vapor pressure $\left(P_r^{sat} = \dfrac{P^{sat}}{P_c} \right)$:

$$\ln\left(P_r^{sat}\right) = 5.37270\left[1-\frac{1}{T_r}\right] + \omega\left(7.49408 - 11.18177\,T_r^3 + 3.68769\,T_r^6 + 17.92998\,\ln T_r\right) \quad (6.11)$$

T_r is the reduced temperature (T/T_c) and ω the acentric factor defined in Equation 1.1.

6.2 MIXTURE SATURATION POINTS FROM CUBIC EQUATIONS OF STATE

If a single component is not at its vapor pressure, only one phase exists at equilibrium. With two or more components present, the determination of the number of phases is less trivial because the equilibrium phase compositions are unknown. Before considering the general PT-flash problem, the problem of locating mixture saturation pressures is first to be considered. For a mixture initially in liquid form, the saturation point pressure is detected as a pressure at which the first gas bubble is seen to form in the liquid. A saturation point of a liquid is therefore also called a *bubble point*. For a mixture initially in gaseous form, the saturation point is the pressure at which the first liquid drop is formed. The saturation point of a gas is therefore also known as a *dew point*. As compared to the general PT-flash calculation, bubble and dew point calculations are simpler, in the sense that one of the equilibrium phases equals the feed composition.

At the bubble point pressure, the vapor mole fraction β equals zero, and Equation 6.8 can be simplified to

$$F = \sum_{i=1}^{N} z_i\left(K_i - 1\right) = 0 \quad (6.12)$$

For a given estimate of the bubble point pressure, a K-factor estimate may be obtained from the K-factor approximation (Wilson 1969):

$$\ln K_i = \ln\frac{P_{ci}}{P} + 5.373(1+\omega_i)\left(1 - \frac{T_{ci}}{T}\right) \quad (6.13)$$

The liquid phase equals the feed composition and an initial estimate of the vapor phase composition at the bubble point may be obtained from Equation 6.5 with K-factors from Equation 6.13.

The bubble point pressure may be determined by following the iterative scheme in Table 6.1.

Bubble and dew points may also be calculated for a specified pressure, in which case the temperature is the unknown parameter to be determined. A dew point temperature may be calculated as outlined in Table 6.2.

TABLE 6.1
Bubble Point Pressure Calculations

1. Provide an estimate of the bubble point pressure
2. Estimate the K-factors using Equation 6.13
3. Estimate vapor phase composition from $y_i^{j+1} = z_i K_i^j, \quad i = 1,2,\ldots,N,$ where j is an iteration counter
4. Calculate the vapor and liquid phase fugacity coefficients $\left(\left(\varphi_i^V, i=1,2,\ldots,N\right) \text{ and } \left(\varphi_i^L, i=1,2,\ldots,N\right)\right)$ using current estimates for bubble point pressure and vapor phase composition. The liquid composition equals the feed composition. SRK and PR fugacity coefficients may be calculated from, respectively, Equations 4.64 and 4.65
5. Calculate new K-factors from Equation 6.7
6. Evaluate $F = \sum_{i=1}^{N} z_i K_i - 1$
7. Evaluate $\dfrac{dF}{dP} = \sum_{i=1}^{N} z_i K_i \left(\dfrac{\partial \ln \varphi_i^L}{\partial P} - \dfrac{\partial \ln \varphi_i^V}{\partial P} \right)$
8. Calculate the (j + 1)th estimate of the bubble point pressure from $P^{j+1} = P^j - \dfrac{F^j}{\dfrac{dF^j}{dP}}$
9. If not converged, return to 3

TABLE 6.2
Dew Point Temperature Calculation

1. Provide an estimate of the dew point temperature
2. Estimate the K-factors using Equation 6.13
3. Estimate liquid phase composition from $x_i^{j+1} = \dfrac{z_i}{K_i^j} \, i = 1,2,\ldots, N,$ where j is an iteration counter
4. Calculate the vapor and liquid phase fugacity coefficients $\left(\left(\varphi_i^V, i=1,2,\ldots, N\right) \text{ and } \left(\varphi_i^L, i=1,2,\ldots, N\right)\right)$ using current estimates for dew point temperature and liquid phase composition. The vapor composition equals the feed composition. SRK and PR fugacity coefficients may be calculated from, respectively, Equations 4.64 and 4.65
5. Calculate new K-factors from Equation 6.7
6. Evaluate $F = \sum_{i=1}^{N} \dfrac{z_i}{K_i} - 1$
7. Evaluate $\dfrac{dF}{dT} = \sum_{i=1}^{N} \dfrac{z_i}{K_i} \left(\dfrac{\partial \ln \varphi_i^V}{\partial T} - \dfrac{\partial \ln \varphi_i^L}{\partial T} \right)$
8. Calculate the (j + 1)th estimate of the dew point temperature from $T^{j+1} = T^j - \dfrac{F^j}{\dfrac{dF^j}{dT}}$
9. If not converged, return to 3

Though in principle simpler than PT-flash calculations, bubble and dew point calculations are complicated by the fact that it is not generally known in advance whether the mixture considered really has a bubble or a dew point at the specified P or T. Figure 1.6 shows a phase envelope for a natural gas mixture. The bubble point line ends in the critical point (CP) at a temperature of around −60°C. A bubble point calculation for a higher temperature should therefore give the answer that no bubble point can be located. It can, however, be quite hard to distinguish cases with no saturation point from cases for which the saturation point calculation is causing numerical problems.

Figure 1.6 also reveals that the natural gas considered has two dew point pressures in a temperature interval above the critical temperature. This may cause convergence problems in a saturation point calculation, and either the upper or lower dew point will be located, at best. A phase envelope calculation as outlined in Section 6.5 will track all the saturation points existing for a given temperature or pressure.

6.3   FLASH CALCULATIONS

Michelsen and Mollerup (2007) have given a very thorough description of flash calculation techniques. They have also dealt with how to derive the thermodynamic quantities and derivatives needed to make a flash algorithm fast and robust.

6.3.1   STABILITY ANALYSIS

A flash calculation presents the problem that the number of phases is generally not known in advance. An important element of a flash calculation is therefore the determination of the number of phases present. This may be accomplished by carrying out a stability analysis (Michelsen 1982a).

As explained in Appendix A, a closed system will try to arrange its molecules in the position that minimizes its Gibbs free energy, G. Consider two samples of pure components, named 1 and 2, being introduced into a closed cell kept at a fixed pressure and temperature. The two samples will mix if the mixing process leads to a decrease in G. Figure 6.4 exemplifies ΔG of mixing (ΔG^{mix}) for a binary mixture. If the two substances are miscible in all proportions, the ΔG of the mixing curve may look like the one marked I. For a zero-mole fraction of component 1, the "mixture" will consist of pure 2, and the ΔG of mixing will be zero. With component 1 introduced in a nonzero amount, the ΔG of mixing will initially decrease and subsequently pass through a minimum. For pure component 1, the ΔG of mixing is again zero. Consider a mixture with a mole fraction of component 1 equal to x_1^m. Imagine that this mixture splits into two phases, A and B, in which the mole fractions of component 1 are x_1^A and x_1^B, respectively. After this phase split, the total system will have a ΔG of mixing determined by the intersection between a vertical line through x_1^m (not shown in figure) and the dashed line connecting A and B on curve I. This phase split will lead to an increase in ΔG of mixing and will therefore not take place. The mixture is said to be stable. This will be true for any mixture on curve I.

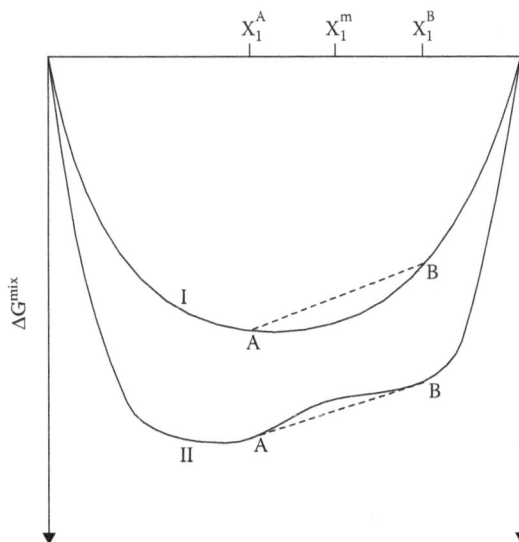

FIGURE 6.4   Principle of stability analysis for a binary mixture.

The curve marked II illustrates a situation with limited miscibility between the two substances. Consider again a mixture with a mole fraction of component 1 equal to x_1^m. Imagine that this mixture splits into two phases, A and B, and that the mole fractions of component 1 in the two phases are x_1^A and x_1^B, respectively. Similar to the situation on curve I, after the phase split, the total system will have a ΔG of mixing determined by the intersection point between a vertical line through x_1^m and the dashed line connecting A and B on curve II. In this case, the phase split leads to a decrease in ΔG of mixing. In other words, a mixture with the mole fraction of component 1 equal to x_1^m will spontaneously split into two separate phases, A and B. The mole fraction of component 1 in the two phases will be x_1^A and x_1^B, respectively. The mixture is said to be unstable. The ratio

$$\frac{x_1^m - x_1^A}{x_1^B - x_1^m}$$

gives the relative molar amounts formed of phases B and A.

The equation for the tangent to the ΔG of mixing curve in point A is

$$\Delta G^{mix}(x_1) = \Delta G^{mix}\left(x_1^A\right) + \left[\frac{d\left(\Delta G^{mix}\left(x_1^A\right)\right)}{dx_1}\right]\left(x_1 - x_1^A\right) \tag{6.14}$$

and the equation for the tangent to the ΔG of mixing curve in point B is

$$\Delta G^{mix}(x_1) = \Delta G^{mix}\left(x_1^B\right) + \left[\frac{d\left(\Delta G^{mix}\left(x_1^B\right)\right)}{dx_1}\right]\left(x_1 - x_1^B\right) \tag{6.15}$$

The derivative of ΔG of mixing with respect to the mole fraction of component 1 may be expressed as

$$\frac{d\Delta G^{mix}}{dx_1} = \frac{\partial G}{\partial n_1}\frac{dn_1}{dx_1} + \frac{\partial G}{\partial n_2}\frac{dn_2}{dx_1} = \mu_1 - \mu_2 \tag{6.16}$$

where n_1 and n_2 are the moles of components 1 and 2, respectively, and μ_1 and μ_2 are the chemical potentials of components 1 and 2 evaluated in the same point as the derivative of ΔG of mixing. The equation for the tangent to the ΔG of mixing curve in point A may then be rewritten to

$$\Delta G^{mix}(x_1) = \Delta G^{mix}\left(x_1^A\right) + \left(\mu_1^A - \mu_2^A\right)\left(x_1 - x_1^A\right) \tag{6.17}$$

or

$$\begin{aligned}\Delta G^{mix}(x_1) &= \mu_1^A x_1^A + \mu_2^A\left(1 - x_1^A\right) + \left(\mu_1^A - \mu_2^A\right)\left(x_1 - x_1^A\right)\\ &= \mu_1^A x_1 + \mu_2^A\left(1 - x_1\right)\end{aligned} \tag{6.18}$$

and the equation for the tangent to the ΔG of mixing curve in point B to

$$\Delta G^{mix}(x_1) = \mu_1^B x_1 + \mu_2^B(1 - x_1) \tag{6.19}$$

In case the two tangents coincide, the following relation will apply:

$$\mu_1^A x_1 + \mu_2^A(1 - x_1) = \mu_1^B x_1 + \mu_2^B(1 - x_1) \tag{6.20}$$

This relation will only be true in general if the chemical potentials of the two components are the same in positions A and B. Equality in the chemical potentials of each component implies phase equilibrium, and it can be concluded that two phases in equilibrium are located on the common tangent to the ΔG of mixing curve.

This observation may be generalized to a criterion saying that a mixture will remain single phase (be stable) if the tangent to the ΔG of mixing curve in the point of the feed composition does not intersect the ΔG of mixing curve anywhere. If it intersects the ΔG of mixing curve, two or more phases will be present.

Figure 6.5 shows a phase envelope for a mixture consisting of 40 mole percentage methane (C_1) and 60 mole percentage carbon dioxide (CO_2). The phase envelope is calculated using the Soave–Redlich–Kwong equation of state (Equation 4.20). From Figures 6.6 through 6.8, it is possible to study the ΔG of mixing for this system at different conditions. Figure 6.6 shows ΔG^{mix} for $-42°C$

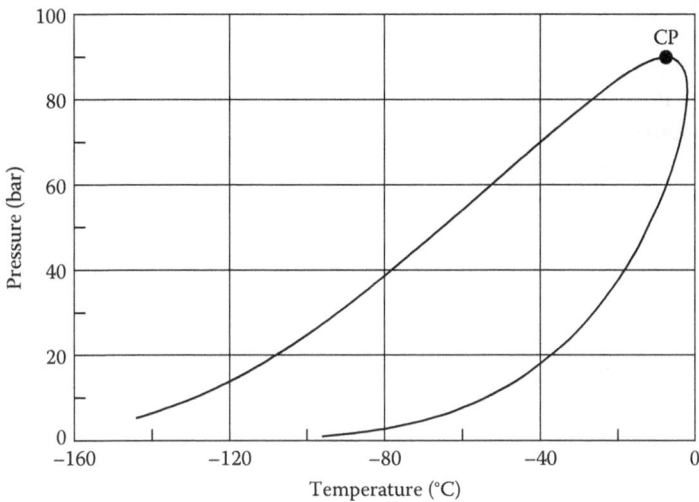

FIGURE 6.5 Phase envelope for mixture consisting of 40 mole percentage C_1 and 60 mole percentage CO_2. CP is the critical point.

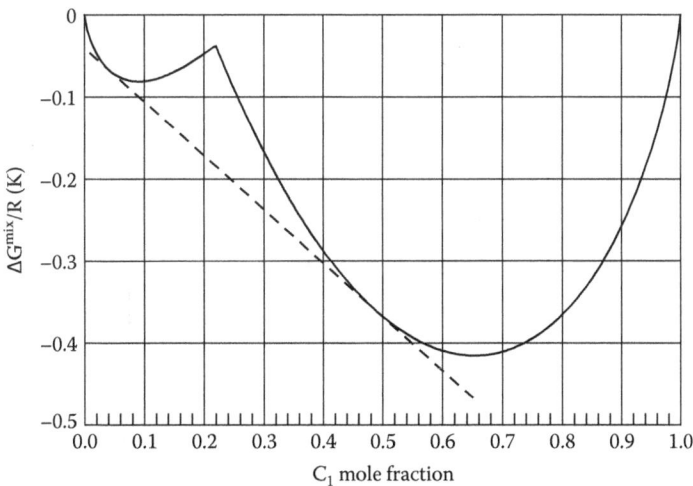

FIGURE 6.6 ΔG^{mix} for binary mixtures of C_1 and CO_2 at $-42°C$ and 20 bar.

and 20 bar. From the phase envelope in Figure 6.5, it is seen that the mixture at these conditions will split into two phases. A PT-flash calculation for −42°C and 20 bar gives the results presented in Table 6.3. The methane mole fractions contained in the liquid and gas phases in Table 6.3 are seen to be identical to the methane mole fractions in Figure 6.6, for which the ΔG^{mix} curve has a common tangent. It is further possible to conclude from Figure 6.6 that a binary mixture of C_1 and CO_2 containing more than approximately 49 mole percentage of methane will be single phase at −42°C and 20 bar. For higher C_1 mole fractions, the tangents to the ΔG^{mix} curve will not intersect with the curve.

Figure 6.7 shows the ΔG of mixing curve of C_1–CO_2 mixtures for −30°C and 26 bar. The tangent to the curve for a C_1 mole fraction of 0.40 (~40 mole percentage) does not intersect the curve anywhere, indicating that a considered mixture is single phase at these conditions. This is in accordance with the phase envelope in Figure 6.5.

Figure 6.8 shows the ΔG of mixing curve of C_1–CO_2 mixtures for −7.6°C and 90 bar, at which conditions a mixture of 40 mole percentage of C_1 and 60 mole percentage of CO_2 is at its critical point (CP). The ΔG^{mix} is very flat in the vicinity of a C_1 mole fraction of 0.40 and almost coincides with the tangent to the curve in this point. At the CP, the gas and liquid phases become equal. When approaching the CP from the two-phase side, the two minima in the ΔG^{mix} curve such as those in Figure 6.6 will approach each other and finally merge right in the CP. When moving away from the CP on the single-phase side, the flat part of the ΔG of mixing curve will gradually become more rounded and take a shape similar to that in Figure 6.7.

TABLE 6.3

Flash Calculation Results for Mixture of C_1 and CO_2 at −42°C and 20 bar

Component	Feed (Mol%)	Liquid (Mol%)	Gas (Mol%)
C_1	40.0	4.2	48.8
CO_2	60.0	95.8	51.2
Phase mol%	100.0	80.3	19.7

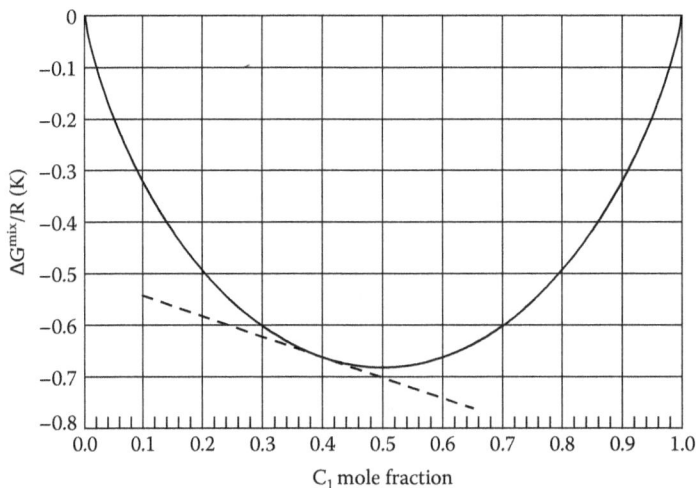

FIGURE 6.7 ΔG^{mix} for binary mixtures of C_1 and CO_2 at −30°C and 26 bar.

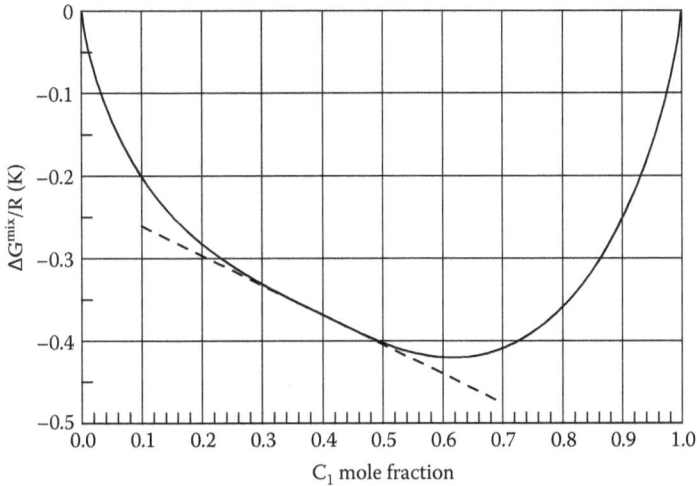

FIGURE 6.8 ΔG^{mix} for binary mixtures of C_1 and CO_2 at $-7.6°C$ and 90 bar. In these conditions, a mixture of 40 mol% C_1 and 60 mol% CO_2 is at its critical point.

The stability analysis as outlined in the preceding text for a binary mixture can be extended to multicomponent mixtures. For a multicomponent mixture, the ΔG of mixing curve will change to a ΔG of mixing surface and the tangent line to a tangent plane. A mixture will remain single phase if the tangent plane to the ΔG of mixing surface in the point of a molar composition does not intersect the ΔG of mixing surface anywhere. Intersection with the ΔG of mixing surface indicates the presence of two or more phases. Consider the tangent plane to the ΔG of mixing surface in the point of a feed molar composition equal to $(z_1, z_2, \ldots, z_N)$. The distance (tangent plane distance or TPD) from this plane to the ΔG of mixing surface evaluated for a trial y-phase $(y_1, y_2, \ldots, y_N)$ equals

$$\text{TPD}(y_1, y_2, \ldots, y_N) = \sum_{i=1}^{N} y_i \left(\mu_i^{y-\text{phase}} - \mu_i^{\text{feed}} \right) \tag{6.21}$$

If the tangent plane distance is non-negative for all possible trial phases $(y_1, y_2, \ldots, y_N)$, the feed mixture will only form one phase. If on the other hand a trial phase can be located that gives a negative value of TPD, the mixture will split into two or more phases.

6.3.2 Solving the Flash Equations

If the stability analysis reveals that two phases are present, the ratios between the component mole fractions in the trial phase $(y_1, y_2, \ldots, y_N)$ and in the feed mixture $(z_1, z_2, \ldots, z_N)$ may be used as an initial estimate of the K-factors. The Rachford–Rice equation

$$F(\beta) = \sum_{i=1}^{N} \frac{z_i (K_i - 1)}{1 + \beta (K_i - 1)} = 0 \tag{6.22}$$

is solved for the vapor mole fraction β and the following iteration scheme may be followed:

$$\beta_{j+1} = \beta_j - \frac{F_j}{\left(\dfrac{dF}{d\beta} \right)_j} \tag{6.23}$$

where j is an iteration counter. The derivative of F with respect to β equals

$$\frac{dF}{d\beta} = -\sum_{i=1}^{N} \frac{z_i(K_i-1)^2}{(1+\beta(K_i-1))^2} \quad (6.24)$$

Having determined β corresponding to the assumed K-factors, new estimates of the phase mole fractions may be determined from Equations 6.5 and 6.6. The fugacity coefficients of these compositions are determined from the cubic equation of state, and a new K-factor estimate is obtained from Equation 6.4. A new β is determined using Equations 6.22 through 6.24. The successive substitution outlined here can be quite time consuming, especially for near-critical mixtures, that is, mixtures for which all the K-factors approach 1. Convergence may be accelerated as outlined by Michelsen (1982b, 1998).

For some systems, as outlined by Heidemann and Michelsen (1995), successive substitution may also lead to convergence problems, and different techniques must be applied. Successive substitution is, for example, inappropriate when gas hydrate phases are to be considered. Gas hydrate flash techniques will be further dealt with in Chapter 14.

Much computer time can be saved in a two-phase flash calculation using a cubic equation of state with all binary interaction coefficients set to zero. Irrespective of the number of components, a flash calculation with no nonzero interaction parameters can be reduced to involve only three parameters (Pedersen et al. 1985; Michelsen 1986a). Time can also be saved by dividing the component pairs into those with $k_{ij} = 0$ and those with $k_{ij} \neq 0$ (Hendriks 1987; Jensen and Fredenslund 1987).

In transient compositional simulations, as in reservoir simulations or in dynamic flow or process simulations, the flash calculation time can be reduced considerably by taking advantage of the fact that phase compositions may only change moderately in time and position. Suppose the flash result at position x is a single-phase liquid at time t, and the liquid is highly undersaturated. Unless pressure, temperature, or composition change substantially from time t to t + Δt, it is unlikely that a gas phase will be present at position x at time t + Δt. Rasmussen et al. (2006) have suggested a procedure for keeping track of the degree of undersaturation, and as long as the system is clearly single phase, the fairly time-consuming stability analysis can be skipped. Rasmussen et al. also outline procedures for saving computation time in the two-phase region by using the flash result from the previous time step as initial estimate.

6.3.3 MULTIPHASE PT-FLASH

The stability analysis outlined in Section 6.3.1 may be extended to test for the possible presence of three or more phases (Michelsen 1982a). The complexity of the stability analysis is somewhat increased when more than two phases are to be considered. The upper part of Figure 6.9 shows a system that, as an anticipated single phase, has liquid-like properties and a Gibbs free energy of G^0. The system may be checked for stability by splitting off a vapor phase. If the total Gibbs energy of the system can be reduced by splitting the mixture into a gas–liquid system ($G' < G^0$), the correct flash solution will consist of two or more phases. In a multiphase flash calculation, the stability analysis is continued to look for a second liquid phase, which may reveal that Gibbs energy can be further reduced by splitting the mixture into a gas–liquid–liquid system ($G'' < G'$). It could also be that a liquid–liquid system with no gas phase would have an even lower Gibbs energy. This is investigated as part of a multiphase flash calculation but not dealt with in Figure 6.9. The stability analysis may be continued searching for a third liquid phase, and so on. As indicated in Figure 6.9, the fourth phase could also be a solid phase—for example, solid wax, as is further dealt with in Chapter 11. A solid phase will form if this reduces the Gibbs free energy as compared with the three-phase solution ($G''' < G''$). The lower part of Figure 6.9 illustrates a similar series of stability analyses starting with a mixture, which as a single phase has vapor-like properties.

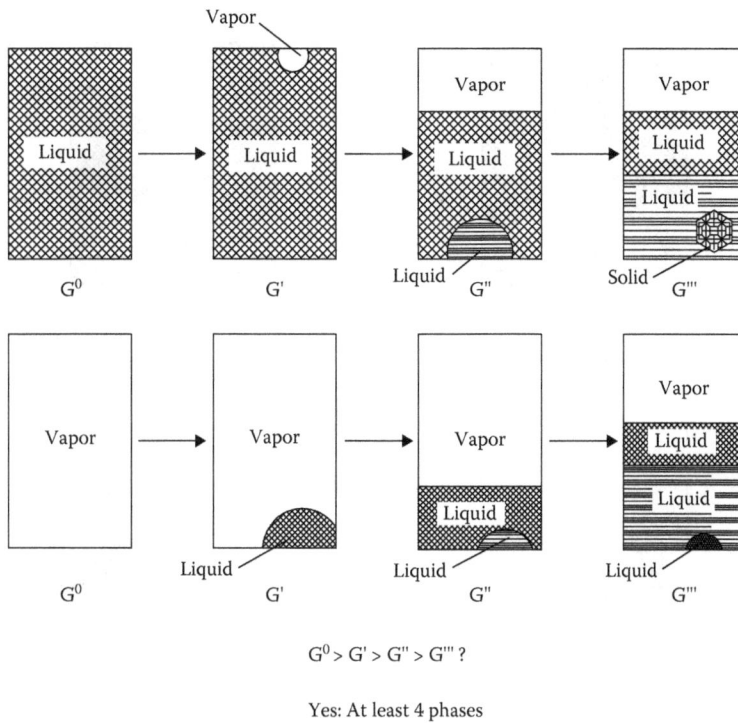

FIGURE 6.9 Stability analysis carried out as part of general multiphase flash. G is the total Gibbs free energy. $G = G^0$ for one phase, $G = G'$ for two phases, and so on.

The calculation time needed for multiphase flash calculations may be reduced if, for example, it is known that one of the phases is likely to be aqueous. The upper part of Figure 6.10 shows a mixture of hydrocarbon and water (aqueous). An initial stability analysis has revealed that a water-rich liquid phase will split off. As a single phase, the remaining hydrocarbon mixture has liquid-like properties. If no other phases are considered than one vapor, one hydrocarbon liquid, and one aqueous phase, all that is needed in Figure 6.10 is a stability analysis for a possible hydrocarbon vapor phase. If the Gibbs free energy (G') of the three-phase system is lower than that of the two-phase system (G^0), the flash solution is a three-phase system. The lower part of Figure 6.10 shows a mixture consisting of water and a vapor-like hydrocarbon mixture that is being tested for stability with respect to the precipitation of a liquid hydrocarbon phase. A flash calculation in which the phases considered are limited to gas, hydrocarbon liquid, and liquid aqueous water, and in which it is further assumed that only one phase of a given type can exist, is much simpler and therefore less time consuming than a general multiphase flash calculation. It is therefore recommended to take advantage of any previous knowledge about the system when designing or choosing the flash algorithm for a particular purpose.

For a system consisting of J phases, the analogue of the Rachford–Rice equation (Equation 6.3) is

$$\sum_{i=1}^{N} \frac{z_i \left(K_i^m - 1 \right)}{H_i} = 0 \; m = 1, 2, \; \ldots, \; J-1 \tag{6.25}$$

where

$$H_i = 1 + \sum_{m=1}^{J-1} \beta^m \left(K_i^m - 1 \right) \tag{6.26}$$

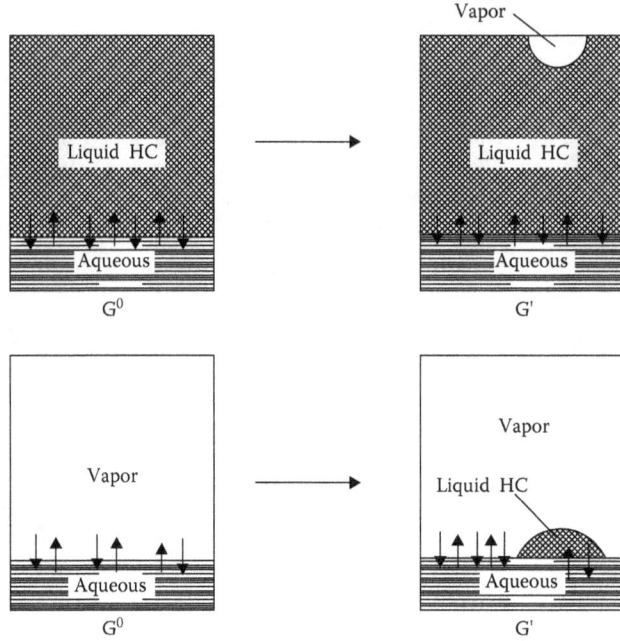

FIGURE 6.10 Stability analysis carried out as part of three-phase flash calculation where the phases considered are gas, hydrocarbon liquid, and aqueous liquid (no more than one phase of each type). G is the total Gibbs free energy. $G = G^0$ for one hydrocarbon (HC) phase, and $G = G'$ for two hydrocarbon phases.

β^m is the molar fraction of phase m. K_i^m equals the ratio of mole fractions of component i in phase m and phase J. With initial estimates of the K-factors, the molar phase fractions β^1 to β^J can be found from Equations 6.25 and 6.26, using a similar procedure as that outlined for two phases in Section 6.3.2. The phase compositions may subsequently be found from

$$y_i^m = \frac{z_i K_i^m}{H_i} \quad i = 1, \ 2, \ ..., \ N; \ m = 1, \ 2, \ ..., \ J-1 \tag{6.27}$$

$$y_i^J = \frac{z_i}{H_i} \quad i = 1, \ 2, \ ..., \ N \tag{6.28}$$

where y_i^m and y_i^J are the mole fractions of component i and phase m and J, respectively. As is the case with two-phase flash calculations, multiphase flash calculations may be accelerated considerably (Michelsen 1982b).

In reservoir and process simulations, water-free oil and gas mixtures are usually regarded as systems that are unlikely to form more than two fluid phases. A multiphase flash algorithm may be used to investigate the validity of this assumption. Table 6.4 shows the composition of a light gas condensate. This gas condensate mixture was characterized for the PR equation using the characterization procedure of Pedersen et al. in Chapter 5. The characterized composition is shown in Table 6.5. A multiphase flash calculation is performed using the PR equation of state. Table 6.6 shows the results of a PT-flash for this mixture carried out for 52 bar and −72°C. It is seen that the

TABLE 6.4
Molar Composition of Gas Condensate

Component	Mol%	Molecular Weight	Density (g/cm³) at 15°C and 1.01 bar
N_2	0.08	—	—
CO_2	2.01	—	—
C_1	82.51	—	—
C_2	5.81	—	—
C_3	2.88	—	—
iC_4	0.56	—	—
nC_4	1.24	—	—
iC_5	0.52	—	—
nC_5	0.60	—	—
C_6	0.72	—	—
C_{7+}	3.06	140.3	0.774

TABLE 6.5
Gas Condensate Mixture in Table 6.4 after Characterization for PR Equation of State

Component	Mol%	T_c (°C)	P_c (bar)	Acentric Factor
N_2	0.08	−147.0	33.94	0.040
CO_2	2.01	31.1	73.76	0.225
C_1	82.51	−82.6	46.00	0.008
C_2	5.81	32.3	48.84	0.098
C_3	2.88	96.7	42.46	0.152
iC_4	0.56	135.0	36.48	0.176
nC_4	1.24	152.1	38.00	0.193
iC_5	0.52	187.3	33.84	0.227
nC_5	0.60	196.5	33.74	0.251
C_6	0.72	234.3	29.69	0.296
C_7–C_9	1.66	280.4	26.72	0.373
C_{10}–C_{13}	0.91	352.5	21.29	0.518
C_{14}–C_{55}	0.49	473.1	16.67	0.803

TABLE 6.6
Multiphase PT-Flash Calculation for 52 bar and –72°C for Gas Condensate Mixture in Table 6.5 Using PR Equation of State

Component	Feed (Mol%)	Gas (Mol%)	Liquid I (Mol%)	Liquid II (Mol%)
N_2	0.08	0.18	0.08	0.05
CO_2	2.01	1.08	1.88	2.36
C_1	82.51	96.45	87.95	75.66
C_2	5.81	1.86	5.28	7.28
C_3	2.88	0.33	2.17	4.00

(Continued)

TABLE 6.6 *(Continued)*

Multiphase PT-Flash Calculation for 52 bar and −72°C for Gas Condensate Mixture in Table 6.5 Using PR Equation of State

Component	Feed (Mol%)	Gas (Mol%)	Liquid I (Mol%)	Liquid II (Mol%)
iC_4	0.56	0.03	0.38	0.81
nC_4	1.24	0.05	0.76	1.83
iC_5	0.52	0.01	0.28	0.79
nC_5	0.60	0.01	0.29	0.93
C_6	0.72	0.00	0.29	1.14
$C_7–C_9$	1.66	0.00	0.48	2.73
$C_{10}–C_{13}$	0.91	0.00	0.14	1.54
$C_{14}–C_{55}$	0.49	0.00	0.01	0.87
Phase mole percentage	100	17.51	26.15	56.33

TABLE 6.7

Molar Composition of Oil Mixture for Which Phase Equilibrium Studies Have Been Carried Out with CO_2 Mixed In. The C_{7+} Molecular Weight Is 271 and the C_{7+} Density Is 0.886 g/cm³

Component	Mol%
N_2	0.159
CO_2	0.168
C_1	24.845
C_2	5.125
C_3	6.922
iC_4	2.014
nC_4	4.895
iC_5	2.752
nC_5	2.054
C_6	3.422
C_{7+}	47.644

mixture at these conditions splits into one gas phase and two liquid phases. A temperature as low as −72°C is usually not experienced during oil and gas production. The presence of a third phase at these conditions therefore has little practical importance. As is outlined in Section 6.5, the presence of a three-phase region does, however, have some implications for phase envelope calculations.

Lindeloff et al. (2013) have studied systems consisting of a Middle East oil mixed with CO_2 at both reservoir temperature and lower temperatures. The oil composition is shown in Table 6.7. It was characterized for the SRK-Peneloux equation of state, and the characterized fluid composition is shown in Table 6.8 with nonzero binary interaction parameters in Table 6.9. Mixed with CO_2 the fluid at a temperature of 26.7°C splits into three phases, two liquid phases and a gas phase. The experimental conditions and the experimental and simulated phase compositions can be seen in Table 6.10.

Reservoir simulation studies will usually only consider two non-aqueous phases, gas and oil. The existence of a third phase may disturb a PT-flash calculation carried out with a two-phase flash algorithm. A two-phase algorithm will not search for a third phase. Once the stability analysis reveals

TABLE 6.8

EoS Model for the Oil Composition in Table 6.7 Developed for the Volume-corrected SRK Equation of State. NonZero Binary Interaction Parameters Can Be Seen from Table 6.9

Component	Mol%	Molecular Weight	T_c °C	P_c bar	Acentric Factor	Volume Correction cm³/mol
CO_2	0.168	44.01	31.05	73.76	0.2250	3.03
N_2+C_1	25.02	16.119	−83.36	45.87	0.0084	0.63
C_2+C_3	12.088	38.146	78.14	44.59	0.1339	4.03
C_4+C_6	15.146	68.427	182.66	34.11	0.2345	11.22
C_7-C_{14}	23.626	137.938	326.83	20.71	0.6041	35.43
$C_{15}-C_{23}$	12.827	252.796	497.12	17.15	0.8897	32.41
$C_{24}-C_{47}$	7.546	427.687	691.57	16.11	1.2105	−65.48
$C_{48}-C_{80}$	3.578	891.166	893.08	15.99	1.0960	−224.19

TABLE 6.9

NonZero Binary Interaction Parameters for Use with the EoS Model in Table 6.8

k_{ij}	CO_2	N_2+C_1
N_2+C_1	0.1190	
C_2+C_3	0.1200	0.0004
C_4+C_6	0.1200	0.0005
C_7-C_{14}	0.1339	0.0005
$C_{15}-C_{23}$	0.0762	0.0005
$C_{24}-C_{47}$	0.0849	0.0005
$C_{48}-C_{80}$	0.1081	0.0005

TABLE 6.10

Experimental and Simulated Phase Compositions in the Region with Observed VLLE Behavior for Mixtures of CO_2 and the Oil Composition in Table 6.7 Characterized as Shown in Table 6.8. The 1st Section of the Table is for a Molar Ratio of CO_2/Oil of 2.25, a Pressure of 76.9 bar and a Temperature of 26.7°C. The 2nd Section of the Table is for a Molar Ratio of CO_2/Oil of 2.25, a Pressure of 74.1 bar and a Temperature of 26.7°C.

Component	Liq 1 Exp Mol%	Liq 1 Sim Mol%	%Dev Mol%	Liq 2 Exp Mol%	Liq 2 Sim Mol%	%Dev Mol%	Vap Exp Mol%	Vap Sim Mol%	%Dev Mol%
CO_2	54.99	57.83	5.2	88.46	88.57	0.1	86.28	86.63	0.4
N_2+C_1	4.26	4.274	0.3	5.61	6.602	17.7	11.68	11.649	−0.3
C_2+C_3	1.61	1.672	3.9	1.31	1.202	−8.2	1.07	0.946	−11.6
C_4-C_6	4.06	4.369	7.6	2.24	1.788	−20.2	0.8	0.68	−15.0
C_{7+}	35.08	31.855	−9.2	2.38	1.84	−22.7	0.17	0.095	−44.1

(Continued)

TABLE 6.10 (*Continued*)

Experimental and Simulated Phase Compositions in the Region with Observed VLLE Behavior for Mixtures of CO_2 and the Oil Composition in Table 6.7 Characterized as Shown in Table 6.8. The 1st Section of the Table is for a Molar Ratio of CO_2/Oil of 2.25, a Pressure of 76.9 bar and a Temperature of 26.7°C. The 2nd Section of the Table is for a Molar Ratio of CO_2/Oil of 2.25, a Pressure of 74.1 bar and a Temperature of 26.7°C.

Component	Liq 1 Exp Mol%	Liq 1 Sim Mol%	%Dev Mol%	Liq 2 Exp Mol%	Liq 2 Sim Mol%	%Dev Mol%	Vap Exp Mol%	Vap Sim Mol%	%Dev Mol%
CO_2	53.47	58.47	9.4	89.48	89.73	0.3	88.06	88.42	0.4
N_2+C_1	4.47	3.52	−21.3	4.38	5.32	21.5	9.99	9.96	−0.3
C_2+C_3	1.78	1.61	−9.4	1.25	1.158	−7.4	1.02	0.90	−11.8
C_4-C_6	4.07	4.41	8.3	2.22	1.831	−17.5	0.83	0.64	−22.5
C_{7+}	36.21	31.99	−11.7	2.67	1.964	−26.4	0.10	0.08	−21.0

Source: Lindeloff, N., Mogensen, K. M., Pedersen, K. S., and Tybjerg, P., Investigation of miscibility behavior of CO_2 rich hydrocarbon systems—with application for gas injection EOR, SPE 166270-MS, presented at *SPE ATCE*, New Orleans, LA, September 30–October 2, 2013.

that more than one phase exists, the algorithm will start looking for a two-phase solution, that is, a solution fulfilling Equation 6.1. Any such solution will be regarded as the correct one. For a mixture that actually forms three phases, there may be more than one solution to Equation 6.1, and the iteration for a two-phase solution may lead to oscillations between two nearby phase compositions, both fulfilling Equation 6.1. If the flash algorithm converges to a result, it will be false.

Paterson et al. (2018) have presented a multiflash algorithm based on the so-called RAND method, which is one well suited when not only phase equilibria but also chemical equilibria are to be considered.

6.3.4 THREE PHASE PT-FLASH WITH A PURE WATER PHASE

Water is often present as a third phase during oil and gas production. The solubility of the gas and oil constituents in the water phase is usually quite limited. It is therefore often acceptable with an approximate PT-flash calculation considering the water phase to be pure water. In that case, the PT-flash calculation can be much simplified (Michelsen 1981) as compared with a general multiphase flash calculation, in which all components can be present in all phases. This is illustrated in Figure 6.11. It is similar to Figure 6.10, but the calculation is further simplified by not having to consider the solubility of other components in the water phase.

In a three-phase PT-flash with a check for pure water, the feed is initially assumed to form only one mixed phase (gas or oil with some dissolved water). It is tested whether pure water will separate from this imaginary phase. This test is done by comparing the chemical potential of water in a pure water phase with the chemical potential of water in the feed. The expression for the chemical potential of pure water takes the form

$$(\mu_w)_{pure} = \mu_w^0 + RT(\ln f_w)_{pure} = \mu_w^0 + RT(\ln P + \ln(\varphi_w)_{pure}) \tag{6.29}$$

and the chemical potential of water in the feed can be expressed as

$$(\mu_w)_{mix} = \mu_w^0 + RT(\ln P + \ln(\varphi_w)_{mix} + \ln z_w) \tag{6.30}$$

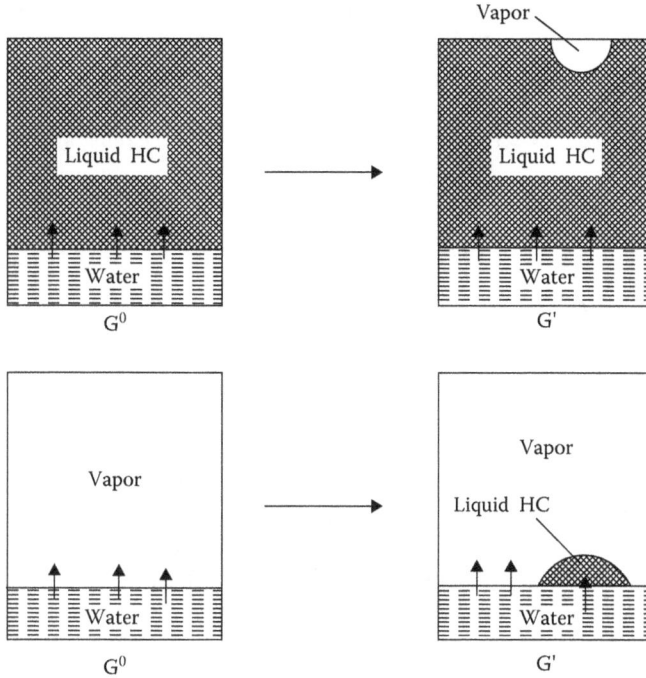

FIGURE 6.11   Stability analysis carried out as part of three-phase flash calculation where the phases considered are gas, hydrocarbon liquid, and pure liquid water (no more than one phase of each type). G is the total Gibbs energy. $G = G^0$ for one hydrocarbon phase, and $G = G'$ for two hydrocarbon phases.

z_w is the mole fraction of water in the total mixture. The terms *chemical potential*, *fugacity*, and *fugacity coefficients* are explained in Appendix A. A pure liquid water phase will precipitate if water has a lower chemical potential in pure form than mixed into the feed composition:

$$(\mu_w)_{pure} < (\mu_w)_{mix} \qquad (6.31)$$

This relation may be rewritten in terms of fugacity coefficients

$$\ln(\varphi_w)_{mix} + \ln z_w > \ln(\varphi_w)_{pure} \qquad (6.32)$$

If liquid water does precipitate, the mole fraction x_w of water in the mixed (hydrocarbon + water) phase is found from the following equation:

$$\ln(\varphi_w)_{mix} + \ln x_w - \ln(\varphi_w)_{pure} = 0 \qquad (6.33)$$

The remaining water is found in a pure liquid water phase.

A two-phase flash calculation is carried out for the remaining mixed hydrocarbon phase with the mole fraction of water from Equation 6.33. If this mixture splits into two phases, the amount of pure water is subsequently adjusted, considering that Equation 6.33 must be fulfilled for each mixed phase.

When a pure water phase does not separate from the feed mixture, that is, Equation 6.32 is not fulfilled, a usual two-phase flash calculation is performed on the total feed mixture. In each iterative step for the equilibrium phase compositions, it is checked whether a pure water phase will form.

6.3.5 Pure Inorganic Solids

Most reservoir fluids and natural gas mixtures contain carbon dioxide (CO_2) and may also carry water (H_2O). CO_2, H_2O and other inorganics may precipitate as pure solid phases if the fluid is cooled. The solid form of water is ice, but when gas components are present, water might instead form part of a gas hydrate structure as dealt with in Chapter 13.

To decide whether it is thermodynamically favorable for a pure component to be in liquid or solid form, one may consider the change in Gibbs free energy, G, as a result of the phase transition from solid to liquid form. If a pure component is in solid form at a given P and T but can reduce its G by melting (fusion), its thermodynamic equilibrium state will be liquid. If fusion makes G increase, it is thermodynamically favorable for the component to remain in solid form. No change in the Gibbs free energy by a possible fusion indicates that the liquid and solid states are equally favorable and that the component is right at its melting point. The following general thermodynamic relation exists for dG (combining Equations A.6 and A.7 in Appendix A):

$$dG = dH - TdS \tag{6.34}$$

where dH and dS are the changes in enthalpy and entropy, respectively. Applied to the fusion of a pure component, this relation becomes

$$\Delta G^f = \Delta H^f - T\Delta S^f \tag{6.35}$$

where the superscript f stands for fusion, ΔH^f is enthalpy (or heat) of fusion, and ΔS^f is entropy of fusion. If melting of the component takes place right at the melting temperature, where $\Delta G^f = 0$, the entropy of fusion will be

$$\Delta S^f = \frac{\Delta H^f}{T^f} \tag{6.36}$$

where T^f is the melting temperature of the component.

An expression for the change in Gibbs free energy as a result of melting a pure component at a temperature $T \neq T^f$ may be obtained by considering the hypothetical process sketched in Figure 6.12. The component considered is initially in the solid state, a, at the temperature, T. The component is to be converted to the liquid state, d, at the same temperature T. Instead of following the path a –> d, it follows the path a –> b, b –> c, and finally c –> d. While the component is kept in a solid state, the temperature is initially changed to T^f, which is associated with a change in enthalpy of

$$\Delta H_{ab} = \int_T^{T^f} C_P^S dT \tag{6.37}$$

where C_P is the heat capacity at constant pressure, and the superscript S stands for solid. At $T = T^f$, the component undergoes a phase transition from solid to liquid form (melting). This transition is associated with an enthalpy change of

$$\Delta H_{bc} = \Delta H^f \tag{6.38}$$

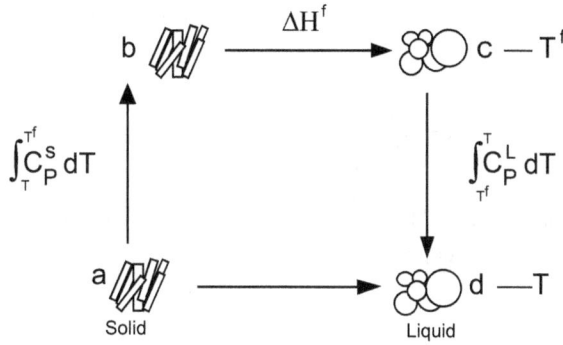

FIGURE 6.12 Hypothetical melting process at temperature T.

Finally, the temperature of the liquid is changed to T, giving rise to the following enthalpy change:

$$\Delta H_{cd} = \int_{T^f}^{T} C_P^L dT \tag{6.39}$$

where C_P^L is the heat capacity of component i in liquid form. The total enthalpy change associated with the transition from solid to liquid at temperature T using Equations 6.37 through 6.39 becomes

$$\Delta H_{ad} = \Delta H_{ab} + \Delta H_{bc} + \Delta H_{cd} = \Delta H^f + \int_{T}^{T^f} \Delta C_P dT \tag{6.40}$$

where ΔC_P is the difference between the solid and liquid state heat capacities for the considered component. In a quite similar manner, the following expression may be derived for the changes in entropy and Gibbs free energy from state a to d:

$$\Delta S_{ad} = \frac{\Delta H^f}{T^f} + \int_{T}^{T^f} \frac{\Delta C_P}{T} dT \tag{6.41}$$

$$\Delta G_{ad} = \Delta H^f \left(1 - \frac{T}{T^f}\right) + \int_{T}^{T^f} \Delta C_P dT - \int_{T}^{T^f} \frac{\Delta C_P}{T} dT \tag{6.42}$$

Phase equilibrium calculations require knowledge of component fugacities (f) or component fugacity coefficients (φ). These terms are defined in Appendix A. For a pure component i, the change in Gibbs free energy is related to the change in fugacity through the relation (Equation A.23 in Appendix A)

$$dG_i = -RT d \ln f_i \tag{6.43}$$

The fugacity of a pure component at a reference pressure is named *reference fugacity*. The term $f_i^{\circ L}(P_{ref})$ is used for the reference fugacity of component i in liquid state and $f_i^{\circ S}(P_{ref})$ for the reference fugacity of component i in solid state. Using Equation 6.43, the change in G as a result of the melting process sketched in Figure 6.12 and taking place at P_{ref} may be written as

$$\Delta G_{ad} = -RT \left[\ln\left(f_i^{\circ L}(P_{ref})\right) - \ln\left(f_i^{\circ S}(P_{ref})\right) \right] = RT \ln \frac{f_i^{\circ S}(P_{ref})}{f_i^{\circ L}(P_{ref})} \tag{6.44}$$

This expression combined with Equation 6.42 may be used to generate a relation between the fugacities of component i in the solid and liquid states at temperature T and pressure P_{ref}

$$f_i^{\circ S}\left(P_{ref}\right)=f_i^{\circ L}\left(P_{ref}\right)\exp\left(\frac{\Delta H_i^f}{RT}\left(1-\frac{T}{T_i^f}\right)+\frac{1}{RT}\int_T^{T_i^f}\Delta C_{Pi}dT-\frac{1}{RT}\int_T^{T_i^f}\frac{\Delta C_{Pi}}{T}dT\right) \qquad (6.45)$$

The pressure dependence of the fugacity of component i, assuming that the liquid and solid molar volumes, V_i^L and V_i^S, are independent of pressure, can be seen from Equations A.27 and A.28 in Appendix A:

$$f_i^L\left(P\right)=f_i^{\circ L}\left(P_{ref}\right)\exp\frac{V_i^L\left(P-P_{ref}\right)}{RT} \qquad (6.46)$$

$$f_i^S\left(P\right)=f_i^{\circ S}\left(P_{ref}\right)\exp\frac{V_i^S\left(P-P_{ref}\right)}{RT} \qquad (6.47)$$

These two equations combined with Equation 6.45 may be used to establish the following relation between the fugacity of component i at pressure P in pure solid and in pure liquid form:

$$f_i^{\circ S}\left(P\right)=f_i^{\circ L}\left(P\right)\exp\left(\begin{array}{c}\dfrac{\Delta H_i^f}{RT}\left(1-\dfrac{T}{T_i^f}\right)+\dfrac{1}{RT}\displaystyle\int_T^{T_i^f}\Delta C_{Pi}dT-\dfrac{1}{RT}\displaystyle\int_T^{T_i^f}\dfrac{\Delta C_{Pi}}{T}dT \\[2mm] +\dfrac{\Delta V_i\left(P-P_{ref}\right)}{RT}\end{array}\right) \qquad (6.48)$$

where the term ΔV_i is used for the difference between the solid and liquid phase molar volumes of component i.

Table 6.11 shows parameters required to calculate the fugacity of pure solid H_2O and CO_2. The pure component liquid fugacity is calculated from the applied cubic equation of state, and so are the fugacities of the components in the fluent phase(s) in equilibrium with the pure solid.

When no data is available for ΔC_p and ΔV of an inorganic component, the following approximate expression may be used for the solid-state fugacity

$$f_i^{\circ S}\left(P\right)=f_i^{\circ L}\left(P\right)\exp\left(\frac{\Delta H_i^f}{RT}\left(1-\frac{T}{T_i^f}\right)\right) \qquad (6.49)$$

TABLE 6.11

Parameters Used to Calculate the Fugacity of Pure Solid H_2O and CO_2

Component	Temperature of Fusion K	Enthalpy of Fusion J/mol	ΔC_p J/(mol K)	ΔV cm³/mol	P_{ref} bar
H_2O	273.15	6009	23.6514	1.600	0.0
CO_2	216.58	9020	-	−6.2547	5.17

TABLE 6.12

CO_2 Rich Gas Mixture Flashed to 80 bar and −100°C. The SRK Equation Is Used for the Liquid Phase. The Fugacity of CO_2 in the Solid Phase Is Calculated from Equation 6.48 Using the CO_2 Parameters in Table 6.11

Component	Feed Mol%	Liquid Mol%	Solid Mol%
N_2	0.170	0.329	
CO_2	50.420	4.049	100.000
C_1	45.200	87.475	
C_2	2.599	5.031	
C_3	1.030	1.993	
iC_4	0.180	0.348	
nC_4	0.275	0.532	
iC_5	0.070	0.135	
nC_5	0.049	0.094	
nC_6	0.007	0.014	
Total	100.000	51.672	48.328

CO_2 is often used as injection gas for EOR purposes (Chapter 15). The gas produced from a field with CO_2 gas injection will after some time be rich in CO_2. One way to reduce the CO_2 concentration in the gas could be to freeze the CO_2 out. Table 6.12 shows a gas composition containing 50.42 mol% CO_2 and the result of a PT flash calculation at 80 bar and −100°C. At these conditions a pure solid CO_2 phase is formed, which is in equilibrium with a liquid phase with a CO_2 mol% of 4.049. This is a reduction of the CO_2 concentration in the fluent phase by more than a factor of 10.

6.3.6 OTHER FLASH SPECIFICATIONS

P and T are not always the most convenient flash specification variables. Some of the processes taking place during oil and gas production are not at a constant P and T. Passage of a valve may, for example, be approximated as a constant enthalpy (H) process and a compression as a constant entropy (S) process. The temperature after a valve may, therefore, be simulated by initially performing a PT-flash at the conditions at the inlet to the valve. If the enthalpy is assumed to be the same at the outlet, the temperature at the outlet can be found from a PH-flash with P equal to the outlet pressure and H equal to the enthalpy at the inlet. A PT-flash followed by a PS-flash may similarly be used to determine an approximate temperature after a compressor. A VT-flash may be used to study how pressure varies with temperature in a closed system. It may, for example, be used to simulate the conditions in a pipeline during shutdown.

Figure 6.13 gives an overview of the application areas for four different flash specifications. The alternative flash specification variables (H, S, or V) may replace P and/or T in the flash calculation, or the flash calculation may be performed as an over iteration in a PT-flash. The latter means that P and/or T are guessed, and for each P and T, the solution is checked for correspondence with the flash specification variables. Michelsen (1999) has presented flash calculation techniques for use with alternative state function-based flash specification variables.

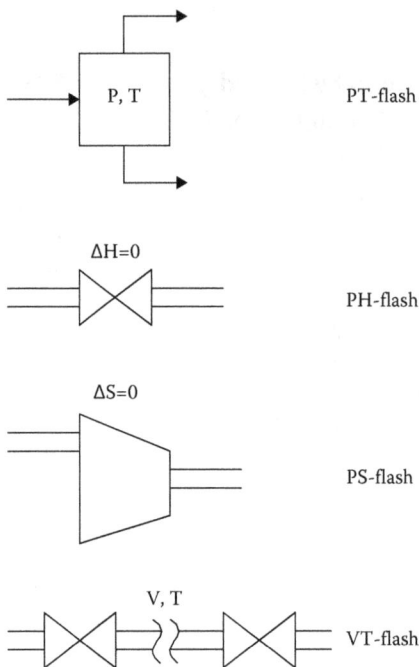

FIGURE 6.13   Application areas for alternative flash specifications.

6.4   ALLOCATION CALCULATIONS

Flash calculations combined with the common EoS technology presented in Section 5.6 can be used in an allocation procedure that takes both process conditions and mixing effects into account. Allocation consists in calculating the volumetric contributions to the total gas and oil (liquid) production at standard (or other reference) conditions from each of the well streams led to a process facility.

To quantify the volumetric contributions of each well stream, the partial molar volumes of each component in the gas and liquid product streams are needed. The partial molar volume, $\bar{V}_i$, of component i in a gas or a liquid phase is the contribution of component i to the overall molar volume, meaning that the overall molar volume can be written

$$V = \sum_{i=1}^{N} z_i \bar{V}_i \qquad (6.50)$$

where z_i is the mole fraction of component i and N the number of components. The partial molar volume of component i in an ideal gas mixture equals the molar volume of the pure component i at the same conditions. For a non-ideal gas or a liquid mixture, the partial molar volume of component i can be found from

$$\bar{V}_i = RT \left(\frac{\partial \ln(\varphi_i)}{\partial P} \right)_T + \frac{RT}{P} \qquad (6.51)$$

To exemplify allocation calculations, a case is considered where three well streams are processed through the same process plant, as sketched in Figure 6.14. A flowmeter measures the flow rate of each well stream. The fluid compositions and the mass, volumetric, and molar flow rates are shown in Table 6.13. Before the inlet to the process plant, the well streams are mixed in a manifold.

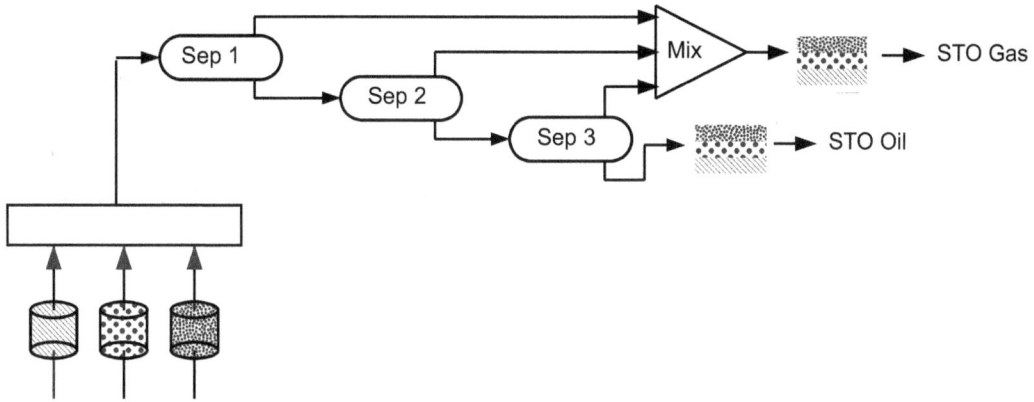

FIGURE 6.14 Separator setup for processing of three mixed well streams. The separator conditions are: 1st 48 bar, 55°C; 2nd 20 bar, 35°C; 3rd 1.01 bar, 15°C. Flowmeters measure the flowrates before the well streams are mixed in the manifold.

TABLE 6.13

Reservoir Fluid Compositions Processed through the Process Plant Sketched in Figure 6.14. The Total Flowrate from Each Well Is Measured by a Flow Meter

Component	Fluid 1 Mol %	Fluid 2 Mol %	Fluid 3 Mol %
N_2	0.528	0.477	0.560
CO_2	3.304	4.040	3.549
C_1	73.036	57.413	45.326
C_2	7.688	9.280	5.478
C_3	4.107	5.621	3.699
iC_4	0.700	1.004	0.700
nC_4	1.426	2.220	1.650
iC_5	0.538	0.831	0.730
nC_5	0.672	1.053	0.870
C_6	0.838	1.346	1.330
C_7	1.265	2.150	2.556
C_8	1.294	2.560	3.143
C_9	0.770	1.472	2.081
C_{10+}	3.835	10.532	28.329
C_{7+} molecular weight	169.5	195.500	254.900
C_{7+} density (g/cm^3)	0.8163	0.8315	0.8760
Fluid molecular weight	32.45	53.54	107.86
Meter mass flow rates (kg/hr)	2,500	2,500	5,000
Meter molar flow rates (kmol/hr)	77.04	46.69	46.36
Mole fraction of total	0.4530	0.2745	0.2725

A common EoS model as described in Section 5.6 is developed for the three fluid compositions in Table 6.13. The applied fluid weights (*Wgt* in Equations 5.49–5.53) are the respective feed mole fractions in Table 6.13. The common EoS model is shown in Table 6.14 with nonzero binary interaction parameters in Table 6.15.

TABLE 6.14

Common EoS Model for SRK-Peneloux for the Three Fluid Compositions in Table 6.13. NonZero Binary Interaction Parameters Are Shown in Table 6.15

Molecular Weight	Critical Temperature °C	Critical Pressure bar	Acentric Factor	Volume Shift cm³/mol
28.01	−146.95	33.94	0.0400	0.92
44.01	31.05	73.76	0.2250	3.03
16.04	−82.55	46.00	0.0080	0.63
30.07	32.25	48.84	0.0980	2.63
44.10	96.65	42.46	0.1520	5.06
58.12	134.95	36.48	0.1760	7.29
58.12	152.05	38.00	0.1930	7.86
72.15	187.25	33.84	0.2270	10.93
72.15	196.45	33.74	0.2510	12.18
86.18	234.25	29.69	0.2960	17.98
96.00	263.10	32.44	0.4679	8.69
107.00	281.96	29.44	0.4998	14.54
121.00	303.05	26.30	0.5398	22.07
146.40	337.93	22.66	0.6128	32.48
182.12	378.23	19.41	0.7072	42.56
220.62	416.62	17.32	0.8056	47.01
262.21	453.77	16.01	0.9034	44.13
309.63	492.94	15.08	1.0068	34.49
364.78	534.77	14.39	1.1123	17.15
433.20	583.27	13.86	1.2205	−10.30
538.65	653.06	13.42	1.3299	−61.26
757.38	791.46	13.14	1.2684	−181.07

TABLE 6.15

NonZero Binary Interaction Parameters for Use with the Common EoS in Table 6.14

	N_2	CO_2
CO_2	−0.0315	
C_1	0.0278	0.1200
C_2	0.0407	0.1200
C_3	0.0763	0.1200
iC_4	0.0944	0.1200
nC_4	0.0700	0.1200
iC_5	0.0867	0.1200
nC_5	0.0878	0.1200
C_6	0.0800	0.1200
C_{7+}	0.0800	0.1000

TABLE 6.16

Molar Compositions of the Three Fluids and of the Mixture Fed to the Process Plant in Figure 6.14. The Last Three Columns Show the Fraction Each Inlet Fluid Contributes to the Total Number of Moles of a Given Component in the Mixed Fluid

Component	Fluid 1 Mol %	Fluid 2 Mol %	Fluid3 Mol %	Mixed Fluid Mole%	Fluid 1 Fraction of mix	Fluid 2 Fraction of mix	Fluid 3 Fraction of mix
N_2	0.528	0.477	0.560	0.523	0.4575	0.2506	0.2919
CO_2	3.304	4.040	3.549	3.573	0.4189	0.3104	0.2707
C_1	73.036	57.413	45.326	61.196	0.5406	0.2575	0.2018
C_2	7.688	9.280	5.478	7.523	0.4629	0.3386	0.1984
C_3	4.107	5.621	3.699	4.412	0.4218	0.3498	0.2285
iC_4	0.700	1.004	0.700	0.783	0.4048	0.3518	0.2434
nC_4	1.426	2.220	1.650	1.705	0.3788	0.3575	0.2637
iC_5	0.538	0.831	0.730	0.671	0.3633	0.3402	0.2965
nC_5	0.672	1.053	0.870	0.830	0.3665	0.3481	0.2854
C_6	0.838	1.346	1.330	1.111	0.3415	0.3325	0.3260
C_7	1.265	2.150	2.556	1.860	0.3082	0.3173	0.3745
C_8	1.294	2.560	3.143	2.146	0.2732	0.3276	0.3992
C_9	0.770	1.472	2.081	1.320	0.2643	0.3061	0.4296
$C_{10}-C_{12}$	1.355	3.120	6.227	3.167	0.1938	0.2704	0.5358
$C_{13}-C_{14}$	0.625	1.548	3.373	1.627	0.1741	0.2611	0.5648
$C_{15}-C_{17}$	0.655	1.737	4.123	1.897	0.1565	0.2514	0.5921
$C_{18}-C_{20}$	0.424	1.223	3.219	1.405	0.1367	0.2389	0.6244
$C_{21}-C_{24}$	0.342	1.087	3.223	1.331	0.1164	0.2241	0.6596
$C_{25}-C_{28}$	0.191	0.680	2.317	0.905	0.0958	0.2064	0.6978
$C_{29}-C_{34}$	0.141	0.575	2.314	0.852	0.0750	0.1852	0.7398
$C_{35}-C_{44}$	0.078	0.389	2.030	0.695	0.0507	0.1537	0.7956
$C_{45}-C_{80}$	0.024	0.173	1.504	0.468	0.0229	0.1013	0.8759

Molar compositions of the three well streams and of the mixture fed to the process plant are shown in Table 6.16. The table further shows the fraction of each component in the mixed fluid that originates from each individual inlet fluid. These fractions will be the same throughout the process and in the product streams. The fraction of component i originating from Fluid j is calculated as

$$Frac_i^j = \frac{Frac^J \times z_i^j}{z_i^{mix}} \tag{6.52}$$

where $Frac^J$ is the mole fraction of Fluid j of the total (mixed) molar feed to the process plant, z_i^j the mole fraction of component i in Fluid j, and z_i^{mix} the mole fraction of component i in the mixed feed.

A separator calculation is carried out for the mixed stream with the separator stages given in Figure 6.14. The compositions of the produced gas and liquid at standard conditions are shown in Table 6.17. Also shown are the flow rates of each product stream in mass, volumetric, and molar

TABLE 6.17

Produced Gas and Liquid Compositions at Standard Conditions and Flowrates of Each Product Stream (Stock Tank Gas and Stock Tank Liquid) in Mass, Volumetric, and Molar Units. The Last Two Columns Show the Partial Molar Volumes of Each Component in Each of the Two Product Streams

Component	Stock Tank Gas Mol%	Stock Tank Liquid Mol%	Partial Molar Volumes Stock Tank Gas cm³/mol	Partial Molar Volumes Stock Tank Liquid cm³/mol
N_2	0.6657	0.0001	23672.13	50.66
CO_2	4.5242	0.1010	23543.71	48.26
C_1	77.9215	0.1707	23602.12	52.40
C_2	9.3613	0.8149	23467.78	64.38
C_3	4.7808	3.0644	23354.82	79.43
iC_4	0.6563	1.2470	23263.77	95.44
nC_4	1.1972	3.5563	23242.21	94.32
iC_5	0.2789	2.1001	23148.28	108.90
nC_5	0.2837	2.8250	23125.58	109.68
C_6	0.1667	4.5580	23003.29	127.17
C_7	0.0856	8.3333	22937.42	129.73
C_8	0.0524	9.7828	22846.36	140.96
C_9	0.0151	6.0811	22731.50	156.01
$C_{10}-C_{12}$	0.0094	14.6871	22536.04	182.68
$C_{13}-C_{14}$	0.0008	7.5617	22282.68	220.96
$C_{15}-C_{17}$	0.0001	8.8193	22028.05	262.03
$C_{18}-C_{20}$	0.0000	6.5311	21780.83	304.62
$C_{21}-C_{24}$	0.0000	6.1893	21518.55	351.89
$C_{25}-C_{28}$	0.0000	4.2064	21238.93	405.56
$C_{29}-C_{34}$	0.0000	3.9629	20921.85	470.82
$C_{35}-C_{44}$	0.0000	3.2329	23672.13	569.18
$C_{45}-C_{80}$	0.0000	2.1747	23543.71	771.43
Mass flow rate (kg/hour)	2857.49	7142.51		
Volumetric flow rate (m³/hour)	3146.35	8.4944		
Molar flow rate (kmol/hour)	133.52	36.59		

units. The last two columns in Table 6.17 show the partial molar volumes of each component in each of the product streams (calculated from Equation 6.51).

Table 6.18 shows the total number of moles of each component produced per hour and how many of these moles originate from each individual feed stream. The latter numbers are calculated by multiplying the total number of moles of each component by the mole fractions in the last three columns of Table 6.16. Table 6.19 shows the contribution of each component to the total volumetric production per hour and how this is distributed on the individual feed streams. In the bottom row the volumetric contributions from each component are added up to give the total volumetric flowrates and the volumetric flowrates from each feed stream. The latter flow rates are the target numbers of an allocation calculation.

TABLE 6.18

Total Molar Production Rates of Each Component and the Contributions to the Production from Each Feed Stream

Component	Stock Tank Gas Moles/hour				Stock Tank Liquid Moles/hour			
	Total	Fluid 1	Fluid 2	Fluid 3	Total	Fluid 1	Fluid 2	Fluid 3
N_2	0.8889	0.4067	0.2227	0.2595	0.0000	0.0000	0.0000	0.0000
CO_2	6.0408	2.5305	1.8751	1.6352	0.0369	0.0155	0.0115	0.0100
C_1	104.0429	56.2495	26.7941	20.9992	0.0625	0.0338	0.0161	0.0126
C_2	12.4995	5.7865	4.2326	2.4804	0.2982	0.1381	0.1010	0.0592
C_3	6.3835	2.6923	2.2327	1.4585	1.1214	0.4730	0.3922	0.2562
iC_4	0.8764	0.3547	0.3083	0.2133	0.4563	0.1847	0.1605	0.1111
nC_4	1.5986	0.6056	0.5715	0.4215	1.3014	0.4930	0.4652	0.3431
iC_5	0.3723	0.1353	0.1267	0.1104	0.7685	0.2792	0.2614	0.2279
nC_5	0.3788	0.1388	0.1319	0.1081	1.0338	0.3788	0.3599	0.2951
C_6	0.2226	0.0760	0.0740	0.0726	1.6679	0.5696	0.5546	0.5438
C_7	0.1144	0.0352	0.0363	0.0428	3.0495	0.9399	0.9677	1.1419
C_8	0.0700	0.0191	0.0229	0.0280	3.5799	0.9781	1.1727	1.4292
C_9	0.0202	0.0053	0.0062	0.0087	2.2253	0.5880	0.6812	0.9560
C_{10}–C_{12}	0.0126	0.0024	0.0034	0.0067	5.3746	1.0416	1.4533	2.8797
C_{13}–C_{14}	0.0011	0.0002	0.0003	0.0006	2.7671	0.4818	0.7224	1.5629
C_{15}–C_{17}	0.0002	0.0000	0.0000	0.0001	3.2273	0.5050	0.8113	1.9110
C_{18}–C_{20}	0.0000	0.0000	0.0000	0.0000	2.3900	0.3266	0.5711	1.4923
C_{21}–C_{24}	0.0000	0.0000	0.0000	0.0000	2.2649	0.2635	0.5075	1.4938
C_{25}–C_{28}	0.0000	0.0000	0.0000	0.0000	1.5393	0.1474	0.3178	1.0741
C_{29}–C_{34}	0.0000	0.0000	0.0000	0.0000	1.4502	0.1088	0.2685	1.0728
C_{35}–C_{44}	0.0000	0.0000	0.0000	0.0000	1.1830	0.0599	0.1819	0.9412
C_{45}–C_{80}	0.0000	0.0000	0.0000	0.0000	0.7958	0.0182	0.0806	0.6970
Total (kmol/hour)	133.5227	69.0383	36.6387	27.8457	36.5939	8.0245	10.0583	18.5111

TABLE 6.19

Total Volumetric Production Rates for Each Component and Breakdown of the Production Rates on Each Feed Stream. The Last Row Shows the Total Gas and Liquid Production Flowrates and a Breakdown of These Flowrates on Each of the Three Feed Streams

Component	Stock Tank Gas m³/hour				Stock Tank Liquid m³/hour			
	Total	Fluid 1	Fluid 2	Fluid 3	Total	Fluid 1	Fluid 2	Fluid 3
N_2	21.04	9.63	5.27	6.14	0.0000	0.0000	0.0000	0.0000
CO_2	142.22	59.58	44.15	38.50	0.0018	0.0007	0.0006	0.0005
C_1	2455.63	1327.61	632.40	495.63	0.0033	0.0018	0.0008	0.0007
C_2	293.34	135.80	99.33	58.21	0.0192	0.0089	0.0065	0.0038
C_3	149.08	62.88	52.14	34.06	0.0891	0.0376	0.0312	0.0203
iC_4	20.39	8.25	7.17	4.96	0.0436	0.0176	0.0153	0.0106
nC_4	37.15	14.08	13.28	9.80	0.1227	0.0465	0.0439	0.0324

(Continued)

TABLE 6.19 *(Continued)*
Total Volumetric Production Rates for Each Component and Breakdown of the Production Rates on Each Feed Stream. The Last Row Shows the Total Gas and Liquid Production Flowrates and a Breakdown of These Flowrates on Each of the Three Feed Streams

Component	Stock Tank Gas m³/hour				Stock Tank Liquid m³/hour			
	Total	Fluid 1	Fluid 2	Fluid 3	Total	Fluid 1	Fluid 2	Fluid 3
iC_5	8.62	3.13	2.93	2.56	0.0837	0.0304	0.0285	0.0248
nC_5	8.76	3.21	3.05	2.50	0.1134	0.0416	0.0395	0.0324
C_6	5.12	1.75	1.70	1.67	0.2121	0.0724	0.0705	0.0692
C_7	2.62	0.81	0.83	0.98	0.3956	0.1219	0.1255	0.1481
C_8	1.60	0.44	0.52	0.64	0.5046	0.1379	0.1653	0.2015
C_9	0.46	0.12	0.14	0.20	0.3472	0.0917	0.1063	0.1491
C_{10}–C_{12}	0.28	0.06	0.08	0.15	0.9818	0.1903	0.2655	0.5261
C_{13}–C_{14}	0.02	0.00	0.01	0.01	0.6114	0.1065	0.1596	0.3453
C_{15}–C_{17}	0.00	0.00	0.00	0.00	0.8456	0.1323	0.2126	0.5007
C_{18}–C_{20}	0.00	0.00	0.00	0.00	0.7280	0.0995	0.1740	0.4546
C_{21}–C_{24}	0.00	0.00	0.00	0.00	0.7970	0.0927	0.1786	0.5257
C_{25}–C_{28}	0.00	0.00	0.00	0.00	0.6243	0.0598	0.1289	0.4356
C_{29}–C_{34}	0.00	0.00	0.00	0.00	0.6828	0.0512	0.1264	0.5051
C_{35}–C_{44}	0.00	0.00	0.00	0.00	0.6734	0.0341	0.1035	0.5357
C_{45}–C_{80}	0.00	0.00	0.00	0.00	0.6139	0.0140	0.0622	0.5377
Total (m³/hour)	3146.35	1627.33	863.01	656.01	8.4944	1.3895	2.0451	5.0599

6.5 PHASE ENVELOPE CALCULATIONS

A phase envelope, as presented in Chapter 1, may, in principle, be calculated by performing a series of saturation point calculations as outlined in Section 6.2, but if the complete phase envelope is needed, this method is not to be recommended. It is both time consuming and likely to cause convergence problems at higher pressures and near the critical point (CP). The procedure outlined by Michelsen (1980) may be used instead. The phase envelope calculation is started at a moderate pressure (<20 bar) from either the dew point or the bubble point side. At the starting pressure, the saturation point temperature is calculated as outlined in Section 6.2. Because the pressure is moderate, convergence is easily obtained. A second saturation pressure is calculated at a slightly higher pressure. The third point and subsequent saturation points are calculated making use of the K-factors, pressures, and temperatures in each of the two previous points on the phase envelope. This ensures a reasonable initial estimate and, using the procedure outlined by Michelsen, it creates no problems to locate and pass the CP. The CP may alternatively be located as described by Michelsen and Heidemann (1981).

Michelsen's technique for construction of phase envelopes is not limited to dew and bubble point lines. It may also be used to construct inner lines in a phase envelope, that is, the PT values for which the vapor mole fraction equals a specified value. Figure 6.15 shows the phase envelope of the oil mixture in Table 6.20 calculated using the SRK equation of state. The characterized fluid composition is shown in Table 6.21. It is seen that the dew and bubble point lines as well as the inner lines meet at the CP at which the gas and liquid phases are indistinguishable and the vapor mole fraction β may therefore be assigned any value between 0 and 1.

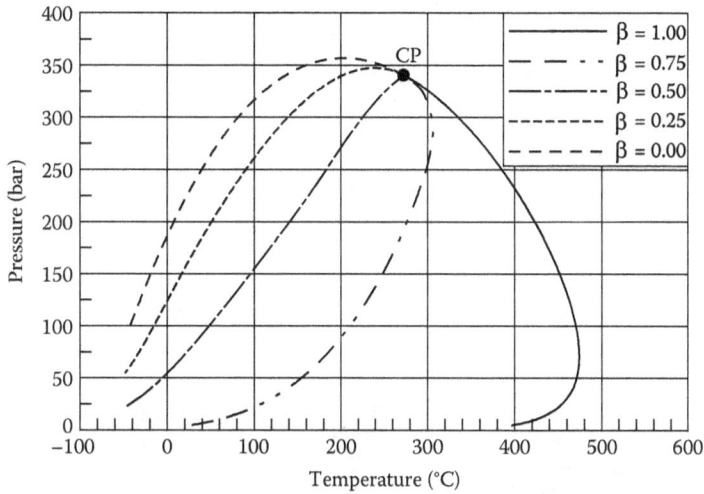

FIGURE 6.15 Phase envelope of oil mixture of Table 6.21 calculated using SRK equation of state. CP stands for critical point and β for vapor mole fraction.

TABLE 6.20
Molar Composition of Reservoir Oil Mixture

Component	Mol%	Molecular Weight	Density at 15°C and 1.01 bar (g/cm³)
N_2	0.546	—	—
CO_2	2.826	—	—
C_1	55.565	—	—
C_2	8.594	—	—
C_3	5.745	—	—
iC_4	1.009	—	—
nC_4	2.435	—	—
iC_5	0.895	—	—
nC_5	1.240	—	—
C_6	1.581	—	—
C_7	2.552	91.5	0.738
C_8	2.747	101.2	0.765
C_9	1.699	119.1	0.781
C_{10+}	12.564	254.9	0.870

Figure 6.16 shows the results of phase envelope calculations performed for the gas condensate mixture in Table 6.5. No CP is located. This may at first sight look like a simulation failure but does in fact have a more sensible explanation. It has already been shown through the PT-flash results in Table 6.6 that the mixture considered forms three phases in a PT region at low temperatures. The three-phase area has been located (Michelsen 1986b) and is shown in Figure 6.16 and in an enlarged scale in Figure 6.17. This example illustrates the fact that a hydrocarbon mixture will not always have a CP.

TABLE 6.21

Oil Composition in Table 6.20 after Characterization

Component	Mol%	T_c (°C)	P_c (bar)	Acentric Factor
N_2	0.546	−147.0	33.94	0.040
CO_2	2.826	31.1	73.76	0.225
C_1	55.566	−82.6	46.00	0.008
C_2	8.594	32.3	48.84	0.098
C_3	5.745	96.7	42.46	0.152
iC_4	1.009	135.0	36.48	0.176
nC_4	2.435	152.1	38.00	0.193
iC_5	0.895	187.3	33.84	0.227
nC_5	1.240	196.5	33.74	0.251
C_6	1.581	234.3	29.69	0.296
C_7–C_{13}	11.483	316.5	24.96	0.442
C_{14}–C_{22}	5.089	466.1	16.56	0.792
C_{23}–C_{80}	2.990	676.5	13.28	1.137

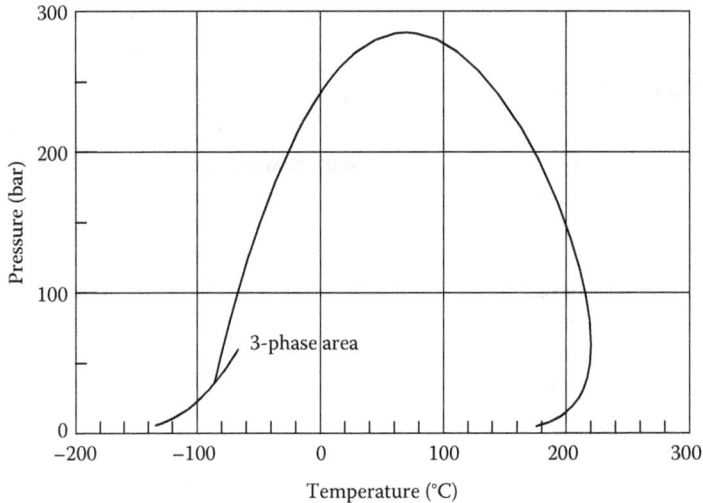

FIGURE 6.16 Phase envelope of gas condensate mixture of Table 6.5 calculated using PR equation of state.

The oil mixture in Table 6.20 has been characterized for the SRK equation, and the characterized mixture is shown in Table 6.21. Figure 6.18 shows the phase envelope for a mixture of CO_2 and the oil in Table 6.21 mixed in the molar ratio 2.50:1.0. It has a three-phase area of extending to temperature above 0°C. At a temperature of around 30°C, the phase boundary increases almost vertically in pressure. This is an indication of a liquid–liquid equilibrium at the low temperature side of the phase boundary. Table 6.22 shows the results of a PT-flash calculation for 900 bar and 30°C. Two liquid phases of approximately the same composition are formed. The density is above 0.9 g/cm³ for both phases.

Figure 6.19 shows experimental saturation points measured for a three-component mixture consisting of 9.87 mole percentage CO_2, 40.23 mole percentage H_2S, and 49.90 mole percentage C_1

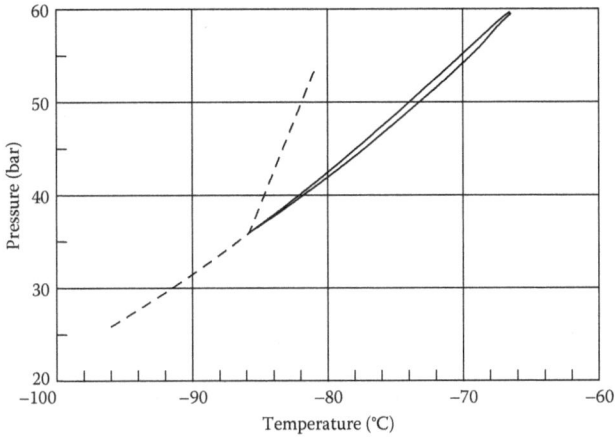

FIGURE 6.17 Close-up of low-temperature part of phase envelope in Figure 6.16. The solid line surrounds the three-phase area.

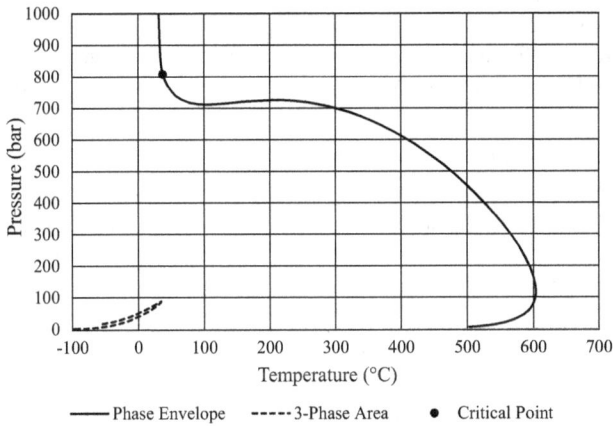

FIGURE 6.18 Phase envelope of oil mixture in Table 6.8 mixed with CO_2 in molar ratio 1:2.5.

TABLE 6.22

PT-Flash Calculation for 900 bar and 30°C for the Oil Mixture in Table 6.21 Mixed with CO_2 in the Molar Ratio 1:2.5. Phase Compositions and Phase Densities Are Calculated Using the SRK-Peneloux Equation

Component	Feed Mol%	Liquid 1	Liquid 2
CO_2	71.477	70.760	83.245
$N_2 + C_1$	7.149	7.176	6.697
$C_2 + C_3$	3.454	3.499	2.715
$iC_4 - C_6$	4.327	4.415	2.888
$C_7 - C_{14}$	6.750	6.999	2.669
$C_{15} - C_{23}$	3.665	3.802	1.408
$C_{24} - C_{47}$	2.156	2.267	0.326
$C_{48} - C_{80}$	1.022	1.081	0.053
Phase mole percentage	100	94.26	5.74
Density (g/cm³)		0.927	0.970

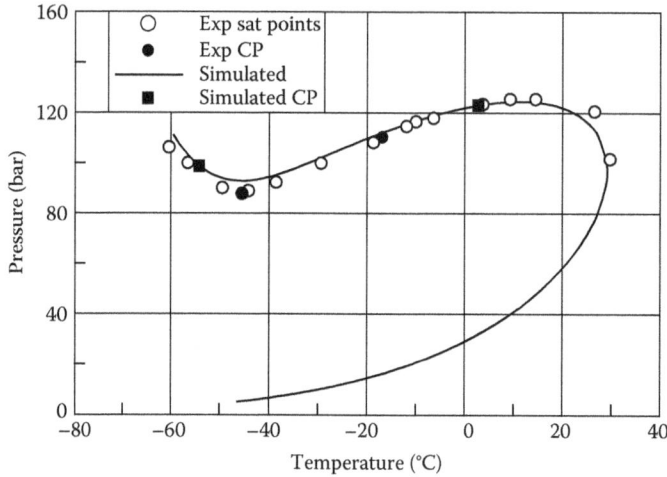

FIGURE 6.19 Experimental saturation points (circles) measured for a three-component mixture consisting of 9.87 mole percentage CO_2, 40.23 mole percentage H_2S, and 49.90 mole percentage C_1 (Robinson et al. 1981). The solid circles are experimentally determined critical points (CP). The full-drawn line shows the phase envelope simulated using the PR equation of state with binary interaction parameters of: CO_2–H_2S: 0.0974, CO_2–C_1: 0.110, and H_2S–C_1: 0.069. The two solid squares are simulated critical points.

(Robinson et al. 1981). As can be seen from the figure (solid circles), two CPs have been detected experimentally for this mixture. As can also be seen from the figure, the existence of two CPs is reproduced (solid squares) in a phase envelope calculation on the same mixture using the PR equation of state. The locations of the simulated CPs deviate slightly from those observed experimentally but considering that a CP is surrounded by a region with near-critical phase behavior, the agreement is quite good.

Phase envelope calculations for mixtures containing water and possibly hydrate inhibitors are complicated by the fact that the hydrocarbon phase boundary may be quite affected by the presence of aqueous components and because an additional (aqueous) phase must be accounted for. Lindeloff and Michelsen (2003) have outlined a procedure for how to make phase envelope calculations for hydrocarbon–water mixtures. Chapter 16 shows examples of the application of this algorithm.

6.6   PHASE IDENTIFICATION

If a PT-flash calculation for an oil or gas mixture shows the presence of two phases, the one with lower density is usually assumed to be gas or vapor, and the one with higher density is assumed to be liquid or oil. In the case of a single-phase solution, it is less obvious whether to consider this single phase a gas or a liquid. No generally accepted definition exists to distinguish a gas from a liquid. Because the terms *gas* and *oil* are very much used in the oil industry, it is, however, of interest to try to establish a reasonable criterion for distinguishing between the two types of phases. Figure 6.20 shows the phase envelope of a volatile oil. Four single-phase conditions are marked on the figure (points 1 to 4). Point 1 is just outside the two-phase region on the bubble point side. Therefore, it is natural to classify the mixture at these conditions as being a liquid. Point 4 is also just outside the two-phase region but on the dew point side, suggesting that the mixture is gaseous at these conditions. At the conditions of points 2 and 3, it is less obvious whether the mixture is to be considered a gas or a liquid. Point 2 is located at a temperature lower than the critical temperature. This could suggest that the mixture at point 2 is a liquid. Similarly, point 3 is at a temperature higher than the critical temperature, suggesting that the fluid in point 3 is a gas. This leads to the following suggestion for a phase identification criterion (as illustrated in Figure 6.21).

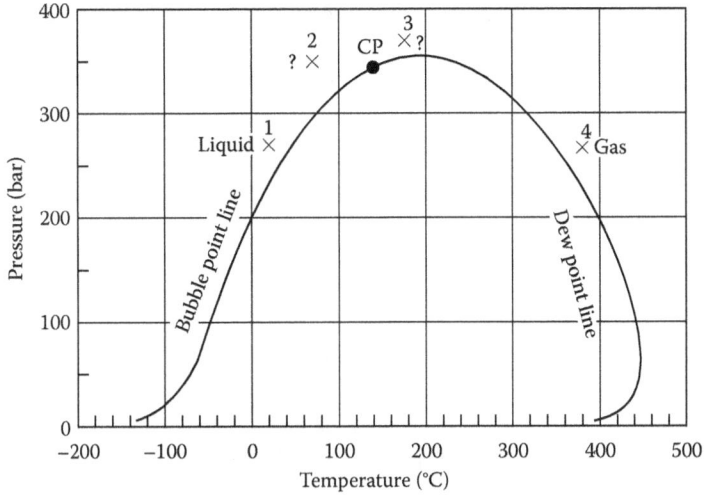

FIGURE 6.20 Illustration of the problem with phase identification of single-phase mixtures.

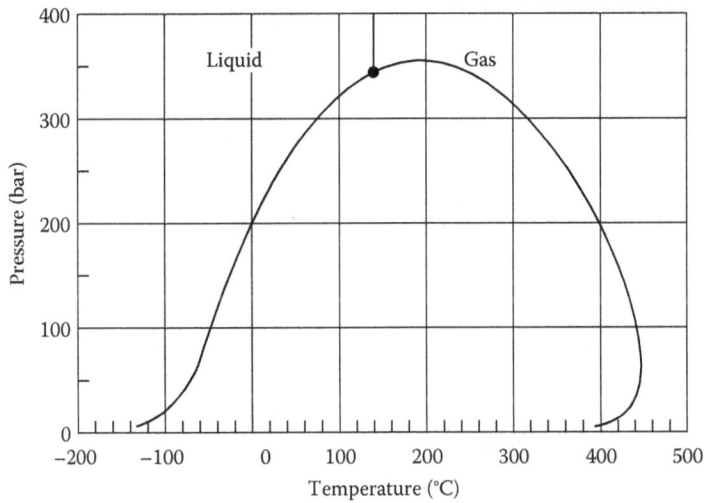

FIGURE 6.21 Possible phase identification criterion.

Liquid

1. If the pressure is lower than the critical pressure and the temperature lower than the bubble point temperature.
2. If the pressure is higher than the critical pressure and the temperature lower than the critical temperature.

Gas

1. If the pressure is lower than the critical pressure and the temperature higher than the dew point temperature.
2. If the pressure is higher than the critical pressure and the temperature higher than the critical temperature.

This criterion has the disadvantage that the CP and possibly also the dew or bubble point temperature at the actual pressure must be simulated. This is impractical when computation time matters. Therefore, it is often more convenient to work with a simpler criterion such as the following:

Liquid

$$\frac{V}{b} < Const$$

Gas

$$\frac{V}{b} > Const$$

V is the molar volume, b the b parameter of the cubic equation of state, and Const a constant, the value of which depends on the equation of state. For the SRK and PR equations presented in Chapter 4, Const = 1.75 is a convenient choice.

As long as the thermodynamic model applied is the same, independent of phase type, (as is, e.g., the case with a cubic equation of state), the phase identification criterion has no importance for the simulated properties.

REFERENCES

Dong, W.-G. and Lienhard, J.H., Corresponding states correlation of saturated metastable properties, *Can. J. Chem. Eng.* 64, 158–161, 1986.

Heidemann, R.A. and Michelsen, M.L., Instability of successive substitution, *Ind. Eng. Chem. Res.* 34, 958–966, 1995.

Hendriks, E.M., Simplified phase equilibrium equations for multicomponent systems, *Fluid Phase Equilib.* 33, 207–221, 1987.

Jensen, B.H. and Fredenslund, Aa., A simplified flash procedure for multicomponent mixtures containing hydrocarbons and one non-hydrocarbon using two-parameter cubic equations of state, *Ind. Eng. Chem. Res.* 26, 2129–2134, 1987.

Lindeloff, N. and Michelsen, M.L., Phase Envelope calculations for hydrocarbon-water mixtures, SPE 85971, *SPE J.* 298–303, September 2003.

Lindeloff, N., Mogensen, K.M., Pedersen, K.S., and Tybjerg, P., Investigation of miscibility behavior of CO$_2$ rich hydrocarbon systems—with application for gas injection EOR, SPE 166270-MS, presented at the *SPE ATCE*, New Orleans, LA, September 30–October 2, 2013.

Michelsen, M.L., Calculation of phase envelopes and critical points for multicomponent mixtures, *Fluid Phase Equilib.* 4, 1–10, 1980.

Michelsen, M.L., Three-phase envelope and three-phase flash algorithms with a liquid water phase, SEP Report 8123, *Department of Chemical Engineering*, The Technical University of Denmark, 1981.

Michelsen, M.L., The isothermal flash problem. Part I. Stability, *Fluid Phase Equilib.* 9, 1–19, 1982a.

Michelsen, M.L., The isothermal flash problem. Part II. Phase-split calculation, *Fluid Phase Equilib.* 9, 21–40, 1982b.

Michelsen, M.L., Simplified flash calculations for cubic equations of state, *Ind. Eng. Chem. Process Des. Dev.* 25, 184–188, 1986a.

Michelsen, M.L., Some aspects of multiphase calculations, *Fluid Phase Equilib.* 30, 15–29, 1986b.

Michelsen, M.L., Speeding up two-phase PT-flash, with applications for calculation of miscible displacement, *Fluid Phase Equilib.* 143, 1–12, 1998.

Michelsen, M.L., State function based flash specifications, *Fluid Phase Equilib.* 158–160, 617–626, 1999.

Michelsen, M.L. and Heidemann, R.A., Calculation of critical points from cubic two-constant equations of state, *AICHE J.* 27, 521–523, 1981.

Michelsen, M.L. and Mollerup, J., *Thermodynamic Models: Fundamentals and Computational Aspects*, Tie-Line Publications, Holte, Denmark, 2007.

Paterson, D., Michelsen, M.L., Stenby, E.H., and Yan, W., RAND-based formulations for iso-thermal multi-phase flash, *SPE J.* 23, 535–549, 2018.

Pedersen, K.S., Thomassen, P., and Fredenslund, Aa., Thermodynamics of petroleum mixtures containing heavy hydrocarbons. 3. Efficient flash calculation procedures using the SRK-equation of state, *Ind. Eng. Chem. Process. Des. Dev.* 24, 1985, 948–954.

Rachford, H.H. Jr. and Rice, J.D., Procedure for use of electronic digital computers in calculating flash vaporization hydrocarbon equilibrium, *J. Pet. Technol.* 4, 1952, sec. 1, p. 19 and sec. 2, p. 3.

Rasmussen, C.P., Krejbjerg, K., Michelsen, M.L., and Bjurstrøm, K.E., Increasing computational speed of flash calculations with applications for compositional, transient simulations, *SPE 84181, SPE Reserv. Evaluation Eng.* 32–38, February 2006.

Robinson, D.B., Ng, H.-J., and Leu, A.D., The behavior of CH_4-CO_2-H_2S mixtures at sub-ambient temperatures, *Research Report RR-47*, Gas Processors Association, Tulsa, OK, 1981.

Wilson, G.M., A modified Redlich-Kwong equation of state, application to general physical data calculation, Paper No. 15C, presented at the *1969 AIChE 65th National Meeting*, Cleveland, OH, March 4–7, 1969.

7 PVT Simulation

Before a characterized fluid composition is applied, for example, in compositional reservoir simulation studies, it is recommended to perform PVT simulations and compare with measured PVT data. This can be seen as a final quality check (QC) of the fluid composition. The check can be performed by characterizing the fluid without any lumping of the components lighter than C_7, and with the C_{7+} fraction lumped into 10–12 pseudocomponents using the method of Pedersen et al. described in Chapter 5. For an accurate plus composition, such fluid description will generally provide a good match of measured PVT data. Further lumping may be needed for the actual application. If this lumping deteriorates the match of the PVT data, regression may be performed, as outlined in Chapter 9.

If the initial fluid characterization fails to give the desired match of measured PVT data, the compositional analysis and the PVT data should be critically checked. An inaccurate plus molecular weight is a common source of error. Regression to measured PVT data as described in Chapter 9 is an option if major deviations between measured and simulated PVT data persist after a critical evaluation analysis of the measured data. It is described in Section 2.2 how to QC a compositional analysis.

7.1 CONSTANT MASS EXPANSION

The constant mass expansion (CME) experiment is described in Section 3.1.1. Table 3.9 shows CME data for the gas condensate composition in Table 3.8. The composition is characterized for the Soave–Redlich–Kwong equation with temperature-dependent volume shift parameters. The property correlations in Equations 5.1 through 5.3 and the SRK coefficients in Table 5.3 have been used. Table 7.1 shows the characterized mixture lumped with a total of 22 components (10 defined components and 12 pseudocomponents). Lumping starts with C_{10}. Each of C_{10+} fractions contains approximately the same weight amount. A carbon number fraction is not split between two pseudocomponents, which explains the slight variation in weight amounts of the various pseudocomponents. Binary interaction coefficients are shown in Table 7.2. Table 7.3 shows the mixture lumped with a total of six pseudocomponents with binary interaction coefficients shown in Table 7.4. In this case, lumping starts with C_7. Again, the cut points are selected to have approximately the same weight amount in each pseudocomponent. In Table 7.3, N_2 and C_1, as the two most volatile components, have been lumped together. CO_2 is less volatile and has been lumped with C_2 and C_3. Finally, the C_4–C_6 components have been lumped together.

Experimental and simulated CME results are plotted in Figures 7.1 through 7.3. Figure 7.1 shows relative volumes (total volume divided by saturation point volume), Figure 7.2 shows liquid volumes (liquid volume in percentage of saturation point volume), and Figure 7.3 shows gas-phase Z-factors (defined in Equation 3.2). Simulation results are shown for the mixture composition lumped to a total of 22 (pseudo)components (Tables 7.1 and 7.2) as well as for the mixture lumped to a total of 6 pseudocomponents (Tables 7.3 and 7.4). The simulated relative volumes and the simulated Z-factors agree nicely with the experimental results, but some deviations are seen between the experimental and simulated liquid volumes. It is shown in Chapter 9 how the match can be improved by regression.

Table 3.5 shows CME data for the oil mixture in Table 3.6. The composition was characterized for the Soave–Redlich–Kwong and Peng–Robinson equations of state, and a constant (temperature independent) Peneloux volume shift parameter was used. The component properties in Table 7.5 were found using the property correlations in Equations 5.1 through 5.3 and the coefficients in

DOI: 10.1201/9780429457418-7

TABLE 7.1
The Gas Condensate Mixture in Table 3.8 Characterized for the SRK–Peneloux(T) Equation of State[*]

	Mol%	Weight%	Mol Weight	T_c (°C)	P_c (bar)	Acentric Factor	c_0 (cm³/mol)	c_1 (cm³/mol K)
N_2	0.600	0.577	28.0	−147.0	33.94	0.040	0.92	0.0000
CO_2	3.340	5.044	44.0	31.1	73.76	0.225	3.03	0.0100
C_1	74.167	40.831	16.0	−82.6	46.00	0.008	0.63	0.0000
C_2	7.901	8.153	30.1	32.3	48.84	0.098	2.63	0.0000
C_3	4.150	6.280	44.1	96.7	42.46	0.152	5.06	0.0000
iC_4	0.710	1.416	58.1	135.0	36.48	0.176	7.29	0.0000
nC_4	1.440	2.872	58.1	152.1	38.00	0.193	7.86	0.0000
iC_5	0.530	1.312	72.2	187.3	33.84	0.227	10.93	0.0000
nC_5	0.660	1.634	72.2	196.5	33.74	0.251	12.18	0.0000
C_6	0.810	2.395	86.2	234.3	29.69	0.296	17.98	0.0000
C_7	1.200	3.747	91.0	255.6	34.98	0.453	4.74	0.0194
C_8	.1.150	4.104	104.0	279.3	31.23	0.491	10.89	0.0129
C_9	0.630	2.573	119.0	302.6	27.68	0.534	18.23	0.0047
C_{10}	0.500	2.282	133.0	321.1	24.78	0.574	25.88	−0.0050
C_{11}–C_{12}	0.560	2.869	149.3	340.0	21.93	0.619	35.49	−0.0192
C_{13}	0.280	1.614	168.0	362.4	20.39	0.669	39.92	−0.0321
C_{14}–C_{15}	0.390	2.504	187.1	383.6	19.13	0.720	43.16	−0.0469
C_{16}–C_{17}	0.290	2.126	213.7	409.3	17.47	0.788	47.97	−0.0715
C_{18}–C_{19}	0.220	1.829	242.3	434.3	16.09	0.857	51.18	−0.0999
C_{20}–C_{22}	0.171	1.695	288.8	473.0	14.70	0.963	48.87	−0.1500
C_{23}–C_{28}	0.178	2.116	346.4	518.2	13.83	1.080	35.07	−0.2000
C_{29}–C_{80}	0.121	2.026	487.9	624.3	12.85	1.269	−18.13	−0.3100

[*] Fluid is described using a total of 10 defined components and 12 C_{7+} components. c_0 and c_1 are Peneloux volume shift parameters as defined in Equation 5.9. Binary interaction parameters can be seen in Table 7.2.

TABLE 7.2
Nonzero Binary Interaction Coefficients for Use with the Mixture in Table 7.1

	N_2	CO_2
CO_2	−0.032	—
C_1	0.028	0.120
C_2	0.041	0.120
C_3	0.076	0.120
iC_4	0.094	0.120
nC_4	0.070	0.120
iC_5	0.087	0.120
nC_5	0.088	0.120
C_6	0.080	0.120
C_{7+}	0.080	0.100

TABLE 7.3

The Gas Condensate Mixture in Table 3.8 Characterized for the SRK–Peneloux(T) Equation of State*

	Mol%	Weight%	Mol wgt	T_c (°C)	P_c (bar)	Acentric Factor	c_0 (cm³/mol)	c_1 (cm³/ (mol K))
$N_2 + C_1$	74.767	41.406	16.1	−83.45	45.83	0.0085	0.63	0.0000
$CO_2 + C_2-C_3$	15.392	19.478	36.9	52.71	53.24	0.1483	3.37	0.0000
C_4-C_6	4.150	9.630	67.6	182.31	34.42	0.2306	10.82	0.0000
C_7-C_9	2.980	10.424	101.9	276.53	31.70	0.4883	10.12	0.0127
$C_{10}-C_{15}$	1.730	9.269	156.1	351.01	21.61	0.6439	35.96	−0.0247
$C_{16}-C_{80}$	0.980	9.793	291.2	493.08	14.99	0.9936	50.57	−0.1500

* The fluid composition is described using three lumped defined components and three C_{7+} fractions. c_0 and c_1 are Peneloux volume shift parameters as defined in Equation 5.9. Binary interaction parameters can be seen in Table 7.4.

TABLE 7.4

Nonzero Binary Interaction Coefficients for Use with the Mixture in Table 7.3

	$N_2 + C_1$	$CO_2 + C_2-C_3$
$CO_2 + C_2-C_3$	0.0261	—
C_4-C_6	0.0007	0.0260
C_7-C_9	0.0006	0.0217
$C_{10}-C_{15}$	0.0006	0.0217
$C_{16}-C_{80}$	0.0006	0.0217

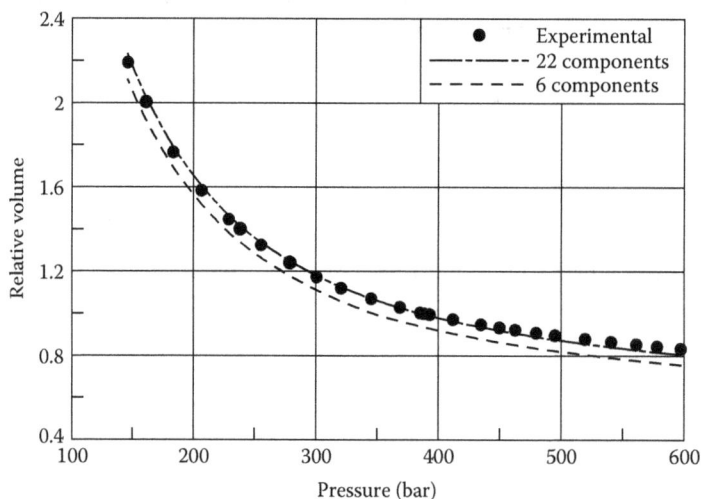

FIGURE 7.1 Measured and simulated relative volumes in a constant mass expansion experiment at 155°C for the gas condensate mixture in Table 3.8. The SRK equation with T-dependent volume correction was used in the simulations. The characterized mixture compositions (containing 22 and 6 (pseudo)components, respectively) are shown in Tables 7.1 and 7.3 and binary interaction parameters in Tables 7.2 and 7.4. The experimental data is shown in Table 3.9.

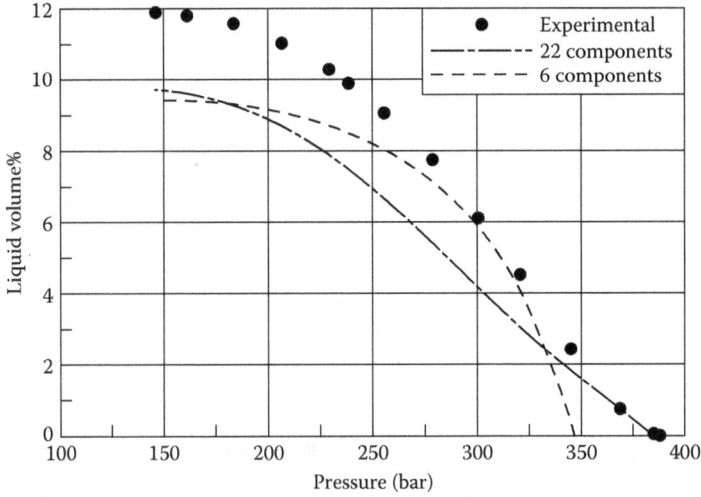

FIGURE 7.2 Measured and simulated liquid volumes (percentage of saturation point volume) in a constant mass expansion experiment at 155°C for the gas condensate mixture in Table 3.8. The SRK equation with T-dependent volume correction was used in the simulations. The characterized mixture compositions (containing 22 and 6 (pseudo)components, respectively) are shown in Tables 7.1 and 7.3 and binary interaction parameters in Tables 7.2 and 7.4. The experimental data is shown in Table 3.9.

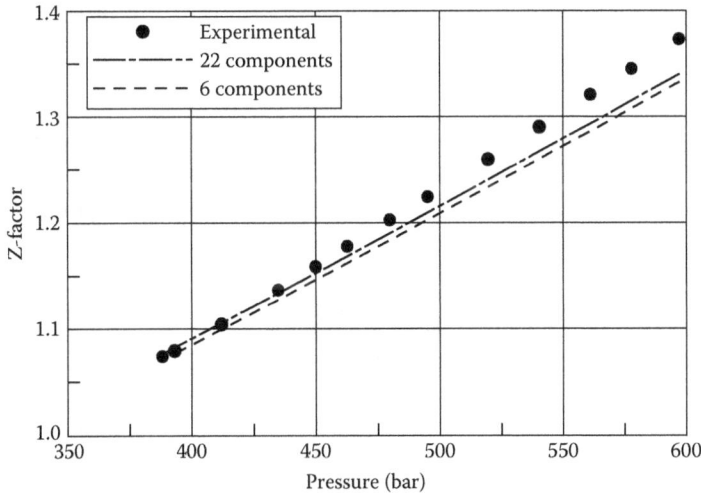

FIGURE 7.3 Measured and simulated gas phase Z-factors in a constant mass expansion experiment at 155°C for the gas condensate mixture in Table 3.8. The SRK equation with T-dependent volume correction was used in the simulations. The characterized mixture compositions (containing 22 and 6 (pseudo) components, respectively) are shown in Tables 7.1 and 7.3 and binary interaction parameters in Tables 7.2 and 7.4. The experimental data is shown in Table 3.9.

Table 5.3. The C_{7+} fraction was represented using three pseudocomponents. Binary interaction coefficients are shown in Table 7.6.

Experimental and simulated CME results are plotted in Figures 7.4 and 7.5. Figure 7.4 shows relative volumes (total volume divided by saturation point volume), and Figure 7.5 shows the isothermal oil compressibilities (defined in Equation 3.4). The simulated relative volumes agree nicely with the

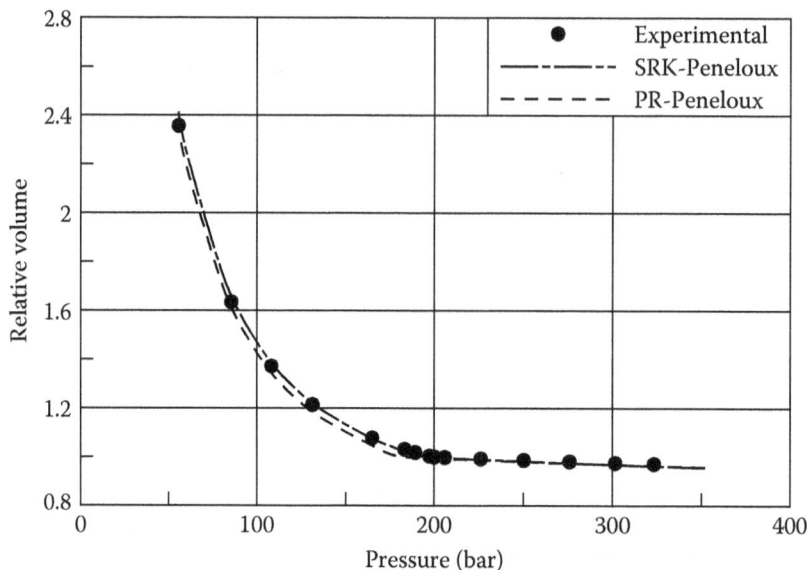

FIGURE 7.4 Measured and simulated relative volumes in a constant mass expansion experiment at 97.5°C for the oil mixture in Table 3.6. The SRK and PR equations with Peneloux volume correction were used in the simulations. The characterized mixture compositions are shown in Table 7.5 and binary interaction parameters in Table 7.6. The experimental data is shown in Table 3.9.

TABLE 7.5
The Oil Mixture in Table 3.6 Characterized for the SRK–Peneloux and PR–Peneloux Equations of State*

Component	Mol%	SRK–Peneloux				PR–Peneloux			
		T_c (°C)	P_c (bar)	Acentric Factor	c_0 (cm3/ mol)	T_c (°C)	P_c (bar)	Acentric Factor	c_0 (cm3/ mol)
N_2	0.39	−147.0	33.94	0.040	0.92	−147.0	33.94	0.040	−4.23
CO_2	0.30	31.1	73.76	0.225	3.03	31.1	73.76	0.225	−1.64
C_1	40.20	−82.6	46.00	0.008	0.63	−82.6	46.00	0.008	−5.20
C_2	7.61	32.3	48.84	0.098	2.63	32.3	48.84	0.098	−5.79
C_3	7.95	96.7	42.46	0.152	5.06	96.7	42.46	0.152	−6.35
iC_4	1.19	135.0	36.48	0.176	7.29	135.0	36.48	0.176	−7.18
nC_4	4.08	152.1	38.00	0.193	7.86	152.1	38.00	0.193	−6.49
iC_5	1.39	187.3	33.84	0.227	10.93	187.3	33.84	0.227	−6.20
nC_5	2.15	196.5	33.74	0.251	12.18	196.5	33.74	0.251	−5.12
C_6	2.79	234.3	29.69	0.296	17.98	234.3	29.69	0.296	1.39
C_7–C_{13}	19.33	311.1	25.42	0.563	25.04	318.8	24.38	0.447	15.18
C_{14}–C_{26}	8.64	449.3	16.20	0.894	48.09	482.5	16.10	0.834	27.16
C_{27}–C_{80}	3.98	679.6	13.32	1.256	−51.49	776.1	12.50	1.120	−24.73

* The fluid is described using a total of ten defined components and three C_{7+} components. c_0 is the Peneloux volume shift parameter as defined in Equation 4.44. Binary interaction parameters can be seen in Table 7.6.

TABLE 7.6

Nonzero Binary Interaction Coefficients for Use with the Oil Mixture in Table 7.5

	SRK–Peneloux		PR–Peneloux	
	N_2	CO_2	N_2	CO_2
CO_2	−0.032	—	−0.017	—
C_1	0.028	0.120	0.031	0.120
C_2	0.041	0.120	0.052	0.120
C_3	0.076	0.120	0.085	0.120
iC_4	0.094	0.120	0.103	0.120
nC_4	0.070	0.120	0.080	0.120
iC_5	0.087	0.120	0.092	0.120
nC_5	0.088	0.120	0.100	0.120
C_6	0.080	0.120	0.080	0.120
C_{7+}	0.080	0.100	0.080	0.100

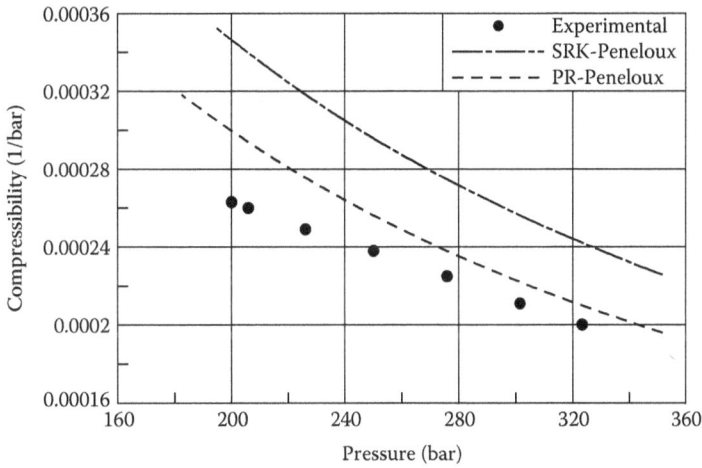

FIGURE 7.5 Measured and simulated isothermal oil compressibilities in a constant mass expansion experiment at 97.5°C for the oil mixture in Table 3.6. The SRK and PR equations with Peneloux volume correction were used in the simulations. The characterized mixture compositions are shown in Table 7.5 and binary interaction parameters in Table 7.6. The experimental data is shown in Table 3.9.

experimental results. Somewhat better results for the isothermal oil compressibilities are obtained with the PR–Peneloux equation than with the SRK–Peneloux equation. The compressibility is determined by the functional form of the equation of state, and it seldom makes sense to try to improve the match by regression to compressibility data.

7.2 CONSTANT VOLUME DEPLETION

The constant volume depletion (CVD) experiment is described in Section 3.1.3. Tables 3.16 and 3.17 show CVD data for the gas condensate composition in Table 3.15. The composition is characterized for the Peng–Robinson equation with constant volume shift parameters. The property correlations in Equations 5.1 through 5.3 and the PR coefficients in Table 5.3 have been used. The

TABLE 7.7

Composition of the Gas Condensate Mixture Whose Constant Volume Depletion Liquid Dropout Curve Is Plotted in Figure 7.6*

Component	Weight% Composition	Mol% Composition with $M_+ = 381.0$	Mol% Composition with $M_+ = 350.5$
N_2	0.57	0.64	0.64
CO_2	4.92	3.53	3.53
C_1	35.94	70.78	70.74
C_2	8.51	8.94	8.94
C_3	7.05	5.05	5.05
iC_4	1.56	0.85	0.85
nC_4	3.09	1.68	1.68
iC_5	1.42	0.62	0.62
nC_5	1.80	0.79	0.79
C_6	2.26	0.83	0.83
C_7	3.09	1.06	1.06
C_8	3.51	1.06	1.06
C_9	2.98	0.79	0.79
C_{10}	2.40	0.57	0.57
C_{11}	1.86	0.38	0.38
C_{12}	1.90	0.37	0.37
C_{13}	1.79	0.32	0.32
C_{14}	1.69	0.27	0.27
C_{15}	1.47	0.23	0.23
C_{16}	1.29	0.19	0.19
C_{17}	1.26	0.17	0.17
C_{18}	1.03	0.13	0.13
C_{19}	1.11	0.13	0.13
C_{20+}	7.48	0.62	0.67

* Molar compositions are shown for two different plus molecular weights (M_+).

mixture composition is lumped to a total of 22 components and pseudocomponents. The simulated liquid dropout curve is shown as a dashed line in Figure 7.6. The general curvature agrees nicely with the experimental data, but the simulated saturation point (starting pressure for liquid dropout curve) is slightly too high. As was mentioned in Section 2.4.2.1, the plus molecular weight (M_+) may be slightly inaccurate, which will result in an inaccurate molar composition because the composition is measured as a weight composition. Table 7.7 shows the measured weight composition (from Table 3.15) and the molar composition based on an M_+ of 381.0 (measured value) and an M_+ of 350.5. The latter M_+ is 8% lower than the experimentally determined value, which percentage is not an unrealistic uncertainty for a plus molecular weight. Both molar compositions in Table 7.7 agree with the measured weight composition.

The adjustment of the plus molecular weight is illustrated graphically in Figure 7.7, which shows mole percentages of each C_{7+} fraction versus molecular weight. The mole percentages of C_7–C_{19} are shown as circles in Figure 7.7. The dashed–dotted line shows the mole percentages of the fractions C_{20}–C_{80} calculated from Equations 5.10 through 5.12, assuming a plus molecular weight (M_+) of 381.0 (measured value). The dashed line shows the mole percentages of the fractions C_{20}–C_{80} calculated from Equations 5.10 through 5.12 but assuming that M_+ is only 350.5, which is the molecular weight that makes the simulated saturation point agree with the measured value. As can be seen

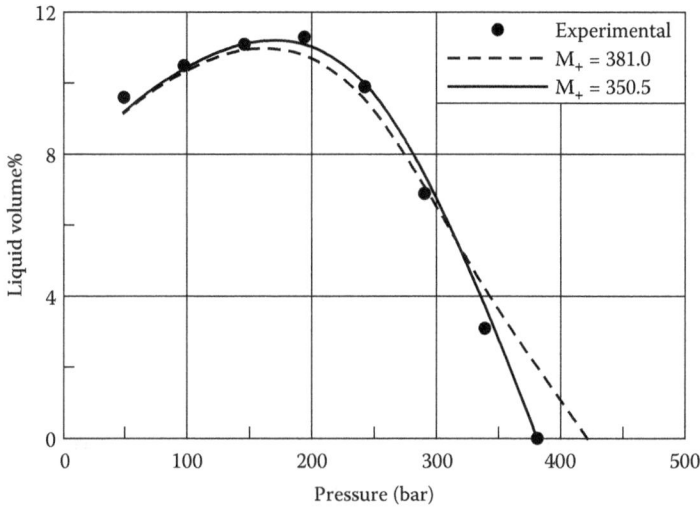

FIGURE 7.6 Measured and simulated liquid volume percentages for a constant volume depletion experiment at 150.3°C for the gas condensate mixture in Table 3.15. The PR equation with Peneloux volume correction was used in the simulations. The dashed line is simulated assuming a plus molecular weight of 380.0 (characterized composition in Table 7.1) and the full-drawn line assuming a plus molecular weight of 350.5. The characterized composition for a plus molecular weight of 350.5 is shown in Table 7.8 and binary interaction parameters in Table 7.9. The experimental data is shown in Table 3.16.

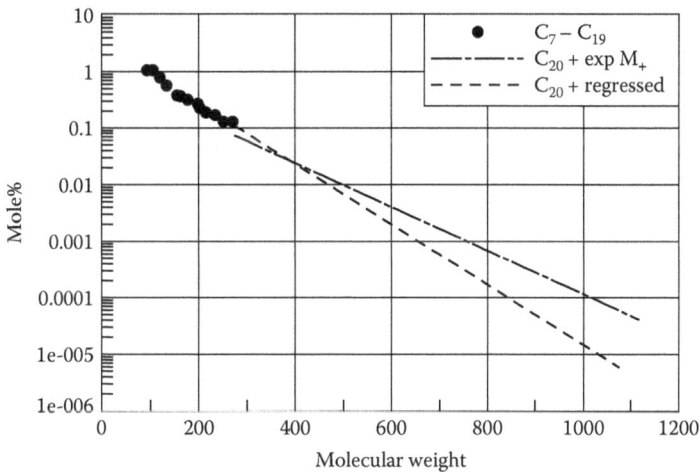

FIGURE 7.7 C_{7+} mole percentages for gas condensate mixture in Table 3.15. C_{20}–C_{80} mole percentages are shown assuming a C_{20+} molecular weight of 381.0 (dashed–dotted line) and assuming a C_{20+} molecular weight of 350.5 (dashed line).

from Figure 7.7, the line for $M_+ = 350.5$ is in better agreement with a best-fit line through the C_7–C_{19} mole percentages than the line for $M_+ = 381.0$. This suggests that an adjustment of the plus molecular weight from 381.0 to 350.5 is justified and not just a convenient fitting tool.

As can be seen from the full-drawn line in Figure 7.6, an almost perfect match of the liquid dropout curve is seen using the molar composition based on a plus molecular weight of 350.5. The characterized composition (for M_+ of 350.5) can be seen in Table 7.8 and binary interaction parameters from Table 7.9. Figure 7.8 shows measured and simulated cumulative mole percentages

TABLE 7.8

The Gas Condensate Mixture in Table 7.7 (M_+ = 350.5) Characterized for the PR–Peneloux Equation of State*

	Mol%	Mol Weight	T_c (°C)	P_c (bar)	Acentric Factor	c_0 (cm³/mol)
N_2	0.640	28.0	−147.0	33.94	0.040	−4.23
CO_2	3.528	44.0	31.1	73.76	0.225	−1.64
C_1	70.742	16.0	−82.6	46.00	0.008	−5.20
C_2	8.935	30.1	32.3	48.84	0.098	−5.79
C_3	5.047	44.1	96.7	42.46	0.152	−6.35
iC_4	0.850	58.1	135.0	36.48	0.176	−7.18
nC_4	1.679	58.1	152.1	38.00	0.193	−6.49
iC_5	0.620	72.2	187.3	33.84	0.227	−6.20
nC_5	0.790	72.2	196.5	33.74	0.251	−5.12
C_6	0.830	86.2	234.3	29.69	0.296	1.39
C_7	1.059	92.2	255.8	30.34	0.325	5.76
C_8	1.059	104.6	280.4	27.84	0.366	9.98
C_9	0.790	119.1	305.0	25.16	0.414	12.85
$C_{10}–C_{11}$	0.949	141.8	341.3	22.32	0.490	18.56
C_{12}	0.370	162.0	368.8	20.43	0.552	21.85
C_{13}	0.320	177.0	389.1	19.45	0.599	24.85
$C_{14}–C_{15}$	0.500	199.8	417.4	18.14	0.668	26.72
$C_{16}–C_{17}$	0.360	224.0	445.3	17.02	0.740	26.29
$C_{18}–C_{19}$	0.260	260.5	484.3	15.74	0.841	22.87
$C_{20}–C_{22}$	0.271	288.6	513.5	15.09	0.915	20.29
$C_{23}–C_{27}$	0.232	340.7	564.4	14.15	1.037	11.17
$C_{28}–C_{77}$	0.170	462.5	682.4	12.87	1.206	−16.43

* Fluid is described using a total of 10 defined components and 12 C_{7+} components. c_0 is the Peneloux volume shift parameter as defined in Equation 4.44. Binary interaction parameters can be seen in Table 7.9.

TABLE 7.9

Nonzero Binary Interaction Coefficients for Use with the Mixture in Table 7.8

	N_2	CO_2
N_2	−0.017	—
CO_2	0.031	0.120
C_1	0.052	0.120
C_2	0.085	0.120
C_3	0.103	0.120
iC_4	0.080	0.120
nC_4	0.092	0.120
iC_5	0.100	0.120
nC_5	0.080	0.120
C_6	0.080	0.100
C_{7+}	0.080	0.100

of gas removed from the cell during the experiment. Figures 7.9 and 7.10 show measured and simulated gas and two-phase Z-factors, respectively. Table 7.10 shows the simulated molar compositions of the gas removed at each pressure stage. The experimentally determined gas compositions can be seen in Table 3.17. All the simulated results are in good agreement with the measured data.

TABLE 7.10
Simulated Molar Compositions (Mol%) of the Gas Depleted in a Constant Volume Depletion Experiment

Pressure (bar)	381.5*	338.9	290.6	242.3	194.1	145.8	97.5	49.3
N_2	0.64	0.65	0.66	0.67	0.68	0.68	0.67	0.66
CO_2	3.53	3.54	3.57	3.60	3.63	3.66	3.69	3.69
C_1	70.74	71.28	72.05	72.90	73.67	74.18	74.24	73.31
C_2	8.94	8.95	8.97	9.00	9.06	9.14	9.25	9.39
C_3	5.05	5.03	5.01	4.98	4.98	5.01	5.10	5.33
iC_4	0.85	0.84	0.84	0.83	0.82	0.82	0.83	0.89
nC_4	1.68	1.66	1.64	1.62	1.60	1.59	1.63	1.75
iC_5	0.62	0.61	0.60	0.58	0.57	0.56	0.57	0.63
nC_5	0.79	0.78	0.76	0.74	0.72	0.70	0.72	0.79
C_6	0.64	0.65	0.66	0.67	0.68	0.68	0.67	0.66
C_{7+}	6.34	5.84	5.13	4.33	3.58	2.98	2.62	2.80
C_{7+} mol. weight. (g/mol)	164	155	146	136	128	122	117	114

Note: The experiment was at 150.3°C for the mixture with the weight% composition in Table 3.15 and the molar composition in Table 7.7, which is for a plus molecular weight of 350.5. The PR equation with Peneloux volume correction was used in the simulations. The characterized composition is shown in Table 7.8 and binary interaction parameters in Table 7.9. The experimentally determined gas compositions can be seen in Table 3.17.
* Saturation point.

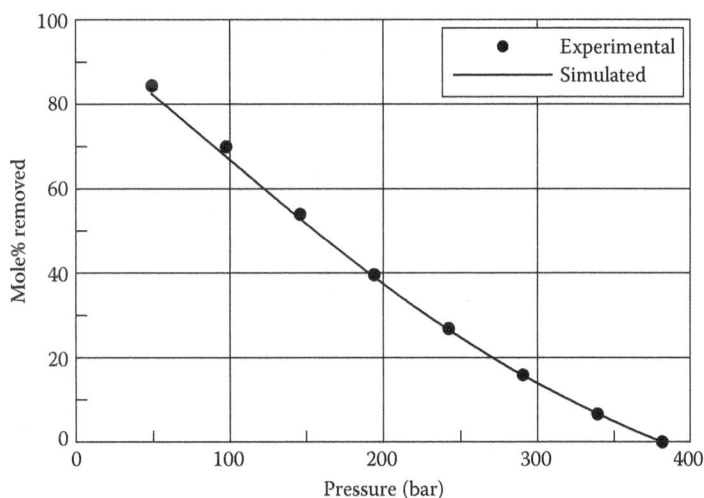

FIGURE 7.8 Measured and simulated cumulative mole percentages removed in a constant volume depletion experiment at 150.3°C for the gas condensate mixture with the weight% composition in Table 3.15 and the molar composition in Table 7.7. The PR equation with Peneloux volume correction was used in the simulations. The characterized composition is shown in Table 7.8 and binary interaction parameters in Table 7.9. The experimental data is shown in Table 3.16.

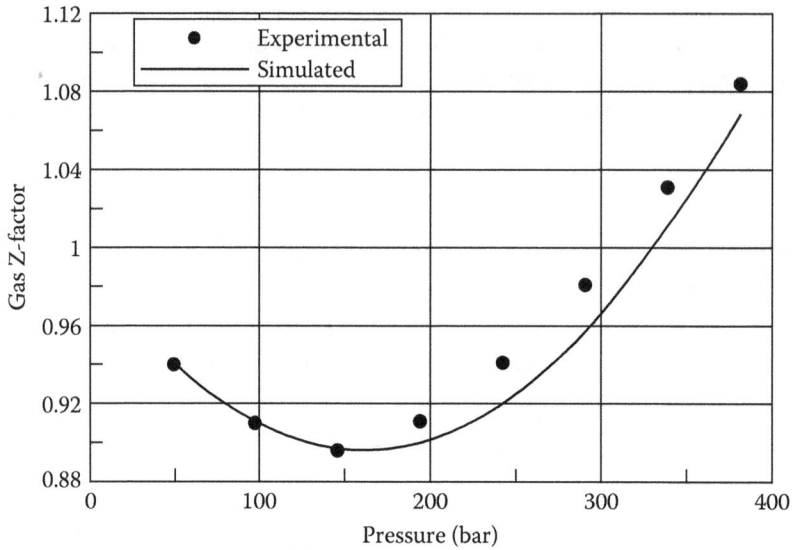

FIGURE 7.9 Measured and simulated gas-phase Z-factors in a constant volume depletion experiment at 150.3°C for the gas condensate mixture with the weight% composition in Table 3.15 and the molar composition in Table 7.7. The PR equation with Peneloux volume correction was used in the simulations. The characterized composition is shown in Table 7.8 and binary interaction parameters in Table 7.9. The experimental data is shown in Table 3.16.

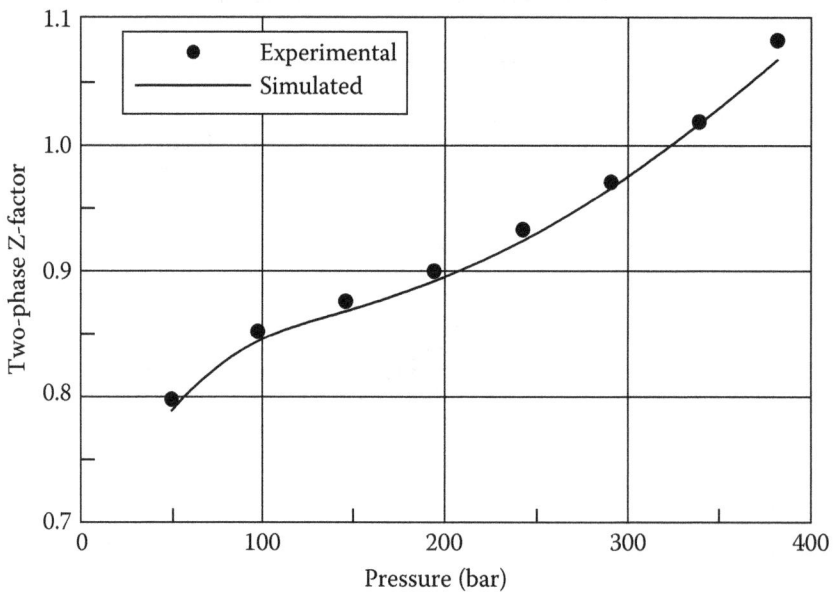

FIGURE 7.10 Measured and simulated two-phase Z-factors in a constant volume depletion experiment at 150.3°C for the gas condensate mixture with the weight% composition in Table 3.15 and the molar composition in Table 7.7. The PR equation with Peneloux volume correction was used in the simulations. The characterized composition is shown in Table 7.8 and binary interaction parameters in Table 7.9. The experimental data is shown in Table 3.16.

7.3 DIFFERENTIAL LIBERATION

Section 3.1.2 describes a differential liberation experiment. Table 3.13 shows differential liberation data for the oil mixture in Table 3.6. The composition was characterized for the Soave–Redlich–Kwong and Peng–Robinson equations of state, and a constant (temperature-independent) Peneloux volume shift parameter was used. The characterized compositions in Table 7.5 were generated using the property correlations in Equations 5.1 through 5.3 and the coefficients in Table 5.3. The C_{7+} fraction was represented using three pseudocomponents. Binary interaction coefficients are shown in Table 7.6.

Experimental and simulated differential liberation results are plotted in Figures 7.11 through 7.16. Figure 7.11 shows B_o factors (defined in Equation 3.9), Figure 7.12 shows solution gas/oil ratios

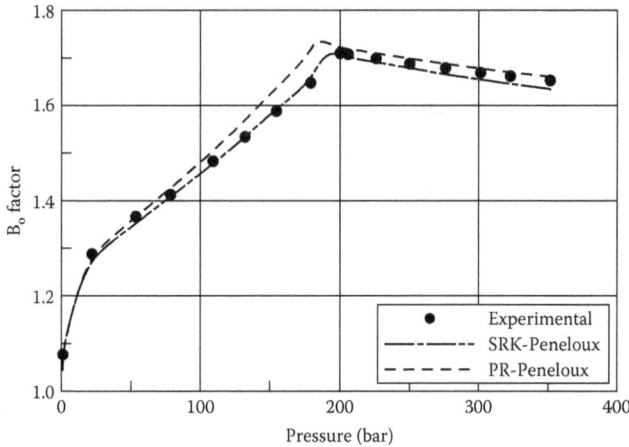

FIGURE 7.11 Measured and simulated oil formation (B_o) factors for differential liberation experiment at 97.5°C for the oil mixture in Table 3.6. The SRK and PR equations with Peneloux volume correction were used in the simulations. The characterized mixture compositions are shown in Table 7.5 and binary interaction parameters in Table 7.6. The experimental data is shown in Table 3.13.

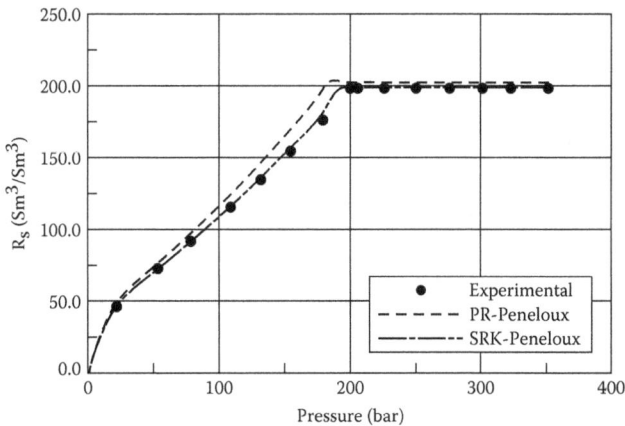

FIGURE 7.12 Measured and simulated solution gas/oil ratios (R_s) for differential liberation experiment at 97.5°C for the oil mixture in Table 3.6. The SRK and PR equations with Peneloux volume correction were used in the simulations. The characterized mixture compositions are shown in Table 7.5 and binary interaction parameters in Table 7.6. The experimental data is shown in Table 3.13.

(defined in Equation 3.10), Figure 7.13 shows B_g (defined in Equation 3.11), Figure 7.14 shows oil densities, Figure 7.15 shows gas-phase Z-factors (defined in Equation 3.2), and Figure 7.16 shows gas gravities (defined in Equation 3.12). For all properties and both applied equations of state, the simulated and experimental results agree nicely.

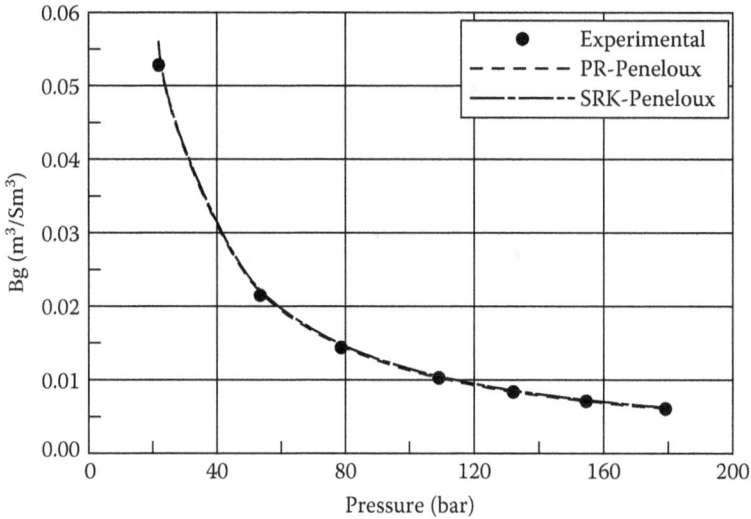

FIGURE 7.13 Measured and simulated gas formation volume factors (B_g) for differential liberation experiment at 97.5°C for the oil mixture in Table 3.6. The SRK and PR equations with Peneloux volume correction were used in the simulations. The characterized mixture compositions are shown in Table 7.5 and binary interaction parameters in Table 7.6. The experimental data is shown in Table 3.13.

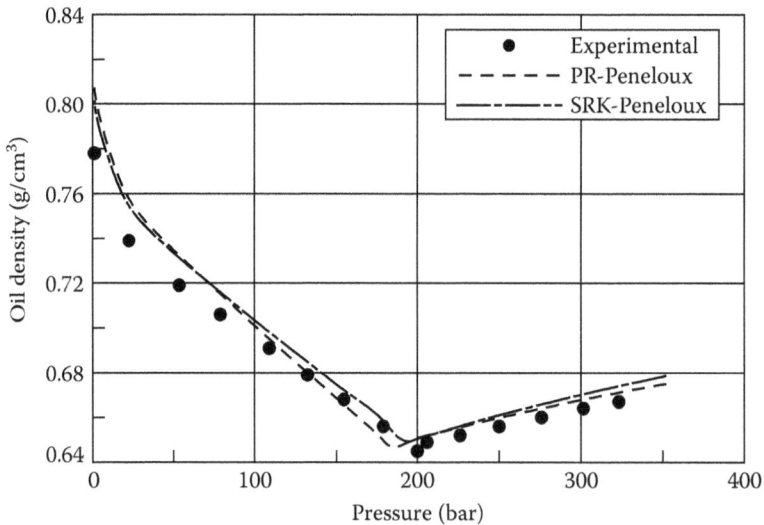

FIGURE 7.14 Measured and simulated oil densities for differential liberation experiment at 97.5°C for the oil mixture in Table 3.6. The SRK and PR equations with Peneloux volume correction were used in the simulations. The characterized mixture compositions are shown in Table 7.5 and binary interaction parameters in Table 7.6. The experimental data is shown in Table 3.13.

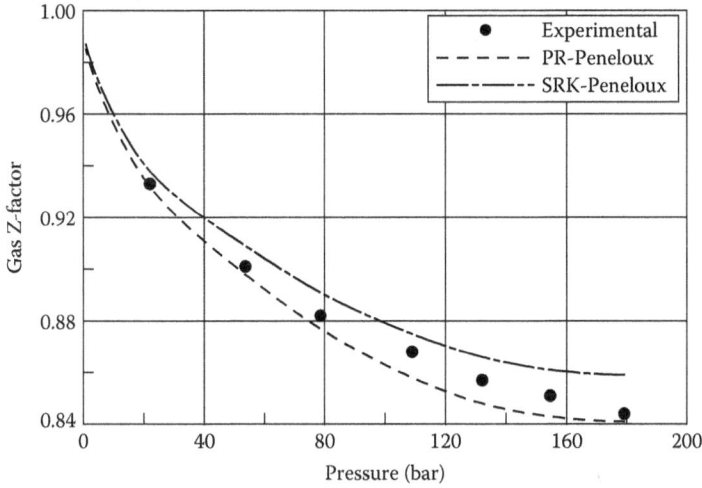

FIGURE 7.15 Measured and simulated gas-phase Z-factors for differential liberation experiment at 97.5°C for the oil mixture in Table 3.6. The SRK and PR equations with Peneloux volume correction were used in the simulations. The characterized mixture compositions are shown in Table 7.5 and binary interaction parameters in Table 7.6. The experimental data is shown in Table 3.13.

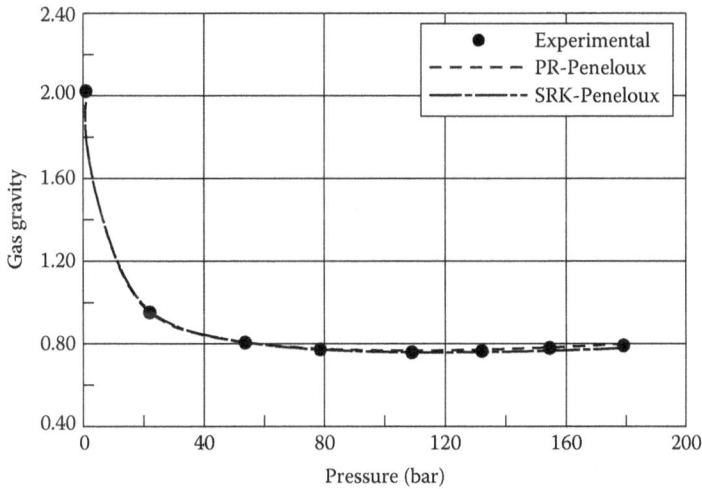

FIGURE 7.16 Measured and simulated gas gravities for differential liberation experiment at 97.5°C for the oil mixture in Table 3.6. The SRK and PR equations with Peneloux volume correction were used in the simulations. The characterized mixture compositions are shown in Table 7.5 and binary interaction parameters in Table 7.6. The experimental data is shown in Table 3.13.

7.4 SEPARATOR TEST

Section 3.1.4 describes how a separator test is performed, and Table 3.21 lists the primary results from the experiment. Table 3.22 shows separator test results for the oil composition in Table 3.23. This composition was characterized for the Soave–Redlich–Kwong and Peng–Robinson equations of state, and a constant (temperature-independent) Peneloux volume shift parameter was used. The characterized compositions in Table 7.11 were generated using the property correlations in Equations 5.1 through 5.3 and the coefficients in Table 5.3. The C_{7+} fraction was represented using four pseudocomponents. Binary interaction coefficients are shown in Table 7.12. Table 7.13 shows simulated separator data for SRK–Peneloux and PR–Peneloux. Simulated separator gas compositions can be seen in Table 7.14. The saturation point simulated using the PR–Peneloux equation is slightly lower than measured. Otherwise, the simulation results agree well with the separator data.

TABLE 7.11
The Oil Mixture in Table 3.23 Characterized for the SRK–Peneloux and PR–Peneloux Equations of State*

Component	Mol%	SRK–Peneloux				PR–Peneloux			
		T_c (°C)	P_c (bar)	Acentric Factor	c_0 (cm³/mol)	T_c (°C)	P_c (bar)	Acentric Factor	c_0 (cm³/mol)
N_2	0.59	−147.0	33.94	0.040	0.92	−147.0	33.94	0.040	−4.23
CO_2	0.36	31.1	73.76	0.225	3.03	31.1	73.76	0.225	−1.64
C_1	40.81	−82.6	46.00	0.008	0.63	−82.6	46.00	0.008	−5.20
C_2	7.38	32.3	48.84	0.098	2.63	32.3	48.84	0.098	−5.79
C_3	7.88	96.7	42.46	0.152	5.06	96.7	42.46	0.152	−6.35
iC_4	1.20	135.0	36.48	0.176	7.29	135.0	36.48	0.176	−7.18
nC_4	3.96	152.1	38.00	0.193	7.86	152.1	38.00	0.193	−6.49
iC_5	1.33	187.3	33.84	0.227	10.93	187.3	33.84	0.227	−6.20
nC_5	2.09	196.5	33.74	0.251	12.18	196.5	33.74	0.251	−5.12
C_6	2.84	234.3	29.69	0.296	17.98	234.3	29.69	0.296	1.39
C_7–C_{10}	14.44	296.1	25.38	0.536	25.51	303.5	25.05	0.416	10.96
C_{11}–C_{17}	9.23	387.5	18.54	0.735	47.08	407.9	18.53	0.648	25.15
C_{18}–C_{31}	5.23	501.1	14.98	1.026	36.31	546.3	14.71	0.986	21.45
C_{32}–C_{80}	2.66	719.3	13.48	1.278	−96.82	826.4	12.26	1.103	−47.07

* The fluid is described using a total of ten defined components and four C_{7+} components. c_0 is the Peneloux volume shift parameter as defined in Equation 4.44. Binary interaction parameters can be seen in Table 7.12.

TABLE 7.12
Nonzero Binary Interaction Coefficients for Use with the Mixture in Table 7.11

	SRK–Peneloux		PR–Peneloux	
	N_2	CO_2	N_2	CO_2
CO_2	−0.032	—	−0.017	—
C_1	0.028	0.120	0.031	0.120
C_2	0.041	0.120	0.052	0.120
C_3	0.076	0.120	0.085	0.120
iC_4	0.094	0.120	0.103	0.120
nC_4	0.070	0.120	0.080	0.120
iC_5	0.087	0.120	0.092	0.120
nC_5	0.088	0.120	0.100	0.120
C_6	0.080	0.120	0.080	0.120
C_{7+}	0.080	0.100	0.080	0.100

TABLE 7.13

Separator Simulation Results for Oil Composition in Table 3.23*

Stage	Pressure (bar)	Temperature (°C)	Gas/Oil Ratio (Sm³/m³)	B_o-Factor (m³/Sm³)
SRK–Peneloux				
Sat. Pt.	196.0	97.8	—	1.562
1	68.9	89.4	105.1	1.255
2	22.7	87.2	33.6	1.152
3	6.9	83.9	15.3	1.095
4	2.0	77.2	10.9	1.044
Std.	1.0	15.0	0.0	1.000
PR–Peneloux				
Sat. Pt.	181.8	97.8	—	1.564
1	68.9	89.4	99.9	1.264
2	22.7	87.2	36.5	1.151
3	6.9	83.9	16.3	1.091
4	2.0	77.2	11.5	1.037
Std.	1.0	15.0	0.0	1.000

* The experimental results can be seen in Table 3.22. The SRK and PR equations with Peneloux volume correction were used in the simulations. The characterized mixture compositions are shown in Table 7.11 and binary interaction parameters in Table 7.12.

TABLE 7.14

Simulated Molar Compositions of Gas from Each Stage of the Separator Train in Table 7.13

Component	Stage 1	Stage 2	Stage 3	Stage 4
SRK–Peneloux				
N_2	1.32	0.56	0.14	0.01
CO_2	0.57	0.70	0.67	0.31
C_1	77.80	64.69	37.06	9.06
C_2	9.47	14.35	19.68	15.06
C_3	6.52	11.96	23.94	34.30
iC_4	0.69	1.30	2.98	5.75
nC_4	1.89	3.59	8.58	18.11
iC_5	0.40	0.73	1.83	4.47
nC_5	0.55	0.98	2.46	6.16
C_6	0.41	0.66	1.61	4.24
C_{7+}	0.41	0.49	1.05	2.54
PR–Peneloux				
N_2	0.90	0.39	0.10	0.01
CO_2	0.48	0.57	0.53	0.24
C_1	77.70	64.86	37.62	9.43
C_2	9.75	14.53	19.80	15.17
C_3	6.52	11.71	23.24	33.28
iC_4	0.67	1.24	2.82	5.45
nC_4	1.94	3.59	8.48	17.86
iC_5	0.42	0.75	1.83	4.47
nC_5	0.57	0.99	2.44	6.09
C_6	0.41	0.63	1.53	4.02
C_{7+}	0.63	0.74	1.60	4.01

* The experimental results can be seen in Table 3.24. The SRK and PR equations with Peneloux volume correction were used in the simulations. The characterized mixture compositions are shown in Table 7.11 and binary interaction parameters in Table 7.12.

7.5   SOLUBILITY SWELLING TEST

A swelling experiment is described in Section 3.2.1. Table 3.38 shows results of a swelling test for the oil and the injection gas in Table 3.39. The oil composition was characterized for the Soave–Redlich–Kwong equation of state, and a constant (temperature independent) Peneloux volume shift parameter was used. The characterized composition in Table 7.15 was generated using the property correlations in Equations 5.1 through 5.3 and the SRK coefficients in Table 5.3. The C_{7+} fraction was represented using 12 pseudocomponents. Binary interaction coefficients are shown in Table 7.16. The fractions C_7–C_9 were kept as separate fractions. The C_{10+} fraction was split into nine pseudo-components of approximately the same weight amount.

To give an idea of the importance of the Peneloux volume correction, C_{7+} pseudocomponent densities at 1.01 bar and 15°C for the characterized mixture in Table 7.15 are in Figure 7.17 plotted against molecular weight. For the lumped fractions, the "experimental" densities are determined from Equations 5.20 and 5.21. Also plotted in Figure 7.17 are the densities of the C_{7+} pseudocomponents calculated using the SRK equation of state (EoS) with and without volume correction. For pseudocomponents with a molecular weight below ~400, the SRK equation simulates too-low densities (too-high molar volumes). For higher molecular weights, the SRK densities are too high (molar volumes too low). This explains why the volume shift parameter in Table 7.15 is positive for the lighter C_{7+} pseudocomponents and negative for the heaviest ones. The volume correction

---

TABLE 7.15
The Oil Composition in Table 3.39 Characterized for the SRK-Peneloux Equation of State

Component	Mol %	Weight%	Molecular Weight	Liquid Density g/cm³	Critical Temperature °C	Critical Pressure bar	Acentric Factor	Volume Shift cm³/mol
N_2	0.137	0.050	28.01		−146.95	33.94	0.0400	0.92
CO_2	0.557	0.320	44.01		31.05	73.76	0.2250	3.03
H_2S	8.743	3.890	34.08		100.05	89.37	0.1000	1.78
C_1	33.707	7.060	16.04		−82.55	46.00	0.0080	0.63
C_2	9.323	3.660	30.07		32.25	48.84	0.0980	2.63
C_3	7.573	4.360	44.10		96.65	42.46	0.1520	5.06
iC_4	1.436	1.090	58.12		134.95	36.48	0.1760	7.29
nC_4	4.032	3.060	58.12		152.05	38.00	0.1930	7.86
iC_5	1.773	1.670	72.15		187.25	33.84	0.2270	10.93
nC_5	2.198	2.070	72.15		196.45	33.74	0.2510	12.18
C_6	2.835	3.190	86.18		234.25	29.69	0.2960	17.98
C_7	3.160	3.960	96.00	0.7266	260.32	31.06	0.4678	11.30
C_8	3.508	4.900	107.00	0.7428	279.10	28.15	0.4997	17.71
C_9	3.007	4.750	121.00	0.7571	300.13	25.12	0.5397	25.96
C_{10}–C_{11}	4.468	8.116	139.15	0.7747	325.17	22.42	0.5913	34.81
C_{12}–C_{13}	2.950	6.453	167.57	0.7968	359.39	19.59	0.6686	45.13
C_{14}–C_{15}	2.307	5.950	197.51	0.8149	391.21	17.66	0.7468	51.57
C_{16}–C_{17}	1.805	5.397	229.04	0.8306	421.47	16.30	0.8251	53.87
C_{18}–C_{20}	1.998	6.835	262.02	0.8475	451.23	15.44	0.9029	50.82
C_{21}–C_{23}	1.382	5.479	303.56	0.8654	485.87	14.70	0.9937	42.31
C_{24}–C_{27}	1.205	5.506	350.07	0.8831	522.23	14.17	1.0856	27.75
C_{28}–C_{34}	1.095	6.052	423.21	0.9057	575.54	13.67	1.2065	−1.78
C_{35}–C_{80}	0.801	6.182	590.89	0.9471	694.07	13.34	1.3201	−85.54

* Binary interaction parameters can be seen from Table 7.16.

TABLE 7.16
Nonzero Binary Interaction Parameters for Use with the Oil Mixture in Table 7.15

	N_2	CO_2	H_2S
CO_2	−0.0315		
H_2S	0.1696	0.0989	
C_1	0.0278	0.1200	0.0800
C_2	0.0407	0.1200	0.0852
C_3	0.0763	0.1200	0.0885
iC_4	0.0944	0.1200	0.0511
nC_4	0.0700	0.1200	0.0600
iC_5	0.0867	0.1200	0.0600
nC_5	0.0878	0.1200	0.0689
C_6	0.0800	0.1200	0.0500
C_{7+}	0.0800	0.1000	

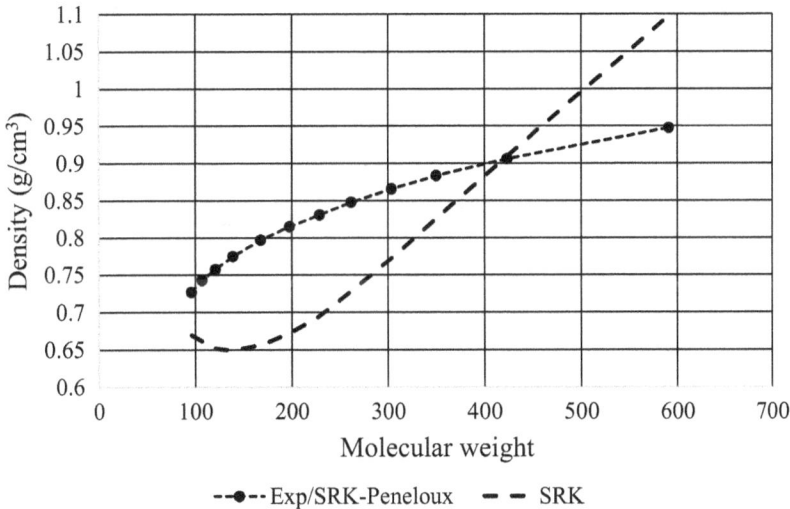

FIGURE 7.17 Experimental densities at 1.01 bar and 15°C of C_{7+} pseudocomponents of the oil mixture in Table 7.15 plotted against molecular weight. Also plotted are the pseudocomponent densities calculated using the SRK equation of state with and without volume correction.

parameter is defined in Equation 4.44 as the difference between the SRK molar volume and the real molar volume. For C_{7+} pseudocomponents, the volume shift parameter is found from Equation 5.6, which, as illustrated in Figure 7.17, will ensure that the SRK–Peneloux densities at standard conditions agree with the experimental densities.

Experimental and simulated swelling results are plotted in Figures 7.18 and 7.19. Figure 7.18 shows how the saturation pressure develops with the amount of gas added, and Figure 7.19 shows the swelling volume at each saturation pressure relative to the volume of the saturated oil before any gas was added.

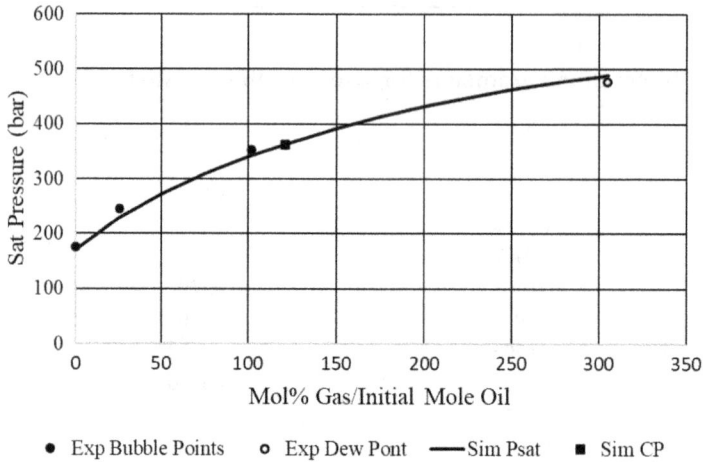

FIGURE 7.18   Measured and simulated saturation points in a swelling experiment at 136.7°C for the oil mixture and the injection gas in Table 3.39. The SRK equation with Peneloux volume correction was used in the simulations. The characterized reservoir fluid composition is shown in Table 7.15 and binary interaction parameters in Table 7.16. The experimental swelling data is shown in Table 3.38.

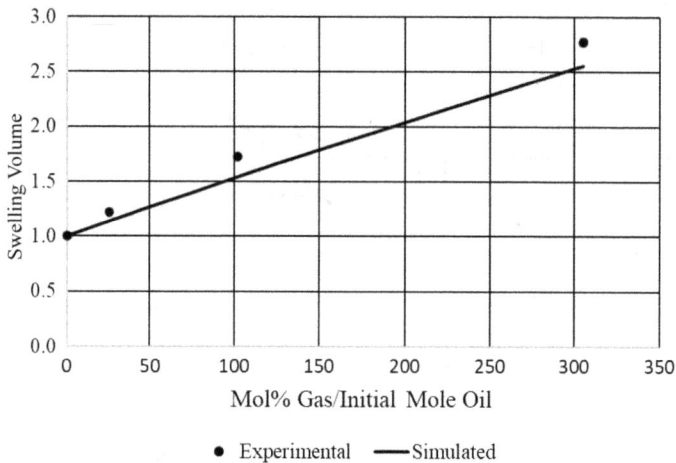

FIGURE 7.19   Measured and simulated swelling volumes for the oil mixture and the injection gas in Table 3.39 in a swelling experiment at 136.7°C. The SRK equation with Peneloux volume correction was used in the simulations. The characterized reservoir fluid composition is shown in Table 7.15 and binary interaction parameters in Table 7.16. The experimental swelling data is shown in Table 3.38.

7.6   GAS REVAPORIZATION

A gas revaporization experiment is described in Section 3.2.5. Table 3.51 shows the results of three gas revaporization experiments carried out on the reservoir fluid in Table 3.50 using three different injection gases, N_2, CO_2, and the hydrocarbon gas in Table 3.50. Kumar et al. (2015) have presented the EoS model in Tables 7.17 and 7.18 for use with the volume-corrected SRK equation, which gave a good match of the experimental gas revaporization data. In Figure 7.20, experimental and simulated liquid volume percents are plotted as a function of the amount of gas added.

TABLE 7.17
EoS Model for Use with the Volume-Corrected SRK Equation for the Reservoir Fluid in Table 3.50. The EoS Model Was Used to Simulate the Gas Revaporization Experiments in Table 3.51. Binary Interaction Parameters Are Shown in Table 7.18

Component	Mol %	Molecular Weight	Crit Temp °C	Crit Pressure bar	Acentric Factor	Volume correction cm³/mol
N_2	0.130	28.0	−147.0	33.94	0.040	0.92
CO_2	2.992	44.0	31.1	73.76	0.225	3.03
H_2S	0.804	34.1	100.1	89.37	0.100	1.78
C_1	74.334	16.0	−82.6	46.00	0.008	0.63
C_2	7.949	30.1	32.3	48.84	0.098	2.63
C_3	3.734	44.1	96.7	42.46	0.152	5.06
C_4	2.506	58.1	146.5	37.50	0.187	7.67
C_5	1.519	72.2	192.2	33.79	0.240	11.60
C_6	0.999	86.2	234.3	29.69	0.296	17.98
$C_7–C_{10}$	3.354	108.6	311.1	19.76	0.362	98.10
$C_{11}–C_{15}$	1.199	166.2	374.1	20.18	0.746	44.12
$C_{16}–C_{25}$	0.427	254.9	463.7	15.94	0.996	50.36
$C_{26}–C_{40}$	0.052	411.5	581.8	13.27	1.204	21.34
$C_{41}–C_{80}$	0.003	629.5	723.9	12.71	1.374	−93.59

Source: Kumar, A., Gohary, M.E., Pedersen, K.S., and Azeem, J., Gas injection as an enhanced recovery technique for gas condensates. A comparison of three injection gases, SPE 177778-MS, presented at the *Abu Dhabi International Petroleum Exhibition and Conference* in Abu Dhabi, UAE, November 9–12, 2015.

TABLE 7.18
NonZero Binary Interaction Parameters k_{ij} for Use with the EoS Model in Table 7.17

	N_2	CO_2	H_2S	C_1
CO_2	−0.0315			
H_2S	0.1696	0.0989		
C_1	0.0278	0.12	0.08	
C_2	0.0407	0.12	0.0852	
C_3	0.0763	0.12	0.0885	
C_4	0.078	0.12	0.0571	
C_5	0.0873	0.12	0.0648	
C_6	0.0044	0.12	0.05	
$C_7–C_{10}$	0.0044	0.1		0.03
$C_{11}–C_{15}$	0.0044	0.1		0.03
$C_{16}–C_{25}$	0.0044	0.1		0.03
$C_{26}–C_{40}$	0.0044	0.1		0.03
$C_{41}–C_{80}$	0.0044	0.1		0.03

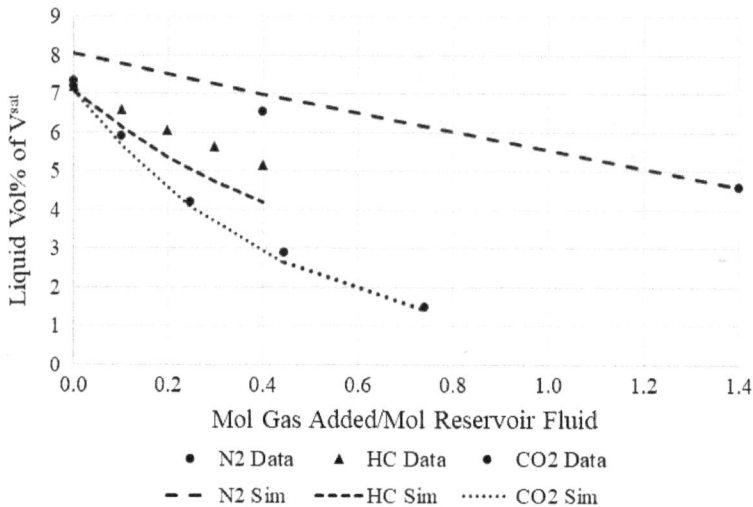

FIGURE 7.20 Experimental and simulated liquid volume percents in the gas revaporation experiment in Table 3.51. The simulations were carried out with the volume-corrected SRK equation using the EoS model in Table 7.17.

7.7 PVT SIMULATIONS WITH PC-SAFT EOS

Table 7.19 shows isothermal oil compressibilities reported by Tybjerg and Pedersen (2017) for the oil mixture in Table 5.22 at a temperature of 123°C. The data is plotted in Figure 7.21 together with simulation results obtained with the PC-SAFT EoS model in Table 5.23 and the SRK-Peneloux and PR-Peneloux equations using the characterization procedure of Pedersen et al. described in Chapter 5. The best match of the experimental compressibility data is seen with the PC-SAFT equation of state. This agrees with the findings of Larsen et al. (2011), who have shown that the PC-SAFT equation provides a better match of the high-pressure oil compressibilities for a Gulf of Mexico reservoir fluid than the SRK–Peneloux equation.

Leekumjorn and Krejbjerg (2013) have presented PC-SAFT PVT simulation results for the fluid in Table 3.44 and see a good match of both routine and PVT data with gas injection reported by Negahban et al. (2010).

7.8 GAS Z-FACTORS

Table 7.20 shows gas Z-factors at a temperature of 41.85°C for a four-component gas measured using a Burnett cell as described in Section 3.1.7 (Mollerup and Angelo 1985). Also shown are Z-factors calculated using the GERG-2008 and the volume-corrected SRK and PR equations. As can be seen from the table, the measured Z-factors are simulated very accurately using the GERG-2008 equation, which is described in Section 4.10.

7.9 MERCURY

Mercury in its elemental form is present in many reservoir fluids. The presence of mercury can lead to operational as well as safety and health problems. It is therefore of interest to be able to calculate the solubility of mercury in reservoir fluids and natural gases. Chapoy et al. (2020a, 2020b) have made a comprehensive review of literature data for the Hg solubility in pure hydrocarbons, inorganic gases, and mixtures of those components and have also presented new experimental data.

TABLE 7.19

Isothermal Compressibilities of Oil Composition in Table 5.22 at 123°C

Pressure bar	Isothermal Compressibility 1/bar
552.6	0.000190
483.7	0.000219
414.9	0.000252
380.2	0.000271
345.8	0.000290
332.0	0.000298
311.6	0.000310
291.0	0.000323
284.1	0.000328

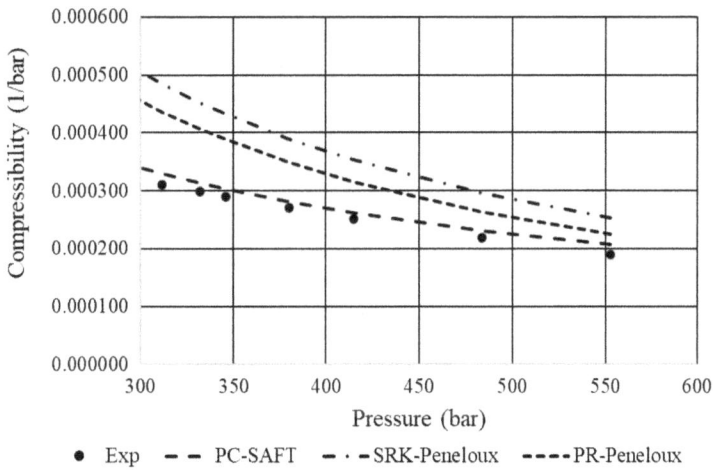

FIGURE 7.21 Measured and simulated isothermal oil compressibilities reported for the oil mixture in Table 5.22 at a temperature of 123°C.

TABLE 7.20

Experimental and Simulated Z-factors of the Gas Composition Consisting of 2.20 mol% N_2, 89.92 mol% C_1, 6.28 mol% C_2, and 1.6 mol% C_3 at a Temperature of 41.85°C

Pressure bar	Experimental Z-Factor	GERG-2008 Z-Factor	%Dev	SRK-Peneloux Z-Factor	%Dev	PR-Peneloux Z-Factor	%Dev
754.2065	1.5443	1.5411	−0.2	1.5115	−2.1	1.5541	0.6
685.2437	1.4469	1.4442	−0.2	1.4221	−1.7	1.4581	0.8
616.7280	1.3500	1.3474	−0.2	1.3333	−1.2	1.3628	0.9
313.8921	0.9357	0.9330	−0.3	0.9566	2.2	0.9596	2.6
260.3572	0.8770	0.8748	−0.3	0.9027	2.9	0.9013	2.8
204.9935	0.8354	0.8333	−0.3	0.8615	3.1	0.8554	2.4
173.7217	0.8257	0.8237	−0.2	0.8491	2.8	0.8402	1.8
144.1796	0.8290	0.8272	−0.2	0.8477	2.3	0.8365	0.9

(Continued)

TABLE 7.20   (*Continued*)

Experimental and Simulated Z-factors of the Gas Composition Consisting of 2.20 mol% N_2, 89.92 mol% C_1, 6.28 mol% C_2, and 1.6 mol% C_3 at a Temperature of 41.85°C

Pressure	Experimental	GERG-2008		SRK-Peneloux		PR-Peneloux	
bar	Z-Factor	Z-Factor	%Dev	Z-Factor	%Dev	Z-Factor	%Dev
124.9679	0.8384	0.8366	−0.2	0.8531	1.8	0.8407	0.3
105.2177	0.8537	0.8520	−0.2	0.8643	1.2	0.8511	−0.3
91.6036	0.8673	0.8658	−0.2	0.8752	0.9	0.8620	−0.6
77.23314	0.8841	0.8827	−0.2	0.8894	0.6	0.8766	−0.8
67.13760	0.8970	0.8958	−0.1	0.9008	0.4	0.8887	−0.9
56.46198	0.9118	0.9107	−0.1	0.9142	0.3	0.9030	−1.0
48.93852	0.9226	0.9217	−0.1	0.9242	0.2	0.9140	−0.9
38.49405	0.9383	0.9375	−0.1	0.9390	0.1	0.9304	−0.8
27.76508	0.9550	0.9544	−0.1	0.9551	0.0	0.9485	−0.7
19.93350	0.9674	0.9670	0.0	0.9673	0.0	0.9624	−0.5
10.17548	0.9832	0.9830	0.0	0.9831	0.0	0.9804	−0.3
5.154022	0.9915	0.9914	0.0	0.9914	0.0	0.9900	−0.2
1.846627	0.9969	0.9969	0.0	0.9964	−0.1	0.9969	0.0

Experimental Data Mollerup, J. and Angelo. P., Measurement and correlation of the volumetric properties of a synthetic natural gas mixture, *Fluid Phase Equilibria* 19, 259–271, 1985.

TABLE 7.21

Critical Properties of Hg for Use with Cubic Equations of State

Equation of state	Critical Temperature[*1)] K	Critical Pressure[*1)] bar	Acentric factor[*2)]
SRK	1764	1670	−0.2100
PR	1764	1670	−0.1805

Source: [*1)] Kozhevnikov, V. Arnold, D., Grodzinskii, E., and Naurzakov, S., Phase transitions and critical phenomena in mercury fluid probed by sound, *Fluid Phase Equilibria* 125, 149–157, 1996.

[*2)] Lim, L. T., Sørensen, H., Lekumjorn, S. and Pottayil, A., A thermodynamic model for prediction of solubility of elemental mercury in natural gas, produced water and hydrate inhibitors, SPE-210631-MS presented at the *SPE Asia Pacific Oil & Gas Conference and Exhibition* in Adelaide, Australia, October 17–19, 2022.

The critical point of Hg is not easily measured. This section uses the T_c and P_c suggested by Kozhevnikov et al. (1996) combined with the acentric factors suggested by Lim et al. (2022). This data is shown in Table 7.21. Using experimental data for the solubility of mercury in inorganic gases and hydrocarbon components, Lim et al. have estimated the binary interaction parameters in Table 7.22 for use with the SRK and the PR equations of state. Table 7.23 shows the solubility of Hg in a gas mixture of the composition in Table 7.24 at temperatures between −30°C and 50°C and varying pressures. This data is plotted in Figure 7.22 with simulation results for the PR-Peneloux equation of state.

TABLE 7.22

Binary Interaction Parameters for Pairs of Hg and Inorganic Gases and Hydrocarbon Components for Use with the Soave–Redlich–Kwong (SRK) and the Peng–Robinson (PR) Equations of State

Component	SRK	PR
N_2	0.1105	0.2218
CO_2	0.3450	0.3450
C_1	0.0690	0.1150
C_2	0.0760	0.0800
C_3	0.0590	0.0550
iC_4	0.0493	0.0432
nC_4	0.0619	0.0535
iC_5	0.0356	0.0356
nC_5	0.0527	0.0461
nC_6	0.0251	0.0115

Source: Lim, L. T., Sørensen, H., Lekumjorn, S. and Pottayil, A., A thermodynamic model for prediction of solubility of elemental mercury in natural gas, produced water and hydrate inhibitors, SPE-210631-MS presented at the *SPE Asia Pacific Oil & Gas Conference and Exhibition* in Adelaide, Australia, October 17–19, 2022.

TABLE 7.23

Solubility of Mercury (yHg) in Parts per Billion (ppb) in the Gas Mixture in Table 7.24

Temperature °C	Pressure bar	yHg ppb
−30	12.3	1.02
−30	20.6	0.67
−30	34.9	0.48
−30	51.7	0.44
0	6.8	44.89
0	19.0	16.86
0	34.6	12.30
0	76.3	8.38
0	160.6	9.49
50	34.5	674.45
50	67.7	479.54
50	151.7	420.33

Source: Chapoy, A., Ahmadi, P., Yamada, J., Kobayashi, Szczepanski, R., Zhang, X. Speranza, A., Elemental mercury partitioning in high pressure fluids part 2: Model validations and measurements in multicomponent systems., *Fluid Phase Equilib.* 523, 112773, 2020b.

TABLE 7.24

Molar Composition of Gas Mixture for Which Hg Solubility Data Is Shown in Table 7.23 and Plotted in Figure 7.22

Component	Mol %
N_2	0.390
CO_2	0.771
C_1	89.24

(Continued)

TABLE 7.24   (*Continued*)

Molar Composition of Gas Mixture for Which Hg Solubility Data Is Shown in Table 7.23 and Plotted in Figure 7.22

Component	Mol %
C_2	6.516
C_3	2.255
iC_4	0.317
nC_4	0.441
iC_5	0.0442
nC_5	0.0299

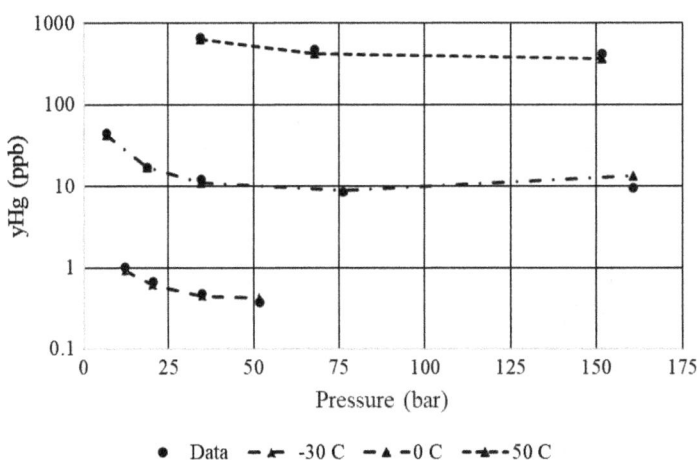

FIGURE 7.22   Measured and simulated solubilities of mercury in the gas mixture in Table 7.24. The simulation results are obtained with the PR-Peneloux equation of state.

7.10   WHAT TO EXPECT FROM A PVT SIMULATION

The preceding examples illustrate the fact that it is generally possible, from an accurate compositional analysis and using a C_{7+} characterization method customized for a particular equation of state, to simulate PVT properties satisfactorily. Inaccuracies in the measured plus molecular weight may have to be assessed, as was done for the fluid composition in Table 3.15 when simulating the CVD experiment. Some deviations can be seen between measured and simulated isothermal oil compressibilities. This is the case with the SRK–Peneloux simulations for the oil mixture in Table 3.6, as seen from Figure 7.5. Problems matching experimental liquid compressibilities is a deficiency in the cubic equation of state itself and not something that is easily cured by adjusting EoS parameters.

Taking into consideration that it is generally possible to predict PVT properties of oil and gas condensate mixtures solely based on an accurate compositional analysis, one may wonder why so much effort is spent on tuning of EoS parameters to match experimental PVT data. There are number of reasons. Compositional analyses are not always of the same quality as those presented in Chapter 3. Most compositional analyses are based on gas chromatography, the type of analyses that will usually lead to less accurate PVT simulation results than simulations based on compositional data originating from a true boiling point (TBP) analysis. The two mentioned analytical techniques are described in Chapter 2. Another reason for tuning the EoS parameters can be the need to lump the components into very few pseudocomponents (e.g., a total of six) to keep down the computation time in compositional reservoir simulations. Regression can also be needed when fluids from different zones of a reservoir are to be represented using the same pseudocomponents, as explained in Section 5.6. Regression to experimental PVT data is dealt with in Chapter 9.

REFERENCES

Chapoy, A., Ahmadi, P., Szczepanski, R., Zhang, X., Speranza, A., Yamada, J., and Kobayashi, A., Elemental mercury partitioning in high pressure fluids part 1: Literature review and measurements in single components, *Fluid Phase Equilib.* 520, 112660, 2020a.

Chapoy, A., Ahmadi, P., Yamada, J., Kobayashi, A., Szczepanski, R., Zhang, X. Speranza, A., Elemental mercury partitioning in high pressure fluids part 2: Model validations and measurements in multicomponent systems, *Fluid Phase Equilib.* 523, 112773, 2020b.

Kozhevnikov, V., Arnold, D., Grodzinskii, E., and Naurzakov, S., Phase transitions and critical phenomena in mercury fluid probed by sound, *Fluid Phase Equlib.* 125, 149–157, 1996.

Kumar, A., Gohary, M.E., Pedersen, K.S., and Azeem, J., Gas injection as an enhanced recovery technique for gas condensates. A comparison of three injection gases, SPE 177778-MS, presented at the *Abu Dhabi International Petroleum Exhibition and Conference*, Abu Dhabi, November 9–12, 2015.

Larsen, J., Sørensen, H., Yang, T., and Pedersen, K.S., EOS and viscosity modeling for a highly undersaturated Gulf of Mexico reservoir fluid, SPE-147075-PP, presented at the *SPE ATCE*, Denver, CO, October 30–November 2, 2011.

Leekumjorn, S. and Krejbjerg, K., Phase behavior of reservoir fluids: Comparisons of PC-SAFT and cubic equation of state simulations, *Fluid Phase Equilib.* 359, 17–23, 2013.

Lim, L.T., Sørensen, H., Lekumjorn, S., and Pottayil, A., A thermodynamic model for prediction of solubility of elemental mercury in natural gas, produced water and hydrate inhibitors, SPE-210631-MS, presented at the *SPE Asia Pacific Oil & Gas Conference and Exhibition*, Adelaide, Australia, October 17–19, 2022.

Mollerup, J. and Angelo, P., Measurement and correlation of the volumetric properties of a synthetic natural gas mixture, *Fluid Phase Equilibria* 19, 259–271, 1985.

Negahban, S., Pedersen, K.S., Baisoni, M.A., Sah, P., and Azeem, J.S., An EoS model for a middle east reservoir fluid with an extensive EOR PVT data material, SPE-136530-PP presented at the *Abu Dhabi International Petroleum Exhibition & Conference*, Abu Dhabi, November 1–4, 2010.

Tybjerg, P. and Pedersen, K.S., Reservoir fluid characterization procedure for the PC-SAFT equation of state, SPE-187170-MS, presented at the *SPE ATCE*, San Antonio, TX, October 9–11, 2017.

8 Thermal Properties

This chapter deals with fluid properties that cannot be calculated solely from an equation of state. An equation of state considers molecular interactions. Thermal properties are also affected by the energy carried inside the molecules, which are not considered by an equation of state. A molecule with more than one atom is not a rigid structure. It has so-called internal degrees of freedom that, as sketched in Figure 8.1, enable the atoms connected in a molecule to rotate and vibrate.

8.1 ENTHALPY

Enthalpy (H) is defined in Equation A.6 in Appendix A and used to describe the total heat content of a system. It has two contributions, the ideal gas contribution and the residual contribution.

$$H = H^{id} + H^{res} \tag{8.1}$$

The highest enthalpy a fluid can have at a given temperature is that of an ideal gas for which the residual enthalpy is zero. The residual enthalpy accounts for molecular attractions and will contribute to the total enthalpy with a negative term. The higher the molecular attractions, the lower the residual enthalpy. A gas passing through a valve is an example of a process with no enthalpy change. That, however, does not mean that the ideal gas and the residual terms are unaffected by the passage of the valve. Because of the pressure drop, the distance between the molecules increases and the intermolecular attractions decrease. That will make the residual enthalpy increase (get closer to zero). To keep the enthalpy constant, the ideal gas enthalpy must decrease, resulting in a temperature drop over the valve. Condensation results in a large drop in residual enthalpy. Figure 8.2 illustrates three different levels of enthalpy, that of an ideal gas for which there is limited interaction between the molecules and the enthalpy is determined by the energy carried inside the molecules. The next (lower) level is that of a gas at a higher pressure for which the attraction between the molecules starts to influence the enthalpy with a negative residual term. Finally, the third level is that in the liquid state in which the attraction between the molecules is high and it would require much energy to bring the fluid back to the ideal gas state.

In the liquid state, the residual enthalpy increases slightly with increasing pressure (becomes less negative). That is caused by the repulsive intermolecular forces acting in the high-pressure liquid state. For a liquid at high pressure, a pressure drop will result in a relaxation of the repulsive forces, which will make the residual enthalpy decrease. A pressure drop at a constant enthalpy (passage of a valve) will for a liquid make the temperature increase. To keep a constant enthalpy with a decreasing residual enthalpy, the ideal gas enthalpy must increase, which will result in a temperature increase. At high pressure (usually over 400 bar), the residual enthalpy for a gas is also seen to increase with increasing pressure, and an isenthalpic expansion will make the temperature increase, as is the case for a liquid.

For an N-component mixture, the ideal gas enthalpy is found from

$$H^{id} = \sum_{i=1}^{N} z_i H_i^{id} \tag{8.2}$$

where z_i is the mole fraction of component i, and H_i^{id} is the molar ideal gas enthalpy of component i. The ideal gas contribution is found from

$$H_i^{id} = \int_{T_{ref}}^{T} C_{P_i}^{id} \, dT \tag{8.3}$$

DOI: 10.1201/9780429457418-8

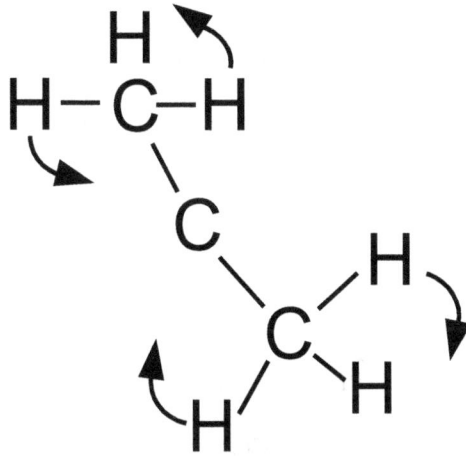

FIGURE 8.1 Illustration of internal molecular degrees of freedom.

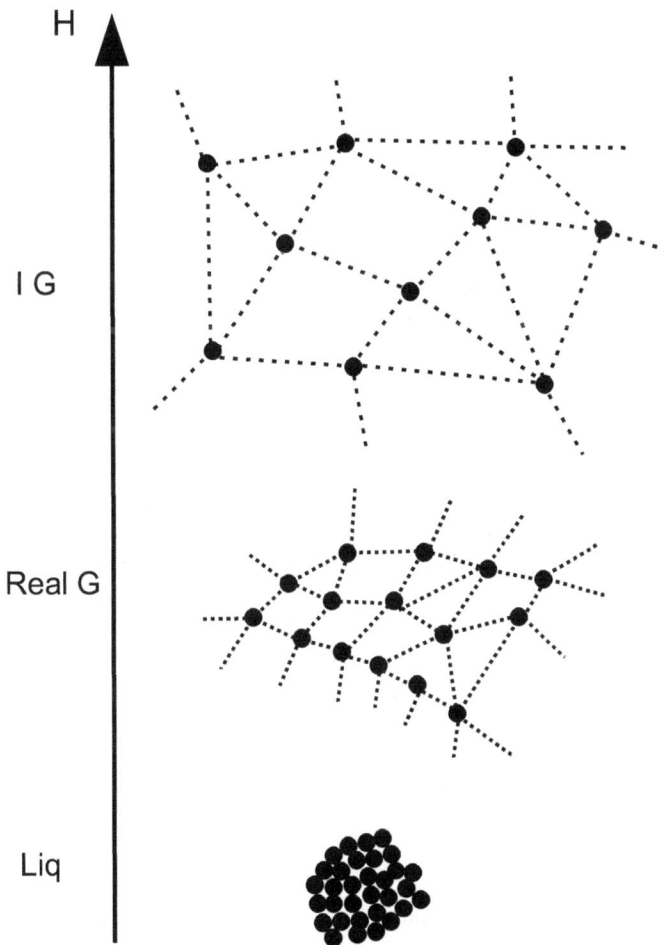

FIGURE 8.2 Enthalpy level of ideal gas (IG), real gas (Real G) and liquid (Liq).

T_{ref} is a reference temperature for which 273.15 K (0°C) is a convenient choice. With $T_{ref} = 273.15$ K, H in Equation 8.1 is the enthalpy relative to the ideal gas enthalpy at 273.15 K. C_{Pi}^{id} is the molar ideal gas heat capacity of component i, which may be calculated from a third-degree polynomial in temperature

$$C_{Pi}^{id} = C_{1,i} + C_{2,i}T + C_{3,i}T^2 + C_{4,i}T^3 \tag{8.4}$$

Poling et al. (2000) have tabulated C_1–C_4 for defined components. For heavy hydrocarbons and T in R, the coefficients C_1–C_4 for heat capacities in Btu/lb may be calculated from (Kesler and Lee 1976):

$$C_1 = -0.33886 + 0.02827\,K - 0.26105\,CF + 0.59332\,\omega\,CF \tag{8.5}$$

$$C_2 = (-0.9291 - 1.1543\,K + 0.0368\,K^2) \times 10^{-4} + CF(4.569.48\,\omega) \times 10^{-4} \tag{8.6}$$

$$C_3 = -1.6658 \times 10^{-7} + CF(0.536 - 0.6828\,\omega) \times 10^{-7} \tag{8.7}$$

$$C_4 = 0 \tag{8.8}$$

where ω is the acentric factor defined in Equation 1.1 and CF is defined as

$$CF = \left(\frac{(12.8 - K)(10 - K)}{10\,\omega}\right)^2 \tag{8.9}$$

K is the Watson characterization factor:

$$K = \frac{T_B^{1/3}}{SG} \tag{8.10}$$

T_B is the normal boiling point in R and SG the specific gravity, which is defined in Equation 5.55 and is approximately equal to the liquid density in g/cm³.

The residual term of H can be derived from the applied equation of state using the following relation

$$H^{res} = -RT^2 \sum_{i=1}^{N} z_i \frac{\partial \ln\varphi_i}{\partial T} \tag{8.11}$$

Where φ_i is the fugacity coefficient and z_i the mole fraction of component i. The derivative is for a constant composition. The enthalpy is, unlike equilibrium phase compositions, influenced by a possible Peneloux volume correction (Section 4.5). The following relation exists between the enthalpy calculated from a Peneloux volume-corrected (Pen) SRK or PR equation and the enthalpy from the same equation with no volume correction (SRK/PR)

$$H_{Pen} = H_{SRK/PR} - c \times P \tag{8.12}$$

where P is the pressure and c the volume correction.

8.2 INTERNAL ENERGY

The internal energy, U, is related to the enthalpy, H, through

$$U = H - PV \tag{8.13}$$

where P is the pressure and V the molar volume.

8.3 ENTROPY

The entropy defines the amount of heat (Q) that can be converted into work (W). The terms Q and W are defined in Appendix A. Figure 8.3 illustrates a fluid system at two different temperatures, T_{high} and T_{low}, with corresponding thermal energies of Q_{high} and Q_{low}, respectively. Assume that the total loss of thermal energy between the high temperature and low temperature state is used to generate work (W), and the process is adiabatic (no heat exchange with surroundings) and reversible; then it is a process of a constant entropy, which means

$$\frac{Q_{max}}{T_{max}} = \frac{Q_{min}}{T_{min}} \tag{8.14}$$

Entropy is a useful quantity to evaluate how much work can be produced from a given amount of heat (Q). For the process in Figure 8.3, which ends at a temperature of T_{min}, the maximum work that can be generated is

$$W_{max} = Q_{max} - Q_{min} \tag{8.15}$$

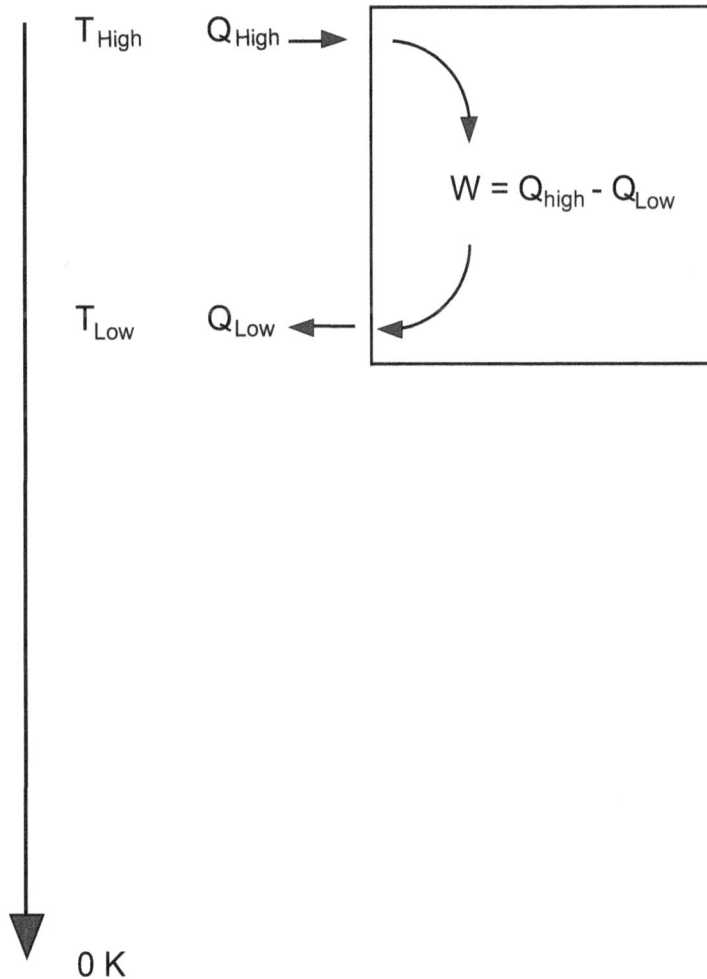

FIGURE 8.3 Conversion of heat to mechanical work.

Had it been possible to continue the process to the absolute zero (0 K or −273.15°C), all heat could have been converted to work.

In the oil industry, entropy is important because there are some processes of approximately constant entropy. That is the case with compression processes, although there will always be some heat loss to the surroundings.

The entropy can be calculated as the sum of two contributions, an ideal gas entropy and a residual entropy:

$$S = \sum_{i=1}^{N} z_i S_i^{id} + S^{res} \tag{8.16}$$

The ideal gas entropy of component i at temperature T may be calculated from

$$S_i^{id} = \int_{T_{ref}}^{T} \frac{C_{pi}^{id}}{T} dT - T \ln \frac{P}{P_{ref}} - R \ln z_i \tag{8.17}$$

P_{ref} is a reference pressure for which 1 atm (1.01325 bar) is a convenient choice. If $T_{ref} = 273.15$ K and $P_{ref} = 1$ atm, S in Equation 8.16 becomes the entropy relative to the entropy of the same composition as an ideal gas at 273.15 K and 1 atm. C_{Pi}^{id} is the molar ideal gas heat capacity of component i and may be calculated from Equation 8.4.

The residual entropy can be derived from

$$S^{res} = \frac{H^{res}}{T} - R \sum_{i=1}^{N} z_i \ln \varphi_i \tag{8.18}$$

where the residual enthalpy (H^{res}) can be found from Equation 8.11.

8.4 HEAT CAPACITY

The heat capacity at constant pressure, C_P, is the temperature derivative of the enthalpy at a constant pressure

$$C_P = \left(\frac{\partial H}{\partial T} \right)_P \tag{8.19}$$

where the enthalpy may be derived as described in Section 8.1.

The heat capacity at constant volume, C_V, is related to the heat capacity at constant pressure as follows:

$$C_V = C_P - T \left(\frac{\partial V}{\partial T} \right)_P \left(\frac{\partial P}{\partial T} \right)_V \tag{8.20}$$

The derivatives $\left(\frac{\partial V}{\partial T} \right)_P$ and $\left(\frac{\partial P}{\partial T} \right)_V$ may be evaluated from the applied equation of state.

8.5 JOULE–THOMSON COEFFICIENT

The Joule–Thomson coefficient is defined as the pressure derivative of the temperature for constant enthalpy and can be related to C_P and enthalpy as follows:

$$\mu_{jT} = \left(\frac{\partial T}{\partial P} \right)_H = -\frac{1}{C_p} \left(\frac{\partial H}{\partial P} \right)_T \tag{8.21}$$

8.6   VELOCITY OF SOUND

The velocity of sound can be expressed as

$$u_{sonic} = -\frac{V}{\sqrt{M}}\sqrt{\left(\frac{\partial P}{\partial V}\right)_S} = \frac{V}{\sqrt{M}}\sqrt{\frac{C_P}{C_V}\left(\frac{\partial P}{\partial T}\right)_V\left(\frac{\partial T}{\partial V}\right)_P} \tag{8.22}$$

where M is the molecular weight. C_P may be derived from Equation 8.19, C_V from Equation 8.20, and V and the derivatives from the applied equation of state.

8.7   EXAMPLE CALCULATIONS

8.7.1   Enthalpy

Figure 8.4 shows a plot of the enthalpy for the gas condensate mixture in Table 5.11 as a function of pressure for a series of temperatures. The enthalpy is calculated using the Peneloux-corrected SRK equation of state (Equation 4.43). For a constant enthalpy, the temperature is seen to increase with pressure at low pressures. The highest increase is seen inside the two-phase area. In this region, the enthalpy at a constant temperature decreases with increasing pressure as a result of the condensation taking place when the pressure is increased. Outside the two-phase region, the influence of pressure on enthalpy is less pronounced. At pressures above ~400 bar, the temperature for a constant enthalpy is seen to decrease with increasing pressure. Whether the temperature in the single-phase region will increase or decrease with pressure depends on the Joule–Thomson coefficient defined in Equation 8.21. For positive Joule–Thomson coefficients, the temperature for a constant H will increase with pressure, whereas a temperature decrease will be seen for a fluid of a negative Joule–Thomson coefficient. It can be seen from Figure 8.4 that the fluid composition in Table 5.11 has a negative Joule–Thomson coefficient for pressures higher than ~400 bar.

Figure 8.5 shows the enthalpy of pure CO_2 as a function of pressure for a series of temperatures calculated using the Span–Wagner equation (Section 4.11).

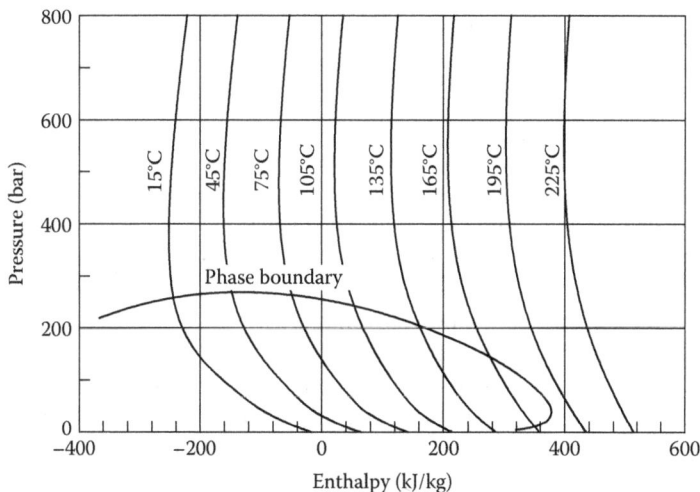

FIGURE 8.4   Isotherms in enthalpy–pressure (HP) diagram for gas condensate mixture in Table 5.11 calculated using SRK equation with Peneloux volume correction.

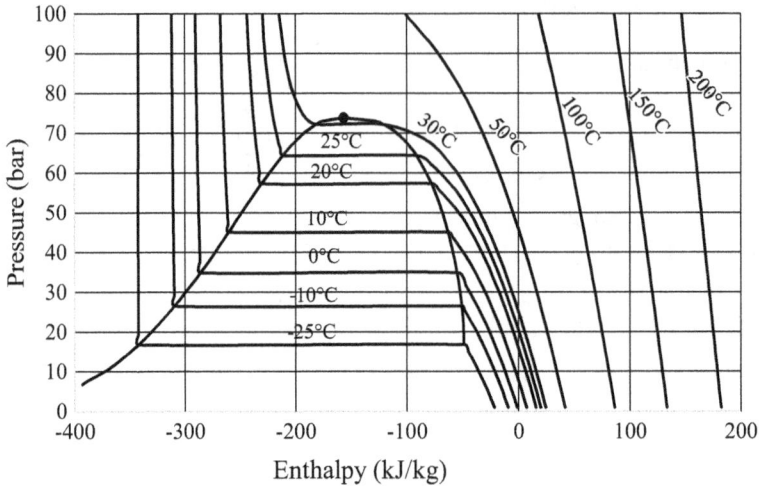

FIGURE 8.5   Isotherms in enthalpy–pressure (HP) diagram for pure CO_2 simulated using the Span–Wagner equation presented in Section 4.11.

8.7.2   ENTROPY

Figure 8.6 shows a plot of the entropy for the oil mixture in Table 3.6 as a function of pressure for a series of temperatures and Figure 8.7 a similar plot for CO_2 calculated using the Span–Wagner equation (section 4.11).

A compression process is often approximated as one of a constant entropy. Figure 8.7 may therefore give an idea about the temperature that can be expected if CO_2 undergoes a compression process. It may, for example, be seen from Figure 8.7 that an outlet temperature of around 100°C can be expected if CO_2 without any heat loss to the surroundings is compressed from 40 bar to 95 bar, starting at a temperature of 30°C.

8.7.3   HEAT CAPACITY

The heat capacity (C_p) of a liquid oil mixture increases with temperature, as can be seen from Figure 8.8 for the oil mixture in Table 3.6, but C_p is not much affected by pressure.

8.7.4   JOULE–THOMSON COEFFICIENT

Figure 8.9 shows Joule–Thomson coefficients for the gas condensate mixture in Table 5.11, calculated for four different temperatures using the PR equation. The Joule–Thomson coefficient is positive at lower pressures, meaning that an increased pressure at a constant enthalpy will result in a higher temperature. The Joule–Thomson coefficient will become zero for some pressure in the range from 350 to 500 bar, depending on temperature. At even higher pressures, the temperature will decrease if the pressure is further increased at a constant enthalpy.

Table 8.1 shows experimental (Wang et al. 2017) and simulated Joule–Thomson coefficients for carbon dioxide (CO_2) and Table 8.2 experimental Joule–Thomson coefficients (Sage et al. 1936) for propane (C_3). Figure 8.10 shows a plot of the experimental Joule–Thomson coefficients of propane for three different temperatures together with values simulated using the PR equation of state.

Meisingset and Pedersen (2016) have outlined how the Joule–Thompson coefficient can be derived from well test analysis data when the reservoir temperature is known. One of the fluids considered is a lean gas condensate of the composition in Table 8.3. During a well test, there is

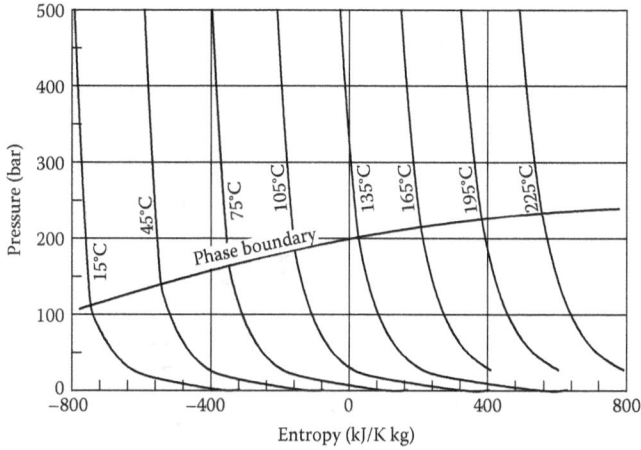

FIGURE 8.6 Isotherms in entropy–pressure (SP) diagram for oil mixture in Table 3.6 calculated using PR equation with Peneloux volume correction.

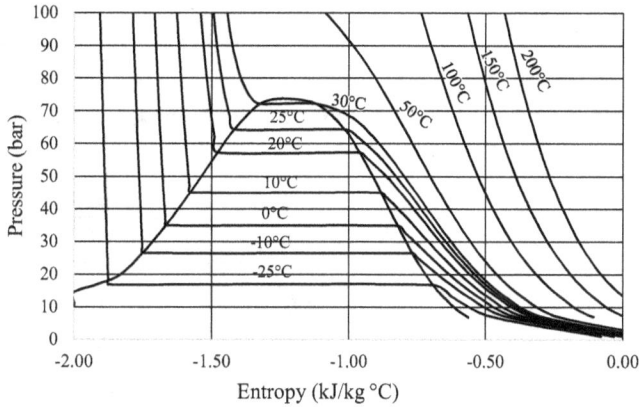

FIGURE 8.7 Isotherms in entropy–pressure (SP) diagram for pure CO_2 simulated using the Span–Wagner equation presented in Section 4.11.

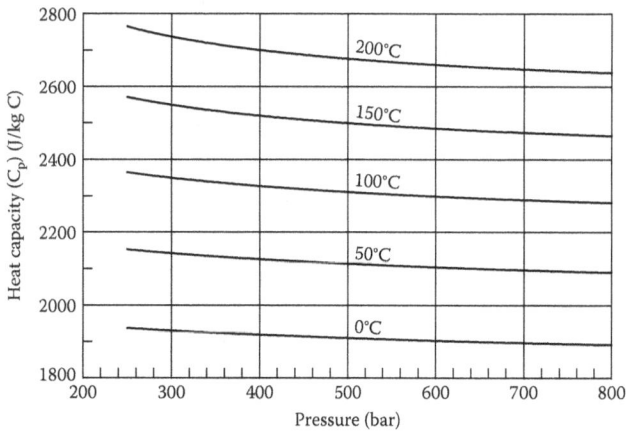

FIGURE 8.8 Heat capacities (C_p) of oil mixture in Table 3.6 as a function of pressure at four different temperatures calculated using PR equation with Peneloux volume correction.

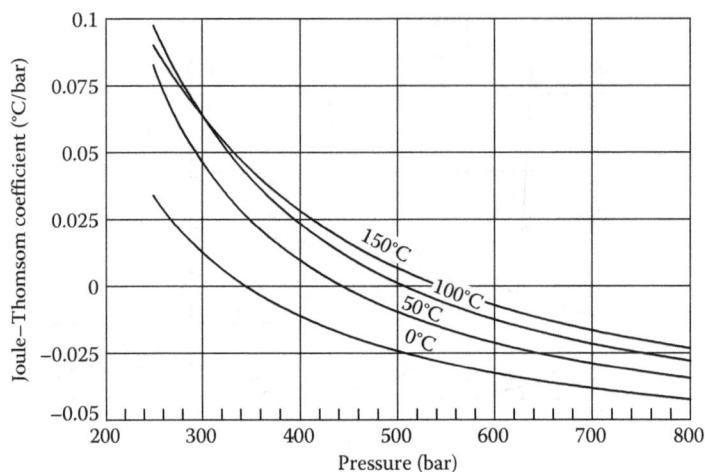

FIGURE 8.9 Joule–Thomson coefficients for gas condensate mixture in Table 5.11 as a function of pressure at four different temperatures calculated using PR equation with Peneloux volume correction.

TABLE 8.1

Experimental and Simulated Joule–Thomson Coefficients for CO_2. The Equations of State Referred to Are Described in Chapter 4

Pressure bar	Temp °C	Exp JT coeff °C/bar	PR JT coeff °C/bar	%Dev	SRK JT coeff °C/bar	%Dev	Span–Wagner JT coeff °C/bar	%Dev	GERG-2008 JT coeff °C/bar	%Dev
74.0	49.85	0.8762	0.8587	−2.0	0.8546	−2.5	0.8151	−7.0	0.8158	−6.9
101.3	49.85	0.5017	0.5273	5.1	0.5297	5.6	0.5365	6.9	0.5369	7.0
141.9	49.85	0.1987	0.1958	−1.5	0.1964	−1.2	0.1660	−16.5	0.1958	−1.5
182.4	49.85	0.1098	0.1161	5.7	0.1172	6.7	0.0913	−16.8	0.1161	5.7
202.7	49.85	0.0901	0.0945	4.8	0.0957	6.2	0.0723	−19.8	0.0945	4.8
74.0	99.85	0.5964	0.6139	2.9	0.6010	0.8	0.5794	−2.9	0.5793	−2.9
101.3	99.85	0.5351	0.5439	1.6	0.5385	0.6	0.5319	−0.6	0.5315	−0.7
141.9	99.85	0.4110	0.4097	−0.3	0.4112	0.0	0.4217	2.6	0.4214	2.5
182.4	99.85	0.2922	0.2835	−3.0	0.2852	−2.4	0.2915	−0.2	0.2915	−0.2
202.7	99.85	0.2435	0.2347	−3.6	0.2357	−3.2	0.2362	−3.0	0.2359	−3.1
74.0	149.85	0.4302	0.4560	6.0	0.4401	2.3	0.4256	−1.1	0.4256	−1.1
101.3	149.85	0.4103	0.4152	1.2	0.4045	−1.4	0.4004	−2.4	0.4002	−2.5
141.9	149.85	0.3528	0.3486	−1.2	0.3441	−2.5	0.3519	−0.3	0.3514	−0.4
182.4	149.85	0.2907	0.2886	−0.7	0.2819	−3.0	0.2944	1.3	0.2942	1.2
202.7	149.85	0.2606	0.2522	−3.2	0.2524	−3.1	0.2648	1.6	0.2648	1.6

Data Source: Wang, J., Wang, Z. and Sun, B., Improved equation of CO_2 Joule–Thomson Coefficient, *Journal of CO_2 Utilization* 19, 296–307, 2017.

a pressure drop from the reservoir to the well, as sketched in Figure 8.11. When the well test has gone on for some time, the heat loss to the surroundings can be neglected. If height differences are disregarded, the enthalpy of the fluid in the reservoir will be the same as that of the fluid at the bottom hole (BH). Figure 8.12 shows the BH pressure and temperature profiles during the well test for

TABLE 8.2

Experimental Joule–Thomson Coefficients for Propane

Pressure bar	Temperature °C					
	21.1	37.8	54.4	71.1	87.8	104.4
	Joule–Thomson Coefficients °C/bar					
1.72	1.732	1.470	1.242	1.072	0.9347	0.8058
3.45	1.941	1.608	1.336	1.155	0.9927	0.8380
6.89	2.281	1.807	1.500	1.280	1.080	0.8863
10.34		1.994	1.653	1.401	1.157	0.9331
13.79			1.815	1.493	1.213	0.9951
17.24			1.989	1.600	1.269	1.038
20.68				1.734	1.332	1.082
24.13				1.896	1.398	1.136
27.58					1.465	1.175
31.03					1.551	1.209
34.47					1.676	1.241
37.92						1.255

Source: Sage, B.H., Kennedy, E.R. and Lacey, W.N., Phase Equilibria in Hydrocarbon Systems. XIII. Joule–Thomson Coefficients of Propane, *Industrial and Engineering Chemistry* 28, 601–604, 1936.

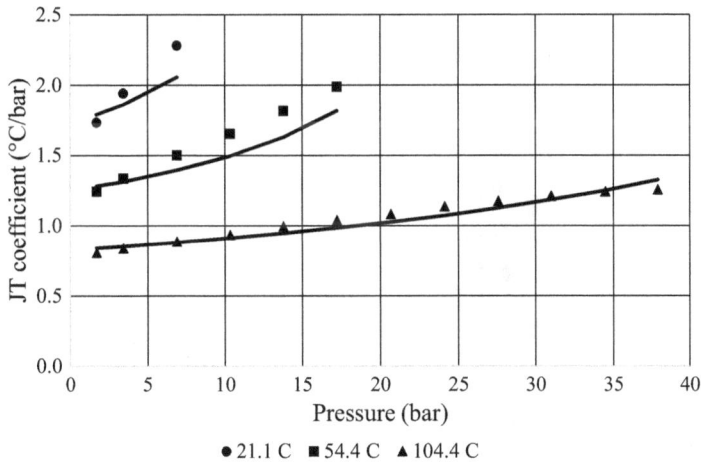

FIGURE 8.10 Experimental Joule–Thomson coefficients of propane for three different temperatures together with values simulated using the PR equation of state. The experimental data is shown in Table 8.2.

the fluid in Table 8.3. Each of the sections, 1–4, represents a period of a constant flow rate. When changing from one stage to the next, the flow rate increases, which gives rise to a lower BH pressure. If the pressure change from one stage to the next is ΔP, it will for a constant enthalpy single phase system result in a temperature change, ΔT, of

$$\Delta T = \Delta P \times \mu_{JT} \tag{8.23}$$

TABLE 8.3

Lean Gas Condensate Fluid Compositions for Which Well Test Data Is Plotted in Figure 8.12

Component	Mol%
N_2	1.57
CO_2	0.30
C_1	92.58
C_2	3.73
C_3	0.62
iC_4	0.36
nC_4	0.09
iC_5	0.08
nC_5	0.03
C_6	0.11
C_{7+}	0.53
C_{7+} molecular weight	114.5
C_{7+} density (g/cm³)	0.784

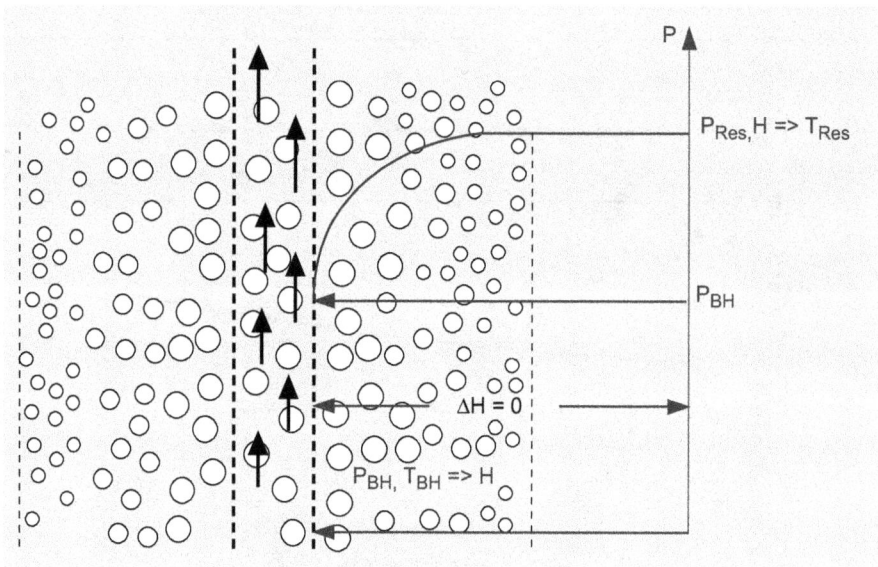

FIGURE 8.11   Sketch of near-well bore area during well test.

Assuming μ_{JT} is (almost) constant in the pressure range covered, the reservoir temperature (T^{res}) can after stage j be calculated as

$$T^{res} = T_{BH} - \sum_{i=1}^{j} \left(\Delta P_i \times \mu_{JT} \right)$$ (8.24)

The dashed line in Figure 8.12 shows the reservoir temperature calculated using Equation 8.24 assuming a μ_{JT} of 0.19°C/bar, which value corresponds with the μ_{JT} calculated for the fluid in

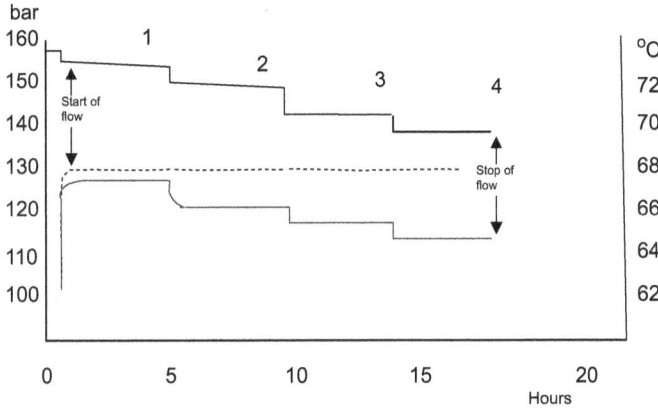

FIGURE 8.12 Variation in bottom hole pressure (upper full drawn line) and temperature (lower full drawn line) during well test for the lean gas condensate reservoir fluid in Table 8.3. The dashed line is the simulated reservoir temperature.

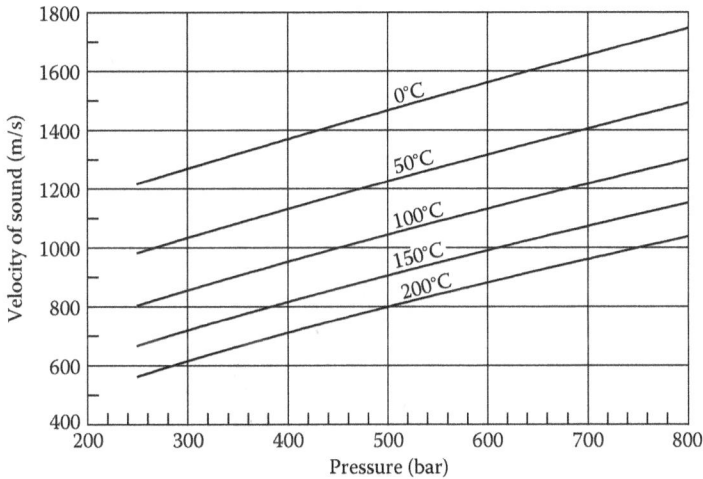

FIGURE 8.13 Velocities of sound in oil mixture of the composition in Table 3.6 as a function of pressure at four different temperatures calculated using PR equation with Peneloux volume correction.

Table 8.3 at the reservoir conditions of 69°C and 158 bar. Had the reservoir temperature not been known, it could have been derived from Equation 8.24. With the reservoir temperature, bottom hole temperature, and pressure stages known, Equation 8.23 can be used to determine the Joule–Thompson coefficient. When the reservoir temperature is unknown or uncertain, the equation may instead be used to determine the reservoir temperature using a Joule–Thompson coefficient derived from an equation of state.

8.7.5 VELOCITY OF SOUND

Figure 8.13 shows a plot of the velocity of sound in the oil mixture in Table 3.6 calculated from Equation 8.22 using the volume-corrected PR equation. For a constant temperature, the sound velocity is seen to increase with increasing pressure, whereas it decreases with increasing temperature for a constant pressure.

REFERENCES

Kesler, M.G. and Lee, B.I., Improve prediction of enthalpy of fractions, *Hydrocarb. Process.* 55, 153–158, 1976.

Meisingset, K.K. and Pedersen, K.S., Joule–Thomson coefficients from well test analysis data, SPE-181622-MS, presented at the *SPE ATCE*, Dubai, September 26–28, 2016.

Poling, B.E., Prausnitz, J.M., and O'Connell, J.P., *The Properties of Gases and Liquids*, McGraw-Hill, New York, 2000.

Sage, B.H., Kennedy, E.R., and Lacey, W.N., Phase equilibria in hydrocarbon systems. XIII. Joule–Thomson coefficients of propane, *Ind. Eng. Chem.* 28, 601–604, 1936.

Wang, J., Wang, Z., and Sun, B., Improved equation of CO_2 Joule–Thomson coefficient, *J. CO_2 Util.* 19, 296–307, 2017.

9 Regression to Experimental PVT Data

It is common practice in the oil industry to try to eliminate deviations between measured and simulated PVT data by adjustment of model parameters. This tradition dates back to a time when extended compositional analyses were rare. With a composition ending at C_{7+} or even C_{6+}, and no established procedure to split up the plus fraction, it was unlikely to see accurate predictions of PVT properties. Coats (1982) has presented an extensive work on how to perform pseudorization with a composition to C_{6+} or C_{7+} as the starting point.

In the 1980s, it became common to have TBP analyses or other extended compositional analyses carried out (see Chapter 2). That type of analytical information was used to develop the procedures presented in Chapter 5 for splitting up a plus fraction. The combination of improved analytical techniques and proper C_{7+} characterization procedures served to improve the quality of PVT simulation results to often not require any parameter tuning.

Nevertheless, regression is still in common use, one reason being that heavy lumping may be required in compositional reservoir simulation studies, and heavy lumping will often deteriorate the quality of the simulation results. Regression may also be needed to get an acceptable match of PVT data for multiple compositions, characterized to the same pseudocomponents using the procedures outlined in Section 5.6.

Also, for near-critical fluids, there may be a need for regression, as the match of the PVT data is quite dependent on the critical point of the fluid being matched.

PVT data as presented in Chapter 3 is the typical source of a parameter regression. The data material may comprise saturation points, phase densities, relative gas and oil volumes, phase compositions, composition of critical mixture (swelling test), and minimum miscibility pressures (slim tube tests).

Typically, the experimental PVT data available for regression will primarily comprise data measured at the reservoir temperature. The parameters estimated by regression will be those for which the model most closely reproduces the measured PVT data, which essentially means the parameters that best represent the phase behavior at the reservoir temperature. This does not necessarily ensure that the estimated parameters will be valid at temperatures not covered by experimental data. For example, it may lead to erroneous results when a PVT program, after regression to conventional PVT data, is used to generate property tables for subsea pipeline conditions or is used to generate properties for a reservoir fluid undergoing gas injection for enhanced oil recovery purposes. These potential dangers of a parameter regression are illustrated in Figure 9.1.

9.1 REGRESSION PARAMETERS

It is good practice not to modify the equation of state parameters for components lighter than C_7. The parameters left for regression are T_cs P_cs, acentric factors, and volume corrections of the C_{7+} components and the binary interaction coefficients (k_{ij}) between C_{7+} and lighter components. Pedersen et al. (1988) warn against the use of nonzero binary hydrocarbon–hydrocarbon interaction coefficients, as it may result in simulation of false liquid–liquid splits. The c parameters of the volume-corrected SRK and PR equations (Equations 4.43 and 4.48) influence the density

DOI: 10.1201/9780429457418-9

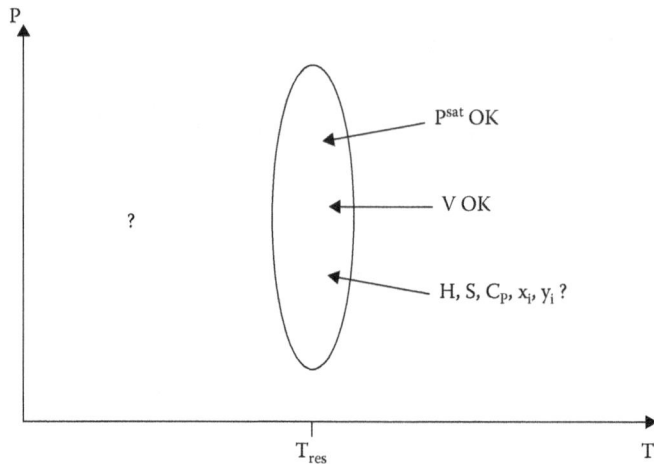

FIGURE 9.1 Dangers of regression to experimental PVT data. Regression will in general provide an improved match of volumetric properties and saturation points for temperatures near the reservoir temperature (T_{res}). An improved match cannot be expected for any other properties. The value of the regression is also questionable for volumetric properties and saturation points at temperatures far from that in the reservoir.

without affecting phase equilibrium results (saturation points, phase compositions, and phase amounts). For C_{7+} pseudocomponents, the c parameter may be determined as the difference between the real molar volume and the molar volume calculated using the SRK/PR equation with no volume correction. The former volume may be calculated as the ratio of the component molecular weight and the density at standard conditions, whose properties for the carbon number fractions are available from the measured composition or from the C_{7+} characterization. The formula used to determine the c parameter (Peneloux volume correction) is shown as Equation 5.6. By determining the C_{7+} c parameters in this manner, it is implicitly assumed that the difference between the real molar volume and that calculated using the SRK/PR equation is constant, independent of T, P, and mixture composition. This is not necessarily a valid assumption. The difference may vary with temperature, which may be accounted for by making the volume correction a linear function of temperature as expressed in Equation 5.9 or by treating the c parameter as a regression parameter.

Experimental data shows that the critical temperatures of C_{7+} components increase with carbon number, while the critical pressures decrease with carbon number. The acentric factors increase with carbon number up to around C_{50}. For higher carbon numbers, a slight decrease is seen with increasing carbon number as ring-shaped and branched molecules of lower acentric factors dominate over n-paraffins. The trends for T_c, P_c, and acentric factor are shown by fully drawn lines in Figure 9.2. The T_cs and P_cs are by far the most efficient regression parameters.

Christensen (1999) has outlined a regression procedure for matching of specific PVT data while limiting the impact on properties not covered in the regression. Instead of regressing on individual component properties (typically T_c, P_c, and acentric factor of pseudocomponents), the coefficients in the T_c, P_c, and m correlations (Equations 5.1 to 5.3) are used as regression parameters. This ensures a smooth development in T_c, P_c, and acentric factor with molecular weight, which would in general not be the case if properties of individual pseudocomponents were tuned on. Essentially, Christensen's procedure consists in rotating the T_c, P_c, and acentric factor curves for the C_{7+} carbon number fractions around the value of C_7, as sketched with dashed lines in Figure 9.2.

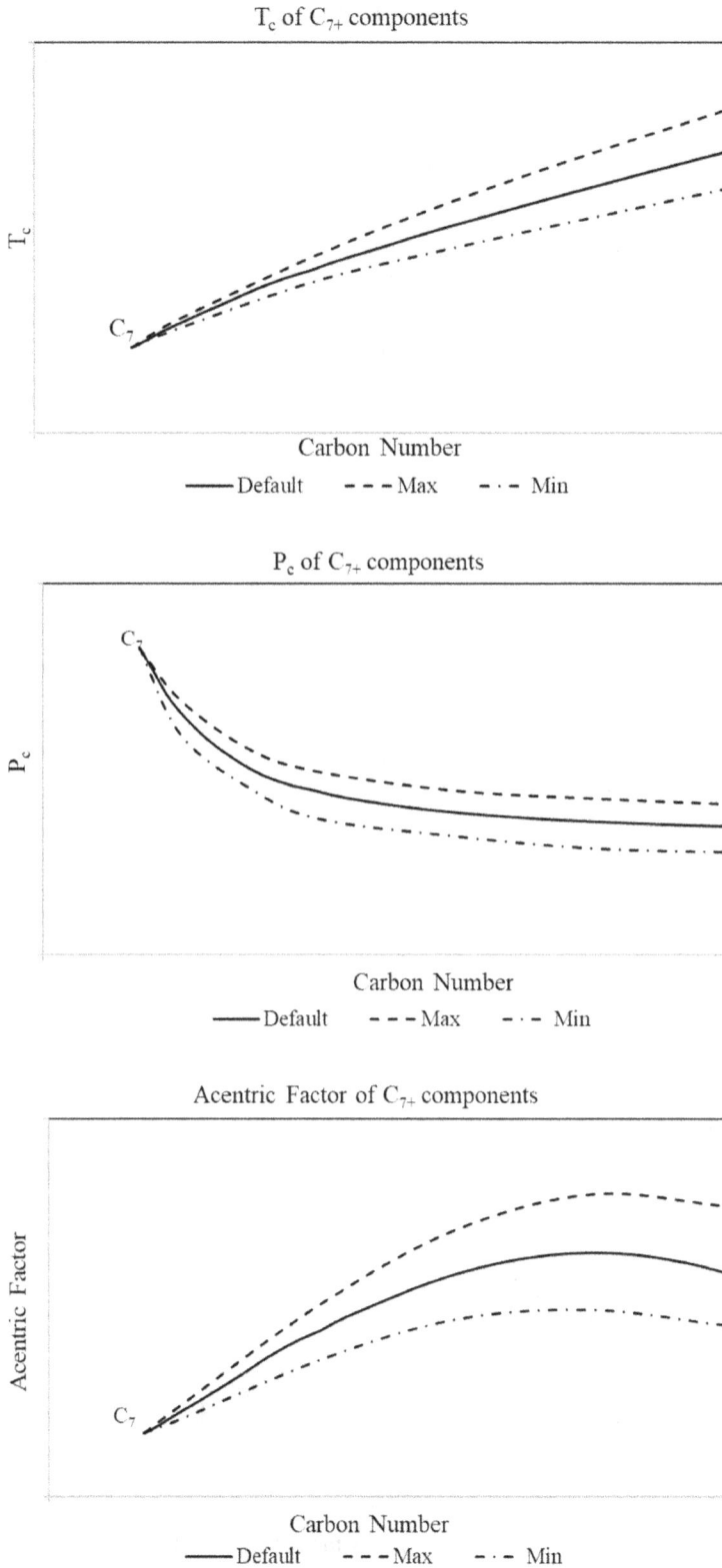

FIGURE 9.2 Qualitative developments of T_c, P_c, and acentric factor with carbon number.

9.2   OBJECT FUNCTION AND WEIGHT FACTORS

The object function (OBJ) to be minimized during a regression calculation may be defined as

$$OBJ = \sum_{j=1}^{NOBS} \left(\frac{r_j}{w_j} \right)^2 \tag{9.1}$$

where NOBS is the number of experimental observations used in the regression, w_j is the weight factor for the j-th observation, and r_j is the residual of the j-th observation:

$$r_j = \frac{OBS_j - CALC_j}{OBS_j} \tag{9.2}$$

OBS stands for observed and CALC for calculated. For liquid dropout curves from a constant mass expansion or a constant volume depletion experiment, adding a constant to all OBS and CALC values is recommended. This constant is added to reduce the importance assigned to data points with small liquid dropouts relative to data points with larger liquid dropouts. A convenient value for the constant could be the maximum liquid dropout divided by three. The weight factor (w_j) and the user-specified weight $(WOBS_j)$ to be assigned to the j-th observation are interrelated as follows:

$$WOBS_j = \frac{1}{w_j^2} \tag{9.3}$$

A minimization algorithm based on the principle of Marquardt (1963) is suited to determine the parameters minimizing the object function in Equation 9.1.

The number of regression parameters must not exceed the number of data points. To ensure that the regressed parameters are generally applicable, it is recommended to limit the number of regression parameters (NPAR) to

$$NPAR = 1 + \ln(NOBS) \tag{9.4}$$

9.3   EXAMPLE OF REGRESSION FOR GAS CONDENSATE

Figure 7.2 reveals some deviations between a measured (Table 3.9) and two simulated liquid dropout curves for the gas condensate mixture in Table 3.8. One simulation was based on the fluid being characterized to a total of 22 components (Table 7.1) and the other on the fluid being characterized to a total of 6 components (Table 7.3).

A regression to the constant mass expansion (CME) data in Table 3.9 was carried out following the procedure in Section 9.1 with the fluid composition in Table 3.8 characterized to a total of 22 (pseudo)components (10 defined components and 12 pseudocomponents). The coefficients c_2, c_3, and d_2 in the correlations used to find T_c and P_c of the C_{7+} fractions (Equations 5.1 and 5.2) were regressed on. The coefficients c_1 in Equation 5.1 and d_1 in Equation 5.2 were modified to keep a constant T_c and P_c of a typical C_7 component (defined as one with a molecular weight of 94 and a density at standard conditions of 0.745 g/cm³). Using this procedure, the regression will rotate the curves for T_c and P_c versus molecular weight around the default values of a C_7 component, as sketched with dashed lines in Figure 9.2. Table 9.1 shows the coefficients to be input in Equations 5.1 and 5.2 after regression. A maximum 20% adjustment of the T_cs and P_cs was accepted relative to the T_cs and P_cs, found using the standard coefficients in Table 5.3. The resulting regressed fluid can be seen in Table 9.2. The mole percentages differ slightly from those in Table 7.1 because the C_{20+}

TABLE 9.1

Coefficients in the Correlations in Equations 5.1 through 5.3 for Gas Condensate Mixture in Table 3.8 after Regression to CME Data in Table 3.9. SRK–Peneloux(T) Equation of State Is Used (the Resulting Characterized Mixture Can Be Seen from Table 9.2)

Subindex/ Coefficient	1	2	3	4	5
c	-2.9446×10^2	1.6170×10^2	4.36029×10^{-1}	-1.8774×10^3	—
d	2.19628	-6.2191×10^{-1}	2.0846×10^2	-3.9872×10^3	1.0
e	7.4310×10^{-1}	4.8122×10^{-3}	9.6707×10^{-3}	-3.7184×10^{-6}	—

TABLE 9.2

The Gas Condensate Mixture in Table 3.8 Characterized for the SRK–Peneloux(T) Equation of State Using a Total of 22 Components and Pseudocomponents

	Mol%	Weight%	Molecular Weight	T_c (°C)	P_c (bar)	Acentric Factor	c_0 (cm³/ mol)	c_1 (cm³/ mol K)
N_2	0.600	0.577	28.0	−147.0	33.94	0.040	0.92	0.0000
CO_2	3.340	5.044	44.0	31.1	73.76	0.225	3.03	0.0100
C_1	74.170	40.831	16.0	−82.6	46.00	0.008	0.63	0.0000
C_2	7.901	8.153	30.1	32.3	48.84	0.098	2.63	0.0000
C_3	4.151	6.280	44.1	96.7	42.46	0.152	5.06	0.0000
iC_4	0.710	1.416	58.1	135.0	36.48	0.176	7.29	0.0000
nC_4	1.440	2.872	58.1	152.1	38.00	0.193	7.86	0.0000
iC_5	0.530	1.312	72.2	187.3	33.84	0.227	10.93	0.0000
nC_5	0.660	1.634	72.2	196.5	33.74	0.251	12.18	0.0000
C_6	0.810	2.395	86.2	234.3	29.69	0.296	17.98	0.0000
C_7	1.200	3.747	91.0	255.6	34.98	0.453	4.74	0.0194
C_8	1.150	4.104	104.0	278.4	28.97	0.491	22.07	0.0264
C_9	0.630	2.573	119.0	303.7	24.28	0.534	42.24	0.0273
C_{10}	0.500	2.282	133.0	327.4	21.26	0.574	59.73	0.0191
$C_{11}-C_{12}$	0.560	2.869	149.3	354.7	18.77	0.619	78.13	0.0017
C_{13}	0.280	1.614	168.0	377.8	16.49	0.669	104.34	0.0001
$C_{14}-C_{15}$	0.390	2.504	187.1	400.7	15.30	0.720	129.38	−0.0066
$C_{16}-C_{17}$	0.290	2.126	213.7	432.7	13.98	0.788	159.02	−0.0282
$C_{18}-C_{19}$	0.220	1.829	242.3	464.7	12.87	0.857	186.98	−0.0556
$C_{20}-C_{22}$	0.166	1.695	288.9	511.7	11.75	0.963	228.09	−0.0980
$C_{23}-C_{28}$	0.176	2.116	346.6	563.0	11.04	1.080	274.00	−0.1400
$C_{29}-C_{80}$	0.125	2.026	490.5	683.1	10.24	1.270	365.24	−0.2500

Note: T_c and P_c of the C_{7+} fractions have been found from Equations 5.1 and 5.2 using the coefficients in Table 9.1, which have been found by a regression to the CME data in Table 3.9. Binary interaction coefficients are not regressed on and can be seen in Table 7.2.

molecular weight was increased by 0.7% from 362 to 364.6 while keeping a constant weight composition. As mentioned in Section 2.4, a fluid composition is measured as a weight% composition. A correct conversion to a mole% composition is dependent on a correct plus molecular weight. The standard technique used to measure the plus molecular weight (Section 2.4.2.1) is, however, slightly

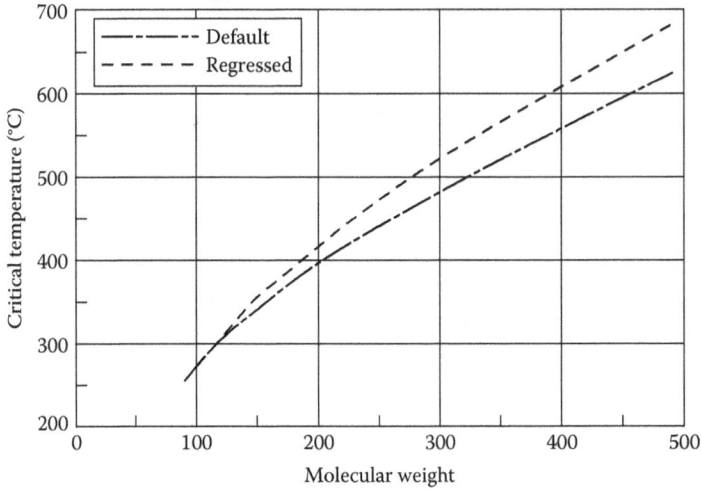

FIGURE 9.3 Critical temperatures of the C_{7+} pseudocomponents for gas condensate mixture in Table 3.8 before and after regression to the CME data in Table 3.9.

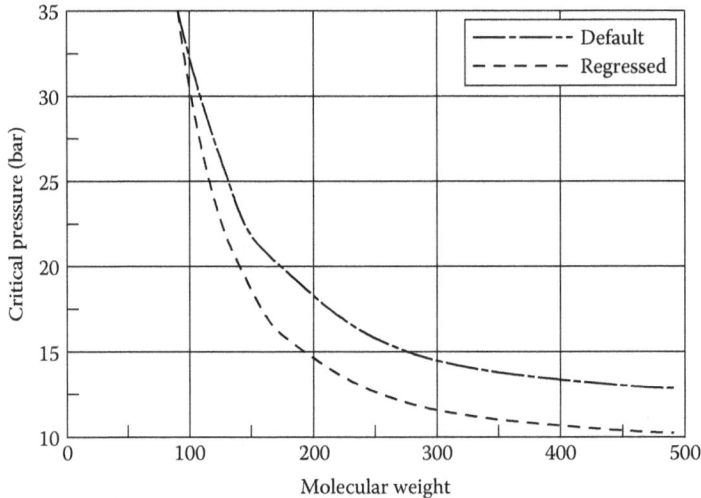

FIGURE 9.4 Critical pressures of the C_{7+} pseudocomponents for gas condensate mixture in Table 3.8 before and after regression to the CME data in Table 3.9.

inaccurate, which justifies a small adjustment of the plus molecular weight. The C_{7+} volume shift parameters (c_0) were decreased by 39% relative to the shift parameters found from Equation 5.6. The critical temperatures and pressures of the C_{7+} pseudocomponents are plotted before and after regression in Figures 9.3 and 9.4, respectively. The qualitative development in T_c and P_c with molecular weight is the same before and after regression. T_c increases and P_c decreases monotonically with molecular weight. Simulated constant mass expansion results before and after regression are shown in Figures 9.5 through 9.7. Figure 9.5 shows that the simulated relative volumes are almost unaffected by the regression, although the regression has markedly improved the match of the liquid dropout curve, as can be seen from Figure 9.6. Finally, Figure 9.7 shows a slightly improved match

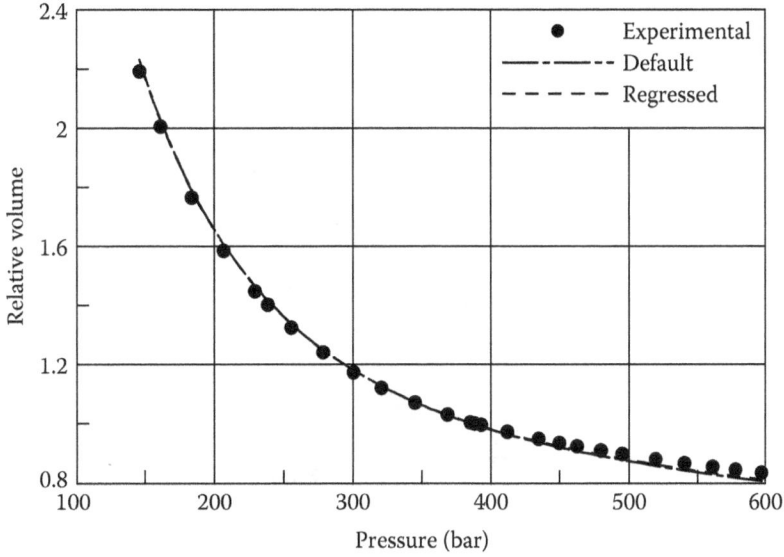

FIGURE 9.5 Measured and simulated relative volumes in a CME experiment at 155°C for the gas conden-sate mixture in Table 3.8. Simulation results are shown for both default and regressed coefficients in T_c and P_c correlations (almost indistinguishable). The SRK equation with T-dependent Peneloux volume correction was used in the simulations, and the fluid was described using a total of 22 components. The characterized regressed mixture composition is shown in Table 9.2. The characterized composition before regression is seen in Table 7.1 and the experimental CME data in Table 3.9.

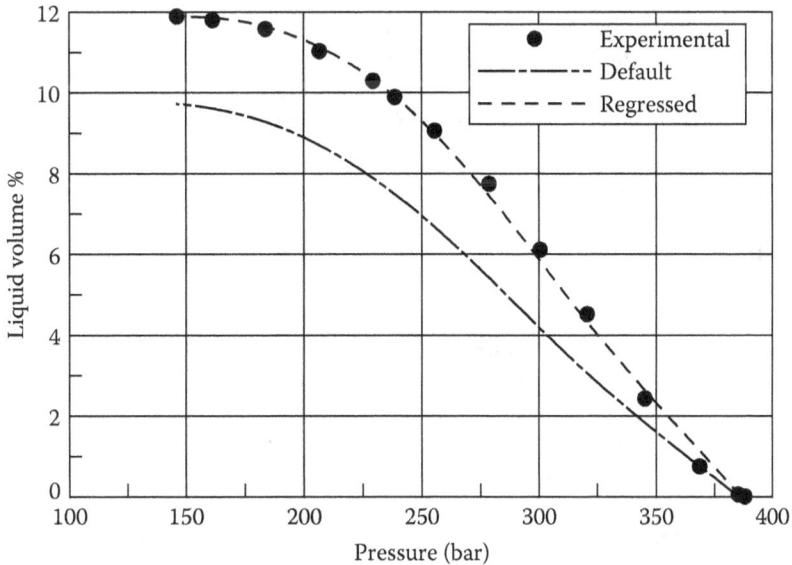

FIGURE 9.6 Measured and simulated liquid volumes (percentage of saturation point volume) in a CME experiment at 155°C for the gas condensate mixture in Table 3.8. Simulation results are shown for both default and regressed T_c and P_c correlations. The SRK equation with T-dependent Peneloux volume correction was used in the simulations, and the fluid was described using a total of 22 components. The characterized regressed mixture composition is shown in Table 9.2. The characterized composition before regression is seen in Table 7.1 and the experimental CME data in Table 3.9.

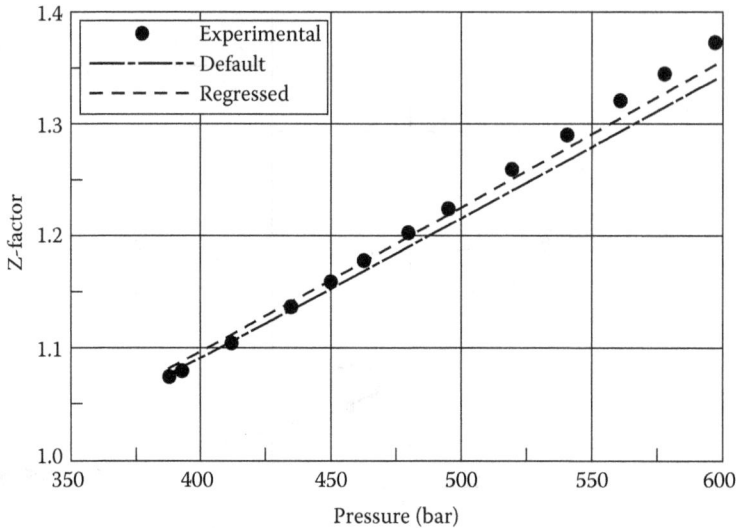

FIGURE 9.7   Measured and simulated gas Z-factors in a CME experiment at 155°C for the gas condensate mixture in Table 3.8. Simulation results are shown for both default and regressed T_c and P_c correlations. The SRK equation with T-dependent Peneloux volume correction was used in the simulations, and the fluid was described using a total of 22 components. The characterized regressed mixture composition is shown in Table 9.2. The characterized composition before regression is seen in Table 7.1 and the experimental CME data in Table 3.9.

of the experimental gas-phase Z factors above the saturation point. Overall, the match of the PVT data is almost perfect after the regression.

As can be seen from Figure 7.2, too small a CME liquid dropout and too low a saturation pressure are simulated using a default six-component description for the fluid composition in Table 3.8 (characterized mixture in Table 7.3). It was recommended in Chapter 5 to lump the C_{7+} components into pseudocomponents of approximately equal weight amounts. Although this recommendation is valid for oil mixtures in general and for gas condensate mixtures when the C_{7+} components are represented using 10–12 pseudocomponents, precautions need to be taken when gas condensate mixtures need to be heavily lumped. A too-small simulated liquid dropout suggests that more importance is to be assigned to the heavy hydrocarbons contained in the gas condensate mixture. With the total C_{7+} fraction of the gas condensate mixture in Table 3.8 described using only three components and an equal-weight-based lumping, the heaviest pseudocomponent will contain all the fractions from C_{16} to C_{80}, as can be seen in Table 7.3. Table 9.3 shows a different lumping (non-regressed) in which the heaviest pseudocomponent contains the C_{51}–C_{80} fractions (binary interaction coefficients in Table 9.4). The dashed–dotted lines in Figures 9.8 through 9.10 show CME simulation results using this lumping. As a result of the changed lumping, the measured liquid dropout curve is better matched than with the six-component description in Figure 7.2. A regression was performed using the same lumping. The resulting characterized mixture can be seen in Table 9.3 (regressed). The pseudocomponent mole percentages differ slightly from those of the non-regressed mixture because the C_{20+} molecular weight was increased by 2.7%, from 362 to 371.9, keeping the weight composition constant. The C_{7+} volume shift parameters (c_0) were decreased by 34% relative to the shift parameter found from Equation 5.6. The relative volumes and the single-phase gas Z-factors are almost unaffected by the regression, as can be seen from Figures 9.8 and 9.10, respectively. The experimental liquid dropout curve is matched much better after regression (Figure 9.9).

TABLE 9.3

The Gas Condensate Mixture in Table 3.8 Characterized for the SRK–Peneloux(T) Equation of State Using a Total of Six Pseudocomponents

	Mol%	Weight%	Molecular Weight	T_c (°C)	P_c (bar)	Acentric Factor	c_0 (cm³/ mol)	c_1 (cm³/ mol K)
Non-Regressed								
$N_2 + C_1$	74.767	41.410	16.1	−83.45	45.83	0.0085	0.63	0.0000
$CO_2 + C_2$–C_3	15.392	19.479	36.9	52.71	53.24	0.148	3.37	0.0015
C_4–C_6	4.150	9.631	67.6	182.3	34.42	0.231	10.82	0.0000
C_7–C_{20}	5.286	24.271	133.8	333.2	25.00	0.613	25.67	−0.0177
C_{21}–C_{50}	0.400	5.101	371.6	545.7	13.62	1.130	30.13	−0.2200
C_{51}–C_{80}	0.004	0.109	791.6	802.0	12.35	1.271	−189.10	−0.5000
Regressed								
$N_2 + C_1$	74.777	41.401	16.1	−83.45	45.83	0.0085	0.63	0.0000
$CO_2 + C_2$–C_3	15.393	19.474	36.9	52.71	53.24	0.148	3.37	0.0015
C_4–C_6	4.151	9.630	67.6	182.3	34.42	0.231	10.82	0.0000
C_7–C_{20}	5.279	24.194	133.6	336.6	23.76	0.612	36.94	−0.0109
C_{21}–C_{50}	0.393	5.109	379.0	576.8	11.39	1.141	128.80	−0.2100
C_{51}–C_{80}	0.007	0.192	798.9	886.0	9.66	1.260	60.53	−0.4900

Note: Coefficients in the correlations used to find T_c and P_c of the C_{7+} fractions (Equations 5.1 and 5.2) were regressed on to improve the match of the CME data in Table 3.9. Binary interaction coefficients were not regressed on and can be seen in Table 9.4.

TABLE 9.4

Nonzero Binary Interaction Coefficients for Use with the Mixture in Table 9.3

	$N_2 + C_1$	$CO_2 + C_2$–C_3
$CO_2 + C_2$–C_3	—	—
C_4–C_6	0.0261	—
C_7–C_{20}	0.0007	0.0260
C_{21}–C_{50}	0.0006	0.0217
C_{51}–C_{80}	0.0006	0.0217

The heaviest pseudocomponent in the regressed composition in Table 9.3 covers the C_{51}–C_{80} fractions but constitutes only 0.007 mol% (0.192 weight percentage) of the total mixture. One may wonder whether a fraction that small will influence the PVT behavior at all. Figure 9.11 shows a simulated liquid dropout curve where the C_{51}–C_{80} fractions have been neglected (dashed–dotted line). The same plot shows the simulated liquid dropout curve where C_{51}–C_{80} is considered (dashed line). It can be seen that the C_{50+} fraction makes the simulated saturation pressure increase by ~50 bar and slightly increases the whole level of liquid dropout. This shows that it is indeed important that the characterization acknowledge the content of C_{50+} components.

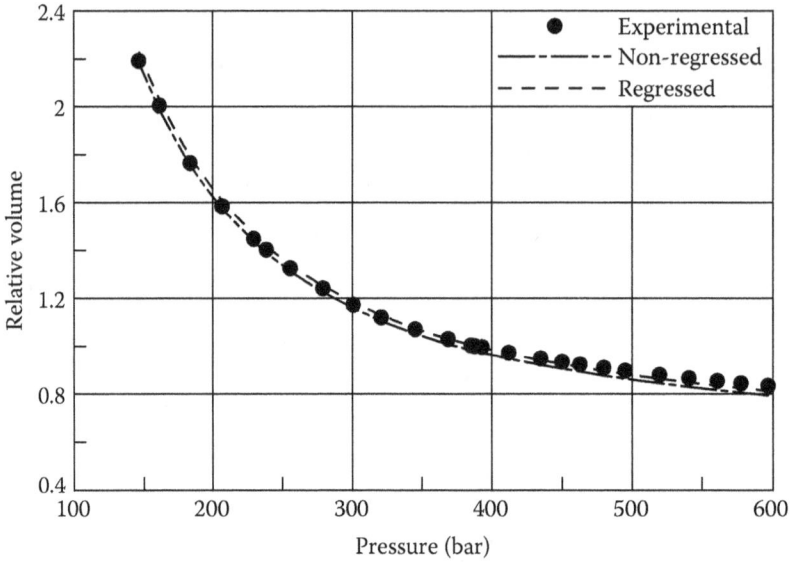

FIGURE 9.8 Measured and simulated relative volumes in a CME experiment at 155°C for the gas condensate mixture in Table 3.8. The mixture was lumped to six pseudocomponents, as shown in Table 9.3. Simulation results are shown for both default and regressed T_c and P_c correlations. The SRK equation with T-dependent Peneloux volume correction was used in the simulations. The experimental CME data can be seen in Table 3.9.

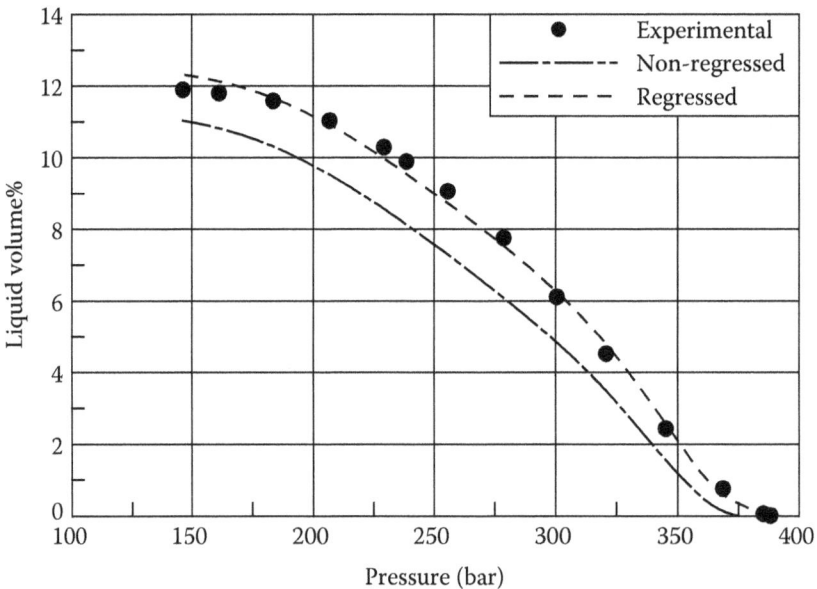

FIGURE 9.9 Measured and simulated liquid volumes (percentage of saturation point volume) in a CME experiment at 155°C for the gas condensate mixture in Table 3.8. The mixture was lumped to six pseudocomponents as shown in Table 9.3. Simulation results are shown for both default and regressed T_c and P_c correlations. The SRK equation with T-dependent Peneloux volume correction was used in the simulations. The experimental CME data can be seen in Table 3.9.

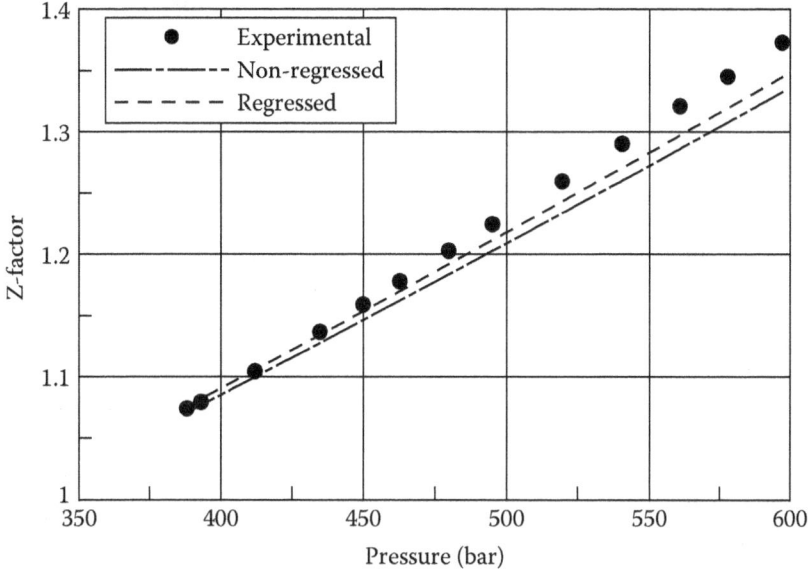

FIGURE 9.10 Measured and simulated gas Z-factors in a CME experiment at 155°C for the gas condensate mixture in Table 3.8. The mixture was lumped to six pseudocomponents as shown in Table 9.3. Simulation results are shown for both default and regressed T_c and P_c correlations. The SRK equation with T-dependent Peneloux volume correction was used in the simulations. The experimental CME data can be seen in Table 3.9.

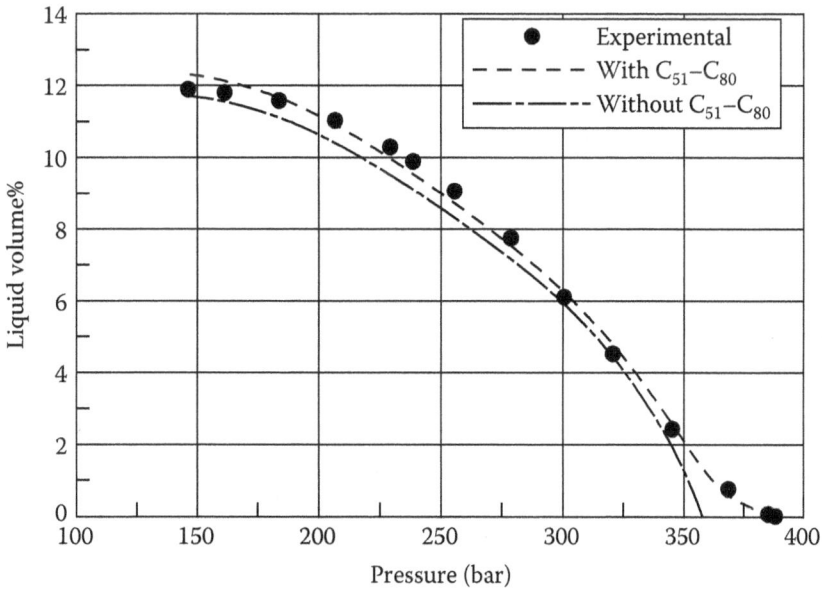

FIGURE 9.11 Measured and simulated liquid volumes (percentage of saturation point volume) in a CME experiment at 155°C for the gas condensate mixture in Table 3.8. The mixture was lumped to six pseudocomponents, as shown in Table 9.3. The dashed line shows the simulation results obtained with the regressed composition in Table 9.3. The dashed–dotted line shows simulation results using the same characterization but neglecting components heavier than C_{50}. The experimental CME data can be seen in Table 3.9.

9.4 TUNING ON SINGLE PSEUDOCOMPONENT PROPERTIES

A possible alternative to the regression procedure outlined in Section 9.1 and exemplified in Section 9.3 would be to tune on properties (T_c, P_c, or acentric factor) of one or more pseudocomponents starting with the properties from the default characterization. Table 7.5 shows a characterized oil mixture (plus fluid in Table 3.6). The heaviest pseudocomponent consists of C_{27}–C_{80} hydrocarbons and is assigned a T_c of 679.6°C when the fluid is characterized default for the SRK–Peneloux equation of state. The phase envelope simulated with the SRK equation of state is shown as a full-drawn line in Figure 9.12. Also shown in Figure 9.12 is the phase envelope simulated having increased T_c of the C_{27}–C_{80} fraction from 679.6°C to 850°C (dashed line). In addition to an increased two-phase area, this change is seen to introduce a three-phase area with one gas and two liquid phases. Table 9.5 shows the results of flash calculations carried out on the oil mixture for a temperature of 100°C and a pressure of 212 bar. Consistent with the phase envelopes in Figure 9.12, a single liquid phase is found when using the default characterization. After having increased the T_c of the C_{27}–C_{80} fraction from 679.6°C to 850°C, the mixture is seen to split into three phases, one gas and two liquid phases.

This example illustrates that adjustment of properties of single pseudocomponents should be used with caution. False phase splits are often seen if T_c and P_c of the C_{7+} pseudocomponents do not develop smoothly and monotonically with molecular weight.

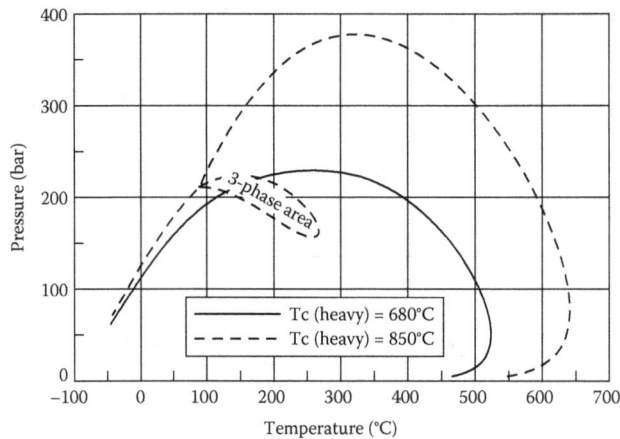

FIGURE 9.12 Simulated phase envelopes of oil mixture in Table 3.6 characterized for the SRK–Peneloux equation of state. The full-drawn line is the phase envelope calculated using the default characterization (Table 7.5). The dashed line shows the phase envelope after increasing the T_c of the C_{27}–C_{80} fraction from 679.6°C to 850°C. In addition to an increased two-phase area, this change is seen to introduce a three-phase area with one gas and two liquid phases, as is further exemplified by the flash calculation result in Table 9.5.

TABLE 9.5

Flash Calculation Results for a Pressure of 212 bar and a Temperature of 100°C for Oil Mixture in Table 3.6 Using the SRK–Peneloux Equation of State

	Mol% (Feed)	Default Characterization Mol% (Liquid)	T_c of C_{27}–C_{80} Increased (from 679.6°C to 850°C) Mol% (Gas)	Mol% (First Liquid)	Mol% (Second Liquid)
N_2	0.39	0.39	1.201	0.387	0.352
CO_2	0.30	0.30	0.399	0.302	0.291
C_1	40.20	40.20	76.592	40.381	37.919

	Mol% (Feed)	Default Characterization Mol% (Liquid)	T_c of C_{27}–C_{80} Increased (from 679.6°C to 850°C)		
			Mol% (Gas)	Mol% (First Liquid)	Mol% (Second Liquid)
C_2	7.61	7.61	8.331	7.688	7.420
C_3	7.95	7.95	6.104	8.051	7.851
iC_4	1.19	1.19	0.740	1.208	1.178
nC_4	4.08	4.08	2.203	4.135	4.073
iC_5	1.39	1.39	0.596	1.411	1.392
nC_5	2.15	2.15	0.846	2.180	2.160
C_6	2.79	2.79	0.832	2.834	2.809
C_7–C_{13}	19.33	19.33	2.055	19.451	20.010
C_{14}–C_{26}	8.64	8.64	0.103	8.637	9.093
C_{27}–C_{80}	3.98	3.98	0.000	3.334	5.451
Total	100.00	100.00	1.77	64.85	33.38

Note: With the default characterization (Table 7.5), a single liquid phase is found for a temperature of 100°C and a pressure of 212 bar. If T_c of the C_{27}–C_{80} fraction is increased from 679.6°C to 850°C, the mixture will at the same conditions split into three phases: one gas and two liquid phases. Nonzero binary interaction parameters can be seen in Table 7.6.

9.5 NEAR-CRITICAL FLUIDS

Table 9.6 shows the molar composition of a reservoir fluid with an experimentally determined critical point of 157°C and 387.5 bar (Pedersen et al. 1989). Table 9.7 shows the fluid characterized for the SRK–Peneloux equation of state with the C_{7+} fraction represented using 12 and 6 pseudocomponents, respectively. Binary interaction coefficients are shown in Table 9.8. Figure 9.13 shows phase envelopes simulated for the two characterized fluids in Table 9.7 using the SRK–Peneloux equation. With 12 C_{7+} pseudocomponents, the critical point is simulated to be 153.6°C and 411.9 bar. With six C_{7+} pseudocomponents, the simulated critical point is 183.9.6°C and 411.4 bar. Though the critical pressure is almost unaffected by the lumping, the critical temperature is seen to increase with the degree of lumping. This is a general observation, and it may complicate regression to PVT data—especially regression to experimental liquid dropout curves for heavily lumped fluid compositions.

CME experiments have been carried out for the fluid in Table 9.6 at four different temperatures—two below and two above the observed critical temperature. The experimental liquid dropout curves can be seen in Table 9.9. When the temperature is lower than the critical temperature, the liquid dropout curve starts with 100% liquid (saturation point is a bubble point), and when the temperature is above the critical temperature, it starts with 0% liquid (saturation point is a dew point). A plot of the four liquid dropout curves can be seen from Figure 9.14. The two dropout curves for temperatures below the mixture critical temperature start with 100% liquid dropout, while the two for a temperature above the critical one start with 0% liquid dropout.

Simulated CME liquid dropout curves can be seen from Figures 9.15 through 9.18. For the two lower temperatures (142.2°C and 151.1°C), the simulated dropout curves using 12 and 6 C_{7+} pseudocomponents both qualitatively agree with the measured liquid dropout. For the two higher temperatures (163.3°C and 171.0°C), the liquid dropout curves simulated using the six C_{7+} pseudocomponent description start with 100% liquid, whereas the experimental data starts with 0% liquid. The reason for this discrepancy is to be sought in the location of the simulated critical point (Figure 9.13), which when the fluid is described using six C_{7+} pseudocomponents is found to be 183.9.6°C. The saturation points for temperatures of 163.3°C and 171.0°C are according to this fluid description on the bubble point branch of the phase envelope where 100% liquid is in equilibrium with an incipient amount of gas.

TABLE 9.6

Molar Composition of Reservoir Fluid Composition with a Critical Point of 157°C and 387.5 bar

Component	Mol%	Molecular Weight (g/mol)	Density at 1.01 bar, 15°C (g/cm³)
N_2	0.46	—	—
CO_2	3.36	—	—
C_1	62.36	—	—
C_2	8.90	—	—
C_3	5.31	—	—
iC_4	0.92	—	—
nC_4	2.08	—	—
iC_5	0.73	—	—
nC_5	0.85	—	—
C_6	1.05	—	—
C_7	1.85	95	0.733
C_8	1.75	106	0.756
C_9	1.40	121	0.772
C_{10}	1.07	135	0.791
C_{11}	0.84	150	0.795
C_{12}	0.76	164	0.809
C_{13}	0.75	177	0.825
C_{14}	0.64	190	0.835
C_{15}	0.58	201	0.841
C_{16}	0.50	214	0.847
C_{17}	0.42	232	0.843
C_{18}	0.42	248	0.846
C_{19}	0.37	256	0.858
C_{20+}	2.63	406	0.897

Source: Pedersen, K.S., Thomassen, P., Fredenslund, Aa., Characterization of gas condensate mixtures, *Advances in Thermodynamics* 1, 157–162, 1989.

A regression was performed using a six–C_{7+} pseudocomponent description to match the experimental critical point of 157°C and 387.5 bar and the four saturation pressures in Table 9.9. The coefficients c_2, c_3, and d_2 in the correlations used for T_c and P_c of the C_{7+} fractions (Equations 5.1 and 5.2) were regressed on. The coefficient c_1 in Equation 5.1 and the coefficient d_1 in Equation 5.2 were modified to keep T_c and P_c of a typical C_7 component constant (see Section 9.1). The C_{20+} molecular weight was increased by 5% keeping the weight composition constant. The regressed composition can be seen in Table 9.10. The simulated critical point after regression was 157°C and 392 bar, meaning that the mixture critical temperature as a result of the regression has been lowered by more than 20°C. That makes the four simulated CME liquid dropout curves start with the right volume percentage of liquid (100% or 0%, depending on temperature), as can be seen from Figure 9.19. It had been difficult to get an equally good match without having the option to regress directly on the mixture critical point.

TABLE 9.7

Reservoir Fluid in Table 9.6 Characterized for the SRK–Peneloux Equation of State Using 12 and 6 C_{7+} Pseudocomponents, Respectively

C_{7+} Fraction Represented Using 12 Pseudocomponents

	Mol%	Molecular Weight	T_c (°C)	P_c (bar)	Acentric Factor	c_0 (cm³/mol)
N_2	0.46	28.0	−147.0	33.94	0.040	0.92
CO_2	3.36	44.0	31.1	73.76	0.225	3.29
C_1	62.36	16.0	−82.6	46.00	0.008	0.63
C_2	8.90	30.1	32.3	48.84	0.098	2.63
C_3	5.31	44.1	96.7	42.46	0.152	5.06
iC_4	0.92	58.1	135.0	36.48	0.176	7.29
nC_4	2.08	58.1	152.1	38.00	0.193	7.86
iC_5	0.73	72.2	187.3	33.84	0.227	10.93
nC_5	0.85	72.2	196.5	33.74	0.251	12.18
C_6	1.05	86.2	234.3	29.69	0.296	17.98
C_7	1.85	95.0	259.7	31.95	0.465	9.63
C_8	1.75	106.0	279.9	29.44	0.497	14.65
C_9	1.40	121.0	302.6	26.08	0.540	22.84
C_{10}–C_{11}	1.91	141.6	331.0	23.03	0.599	31.63
C_{12}–C_{13}	1.51	170.5	365.6	20.28	0.676	40.07
C_{14}–C_{15}	1.22	195.2	392.7	18.89	0.741	42.42
C_{16}–C_{18}	1.34	230.3	425.3	16.86	0.829	47.78
C_{19}–C_{21}	0.85	271.0	460.9	15.64	0.924	44.72
C_{22}–C_{25}	0.72	323.1	501.8	14.53	1.034	35.68
C_{26}–C_{30}	0.57	384.9	547.3	13.73	1.147	17.21
C_{31}–C_{39}	0.52	476.7	610.2	13.05	1.274	−19.73
C_{40}–C_{80}	0.35	678.6	743.1	12.45	1.310	−119.67

C_{7+} Fraction Represented Using 6 Pseudocomponents

	Mol%	Molecular Weight	T_c (°C)	P_c (bar)	Acentric Factor	c_0 (cm³/mol)
N_2	0.46	28.0	−147.0	33.94	0.040	0.92
CO_2	3.36	44.0	31.1	73.76	0.225	3.29
C_1	62.36	16.0	−82.6	46.00	0.008	0.63
C_2	8.90	30.1	32.3	48.84	0.098	2.63
C_3	5.31	44.1	96.7	42.46	0.152	5.06
iC_4	0.92	58.1	135.0	36.48	0.176	7.29
nC_4	2.08	58.1	152.1	38.00	0.193	7.86
iC_5	0.73	72.2	187.3	33.84	0.227	10.93
nC_5	0.85	72.2	196.5	33.74	0.251	12.18
C_6	1.05	86.2	234.3	29.69	0.296	17.98
C_7	1.85	95.0	259.7	31.95	0.465	9.63
C_8	1.75	106.0	279.9	29.44	0.497	14.65
C_9	1.40	121.0	302.6	26.08	0.540	22.84
C_{10}–C_{15}	4.64	165.1	361.8	20.82	0.669	38.44
C_{16}–C_{25}	2.91	265.1	458.9	15.80	0.919	46.44
C_{26}–C_{80}	1.43	488.8	634.9	13.06	1.246	−17.76

Note: No regression has been performed. Binary interaction parameters can be seen in Table 9.8.

TABLE 9.8

Nonzero Binary Interaction Coefficients for Use with the Mixture in Tables 9.7 and 9.10

	N$_2$	CO$_2$
N$_2$	−0.032	–
CO$_2$	0.028	0.120
C$_1$	0.041	0.120
C$_2$	0.076	0.120
C$_3$	0.094	0.120
iC$_4$	0.070	0.120
nC$_4$	0.087	0.120
iC$_5$	0.088	0.120
nC$_5$	0.080	0.120
C$_6$	0.080	0.100
C$_{7+}$	0.080	0.100

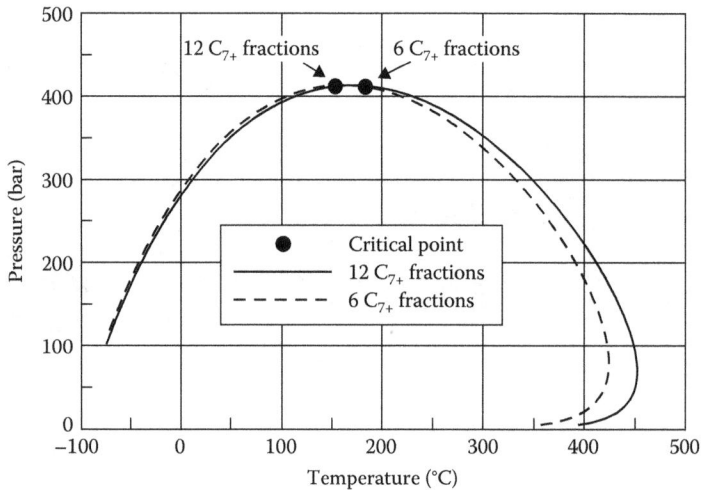

FIGURE 9.13 Simulated phase envelopes for reservoir fluid in Table 9.6. The C$_{7+}$ fraction is represented using 12 pseudocomponents (full-drawn lines) and 6 pseudocomponents (dashed lines). The critical point is found to be at 153.6°C and 411.9 bar with 12 C$_{7+}$ pseudocomponents and at 183.9.6°C and 411.4 with 6 C$_{7+}$ pseudocomponents. The fluid descriptions can be seen in Table 9.7.

TABLE 9.9

Liquid Dropout Curves Measured at Four Different Temperatures for Reservoir Fluid in Table 9.6

Temperature = 142.2°C		Temperature = 151.1°C		Temperature = 163.3°C		Temperature = 170.0°C	
Pressure (bar)	Liquid Volume (%)	Pressure (bar)	Liquid Volume (%)	Pressure (bar)	Liquid Volume (%)	Pressure (bar)	Liquid Volume (%)
449.2	—	449.2	—	449.2	—	449.2	—
435.4	—	435.4	—	435.4	—	435.4	—
421.6	—	421.6	—	421.6	—	421.6	—

Temperature = 142.2°C		Temperature = 151.1°C		Temperature = 163.3°C		Temperature = 170.0°C	
Pressure (bar)	Liquid Volume (%)	Pressure (bar)	Liquid Volume (%)	Pressure (bar)	Liquid Volume (%)	Pressure (bar)	Liquid Volume (%)
414.7	—	407.8	—	407.8	—	407.8	—
407.8	—	403.0	—	403.0	—	403.0	—
403.0	—	400.9	—	400.9	—	394.0	—
394.0	—	394.0	—	394.0	—	387.1	—
390.6*	100.0	389.3*	100.0	387.1	—	383.9*	0.0
380.2	80.0	387.1	83.2	385.7*	0.0	380.2	39.0
373.3	67.2	383.7	57.7	383.7	35.2	378.5	39.7
366.4	58.7	380.2	51.2	381.9	42.2	376.8	41.0
359.5	53.1	373.3	49.3	380.2	43.5	373.3	42.0
345.8	50.6	366.4	48.7	376.8	44.6	369.9	42.8
332.0	49.7	352.6	48.3	373.3	45.0	366.4	43.2
318.2	49.5	338.9	48.4	366.4	45.2	359.5	43.7
290.6	49.1	325.1	48.5	359.5	45.5	352.6	43.9
263.0	48.4	311.3	48.5	352.6	45.5	345.8	43.9
235.6	47.7	283.7	48.2	338.9	45.5	332.0	43.8
194.3	46.4	256.1	47.5	325.1	45.5	318.2	43.7
152.4	45.2	228.5	46.7	311.3	45.5	304.4	43.6
111.3	43.6	201.0	45.7	297.5	45.4	276.8	43.4
89.7	42.7	159.6	44.1	269.9	45.0	249.2	43.0
—	—	122.4	42.2	242.3	44.5	221.6	42.5
—	—	—	—	214.8	43.8	194.1	41.7
—	—	—	—	187.2	43.3	152.7	40.8
—	—	—	—	145.8	42.0	129.9	40.2
—	—	—	—	126.7	41.4	—	—

* Saturation point.

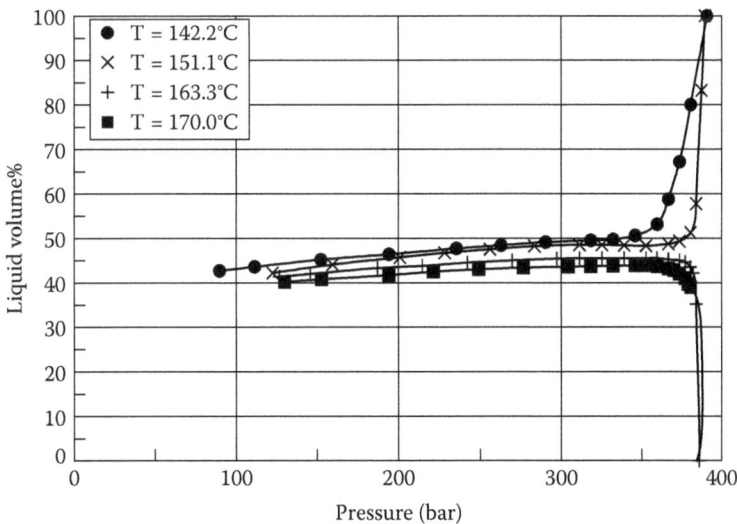

FIGURE 9.14 Measured liquid dropout curves for reservoir fluid in Table 9.6 at four different temperatures, of which two are below and two above the critical temperature of 157°C. The data points are listed in Table 9.9.

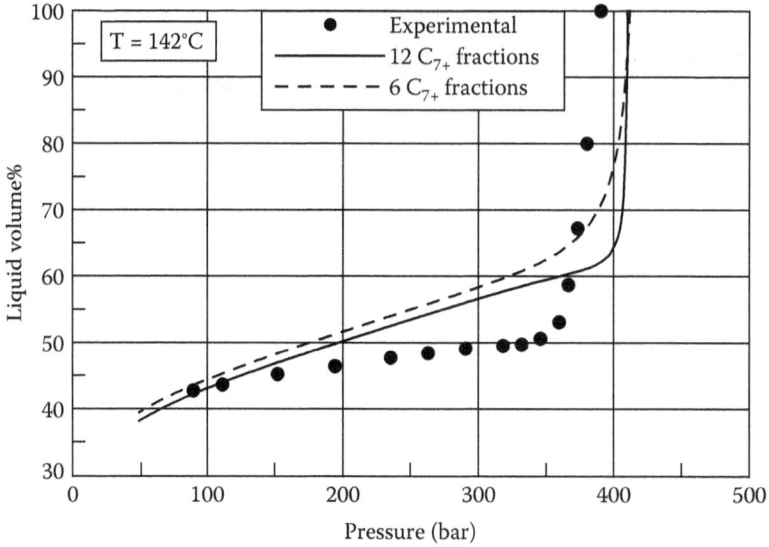

FIGURE 9.15 Measured and simulated liquid dropout curve for reservoir fluid in Table 9.6 at a temperature of 142.2°C. The fluid was represented using 12 (full-drawn lines) and 6 C_{7+} pseudocomponents (dashed lines), respectively. No regression was performed. The fluid description can be seen in Table 9.7. The experimental data can be seen in Table 9.9.

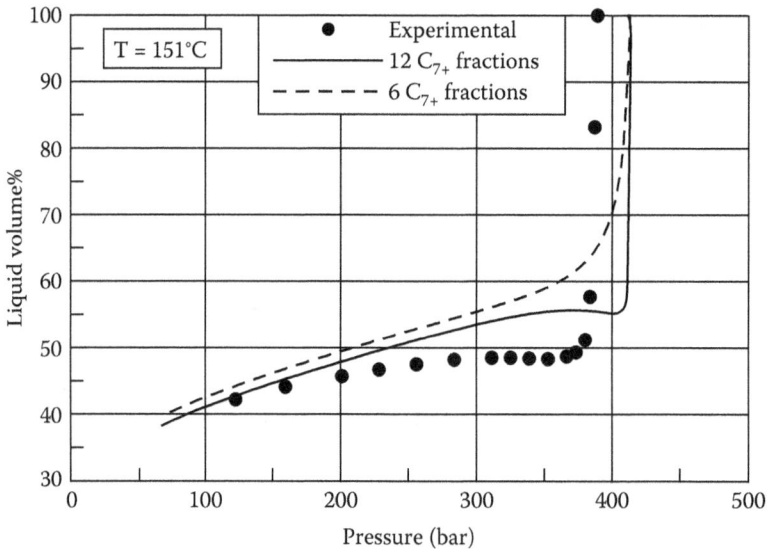

FIGURE 9.16 Measured and simulated liquid dropout curve for reservoir fluid in Table 9.6 at a temperature of 151.1°C. The fluid was represented using 12 (full-drawn lines) and 6 C_{7+} pseudocomponents (dashed lines), respectively. No regression was performed. The fluid description can be seen in Table 9.7. The experimental data can be seen in Table 9.9.

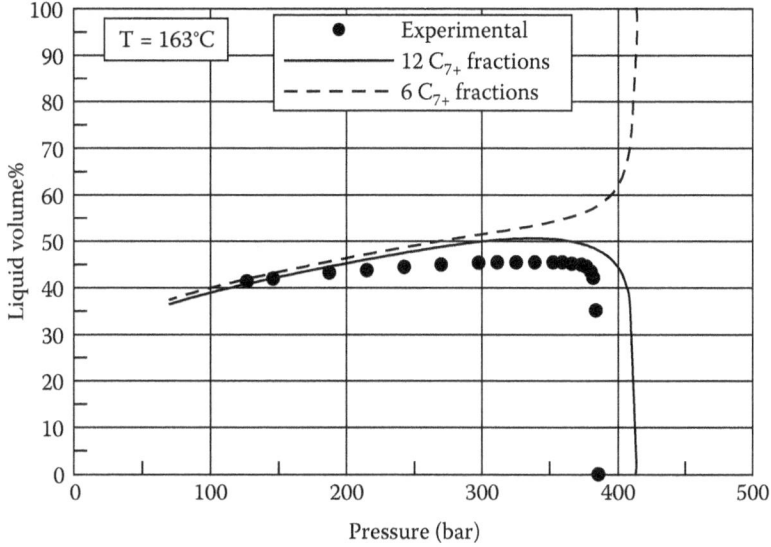

FIGURE 9.17 Measured and simulated liquid dropout curve for reservoir fluid in Table 9.6 at a temperature of 163.3°C. The fluid was represented using 12 (full-drawn lines) and 6 C_{7+} pseudocomponents (dashed lines), respectively. No regression was performed. The fluid description can be seen in Table 9.7. The experimental data can be seen in Table 9.9.

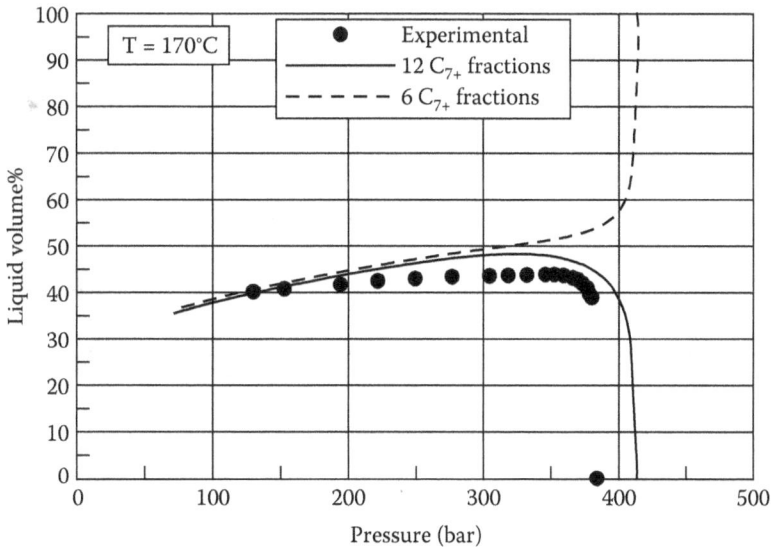

FIGURE 9.18 Measured and simulated liquid dropout curve for reservoir fluid in Table 9.6 at a temperature of 170°C. The fluid was represented using 12 (full-drawn lines) and 6 C_{7+} pseudocomponents (dashed lines), respectively. No regression was performed. The fluid descriptions can be seen in Table 9.7. The experimental data can be seen in Table 9.9.

TABLE 9.10

Reservoir Fluid in Table 9.6 Characterized for the SRK–Peneloux Equation of State Using Six Pseudocomponents to Represent the C_{7+} Fraction

	Mol%	Molecular Weight	T_c (°C)	P_c (bar)	Acentric Factor	c_0 (cm³/mol)
N_2	0.461	28.0	−147.0	33.94	0.040	0.92
CO_2	3.364	44.0	31.1	73.76	0.225	3.29
C_1	62.438	16.0	−82.6	46.00	0.008	0.63
C_2	8.911	30.1	32.3	48.84	0.098	2.63
C_3	5.317	44.1	96.7	42.46	0.152	5.06
iC_4	0.921	58.1	135.0	36.48	0.176	7.29
nC_4	2.083	58.1	152.1	38.00	0.193	7.86
iC_5	0.731	72.2	187.3	33.84	0.227	10.93
nC_5	0.851	72.2	196.5	33.74	0.251	12.18
C_6	1.051	86.2	234.3	29.69	0.296	17.98
C_7	1.852	95.0	259.7	31.95	0.465	9.63
C_8	1.752	106.0	268.3	30.17	0.497	8.6
C_9	1.402	121.0	291.4	27.16	0.540	13.13
C_{10}–C_{15}	4.646	165.1	340.7	22.63	0.669	13.06
C_{16}–C_{26}	2.858	266.8	433.0	17.90	0.924	−7.24
C_{27}–C_{80}	1.362	521.7	565.6	15.27	1.258	−161.12

Note: Regression has been performed to a critical point of 157°C and 387.5 bar. Binary interaction parameters have not been regressed on and can be seen in Table 9.8.

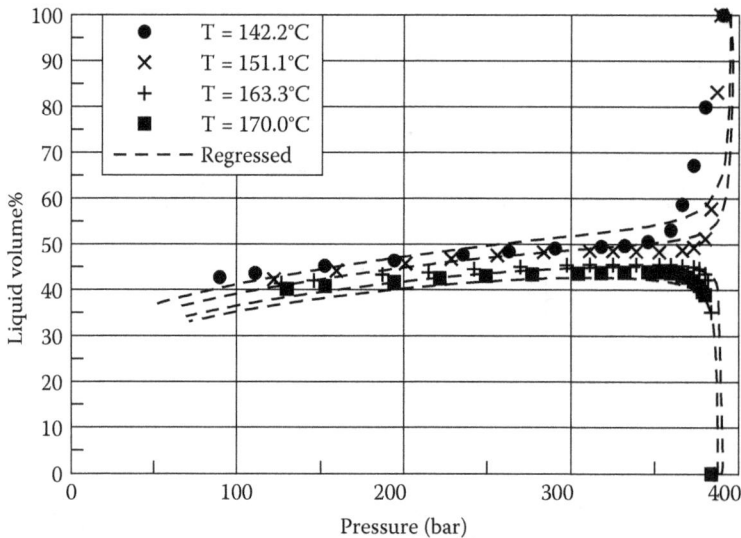

FIGURE 9.19 Liquid dropout curves for reservoir fluid in Table 9.6 at four different temperatures, of which two are below and two above the critical temperature of 157°C. The dashed lines are simulation results obtained with the SRK–Peneloux equation of state using the fluid characterization in Table 9.10. The experimental data can be seen in Table 9.9.

9.6   FLUIDS CHARACTERIZED TO THE SAME PSEUDOCOMPONENTS

Section 5.6 dealt with multiple fluids characterized to the same pseudocomponents. This is a useful way to keep the total number of components low when dealing with multiple streams in a flow line network or when carrying out reservoir simulation studies for a field made up of reservoir zones of different compositions. A typical example could be a reservoir with an upper gas cap and an oil zone beneath separated by a gas–oil contact, as further discussed in Chapter 14. Table 3.6 shows a reservoir oil composition, and Table 3.15 shows a gas condensate composition. An oil zone beneath a gas cap is also called an oil leg. Table 9.11 shows a common characterization (non-regressed parameters) for the two fluid compositions, using the principles of Section 5.6 for characterizing multiple compositions to the same pseudocomponents. Differential liberation simulation results for the oil mixture can be seen as dashed lines in Figures 9.20 through 9.25. Experimental differential liberation results from Table 3.13 are shown in the same plots. The simulated results can be compared with those in Figures 7.11 through 7.16, which were based on a fluid characterization for the oil mixture alone. The simulation results in Figures 9.20 through 9.25 are almost unaffected by the fact that the oil is represented using the same C_{7+} pseudocomponents as a gas condensate mixture.

Constant volume depletion simulation results for the gas condensate mixture in Table 3.15 characterized to the same pseudocomponents as the oil mixture in Table 3.6 can be seen from Figures 9.26 through 9.29, which also show experimental constant volume depletion results from Table 3.16. The simulation results can be compared with those in Figures 7.6 and 7.8 through 7.10, which were achieved based on a fluid characterization for the gas condensate mixture alone. There is a need to improve the match of the liquid dropout curve in Figure 9.26. This will often be the case for gas condensate mixtures when characterized to the same pseudocomponents as one or more oil mixtures.

The coefficients c_2, c_3, d_2, d_3, and e_2 in the correlations used for T_c, P_c, and m of the C_{7+} fractions (Equations 5.1 through 5.3) were regressed on. The coefficients c_1 in Equation 5.1, d_1 in Equation 5.2, and e_1 in Equation 5.3 were modified to keep T_c, P_c, and the acentric factor of a typical C_7 component constant (see Section 9.1). An adjustment of maximum 10% of the plus molecular weights was allowed for both oil and gas condensate mixtures while keeping the weight compositions constant. The resulting fluid characterization common for both fluids can be seen in Table 9.11 (regressed parameters) with binary interaction parameters in Table 9.12. The assumed C_{10+} molecular weight of the oil mixture in Table 3.6 was reduced from 453.0 to 407.7, and the assumed C_{20+} molecular weight of the gas condensate mixture in Table 3.15 was reduced from 381.0 to 342.9. For both compositions, the weight % composition was kept constant. The resulting differential liberation simulation results for the oil mixture can be seen as dashed–dotted lines in Figures 9.20 through 9.25 and the constant volume depletion results for the gas condensate mixture as dashed–dotted lines in Figures 9.26 through 9.29. The experimental data is matched almost perfectly, which is noteworthy considering that both fluid compositions are represented using the same three C_{7+} pseudocomponents.

9.7   PVT DATA WITH GAS INJECTION

The minimum miscibility pressure (MMP) is a key property when gas injection for EOR purposes is considered for a field. The term minimum miscibility pressure was introduced in Section 3.2.4 and is further dealt with in Chapter 15.

Table 9.13 shows a reservoir oil composition for which swelling data is given in Table 9.14. The MMP with CO_2 as injection gas has been determined to 179 bar at the reservoir temperature of 121°C. Table 9.15 shows the reservoir fluid composition lumped to eight pseudocomponents. The fluid was characterized for the SRK–Peneloux equation using the procedure of Pedersen et al. in Chapter 5. CO_2 was kept separate because the injection gas consists of pure CO_2. Nonzero binary interaction parameters are given in Table 9.16. With this fluid description, the CO_2 MMP at 121°C is simulated to 151 bar, which is 28 bar lower than the measured MMP. An EOR PVT study may comprise an extensive amount of data, and the question is what other data to match to improve the match of the MMP while retaining a good description of all other data.

TABLE 9.11

The Oil Mixture in Table 3.6 and the Gas Condensate Mixture in Table 3.15 Characterized to the Same Pseudocomponents

Non-Regressed Parameters

	Oil (Table 3.6)	Gas Condensate (Table 3.15)	T_c (°C)	P_c (bar)	Acentric Factor	c_0 (cm³/mol)
N_2	0.39	0.64	−147.0	33.94	0.040	0.92
CO_2	0.30	3.53	31.1	73.76	0.225	3.03
C_1	40.20	70.79	−82.6	46.00	0.008	0.63
C_2	7.61	8.94	32.3	48.84	0.098	2.63
C_3	7.95	5.05	96.7	42.46	0.152	5.06
iC_4	1.19	0.85	135.0	36.48	0.176	7.29
nC_4	4.08	1.68	152.1	38.00	0.193	7.86
iC_5	1.39	0.62	187.3	33.84	0.227	10.93
nC_5	2.15	0.79	196.5	33.74	0.251	12.18
C_6	2.79	0.83	234.3	29.69	0.296	17.98
$C_7\text{--}C_{12}$	17.74	4.23	302.7	26.41	0.545	22.15
$C_{13}\text{--}C_{23}$	9.25	1.66	426.4	17.04	0.834	48.39
$C_{24}\text{--}C_{80}$	4.96	0.38	651.5	13.38	1.229	−24.40

Regressed Parameters

	Oil (Table 3.6)	Gas Condensate (Table 3.15)	T_c (°C)	P_c (bar)	Acentric Factor	c_0 (cm³/mol)
N_2	0.39	0.64	−147.0	33.94	0.040	0.92
CO_2	0.30	3.53	31.1	73.76	0.225	3.29
C_1	39.91	70.73	−82.6	46.00	0.008	0.63
C_2	7.55	8.93	32.3	48.84	0.098	2.63
C_3	7.89	5.05	96.7	42.46	0.152	5.06
iC_4	1.18	0.85	135.0	36.48	0.176	7.29
nC_4	4.05	1.68	152.1	38.00	0.193	7.86
iC_5	1.38	0.62	187.3	33.84	0.227	10.93
nC_5	2.13	0.79	196.5	33.74	0.251	12.18
C_6	2.77	0.83	234.3	29.69	0.296	18.10
$C_7\text{--}C_{12}$	17.61	4.23	298.5	27.72	0.545	12.80
$C_{13}\text{--}C_{22}$	9.41	1.74	400.3	19.81	0.835	−5.60
$C_{23}\text{--}C_{80}$	5.43	0.39	540.1	16.96	1.202	−115.89

Note: The parameter c_0 is the Peneloux volume shift parameter as defined in Equation 4.44. Binary interaction parameters were not regressed on and can be seen in Table 9.12.

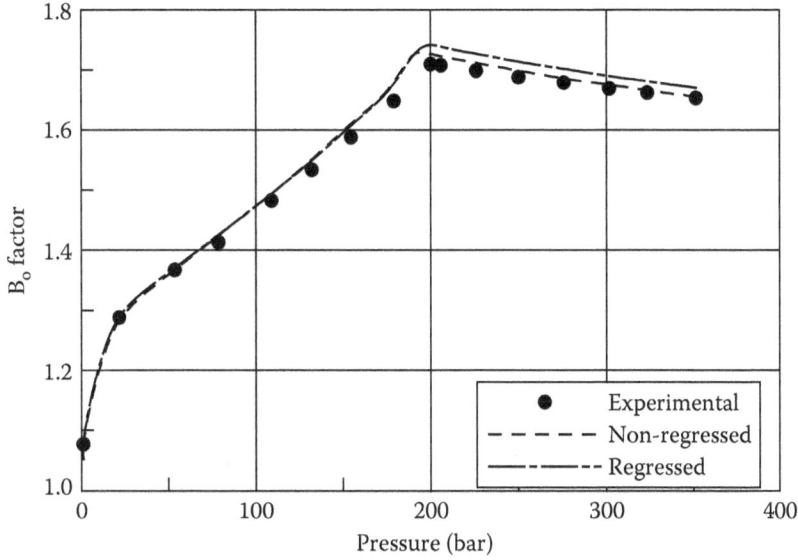

FIGURE 9.20 Measured and simulated oil formation (B_o) factors for differential liberation experiment at 97.5°C for the oil mixture in Table 3.6. Experimental results can be seen in Table 3.13. The simulated results are obtained with the SRK–Peneloux equation of state and using a common fluid characterization for this oil and the gas condensate mixture in Table 3.15, as shown in Table 9.11. Binary interaction parameters are listed in Table 9.12.

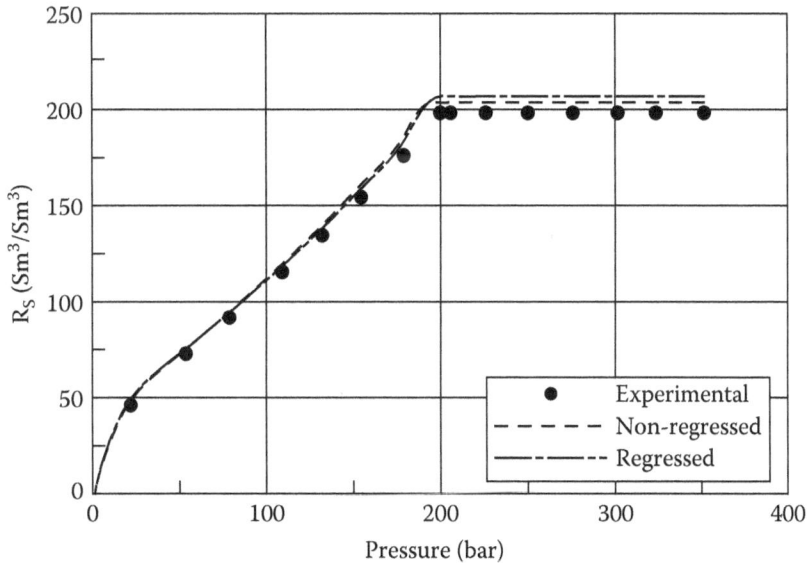

FIGURE 9.21 Measured and simulated solution gas/oil ratios (R_s) for differential liberation experiment at 97.5°C for the oil mixture in Table 3.6. Experimental results can be seen in Table 3.13. The simulated results are obtained with the SRK–Peneloux equation of state and using a common fluid characterization for this oil and the gas condensate mixture in Table 3.15, as shown in Table 9.11. Binary interaction parameters are listed in Table 9.12.

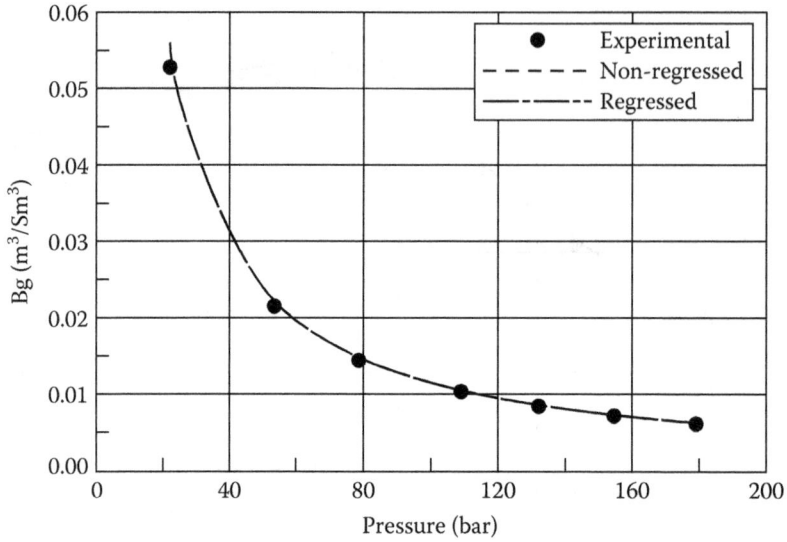

FIGURE 9.22 Measured and simulated gas formation volume factors (B_g) for differential liberation experiment at 97.5°C for the oil mixture in Table 3.6. Experimental results can be seen in Table 3.13. The simulated results are obtained with the SRK–Peneloux equation of state and using a common fluid characterization for this oil and the gas condensate mixture in Table 3.15, as shown in Table 9.11. Binary interaction parameters are listed in Table 9.12.

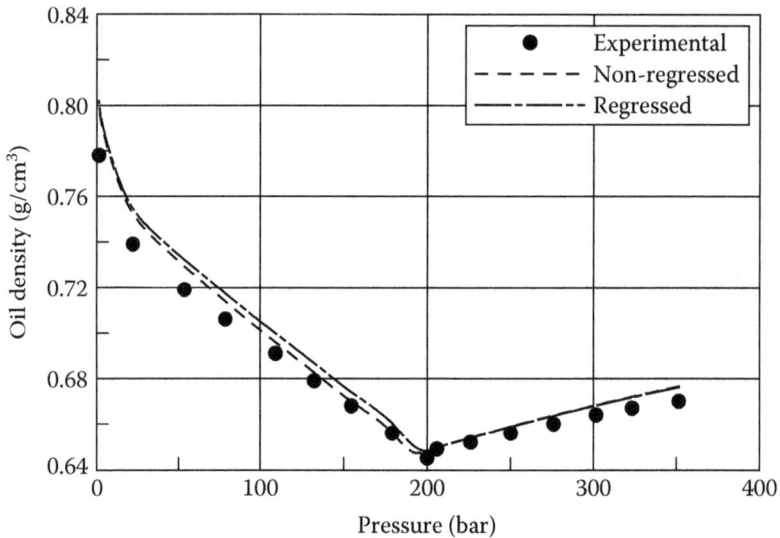

FIGURE 9.23 Measured and simulated oil densities for differential liberation experiment at 97.5°C for the oil mixture in Table 3.6. Experimental results can be seen in Table 3.13. The simulated results are obtained with the SRK–Peneloux equation of state and using a common fluid characterization for this oil and the gas condensate mixture in Table 3.15, as shown in Table 9.11. Binary interaction parameters are listed in Table 9.12.

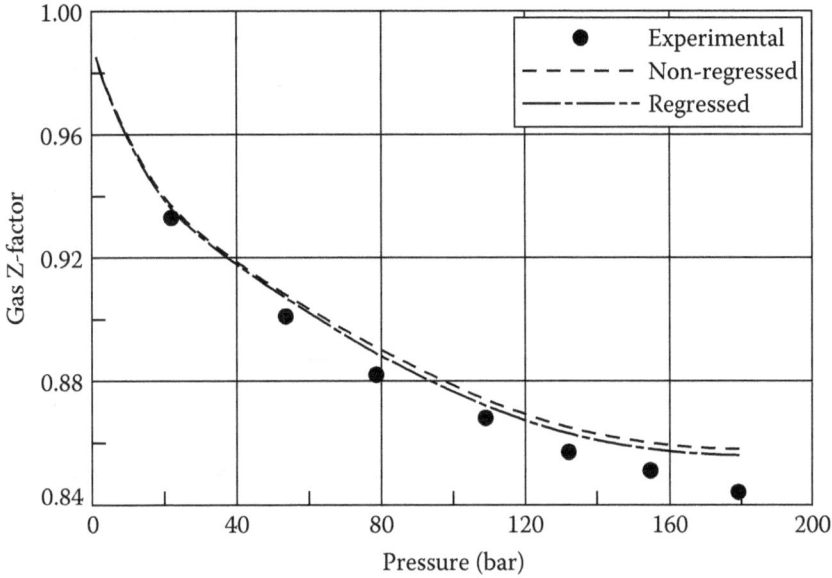

FIGURE 9.24 Measured and simulated gas-phase Z-factors for differential liberation experiment at 97.5°C for the oil mixture in Table 3.6. Experimental results can be seen in Table 3.13. The simulated results are obtained with the SRK–Peneloux equation of state and using a common fluid characterization for this oil and the gas condensate mixture in Table 3.15, as shown in Table 9.11. Binary interaction parameters are listed in Table 9.12.

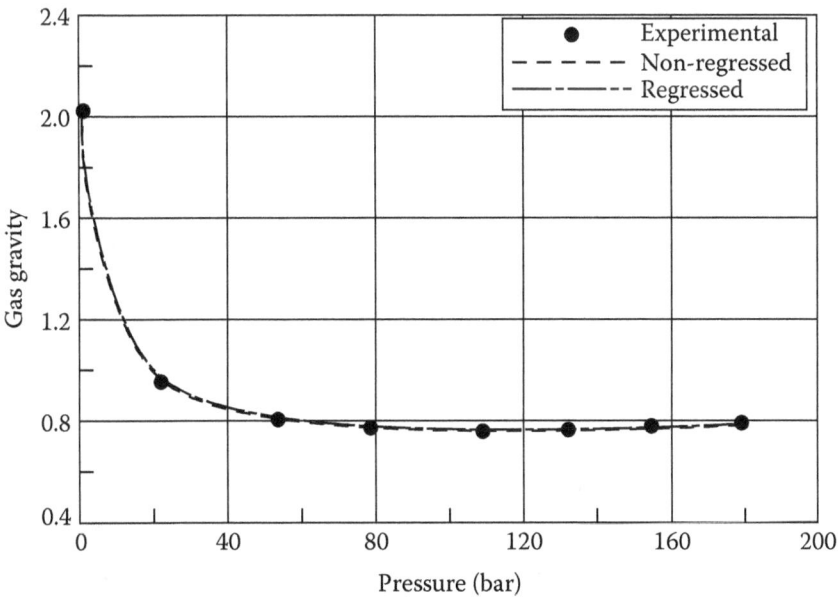

FIGURE 9.25 Measured and simulated gas gravities for differential liberation experiment at 97.5°C for the oil mixture in Table 3.6. Experimental results can be seen in Table 3.13. The simulated results are obtained with the SRK–Peneloux equation of state and using a common fluid characterization for this oil and the gas condensate mixture in Table 3.15, as shown in Table 9.11. Binary interaction parameters are listed in Table 9.12.

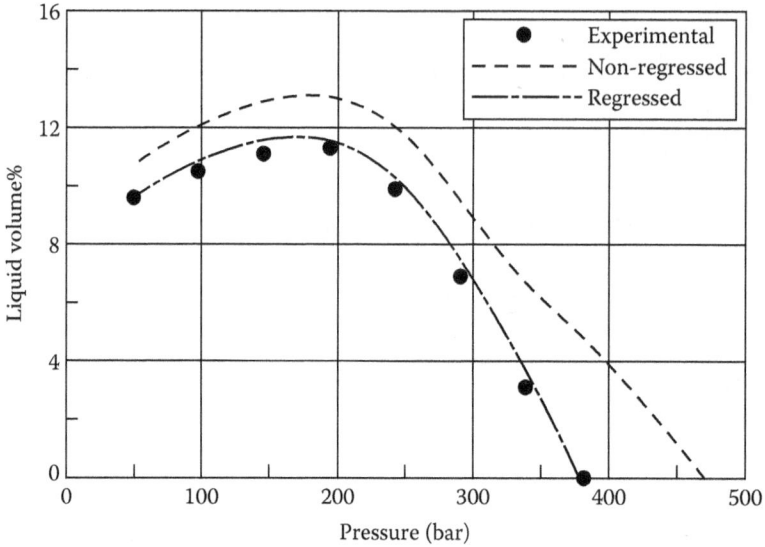

FIGURE 9.26 Measured and simulated liquid volume percentages for a constant volume depletion experiment at 150.3°C for the gas condensate mixture in Table 3.15. Experimental results can be seen in Table 3.16. The simulated results are obtained with the SRK–Peneloux equation of state and using a common fluid characterization for this gas condensate and the oil mixture in Table 3.6, as shown in Table 9.11. Binary interaction parameters are listed in Table 9.12.

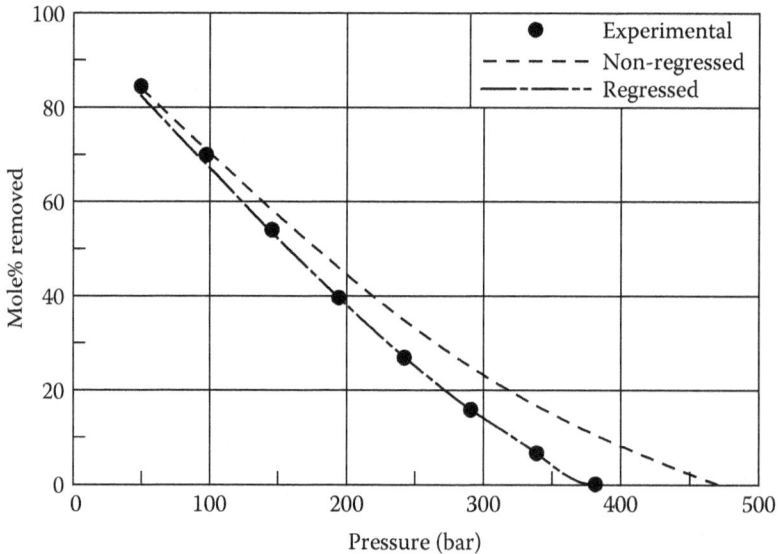

FIGURE 9.27 Measured and simulated cumulative mole percentages removed in a constant volume depletion experiment at 150.3°C for the gas condensate mixture in Table 3.15. Experimental results can be seen in Table 3.16. The simulated results are obtained with the SRK–Peneloux equation of state and using a common fluid characterization for this gas condensate and the oil mixture in Table 3.6 as shown in Table 9.11. Binary interaction parameters are listed in Table 9.12.

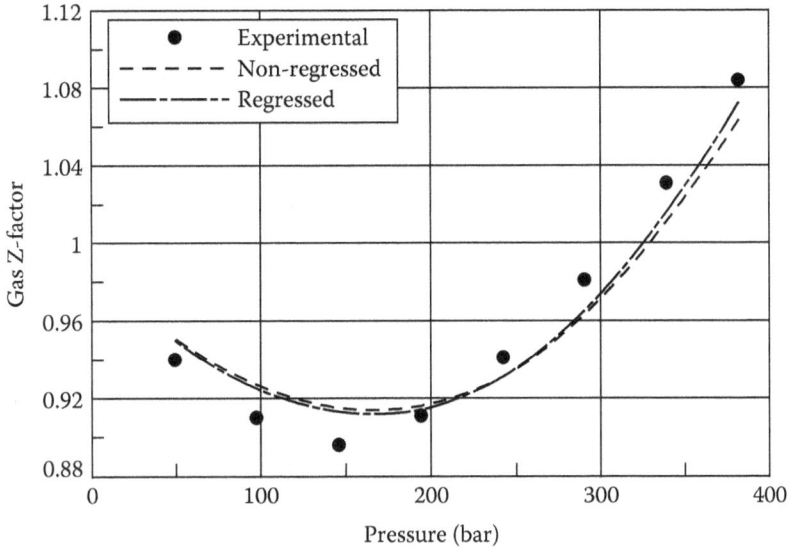

FIGURE 9.28 Measured and simulated gas-phase Z-factors in a constant volume depletion experiment at 150.3°C for the gas condensate mixture in Table 3.15. Experimental results can be seen in Table 3.16. The simulated results are obtained with the SRK–Peneloux equation of state and using a common fluid characterization for this gas condensate and the oil mixture in Table 3.6 as shown in Table 9.11. Binary interaction parameters are listed in Table 9.12.

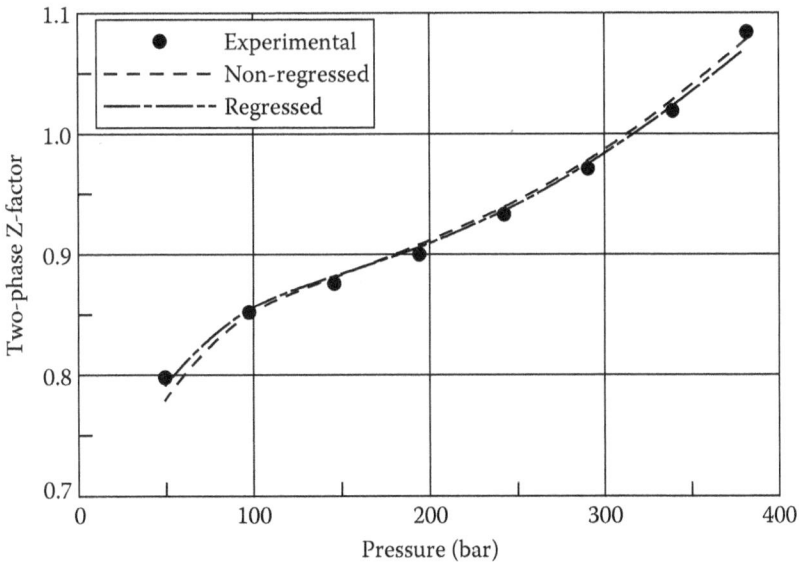

FIGURE 9.29 Measured and simulated two-phase Z-factors in a constant volume depletion experiment at 150.3°C for the gas condensate mixture in Table 3.15. Experimental results can be seen in Table 3.16. The simulated results are obtained with the SRK–Peneloux equation of state and using a common fluid characterization for this gas condensate and the oil mixture in Table 3.6 as shown in Table 9.11. Binary interaction parameters are listed in Table 9.12.

TABLE 9.12

Nonzero Binary Interaction Coefficients for Use with the Compositions in Table 9.11

	SRK–Peneloux	
	N_2	CO_2
CO_2	−0.032	–
C_1	0.028	0.120
C_2	0.041	0.120
C_3	0.076	0.120
iC_4	0.094	0.120
nC_4	0.070	0.120
iC_5	0.087	0.120
nC_5	0.088	0.120
C_6	0.080	0.120
C_{7+}	0.080	0.100

TABLE 9.13

Molar Composition of Reservoir Fluid Composition for Which CO_2 Swelling Data Is Shown in Table 9.14

Component	Mol%	Molecular Weight (g/mol)	Density at 1.01 bar, 15°C (g/cm³)
N_2	0.363	—	—
CO_2	2.991	—	—
C_1	29.066	—	—
C_2	7.163	—	—
C_3	6.577	—	—
iC_4	1.845	—	—
nC_4	4.231	—	—
iC_5	2.274	—	—
nC_5	2.864	—	—
C_6	4.104	—	—
C_7	4.537	98	0.714
C_8	4.451	113	0.736
C_9	3.83	123	0.757
C_{10}	2.828	136	0.774
C_{11}	4.370	151	0.787
C_{12}	3.038	168	0.803
C_{13}	1.926	187	0.814
C_{14}	2.051	206	0.825
C_{15}	1.447	215	0.832
C_{16}	1.288	223	0.843
C_{17}	1.454	245	0.848
C_{18}	1.019	260	0.860
C_{19}	0.736	272	0.870
C_{20+}	5.546	509	0.952

At the point in the reservoir where miscibility develops, the fluid is at a critical point. The composition of the critical fluid at that location is unknown, as miscibility has developed as a result of multiple contacts between gas and oil. The critical composition in the miscibility zone can therefore not be regressed to. Instead regression may be carried out to the critical point on the swelling curve. The swelling experiment in Table 9.14 shows that the saturation point changes from a bubble point to a dew point between 175 and 225 mol% CO_2 added. An estimate of the critical composition could therefore be reservoir oil and CO_2 mixed in a molar ratio of 100:200. With the non-regressed fluid description in Table 9.15, the shift from bubble to dew point is simulated to happen between 50 and 100 mol% CO_2 added. A regression to the swelling data in Table 9.14 was carried out. One

TABLE 9.14

Swelling Data for Oil Composition in Table 9.13 at 121°C with CO_2 Injection Gas

Stage	Mol% Gas/ Initial Mol Oil	GOR Sm³/Sm³	Sat Pressure bar	Saturation Point	Swelling Volume/Initial Oil Volume	Density g/cm³
1	0	0	142.01	Bubble	1.0000	0.6382
2	50	76.9	198.27	Bubble	1.2128	0.6442
3	100	153.7	230.61	Bubble	1.4325	0.6454
4	150	230.7	253.91	Bubble	1.6501	0.6470
5	175	269.1	265.43	Bubble	1.7537	0.6496
6	225	346.1	283.77	Dew	1.9637	0.6530

TABLE 9.15

The Oil Composition in Table 9.13 Characterized Using Non-Regressed and Regressed Parameters. The Binary Interaction Parameters Can Be Seen from Table 9.16

Component	Mol%	Molecular Weight	T_c (°C)	P_c (bar)	Acentric Factor	c_0 cm³/mol
CO_2	2.991	44.01	31.05	73.76	0.2250	3.03
$N_2 + C_1$	29.429	16.19	−83.92	45.74	0.0087	0.63
$C_2 + C_3$	13.740	36.78	69.20	45.18	0.1290	3.79
C_4–C_6	15.318	70.34	191.20	33.67	0.2414	11.77
C_7–C_{11}	20.016	123.06	304.78	24.54	0.5531	28.11
C_{12}–C_{15}	8.462	189.57	385.14	18.40	0.7298	48.12
C_{16}–C_{30}	6.935	279.02	474.37	15.80	0.9547	41.40
C_{31}–C_{80}	3.108	642.34	742.77	13.96	1.2515	−118.70
Non-regressed parameters						
CO_2	2.999	44.01	31.05	73.76	0.2250	3.03
$N_2 + C_1$	29.507	16.19	−83.92	45.74	0.0087	0.63
$C_2 + C_3$	13.777	36.78	69.20	45.18	0.1290	3.79
C_4–C_6	15.359	70.34	191.20	33.67	0.2414	11.77
C_7–C_{11}	20.069	123.06	306.88	27.49	0.5531	8.07
C_{12}–C_{15}	8.484	189.57	368.03	22.08	0.7298	−3.71
C_{16}–C_{32}	6.890	282.92	428.69	18.96	0.9661	−39.43
C_{33}–C_{80}	2.915	683.56	568.21	16.75	1.2394	−324.59

TABLE 9.16

Nonzero Binary Interaction Parameters for Use with the Fluid Description in Table 9.15. The Same Binary Interaction Parameters Were Used with the Non-Regressed and the Regressed Parameters

	CO_2	$N_2 + C_1$
$N_2 + C_1$	0.1181	
$C_2 + C_3$	0.1200	0.0007
C_4–C_6	0.1200	0.0010
C_7–C_{11}	0.1000	0.0010
C_{12}–C_{15}	0.1000	0.0010
C_{16}–C_{30}	0.1000	0.0010
C_{31}–C_{80}	0.1000	0.0010

of the data points tuned to was that 200 mol% CO_2 added should give a critical composition. The coefficients c_1–c_3 in Equation 5.1 and the coefficients d_1–d_2 in Equation 5.2 were used as tuning parameters. The regressed parameters can be seen in Table 9.15 (binary interaction parameters in Table 9.16). The MMP at 121°C calculated using the regressed fluid composition is 175 bar, which is only 4 bar lower than the measured MMP. This example shows that much tedious regression work can be avoided by first evaluating the data material and regressing to the key data. In the actual case, the target was to match the MMP, and the key to matching the MMP was matching the critical point on the swelling curve.

9.8 REGRESSION TARGETING ESTIMATED CRITICAL POINT

Takeshi et al. (2020) have presented a common EoS model for 21 reservoir fluids, 11 of which were oils with swelling data reported. All the swelling experiments for the oils were continued until the bubble point shifted to a dew point. Among the 21 fluids were also 3 near-critical gas condensates with CME and CVD liquid dropout data. The mentioned PVT experiments are described in Chapter 3.

The common EoS model was developed by initially determining the critical point for each of the oils with swelling data and for each of the near-critical gas condensates. For the oils, the assumption was that an EoS model that correctly simulated the critical point on the swelling curve would also simulate the correct critical point of the reservoir fluid before addition of gas. The critical point on the swelling curve was determined by analyzing the CME data for each swelling stage. Figure 9.30 shows schematically CME liquid dropout curves for a reservoir fluid at four different temperatures. When the temperature is much lower than the critical temperature, the liquid dropout curve will start with 100 % liquid at the saturation point, and the liquid shrinkage will be low when the pressure is lowered. Closer to the critical temperature, the liquid shrinkage increases, and right at the critical point, the liquid volume% would decrease from 100% to around 50% by marginally lowering the pressure from that at the saturation point. At temperatures higher than the critical temperature, the liquid dropout will start with zero liquid volume at the saturation point. Close to the critical temperature, the liquid build-up will be high (the liquid dropout curve will be steep). Further away from the critical temperature, the liquid build-up will be lower, and the liquid dropout curve will appear flatter. The CME liquid dropout curves for the fluid mixtures at the various swelling stages will show a similar behavior. As long as the fluid after addition of gas still exhibits a bubble point at the reservoir temperature and the composition is far from critical, the liquid dropout curve will start at 100% liquid and the liquid shrinkage will be low. If, after adding more gas, the fluid still exhibits

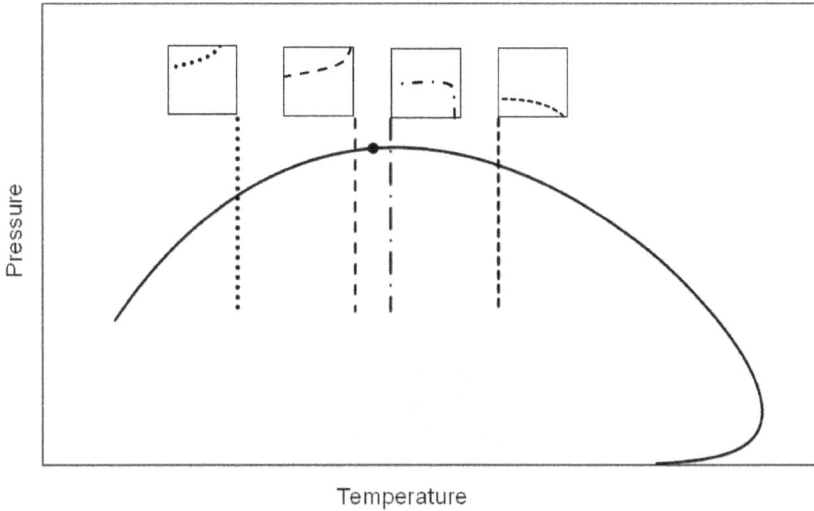

FIGURE 9.30 Schematic liquid dropout curves for pressure depletion experiments performed on a fluid mixture at temperatures at different distances from the critical temperature.

a bubble point, the liquid shrinkage will be higher. Just after passing the critical composition, the liquid will exhibit a dew point, and the liquid build-up will be high when the pressure is reduced from that at the dew point. Had more gas been added (which is usually not the case), the liquid dropout would be lower, and the liquid dropout curve would have been less steep.

Another method was used to determine the critical point of the near-critical gas condensates. Figure 9.31 shows an experimental liquid dropout curve for a gas condensate fluid. Simulated liquid drop-out results are shown with three different fluid characterizations (EoS_1, EoS_2, and EoS_3). The experimental saturation pressure is matched well with all three EoS models, but the liquid dropouts differ. With EoS_1, too little liquid precipitates at pressures below the saturation pressure. EoS_2 provides an almost perfect match of the experimental liquid dropout data, while EoS_3 simulates the saturation point to be a bubble point, while it should have been a dew point.

Phase envelopes simulated using each of the three EoS models are shown in Figure 9.32. The saturation pressure at the reservoir temperature of 150°C is almost the same independent of the model, but the location of the critical point differs. With EoS_1, the critical temperature is simulated to be ~50°C; with EoS_2, it is ~100°C; and with EoS_3, it is ~150°C. If the simulated critical temperature is too low, the liquid dropout curve at the reservoir temperature will be too flat, as is illustrated for EoS_1 in Figure 9.31. If the critical temperature is too high, the EoS model may simulate the saturation point at the reservoir temperature as a bubble point instead of a dew point, as is illustrated for EoS_3 in Figure 9.31. Only with the right critical point will the liquid dropout curve at the reservoir temperature be matched, as is the case for EoS_2. Based on these considerations, it is possible to conclude that the gas condensate fluid considered in Figures 9.31 and 9.32 has a critical temperature of around 100°C.

Pedersen et al. (2023a, 2023b) have elaborated on the ideas brought forward in the paper of Takeshi et al. and found that knowledge of the critical point in addition to the saturation point at the reservoir temperature is sufficient to develop a reliable EoS model for a reservoir fluid. This is a natural extension of the ideas behind a cubic equation for a pure hydrocarbon (Chapter 4), for which parameters a and b and thereby the fluid phase behavior at all conditions is (essentially) determined by the critical temperature and pressure. For the van der Waals equation, no other parameters enter. In the SRK and PR equations, small changes have been introduced through the acentric factor. The procedure of Pedersen et al. is summarized in the following for SRK:

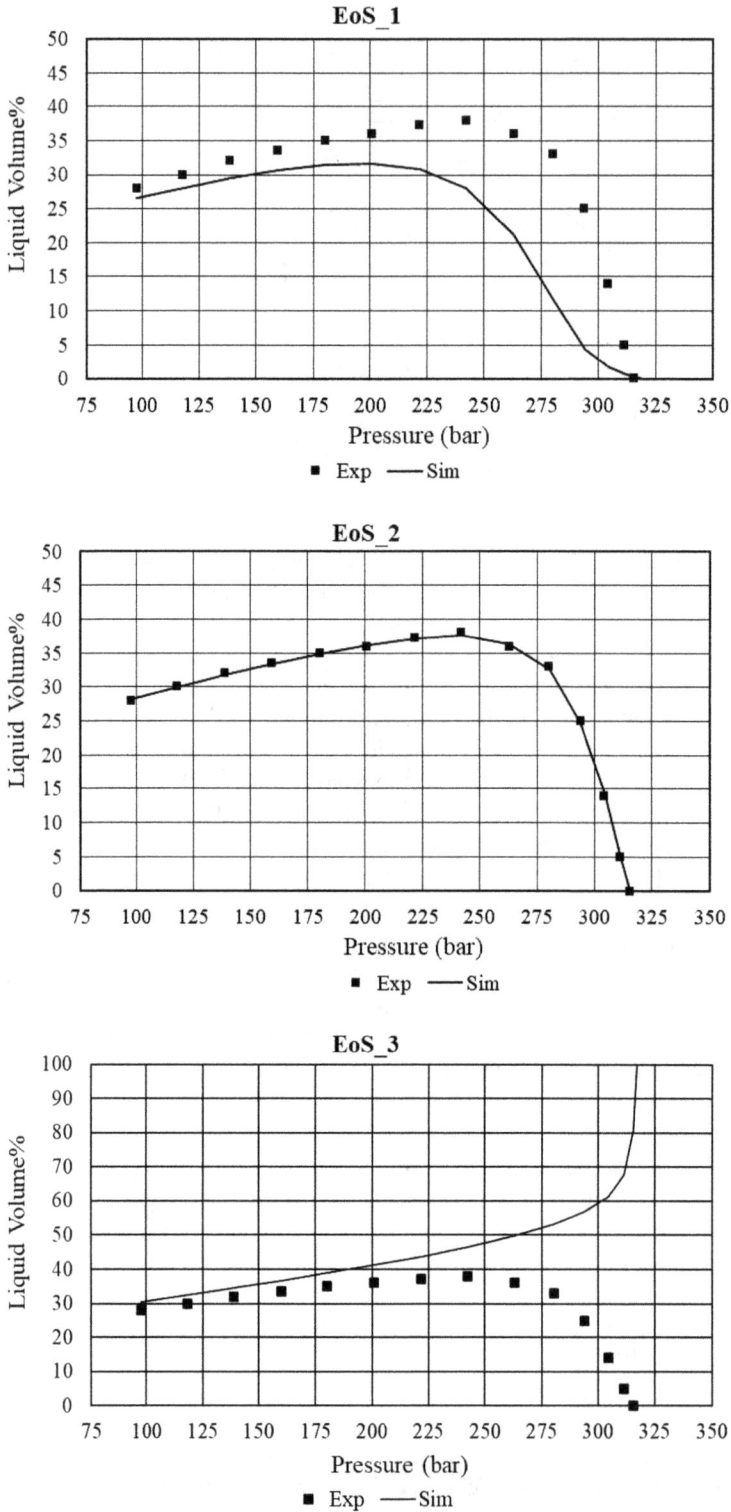

FIGURE 9.31 Experimental liquid dropout data for a gas condensate fluid and simulated liquid dropout curves using three different EoS models. Simulated phase envelopes can be seen from Figure 9.32.

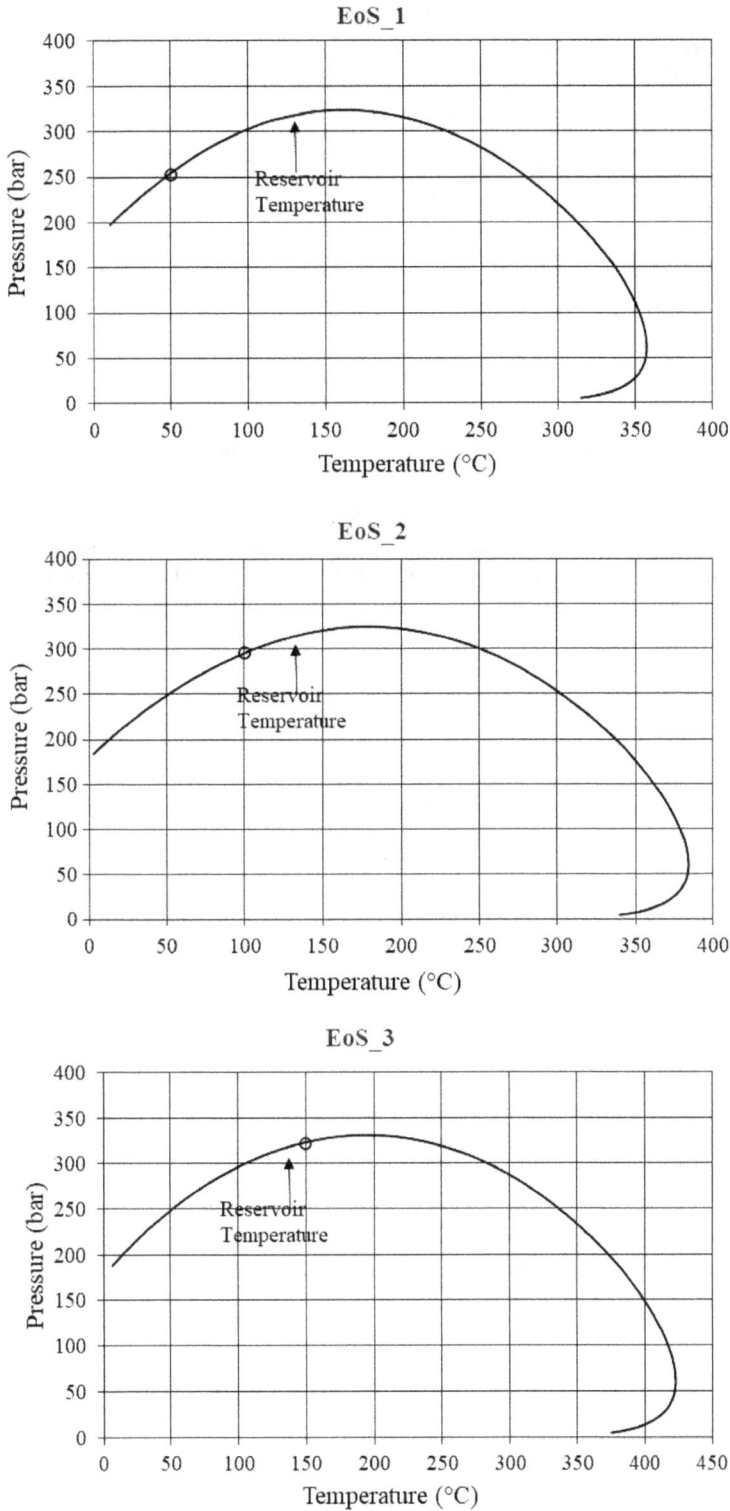

FIGURE 9.32 Simulated phase envelopes for gas condensate fluid using three different EoS models. Experimental and simulated liquid dropout curves can be seen from Figure 9.31.

1. Make a duplicate of the reservoir fluid composition and remove any inorganics (N_2, CO_2, and H_2S) from the duplicate.
2. Calculate the mole% of C_{7+} in the fluid from 1.
3. If the mole% of C_{7+} is below 10, go to 8.
4. Estimate the critical temperature ($T_{c-HCmix}$) in K and critical pressure ($P_{c-HCmix}$) in bar of the hydrocarbon mixture from the following correlations

$$T_{c\_HCmix} = 683.88 + 26.182\ ln\left(\frac{Mole\ C_{7+}}{Mole\ C_1 - C_6)}\right) - 34.061\left(\frac{Mole\ C_{7+}}{Mole\ C_1 - C_6)}\right)^{-1} \qquad 9.5$$

$$P_{c\_HCmix} = 98.035\left(\frac{Mole\ C_{7+}}{Mole\ C_1 - C_6)}\right)^{-0.674} \qquad 9.6$$

5. Characterize the reservoir fluid composition and the fluid composition from 1. to a common EoS (Section 5.6). This will assign the same equation of state parameters to all components and pseudocomponents in both fluid compositions.
6. Regress on the T_cs and P_cs of the C_{7+} components to match the measured saturation pressure for the reservoir fluid and the estimated critical point of the hydrocarbon fluid from 1. The regression is to be performed by rotating the T_c and P_c curves versus carbon number around the value of C_7, as explained in Section 9.1 and illustrated in Figure 9.2.
7. Go to 12.
8. Remove C_1–C_6 from the fluid in 1., leaving only the C_{7+} fraction.
9. Estimate the critical point of the fluid from 8. (C_{7+} fraction) from the following correlations

$$T_{c\_C7+} = 224.92 \times ln\left(C_{7+}\ Molecular\ Weight\right) + 491.08 \qquad 9.7$$

$$P_{c\_C7+} = 11.300 \times ln\left(C_{7+}\ Molecular\ Weight\right) - 24.454 \qquad (9.8)$$

10. Characterize the reservoir fluid composition and the fluid composition from 8. to a common EoS (Section 5.6). This will assign the same equation of state parameters to all components and pseudocomponents in both fluid compositions.
11. Regress on the T_cs and P_cs of the C_{7+} components to match the measured saturation pressure for the reservoir fluid and the estimated critical point of the fluid from 8. (C_{7+} fraction). The regression is to be performed by rotating the T_c and P_c curves versus carbon number around the value of C_7, as explained in Section 9.1 and illustrated in Figure 9.2.
12. Use the developed EoS model for the reservoir fluid. Since reservoir fluids most often contain inorganic components (N_2, CO_2, and H_2S) in addition to hydrocarbons, the reservoir fluid will have a different critical point than the hydrocarbon fluid whose critical point was estimated. Since the inorganic components have well-defined critical properties (T_c, P_c, and acentric factor), it is assumed that the developed EoS model will also predict the correct critical point of the reservoir fluid.

The coefficients in Equations 9.5–9.8 are for use with the SRK equation. Pedersen et al. (2023b) have given coefficients for use with the PR equation of state.

9.8.1 Examples of Critical Point Regression

The critical point regression procedure outlined in the preceding section was tested out on two reservoir fluid compositions, an oil and a gas condensate, whose compositions are shown in Table 9.17. Using Equations 9.5 and 9.6 the critical points of each of the two fluid compositions after removal

TABLE 9.17

Oil Composition for Which Differential Liberation Data Is Shown in Table 9.17 and Gas Condensate Composition for Which CME Liquid Dropout Data Is Shown in Table 9.21

Component	Oil Mol%	Gas Condensate Mol%
N_2	0.27	
CO_2	0.21	5.29
H2S		19.61
C_1	20.11	52.84
C2	6.28	5.00
C_3	6.52	2.96
iC_4	1.23	0.60
nC_4	3.93	1.46
iC_5	1.69	0.65
nC_5	2.49	0.84
C_6	3.85	1.42
C_7	3.82	1.25
C_8	3.88	1.43
C_9	3.43	1.27
C_{10+}	42.29	5.37
C_{7+} molecular weight	252	171
C_{7+} density (g/cm³)	0.8945	0.8122

TABLE 9.18

Experimental Saturation Point Data for the Reservoir Fluid Compositions in Table 9.17 and the Estimated Critical Points of the Same Fluid Compositions After Removal of Inorganics Found Using Equations 9.5 and 9.6

Fluid Composition	Reservoir T °C	Exp Sat P at Reservoir T bar	Estimated Crit T K	Estimated Crit P bar
Oil	77	88.7	658	89
Gas Condensate	135	315	393	365

of inorganics are as shown in Table 9.18, which also shows the experimental saturation points at the reservoir temperature.

A differential liberation experiment has been conducted for the oil in Table 9.17, and the differential liberation data is shown in Table 9.19. The oil has a molar $C_{7+}/(C_1–C_6)$ ratio of 1.158. Using the procedure outlined in the preceding section, EoS model parameters are determined that will provide a fluid description matching the saturation point of the reservoir fluid as well as the critical point data in Table 9.18 for the fluid after removal of inorganics. No other tuning is performed. The resulting EoS model parameters are shown in Table 9.20. The differential liberation experiment was simulated using the tuned reservoir fluid composition in Table 9.20, and the simulation results are plotted in Figure 9.33. A temperature-dependent volume shift parameter (Equation 5.9) was

TABLE 9.19
Experimental Differential Liberation Results for Reservoir Fluid in Table 9.17 at 77°C

Stage	Pressure bar	Oil Formation Volume Factor (FVF) m³/Sm³	Solution Gas/Oil Ratio (Rsd) Sm³/Sm³	Oil Density g/cm³
1	87.54	1.231	60.6	0.7961
2	76.86	1.222	55.7	0.7984
3	66.51	1.209	50.8	0.8029
4	52.72	1.195	44.2	0.8077
5	35.49	1.174	35.1	0.8146
6	18.25	1.151	24.9	0.8210
7	9.70	1.133	17.8	0.8260
8	1.01	1.050	0.0	0.8554

TABLE 9.20
Equation of State Parameters for Oil Composition in Table 9.17 The C_{7+} Parameters Are Determined by Regression to the Data in Table 9.18. NonZero Binary Interaction Parameters Are the Same as in Table 9.8

Comp	Mol%	Crit Temp K	Crit Pres bar	Acen Factor	Volume Shift cm³/mol	T-dependence of Volume Shift cm³/(mol K)
N_2	0.27	126.2	33.94	0.040	0.92	0.000
CO_2	0.21	304.2	73.76	0.225	3.03	0.007
C_1	20.11	190.6	46.00	0.008	0.63	0.000
C_2	6.28	305.4	48.84	0.098	2.63	0.000
C_3	6.52	369.8	42.46	0.152	5.06	0.000
iC_4	1.23	408.1	36.48	0.176	7.29	0.000
nC_4	3.93	425.2	38.00	0.193	7.86	0.000
iC_5	1.69	460.4	33.84	0.227	10.93	0.000
nC_5	2.49	469.6	33.74	0.251	12.18	0.000
C_6	3.85	507.4	29.69	0.296	17.98	0.000
C_7	3.82	524.8	36.18	0.468	−0.62	0.015
C_8	3.88	540.0	33.20	0.500	0.92	0.012
C_9	3.43	556.2	30.05	0.540	2.50	0.007
C_{10}–C_{13}	12.26	589.4	25.61	0.631	2.23	−0.005
C_{14}–C_{16}	6.81	632.7	21.61	0.767	−7.72	−0.026
C_{17}–C_{20}	6.75	670.5	19.67	0.894	−27.40	−0.048
C_{21}–C_{22}	2.60	696.3	18.83	0.981	−46.82	−0.064
C_{23}–C_{25}	3.16	716.6	18.36	1.048	−64.70	−0.077
C_{26}–C_{29}	3.13	744.8	17.85	1.137	−93.35	−0.095
C_{30}–C_{34}	2.67	778.8	17.45	1.231	−133.09	−0.117
C_{35}–C_{42}	2.48	823.4	17.04	1.323	−188.99	−0.144
C_{43}–C_{80}	2.43	921.2	16.70	1.294	−318.91	−0.196

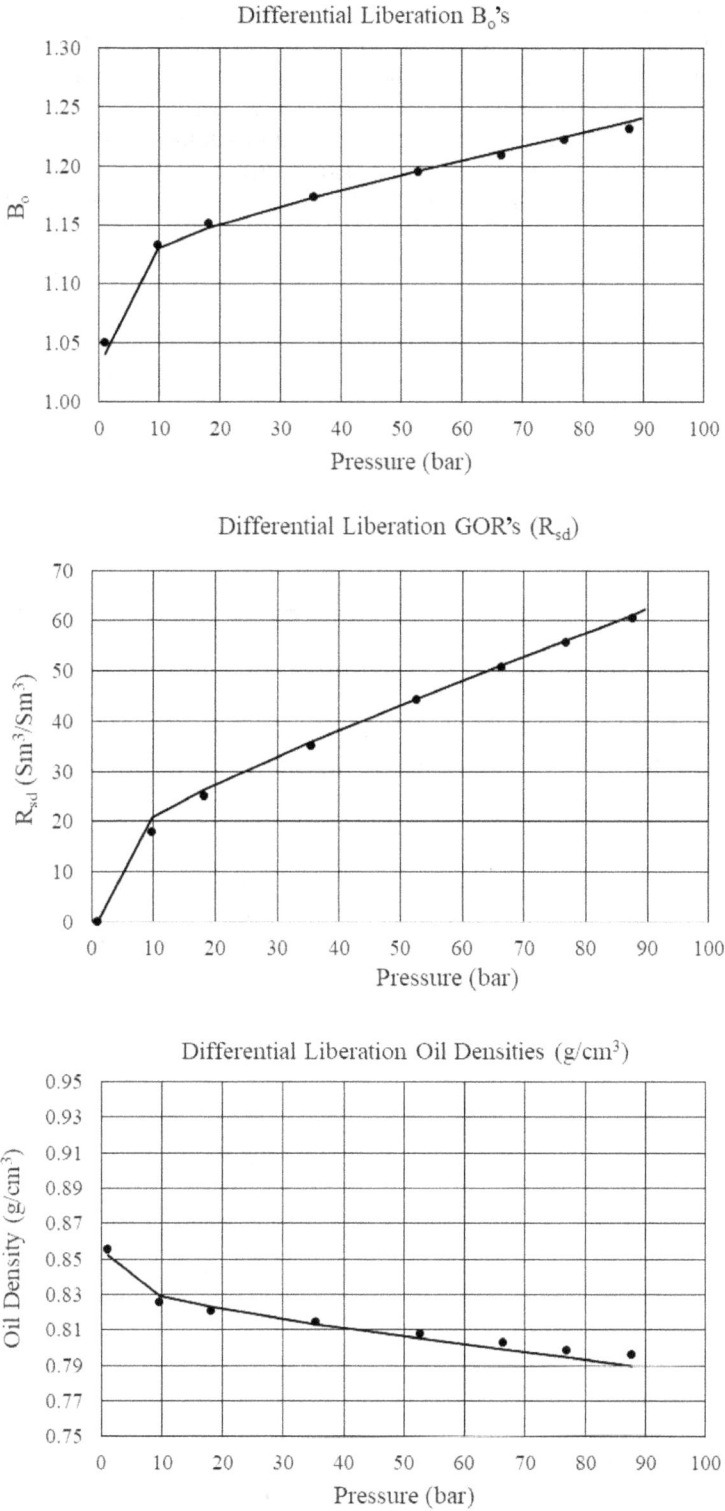

FIGURE 9.33 Experimental and simulated differential liberation data for a temperature of 77°C for the oil composition in Table 9.17. The equation of state parameters have been tuned to the data in Table 9.18. The experimental data is shown in Table 9.19 and the characterized fluid in Table 9.20.

applied when simulating the differential liberation experiment. An almost perfect match is seen even though only a saturation pressure and the estimated critical point in Table 9.18 were tuned to.

A constant mass expansion (CME) experiment has been conducted for the gas condensate fluid in Table 9.17. CME liquid dropout data is shown in Table 9.21. The fluid has a molar $C_{7+}/(C_1-C_6)$ ratio of 0.142. EoS model parameters are determined that match the measured saturation pressure and the estimated critical point of the fluid composition after removal of inorganics. The resulting EoS model parameters are shown in Table 9.22 and binary interaction parameters in Table 9.23. The

TABLE 9.21
Experimental CME Liquid Dropout Data for the Gas Condensate in Table 9.17 at 135°C

Pressure bar	Liquid Volume% of Sat Point Volume
315.5	0.00
311.3	6.84
304.4	18.54
294.0	25.47
276.8	30.55
256.1	34.53

TABLE 9.22
Equation of State Parameters for the Reservoir Fluid Composition in Table 9.17. The C_{7+} Parameters Are Determined by Regression to the Data in Table 9.18. The NonZero Binary Interaction Parameters Can Be Seen from Table 9.23

Component	Res Fluid Mol%	Critical Temp K	Critical Pres bar	Acen Factor	Volume Shift cm³/mol
CO_2	5.29	304.2	73.76	0.225	3.03
H_2S	19.61	373.2	89.37	0.100	1.78
C_1	52.84	190.6	46.00	0.008	0.63
C_2	5.00	305.4	48.84	0.098	2.63
C_3	2.96	369.8	42.46	0.152	5.06
iC_4	0.60	408.1	36.48	0.176	7.29
nC_4	1.46	425.2	38.00	0.193	7.86
iC_5	0.65	460.4	33.84	0.227	10.93
nC_5	0.84	469.6	33.74	0.251	12.18
C_6	1.42	507.4	29.69	0.296	17.98
C_7	1.25	534.0	30.72	0.468	11.76
C_8	1.43	553.2	27.90	0.500	18.57
C_9	1.27	574.6	24.94	0.540	27.23
C_{10}	1.05	593.0	22.95	0.576	34.17
$C_{11}-C_{12}$	1.22	617.0	20.86	0.628	42.29
C_{13}	0.41	643.2	18.99	0.688	49.82
$C_{14}-C_{15}$	0.66	667.4	17.66	0.747	54.81
$C_{16}-C_{18}$	0.70	704.3	16.14	0.841	57.81
$C_{19}-C_{20}$	0.33	734.1	15.37	0.917	54.43
$C_{21}-C_{24}$	0.43	768.1	14.71	1.005	46.62
$C_{25}-C_{31}$	0.36	823.5	14.05	1.139	24.13
$C_{32}-C_{80}$	0.22	938.8	13.58	1.303	−44.52

TABLE 9.23

NonZero Binary Interaction Parameters Used with the Fluid Composition in Table 9.22

	CO_2	H_2S
H_2S	0.10	
C_1	0.12	0.08
C_2	0.12	0.09
C_3	0.12	0.09
iC_4	0.12	0.05
nC_4	0.12	0.06
iC_5	0.12	0.06
nC_5	0.12	0.07
C_6	0.12	0.05
C_{7+}	0.10	0.03

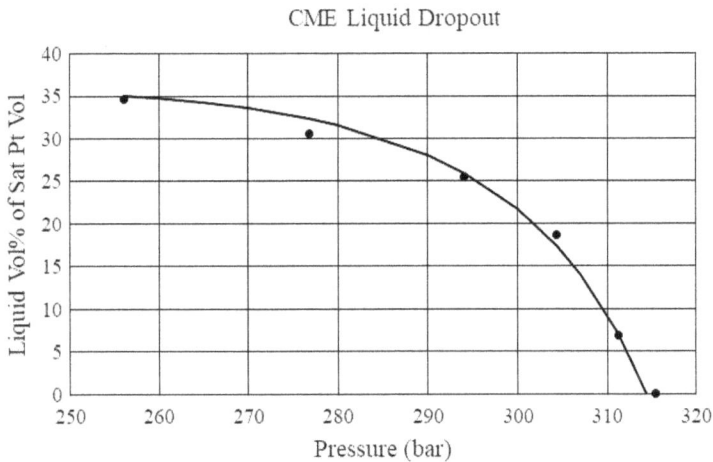

FIGURE 9.34 Experimental and simulated constant mass expension liquid dropout data for the gas condensate reservoir fluid in Table 9.17 for a temperature of 135°C. The equation of state parameters have been tuned to the Data in Table 9.18. The experimental constant mass expansion data is shown in Table 9.21 and the characterized fluid in Table 9.22.

simulated CME liquid dropout curve is shown in Figure 9.34 and is seen to compare well with the experimental data. Again, a match of the saturation pressure of the reservoir fluid and the estimated critical point of the reservoir fluid after removel of inorganics turns out to be sufficient to also match routine PVT data.

REFERENCES

Christensen, P.L., Regression to experimental PVT data, *J. Can. Pet. Technol.* 38, 1–9, 1999.

Coats, K.H., Simulation of gas condensate reservoir performance, SPE paper 10512, presented at the *Sixth SPE symposium on Reservoir Simulation of the Society of Petroleum Engineers of AIME*, New Orleans, LA, January 31–February 3, 1982.

Marquardt, D., An algorithm for least-squares estimation of nonlinear parameters, *SIAM J. Appl. Math.* 11, 431–441, 1963.

Pedersen, K.S., Christensen, P.L., and Shaik, J.A., Use of critical point in equation of state modeling for reservoir fluids, SPE 216776-MS, presented at the *Abu Dhabi International Petroleum Exhibition & Conference*, Abu Dhabi, October 2–5, 2023b.

Pedersen, K.S., Shaikh, J.A., and Christensen, P.L., Importance of the critical point for the phase behavior of a reservoir fluid, *Fluid Ph. Equilibria* 573, 2023a.

Pedersen, K.S., Thomassen, P., and Fredenslund, Aa., On the dangers of tuning equation of state parameters, *Chem. Eng. Sci.* 43, 269–278, 1988.

Pedersen, K.S., Thomassen, P., and Fredenslund, Aa., Characterization of gas condensate mixtures, *Adv. Thermodyn.* 1, 157–162, 1989.

Takeshi, A., Tetsuro, F., Leekumjorn, S., Shaikh, J.A., Pedersen, K.S., Alobeidli, A., and Mogensen, K., A new technique for common EoS model development for multiple reservoir fluids with gas injection, SPE-202923-MS, presented at the *Abu Dhabi International Petroleum Exhibition & Conference*, Abu Dhabi, November 9–12, 2020.

10 Transport Properties

The term transport properties is used about properties that influence and restrict transport of material or heat. The effect of an external force or heat source will not immediately spread throughout the substance upon which it is applied. The rate at which it spreads is determined by transport properties as described in this chapter.

10.1 VISCOSITY

An external stress applied to a portion of a fluid, as illustrated in Figure 10.1, will introduce movement of the molecules of the affected part of the fluid in the direction of the applied stress. The moving molecules will interact with the neighboring molecules, which will start moving, too, but with a lower velocity than that of the molecules exposed to the stress. The dynamic viscosity, η, is defined as

$$\eta = \frac{\tau_{xy}}{\dfrac{\partial v_x}{\partial y}} \tag{10.1}$$

where
τ_{xy} = applied shear stress (force per area unit = F/A)
v_x = velocity of fluid in x-direction
$\dfrac{\partial v_x}{\partial y}$ = shear rate

If the viscosity is independent of shear rate, the fluid is said to be Newtonian in behavior. Figure 10.2 illustrates Newtonian and three kinds of non-Newtonian flow behavior. Pseudoplastic

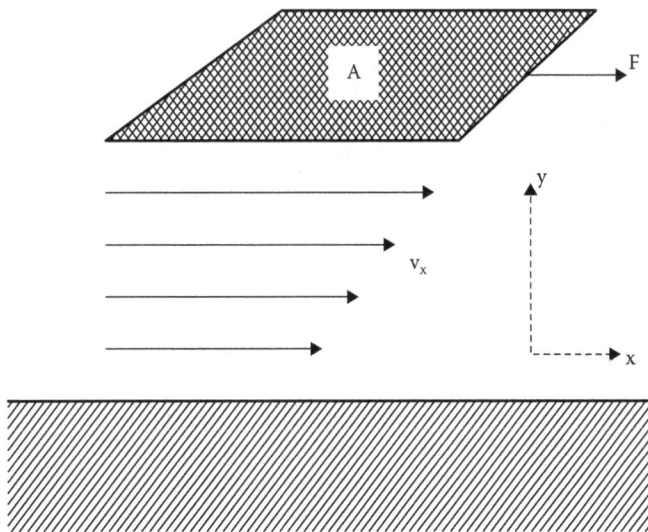

FIGURE 10.1   An external stress applied to a portion of a fluid. The terms on the figure are used in the definition of viscosity (Equation 10.1).

DOI: 10.1201/9780429457418-10

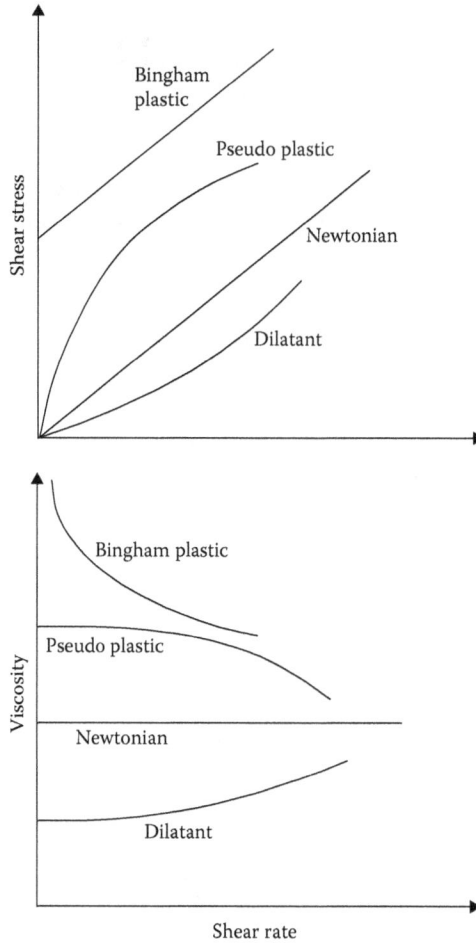

FIGURE 10.2   Newtonian and non-Newtonian flow behaviors.

and dilatant fluids exhibit a nonlinear relationship between shear rate and shear stress. The viscosity of a pseudoplastic fluid decreases with increasing shear rate, whereas the opposite is the case for a dilatant fluid. A Bingham plastic fluid is similar to a Newtonian fluid in the sense that there is a linear relationship between shear stress and shear rate. However, a Bingham plastic fluid differs by requiring a finite shear stress (pressure) to initiate a flow.

Viscosity is a key property for subsurface simulations, well design, and pipeline and process simulations. Several correlations for the viscosity of petroleum systems exist, ranging from simple ones that require information on bulk properties such as API gravity and temperature to more complex ones that rely on the composition of the mixture in question. An example of the simple type of correlations is the Beggs and Robinson correlation (Beggs and Robinson 1975). This chapter will focus on models that are applicable to both gas and liquid phases, and further provide continuous simulation results with T and P in the supercritical region, where it can be difficult to distinguish between gas and liquid. The models should, in other words, not require any information about phase type.

The SI unit for viscosity is N sec/m^2. It is related to other commonly used viscosity units as follows:

$$1 \text{ N sec/m}^2 = 1 \text{ kg/(m sec)} = 1 \text{ Pa sec} = 1000 \text{ mPa sec} = 1000 \text{ centipoise (cP)} = 10 \text{ poise (P)}$$

The *kinematic viscosity* is the ratio of the dynamic viscosity to the density. With the viscosity in P and the density in g/cm^3, the unit of the kinematic viscosity is Stokes or cm^2/sec.

10.1.1   CORRESPONDING STATES VISCOSITY MODELS

The corresponding states principle (CSP) has found application for a number of purposes, a well-known example being the classical Z-factor correlation chart by Standing and Katz (1942). The basic idea is that a given property for any component can be related to that of a well-known reference compound in a corresponding state.

According to the corresponding states principle, the reduced viscosity ($\eta_r = \eta/\eta_c$) can, for example, for a group of substances, be related to the reduced pressure ($P_r = P/P_c$) and temperature ($T_r = T/T_c$) through a unique function, f:

$$\eta_r = f(P_r, T_r) \tag{10.2}$$

Viscosity data at near-critical conditions is rare. Based on considerations for dilute gases (Hirschfelder et al. 1954), the critical viscosity, η_c, can be approximated as

$$\eta_c \approx \frac{P_c^{2/3} M^{1/2}}{T_c^{1/6}} \tag{10.3}$$

M stands for molecular weight. This gives the following expression for reduced viscosity:

$$\eta_r = \frac{\eta(P,T)\, T_c^{1/6}}{P_c^{2/3} M^{1/2}} \tag{10.4}$$

If the function f in Equation 10.2 is known for one component (a reference component) within the group, it will be possible to calculate the viscosity at any (P,T) for any other component within the group. The viscosity of component x at (P,T) can, for example, be expressed as follows:

$$\eta_x(P,T) = \frac{\left(\dfrac{P_{cx}}{P_{co}}\right)^{2/3}\left(\dfrac{M_x}{M_o}\right)^{1/2}}{\left(\dfrac{T_{cx}}{T_{co}}\right)^{1/6}}\, \eta_o\left(\frac{PP_{co}}{P_{cx}}, \frac{TT_{co}}{T_{cx}}\right) \tag{10.5}$$

where o refers to the reference component.

Quite extensive viscosity data has been published for methane, which has enabled Hanley et al. (1975) to make the following correlation for the methane viscosity as a function of density and temperature:

$$\eta(\rho,T) = \eta_o(T) + \eta_1(T)\rho + \Delta\eta'(\rho,T) \tag{10.6}$$

where η_o is the dilute gas viscosity, calculated from

$$\eta_o = \frac{GV(1)}{T} + \frac{GV(2)}{T^{2/3}} + \frac{GV(3)}{T^{1/3}} + GV(4) + GV(5)T^{1/3} + GV(6)T^{2/3} + GV(7)T$$
$$GV(8)T^{4/3} + GV(9)T^{5/3} \tag{10.7}$$

and ρ is the density in mol/L.

The coefficients GV(1) to GV(9) can be seen in Table 10.1. The term η_1 is represented by the following empirical relation:

$$\eta_1(T) = A + B\left(C - \ln\frac{T}{F}\right)^2 \tag{10.8}$$

where the constants A, B, C, and F are given in Table 10.1. Finally, the term $\Delta\eta'$ is given by

$$\Delta\eta'(\rho,T) = \exp\left(j_1 + \frac{j_4}{T}\right)\left[\exp\left[\rho^{0.1}\left(j_2 + \frac{j_3}{T^{3/2}}\right) + \theta\rho^{0.5}\left(j_5 + \frac{j_6}{T} + \frac{j_7}{T^2}\right)\right] - 1.0\right] \tag{10.9}$$

The parameter θ is a function of the actual density, ρ, and the critical density, ρ_c

$$\theta = \frac{\rho - \rho_c}{\rho_c} \tag{10.10}$$

The constants j_1–j_7 can be seen in Table 10.1.

TABLE 10.1
Constants in Equations Expressing Corresponding States Model for Viscosities in 10^{-4} cP

Equation	Constant	Value
10.7	GV(1)	-2.090975×10^5
	GV(2)	2.647269×10^5
	GV(3)	-1.472818×10^5
	GV(4)	4.716740×10^4
	GV(5)	-9.491872×10^3
	GV(6)	1.219979×10^3
	GV(7)	-9.627993×10^1
	GV(8)	4.274152
	GV(9)	-8.141531×10^{-2}
10.8	A	1.696985927
	B	-0.133372346
	C	1.4
	F	168.0
10.9	j_1	-10.3506
	j_2	17.5716
	j_3	-3019.39
	j_4	188.730
	j_5	0.0429036
	j_6	145.290
	j_7	6127.68
10.28	k_1	-9.74602
	k_2	18.0834
	k_3	-4126.66
	k_4	44.6055
	k_5	0.976544
	k_6	81.8134
	k_7	15649.9

The methane density is computed from the Benedict–Webb–Rubin (BWR) equation in the form suggested by McCarty (1974):

$$P = \sum_{n=1}^{9} a_n(T)\rho^n + \sum_{n=10}^{15} a_n(T)\rho^{2n-17}e^{-\gamma\rho^2} \tag{10.11}$$

where a_1–a_{15} and γ are constants shown in Table 10.2.

Equations 10.6 and 10.11 allow the methane viscosity to be calculated at any P and T. This makes methane a convenient choice as reference component in Equation 10.5.

The simple corresponding states principle, as expressed in Equation 10.5, works well for mixtures of light hydrocarbon components such as C_1, C_2, and C_3. If the mixture contains heavy hydrocarbons, a correction is required. Pedersen et al. (1984) have expressed the deviations from the classical corresponding states principle in terms of a parameter, α, and use the following expression for the viscosity of a mixture at pressure P and temperature T:

$$\eta_{mix}(P,T) = \left(\frac{T_{c,mix}}{T_{co}}\right)^{-1/6}\left(\frac{P_{c,mix}}{P_{co}}\right)^{2/3}\left(\frac{M_{mix}}{M_o}\right)^{1/2}\left(\frac{\alpha_{mix}}{\alpha_o}\right)\eta_o(P_o,T_o) \tag{10.12}$$

where

$$P_o = \frac{P\,P_{co}}{P_{c,mix}}\frac{\alpha_o}{\alpha_{mix}};$$

$$T_o = \frac{T\,T_{co}}{T_{c,mix}}\frac{\alpha_o}{\alpha_{mix}} \tag{10.13}$$

The critical temperature and critical molar volume for unlike pairs of molecules (i and j) are found from

$$T_{cij} = \sqrt{T_{ci}\,T_{cj}} \tag{10.14}$$

$$V_{cij} = \frac{1}{8}\left(V_{ci}^{1/3} + V_{cj}^{1/3}\right)^3 \tag{10.15}$$

The critical molar volume of component i may be related to the critical temperature and the critical pressure as follows:

$$V_{ci} = \frac{RZ_{ci}\,T_{ci}}{P_{ci}} \tag{10.16}$$

where Z_{ci} is the compressibility factor of component i at the critical point. Assuming that Z_c is a constant independent of component, the expression for V_{cij} may be rewritten to

$$V_{cij} = \frac{1}{8}\,\text{constant}\left[\left(\frac{T_{ci}}{P_{ci}}\right)^{1/3} + \left(\frac{T_{cj}}{P_{cj}}\right)^{1/3}\right]^3 \tag{10.17}$$

The critical temperature of a mixture is found from

$$T_{c,mix} = \frac{\displaystyle\sum_{i=1}^{N}\sum_{j=1}^{N} z_i z_j T_{cij} V_{cij}}{\displaystyle\sum_{i=1}^{N}\sum_{j=1}^{N} z_i z_j V_{cij}} \tag{10.18}$$

TABLE 10.2

Constants in Equation 10.11 for Pressures (P) in atm (1 atm = 1.01325 bar), Densities (ρ) in mol/l, and Temperature (T) in K. R = 0.08205616 l atm mole^{-1} K^{-1}

Constant	Value/Expression
a_1	RT
a_2	$N_1T + N_2T^{1/2} + N_3 + N_4/T + N_5/T^2$
a_3	$N_6T + N_7 + N_8/T + N_9/T^2$
a_4	$N_{10}T + N_{11} + N_{12}/T$
a_5	N_{13}
a_6	$N_{14}/T + N_{15}/T^2$
a_7	N_{16}/T
a_8	$N_{17}/T + N_{18}/T^2$
a_9	N_{19}/T^2
a_{10}	$N_{20}/T^2 + N_{21}/T^3$
a_{11}	$N_{22}/T^2 + N_{23}/T^4$
a_{12}	$N_{24}/T^2 + N_{25}/T^3$
a_{13}	$N_{26}/T^2 + N_{27}/T^4$
a_{14}	$N_{28}/T^2 + N_{29}/T^3$
a_{15}	$N_{30}/T^2 + N_{31}/T^3 + N_{32}/T^4$
N_1	$-1.8439486666 \times 10^{-2}$
N_2	1.0510162064
N_3	-1.6057820303×10
N_4	8.4844027563×10^2
N_5	$-4.2738409106 \times 10^4$
N_6	$7.6565285254 \times 10^{-4}$
N_7	$-4.8360724197 \times 10^{-1}$
N_8	8.5195473835×10
N_9	$-1.6607434721 \times 10^4$
N_{10}	$-3.7521074532 \times 10^{-5}$
N_{11}	$2.8616309259 \times 10^{-2}$
N_{12}	-2.8685298973
N_{13}	$1.1906973942 \times 10^{-4}$
N_{14}	$-8.5315715698 \times 10^{-3}$
N_{15}	3.8365063841
N_{16}	$2.4986828379 \times 10^{-5}$
N_{17}	$5.7974531455 \times 10^{-6}$
N_{18}	$-7.1648329297 \times 10^{-3}$
N_{19}	$1.2577853784 \times 10^{-4}$
N_{20}	2.2240102466×10^4
N_{21}	$-1.4800512328 \times 10^6$
N_{22}	5.0498054887×10
N_{23}	1.6428375992×10^6
N_{24}	$2.1325387196 \times 10^{-1}$
N_{25}	3.7791273422×10
N_{26}	$-1.1857016815 \times 10^{-5}$
N_{27}	-3.1630780767×10
N_{28}	$-4.1006782941 \times 10^{-6}$
N_{29}	$1.4870043284 \times 10^{-3}$
N_{30}	$3.1512261532 \times 10^{-9}$
N_{31}	$-2.1670774745 \times 10^{-6}$
N_{32}	$2.4000551079 \times 10^{-5}$
Γ	0.0096

where z_i and z_j are mole fractions of components i and j, respectively, and N is the number of components. This expression can be rewritten to

$$T_{c,mix} = \frac{\sum\limits_{i=1}^{N} \sum\limits_{j=1}^{N} z_i z_j \left[\left(\frac{T_{ci}}{P_{ci}} \right)^{1/3} + \left(\frac{T_{cj}}{P_{cj}} \right)^{1/3} \right]^3 \sqrt{T_{ci} T_{cj}}}{\sum\limits_{i=1}^{N} \sum\limits_{j=1}^{N} z_i z_j \left[\left(\frac{T_{ci}}{P_{ci}} \right)^{1/3} + \left(\frac{T_{cj}}{P_{cj}} \right)^{1/3} \right]^3}$$

(10.19)

For the critical pressure of a mixture, $P_{c,mix}$, the following relation is used:

$$P_{c,mix} = constant \frac{T_{c,mix}}{V_{c,mix}}$$

(10.20)

where $V_{c,mix}$ is found from

$$V_{c,mix} = \sum\limits_{i=1}^{N} \sum\limits_{j=1}^{N} z_i z_j V_{cij}$$

(10.21)

The following expression can now be derived for $P_{c,mix}$:

$$P_{c,mix} = \frac{8 \sum\limits_{i=1}^{N} \sum\limits_{j=1}^{N} z_i z_j \left[\left(\frac{T_{ci}}{P_{ci}} \right)^{1/3} + \left(\frac{T_{cj}}{P_{cj}} \right)^{1/3} \right]^3 \sqrt{T_{ci} T_{cj}}}{\left(\sum\limits_{i=1}^{N} \sum\limits_{j=1}^{N} z_i z_j \left[\left(\frac{T_{ci}}{P_{ci}} \right)^{1/3} + \left(\frac{T_{cj}}{P_{cj}} \right)^{1/3} \right]^3 \right)^2}$$

(10.22)

The preceding mixing rules are as recommended by Murad and Gubbins (1977). The mixture molecular weight is found from

$$M_{mix} = 1.304 \times 10^{-4} \left(\overline{M}_w^{2.303} - \overline{M}_n^{2.303} \right) + \overline{M}_n$$

(10.23)

where $\overline{M}_w$ and $\overline{M}_n$ are the weight average and number average molecular weights, respectively:

$$\overline{M}_w = \frac{\sum\limits_{i=1}^{N} z_i M_i^2}{\sum\limits_{i=1}^{N} z_i M_i}$$

(10.24)

$$\overline{M}_n = \sum\limits_{i=1}^{N} z_i M_i$$

(10.25)

The constants in Equation 10.23 are derived empirically from experimental viscosity data. The α parameter of the mixture is found from

$$\alpha_{mix} = 1.000 + 7.378 \times 10^{-3} \rho_r^{1.847} M_{mix}^{0.5173}$$

(10.26)

The same expression is used to find α_o (α value of reference component), except that M_{mix} is replaced by the molecular weight of the reference component (methane). The reduced density, ρ_r, is defined as

$$\rho_r = \frac{\rho_o\left(\dfrac{TT_{co}}{T_{c,mix}}, \dfrac{PP_{co}}{P_{c,mix}}\right)}{\rho_{co}} \tag{10.27}$$

The critical density of methane, ρ_{co}, is equal to 0.16284 g/cm^3.

Viscosity simulations using the corresponding states model will follow the calculation scheme:

1. Mixture T_c and P_c calculated from Equations 10.19 and 10.22.
2. Methane density at $\dfrac{TT_{co}}{T_{c,mix}}$, $\dfrac{PP_{co}}{P_{c,mix}}$ calculated from Equation 10.11 and the reduced density from Equation 10.27.
3. Mixture molecular weight (M_{mix}) calculated from Equation 10.23.
4. Correction factor α_{mix} calculated from Equation 10.26. α_o is calculated from the same expression with M_{mix} replaced by the molecular weight of methane.
5. Methane reference pressure (P_o) and temperature (T_o) calculated from Equation 10.13.
6. Mixture viscosity calculated from Equation 10.12.

Figure 10.3 shows plots of methane pressures as a function of density for temperatures ranging from 20 to 140 K calculated by use of Equation 10.11. The lower the temperature, the steeper the P versus density curve and more inaccurate the density and methane viscosity.

Figure 10.4 shows plots of methane viscosities as a function of temperature for pressures ranging from 100 to 2000 bar. The results shown as dotted lines are calculated using Equation 10.6 and model parameters of Hanley et al. (1975). In the dense liquid region, the left-hand side of Equation 10.6 is mainly governed by the term $\Delta\eta'(\rho,T)$.

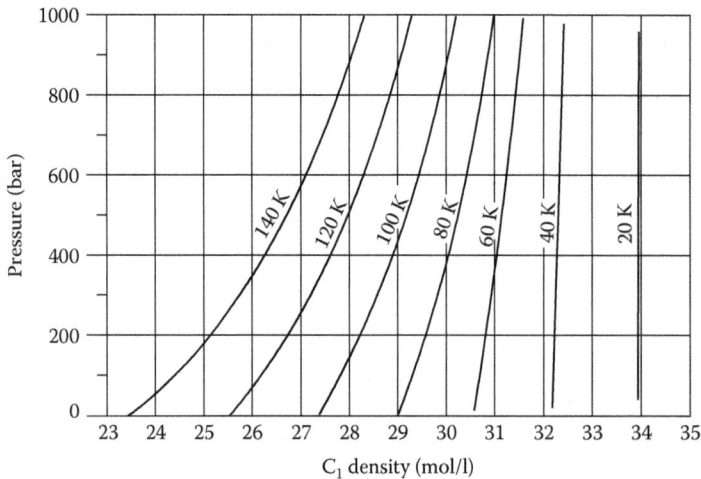

FIGURE 10.3 Methane pressures versus density from model of McCarty (1974) for temperatures ranging from 20 to 140 K.

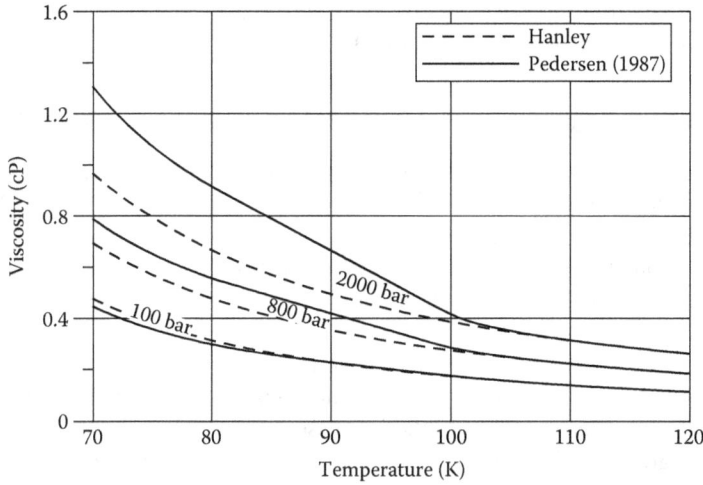

FIGURE 10.4 Simulated methane viscosities versus temperature for pressures ranging from 100 to 2000 bar.

Hanley's model presents some problems when methane is in a solid form at its reference state. This is the case when the methane reference temperature is below approximately 91 K, corresponding to a reduced temperature of 0.48. The melting temperatures of branched paraffins, naphthenes, and aromatics are at a lower reduced temperature than is the case with methane. Pedersen and Fredenslund (1987) suggested overcoming this problem by replacing $\Delta\eta'(\rho,T)$ by the following term when the methane reference temperature (T_0) is below 91 K:

$$\Delta\eta''(\rho,T) = \exp\left(k_1 + \frac{k_4}{T}\right)\left[\exp\left[\rho^{0.1}\left(k_2 + \frac{k_3}{T^{3/2}}\right) + \theta\rho^{0.5}\left(k_5 + \frac{k_6}{T} + \frac{k_7}{T^2}\right)\right] - 1.0\right] \quad (10.28)$$

The constants k_1–k_7 can be seen in Table 10.1.

Continuity between viscosities above and below the freezing point of methane is secured by introducing $\Delta\eta''$ as a fourth term in the viscosity expression as shown in the following:

$$\eta(\rho,T) = \eta_0(T) + \eta_1(T) + F_1\Delta\eta'(\rho,T) + F_2\Delta\eta''(\rho,T) \quad (10.29)$$

$$F_1 = \frac{HTAN+1}{2}; \quad F_2 = \frac{1-HTAN}{2} \quad (10.30)$$

$$HTAN = \frac{\exp(\Delta T) - \exp(-\Delta T)}{\exp(\Delta T) + \exp(-\Delta T)} \quad (10.31)$$

with

$$\Delta T = T - T_F \quad (10.32)$$

where T_F is the freezing point of methane (91 K).

Methane viscosities calculated using the modification of Pedersen and Fredenslund are shown as full-drawn lines in Figure 10.4. For methane reference temperatures lower than 91 K where methane is in solid form, the modified viscosity expression will provide somewhat higher methane viscosities than using the original expressions of Hanley et al. These are in better accordance with viscosity data reported for oil mixtures at conditions in which the reduced temperature is lower than that corresponding to a methane reference temperature of 91 K.

10.1.2 ADAPTATION OF CORRESPONDING STATES VISCOSITY MODEL TO HEAVY OILS

Even though the performance of the corresponding states model in the classical form, with methane as reference component, can be stretched somewhat into the region where methane is in solid form, as suggested by Pedersen and Fredenslund (1987), the classical corresponding model is generally not suited for heavy oil mixtures with viscosities above ~10 cP. As can be seen from Figure 10.3, the curves for the methane density versus pressure are almost vertical for T < 60 K and are therefore unsuited for determining the variation in density from P and T.

Based on viscosity data for eight different heavy oil reservoir fluids, Lindeloff et al. (2004) developed a new heavy oil viscosity model that extends the applicability area of the corresponding states viscosity model. The C_{7+} density at atmospheric pressure and 15°C is above 0.9 g/cm³ for all eight fluids and above 1 g/cm³ for three of the fluids. This is an indication of a high content of aromatic compounds. For three of the fluids, the average molecular weight of the C_{7+} fraction exceeds 500. Being in general rather heavy and aromatic (biodegraded), no wax precipitation takes place, and the oils behave as strictly Newtonian fluids at the temperatures for which viscosity data exist.

Instead of using methane as a reference component and Equation 10.12 to determine the corresponding viscosities, the viscosity of the stabilized crude is used as starting point. Based on viscosity measurements on a wide range of North Sea oils and condensates, Rønningsen (1993) proposed a semi-empirical correlation for stabilized crude viscosities, η_0, at atmospheric pressure in cP:

$$\log_{10}\eta_0 = -0.07995 - 0.1101M - \frac{371.8}{T} + \frac{6.215M}{T} \qquad (10.33)$$

M is the average molecular weight and T the temperature in K. For T > 564.49 K, the sign in front of 0.01101 is changed from − to +. To extend the correlation to be applicable for live oils, Lindeloff et al. proposed the following procedure for evaluating a representative average molecular weight, M:

$$M = \bar{M}_n \left(\frac{1.5}{\text{Visfac3} \times (3^{\text{rd}} \text{ CSP})} \right)^{\text{Visfac4} \times (4^{\text{th}} \text{ CSP})} \qquad \text{for} \left(\frac{\bar{M}_w}{\bar{M}_n} \right) \leq 1.5 \qquad (10.34)$$

$$M = \bar{M}_n \left(\frac{\bar{M}_w}{\text{Visfac3} \times (3^{\text{rd}} \text{ CSP}) \times \bar{M}_n} \right)^{\text{Visfac4} \times (4^{\text{th}} \text{ CSP})} \qquad \text{for} \left(\frac{\bar{M}_w}{\bar{M}_n} \right) > 1.5 \qquad (10.35)$$

where (3rd CSP) and (4th CSP) are tuning parameters, which are 1.0 by default. $\bar{M}_n$ is the number average molecular weight, $\bar{M}_w$ the weight average molecular weight, T is temperature in K, and

$$\text{Visfac3} = 0.2252 \left(\frac{T}{\bar{M}_n} \right) + 0.9738 \qquad (10.36)$$

$$\text{Visfac4} = 0.5354 \times \text{Visfac3} - 0.1170$$

The correlation of Rønningsen applies to systems at atmospheric pressure. To capture pressure effects on the reference fluid, the following pressure dependence is used

$$\eta = \eta^0 e^{0.00384 \frac{P^{0.8226}-1}{0.8226}} \qquad (10.37)$$

for viscosities in cP. η^0 is the viscosity at the actual temperature and atmospheric pressure and P is the actual pressure in atm.

The simulation scheme outlined in the Section 10.1.1 was modified as follows:

1. From 1 to 5: As in Section 10.1.1, except that Equation 10.29 is used for the methane viscosity instead of Equation 10.6.
2. 6: Mixture viscosity calculated from Equation 10.12, with Equation 10.29 as methane viscosity model for $T_0 > 75$ K. For $T_0 < 65$ K, Equations 10.33 and 10.36 are used. For 65 K $< T_0 < 75$ K, the viscosity is calculated as a weighted average between the viscosity results obtained with Equation 10.12 (with Equation 10.29 as methane viscosity model) and with Equations 10.33 and 10.37.

By implementing the heavy oil viscosity correlation in this manner, the classical corresponding states model is still used for higher temperatures and lighter oil mixtures. Averaging the corresponding states and the heavy oil correlation viscosities for methane reference temperatures in the range 65 to 75 K ensures a smooth transition to the heavy oil viscosity correlation.

10.1.3 LOHRENZ–BRAY–CLARK METHOD

The Lohrenz–Bray–Clark (LBC) correlation (1964) expresses gas and oil viscosities as a fourth-degree polynomial in the reduced density, $\rho_r = \rho/\rho_c$

$$[(\eta - \eta^*)\xi + 10^{-4}]^{1/4} = a_1 + a_2\rho_r + a_3\rho_r^2 + a_4\rho_r^3 + a_5\rho_r^4 \tag{10.38}$$

where the constants a_1–a_5 can be seen in Table 10.3.

η^* is the low-pressure gas mixture viscosity. ξ is the viscosity-reducing parameter, which for a mixture is given by

$$\xi = \frac{\left[\sum_{i=1}^{N} z_i T_{ci}\right]^{1/6}}{\left[\sum_{i=1}^{N} z_i M_i\right]^{1/2} \left[\sum_{i=1}^{N} z_i P_{ci}\right]^{2/3}} \tag{10.39}$$

N is the number of components in the mixture, and z_i the mole fraction of component i.

The critical density, ρ_c, is calculated from the critical molar volume

$$\rho_c = \frac{1}{V_c} = \frac{1}{\sum_{i=1}^{N} z_i V_{ci}} \tag{10.40}$$

TABLE 10.3
Parameters in LBC Viscosity Correlation

LBC Parameters	Constants in Equation 10.38
a_1	0.10230
a_2	0.023364
a_3	0.058533
a_4	−0.040758
a_5	0.0093324

Abbreviation: LBC: Lohrenz–Bray–Clark.

For C_{7+} fraction i, the critical molar volume in ft³/lb mole is found from

$$V_{ci} = 21.573 + 0.015122\,M_i - 27.656\rho_i + 0.070615\,M_i \times \rho_i \qquad (10.41)$$

In this expression, M_i is the molecular weight and ρ_i the liquid density in g/cm³ of C_{7+} fraction i. For defined components, literature values are used for the critical molar volumes.

The dilute gas mixture viscosity η^* may be determined from (Herning and Zippener 1936):

$$\eta^* = \frac{\sum_{i=1}^{N} z_i \eta_i^* \sqrt{M_i}}{\sum_{i=1}^{N} z_i \sqrt{M_i}} \qquad (10.42)$$

The following expressions (Stiel and Thodos 1961) can be used to find the dilute gas viscosity, η_i^*, of component i

$$\eta_i^* = 34 \times 10^{-5}\, \frac{1}{\xi_i}\, T_{ri}^{0.94}$$

$$\text{for } T_{ri} < 1.5 \qquad (10.43)$$

$$\eta_i^* = 17.78 \times 10^{-5}\, \frac{1}{\xi_i}\, (4.58\,T_{ri} - 1.67)^{5/8} \qquad (10.44)$$

where ξ_i is given by

$$\xi_i = \frac{T_{ci}^{1/6}}{M_i^{1/2}\, P_{ci}^{2/3}} \qquad (10.45)$$

Because the LBC correlation executes fast in computer code, it is often the preferred choice in compositional reservoir and flow simulation studies. As a predictive model, LBC will usually provide results of low quality. To match experimental data, either V_c of the pseudocomponents or one or more of the coefficients a_1–a_5 will have to be treated as tuning parameters.

Various attempts (e.g., Dandekar et al. 1993) have been made to modify the Lohrenz–Bray–Clark correlation to reduce the need for tuning to measured data. The fundamental assumption of a unique relation between reduced density and viscosity has been retained. Especially for heavy oil mixtures, this is a questionable assumption.

10.1.4 OTHER VISCOSITY MODELS

The gas viscosities reported in a standard PVT report will usually not be measured but calculated values. Many laboratories use the correlation of Lee et al. (1966). It expresses the gas viscosity in cP as follows

$$\eta = 10^{-4} k_v \exp\left[x_v \left(\frac{\rho}{62.4} \right)^{y_v} \right] \qquad (10.46)$$

where

$$x_v = 3.5 + \frac{986}{T} + 0.01M \qquad (10.47)$$

$$y_v = 2.4 - 0.2x_v \qquad (10.48)$$

$$k_v = \frac{(9.4 + 0.02M)T^{1.5}}{209 + 19M + T} \qquad (10.49)$$

T is temperature in R, ρ is gas density in lb/ft^3, and M is molecular weight.

It has been proposed to use the functional form of an equation of state to express the viscosity. A model of Guo et al. (1997) and a model of Quinones-Cisneros et al. (2003) can be mentioned as examples. The latter one is based on the so-called friction theory and expresses the viscosity as the sum of a dilute gas viscosity term, η_0, and a residual friction term, η_f

$$\eta = \eta_0 + \eta_f \tag{10.50}$$

The residual friction term is expressed as

$$\eta_f = \kappa_r P_r + \kappa_a P_a + \kappa_{rr} P_r^2 \tag{10.51}$$

where P_a and P_r are, respectively, the van der Waals (Equation 4.5) attractive and repulsive pressure contributions

$$P_r = \frac{RT}{V-b}; \quad P_a = -\frac{a}{V^2} \tag{10.52}$$

The repulsive and attractive pressure terms may also be calculated from other equations of state, for example, the Soave–Redlich–Kwong (Equation 4.20) or the Peng–Robinson equation (Equation 4.36).

The coefficients κ_r, κ_a, and κ_{rr} in Equation 10.51 are obtained from the following empirical mixing rules:

$$\kappa_r = \sum_{i=1}^{N} \zeta_i \kappa_{ri}; \quad \kappa_a = \sum_{i=1}^{N} \zeta_i \kappa_{ai}; \quad \kappa_{rr} = \sum_{i=1}^{N} \zeta_i \kappa_{rri} \tag{10.53}$$

with

$$\zeta_i = \frac{z_i}{M_i^{0.3} \times MM} \tag{10.54}$$

where z_i is the mole fraction of component i, M_i is the molecular weight of component i, and

$$MM = \sum_{i=1}^{N} \frac{z_i}{M_i^{0.3}} \tag{10.55}$$

The pure component coefficients are calculated from

$$\kappa_{ri} = \frac{\eta_{ci} \hat{\kappa}_{ri}}{P_{ci}}; \quad \kappa_{ai} = \frac{\eta_{ci} \hat{\kappa}_{ai}}{P_{ci}}; \quad \kappa_{rri} = \frac{\eta_{ci} \hat{\kappa}_{rri}}{P_{ci}^2} \tag{10.56}$$

where $\hat{\kappa}_{ri}$, $\hat{\kappa}_{ai}$, and $\hat{\kappa}_{rri}$ are expressed as correlations in the reduced temperature. The 16 constants in the correlations differ for different equations of state. For defined components, the critical viscosity, η_{ci}, may be determined from pure component viscosity data. For C_{7+} pseudocomponents, the following expression is used for η_{ci}:

$$\eta_{ci} = K_c \frac{\sqrt{M_i} \, P_{ci}^{2/3}}{T_{ci}^{1/6}} \tag{10.57}$$

where K_c is an adjustable parameter common for all C_{7+} pseudocomponents. K_c has to be determined from experimental viscosity data, meaning that the application of the model is dependent on the existence of viscosity data for the mixture in question.

For fluids containing solid wax particles, a non-Newtonian viscosity model may be applied, as is described in the section on viscosity of oil–wax suspensions in Section 11.4.

10.1.5 Viscosity Data and Simulation Results

Tables 10.4 through 10.6 show molar reservoir fluid compositions for which viscosity data is given in Table 10.7. The mixture in Table 10.4 is a high-temperature/high-pressure (HT/HP) reservoir fluid. Viscosity simulation results are shown in Figures 10.5 through 10.7 for each of the three mixtures using the corresponding states (CSP) and the LBC viscosity models. As can be seen from Figure 10.7, the LBC model has problems matching the viscosities of the oil in Table 10.6. The corresponding states model, on the other hand, is very accurate for viscosities of this order of magnitude.

Table 10.8 shows the compositions of three heavy oil mixtures as presented by Lindeloff et al. (2004). Viscosity data for these oil mixtures is shown in Table 10.9 (numbering follows that used by Lindeloff et al.). For Oil 5, viscosity data is given for three different temperatures ranging from 38°C to 60°C. A temperature increase of only 22°C makes the viscosity decrease by almost a factor of six. Plots of the viscosity data in Table 10.9 can be seen from Figures 10.8 through 10.12. Also shown in these figures are simulated viscosity results using the corresponding states model, adapted for heavy oil mixtures as proposed by Lindeloff et al. The dashed lines in Figures 10.10 through 10.12 show tuned viscosity simulation results where the multipliers 3rd CSP and 4th CSP in Equations 10.34 and 10.35, which are 1.0 by default, have been treated as tuning parameters. Viscosity data for three temperatures were tuned to at the same time. For viscosities of the order of 1000 cP, viscosity predictions using the extended corresponding states model may deviate as much as a factor 2–3 from the experimental data, but in most cases a good match can be achieved by tuning on the multipliers 3rd CSP and 4th CSP in Equations 10.34 and 10.35, as is exemplified by the dashed lines in Figures 10.10 through 10.12.

Gas viscosities are seldom measured because the commonly used viscosity correlations will, in general, provide results within experimental uncertainty. Kashefi et al. (2013) have, however,

TABLE 10.4

Molar Composition of HT/HP Gas Condensate Mixture. Viscosity Data for This Mixture Is Shown in Table 10.7 and Plotted in Figure 10.5

Component	Mol%	Molecular Weight (g/mol)	Density at 1.01 bar, 15°C (g/cm³)
N_2	0.34	—	—
CO_2	3.59	—	—
C_1	67.42	—	—
C_2	9.02	—	—
C_3	4.31	—	—
iC_4	0.93	—	—
nC_4	1.71	—	—
iC_5	0.74	—	—
nC_5	0.85	—	—
C_6	1.38	—	—
C_7	1.50	109.6	0.6912
C_8	1.69	120.2	0.7255
C_9	1.14	129.5	0.7454
C_{10}	0.80	135.3	0.7864
C_{11+}	4.58	236.2	0.8398

TABLE 10.5
Molar Composition of Oil Mixture. Viscosity Data for This Mixture Is Shown in Table 10.7 and Plotted in Figure 10.6

Component	Mol%	Molecular Weight (g/mol)	Density at 1.01 bar, 15°C (g/cm³)
N_2	0.69	—	—
CO_2	3.14	—	—
C_1	52.81	—	—
C_2	8.87	—	—
C_3	6.28	—	—
iC_4	1.06	—	—
nC_4	2.48	—	—
iC_5	0.87	—	—
nC_5	1.17	—	—
C_6	1.45	—	—
C_7	2.40	91.7	0.741
C_8	2.67	104.7	0.767
C_9	1.83	119.2	0.787
C_{10}	1.77	134.0	0.790
C_{11}	1.19	148.0	0.796
C_{12}	1.16	161.0	0.811
C_{13}	1.01	172.0	0.826
C_{14}	1.04	190.0	0.837
C_{15}	0.89	204.0	0.844
C_{16}	0.73	217.0	0.854
C_{17}	0.63	233.0	0.843
C_{18}	0.71	248.0	0.848
C_{19}	0.59	264.0	0.859
C_{20+}	4.57	425.0	0.909

TABLE 10.6
Molar Composition of Oil Mixture. Viscosity Data (Westvik 1997) for This Mixture Is Shown in Table 10.7 and Plotted in Figure 10.7

Component	Mol%	Molecular Weight (g/mol)	Density at 1.01 bar, 15°C (g/cm³)
N_2	0.291	—	—
CO_2	0.481	—	—
C_1	17.813	—	—
C_2	1.454	—	—
C_3	2.914	—	—
iC_4	1.146	—	—
nC_4	2.750	—	—
iC_5	1.769	—	—
nC_5	2.425	—	—
C_6	3.949	—	—
C_7	4.976	96.2	0.7123
C_8	5.467	109.7	0.7393
C_9	4.387	123.8	0.7583
C_{10+}	50.178	348.6	0.8982

TABLE 10.7

Experimental Viscosity Data for the Compositions in Tables 10.4 through 10.6. The Viscosity Data Is Plotted in Figures 10.5 through 10.7

Table 10.4		Table 10.5		Table 10.6	
140°C		164°C		100°C	
P (bar)	Viscosity (cP)	P (bar)	Viscosity (cP)	P (bar)	Viscosity (cP)
1035.2	0.1052	466.0	0.237	304.0	2.468
966.3	0.0999	427.5	0.222	266.6	2.380
894.0	0.0943	405.0	0.211	231.0	2.277
828.4	0.0891	387.0	0.200	221.0	2.230
759.4	0.0837	365.5	0.195	141.0	2.144
690.5	0.0783	353.0	0.190	111.0	2.092
621.5	0.0729	*330.1	0.180	81.0	2.010
552.6	0.0674	304.3	0.197	*61.5	1.866
483.6	0.0619	275.0	0.221	56.0	1.869
414.7	0.0563	248.0	0.250	51.0	1.893
*396.4	0.0547	207.3	0.296	46.0	1.927
—	—	154.0	0.352	38.0	2.012
—	—	107.5	0.433	28.0	2.115
—	—	49.2	0.579	18.0	2.227
—	—	11.4	0.815	—	—
—	—	1.0	1.034	—	—

* Saturation point.

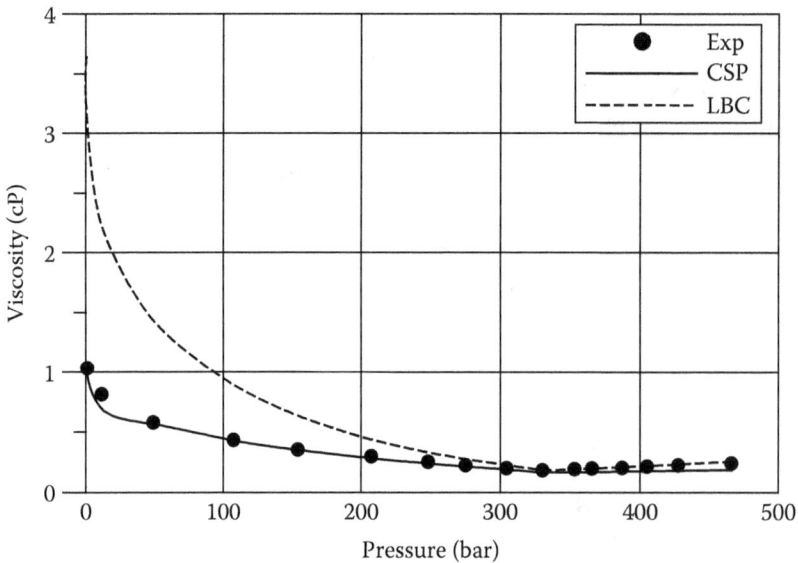

FIGURE 10.5 Measured and calculated viscosities of HT/HP gas condensate at 140°C. The composition is shown in Table 10.4 and the viscosity data in Table 10.7.

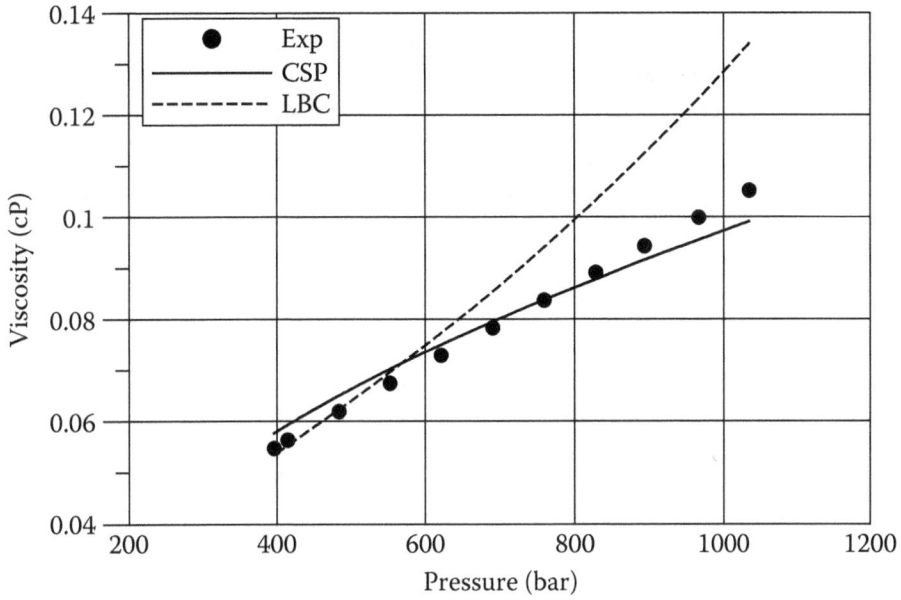

FIGURE 10.6 Measured and calculated viscosities of reservoir oil mixtures at 164°C. The composition is shown in Table 10.5 and the viscosity data in Table 10.7.

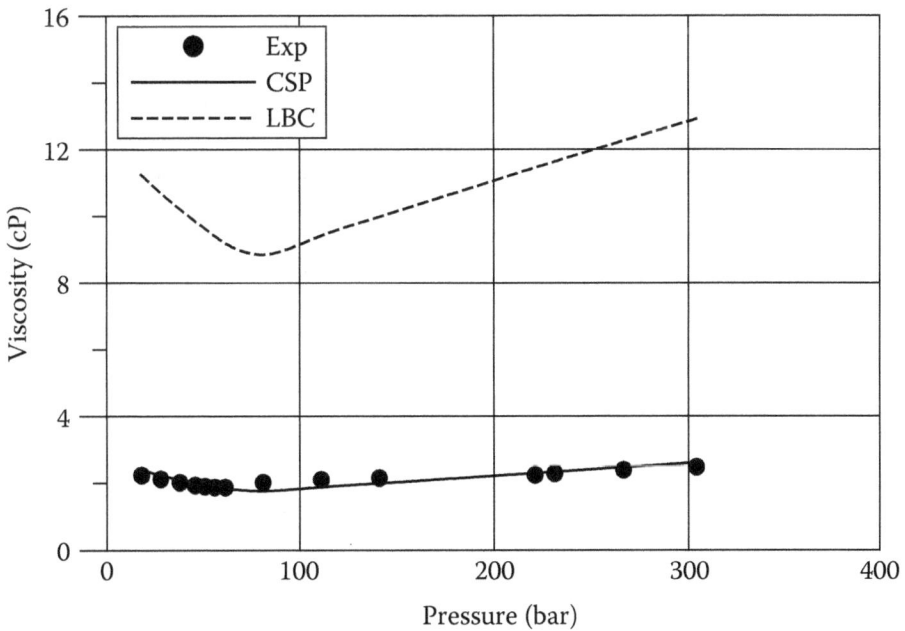

FIGURE 10.7 Measured and calculated viscosities of reservoir oil mixtures at 100°C. The composition is shown in Table 10.6 and the viscosity data in Table 10.7.

TABLE 10.8

Heavy Oil Compositions (Mol%) for Which Viscosity Data Is Shown in Table 10.9. The Density of the C_{7+} Fraction Is in g/cm³ and at 1.01 bar and 15°C

	Oil 1	Oil 2	Oil 5
N_2	0.90	0.31	0.04
CO_2	0.14	0.08	1.21
C_1	38.78	19.43	18.92
C_2	2.03	1.47	0.04
C_3	0.06	0.35	0.04
iC_4	0.01	0.61	0.03
nC_4	0.05	0.29	0.05
iC_5	0.00	0.45	0.05
nC_5	0.00	0.26	0.05
C_6	0.04	0.90	0.23
C_{7+}	57.99	75.86	79.34
M_{C7+}	296	337.5	530.2
Density of C_{7+}	0.955	0.945	1.009

TABLE 10.9

Experimental Viscosity Data for the Compositions in Table 10.8

Oil 1		Oil 2		Oil 5			
55°C		77°C		38°C	49°C	60°C	
P (bar)	Viscosity (cP)	P (bar)	Viscosity (cP)	P (bar)	Viscosity (cP)	Viscosity (cP)	Viscosity (cP)
345.7	8.1	200	11.1	137.9	8500	2268	1505
311.3	7.7	170	10.4	110.3	7756	2085	1348
276.8	7.4	140	9.8	82.7	7011	1898	1167
242.3	7.0	110	9.3	55.2	5945	1735	997
231.0	6.9	*70	8.9	41.4	5541	1760	1061
221.6	6.8	—	—	27.6	5856	2100	1168
214.7	6.7	—	—	13.8	6888	2404	1288
207.9	6.6	—	—	—	—	—	—
*202.6	6.5	—	—	—	—	—	—
173.4	6.9	—	—	—	—	—	—
138.9	7.5	—	—	—	—	—	—
104.4	8.2	—	—	—	—	—	—
70.0	9.0	—	—	—	—	—	—
35.5	10.6	—	—	—	—	—	—
9.8	13.4	—	—	—	—	—	—
1.0	22.7						

Source: Data from Lindeloff, N. et al., *J. Can. Petroleum Technol.* 43, 47–53, 2004.

* Saturation point.

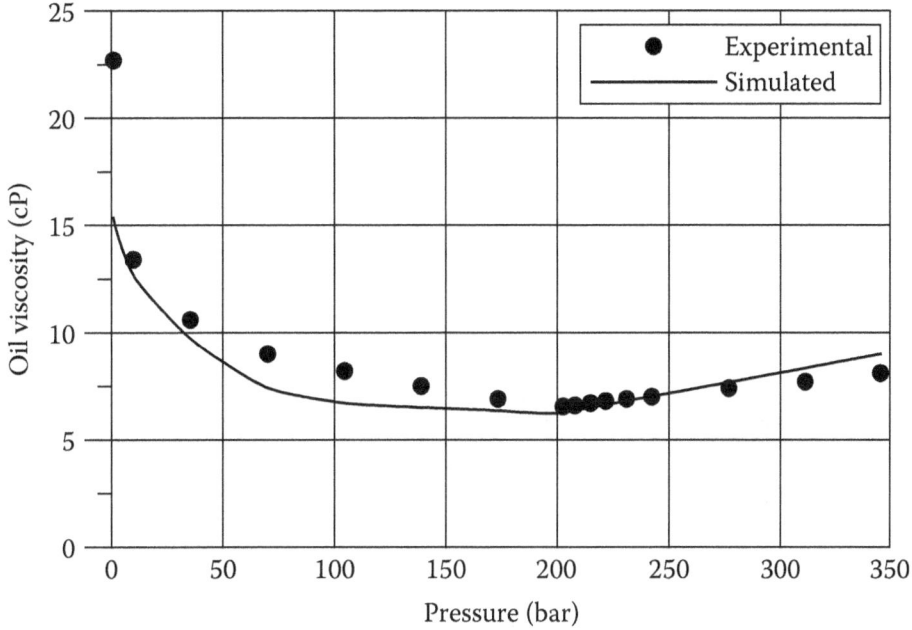

FIGURE 10.8 Measured and simulated viscosities for Oil 1 in Table 10.8 at 55°C. The viscosity data is shown in Table 10.9. The extended corresponding states model of Lindeloff et al. (2004) has been used in the simulations.

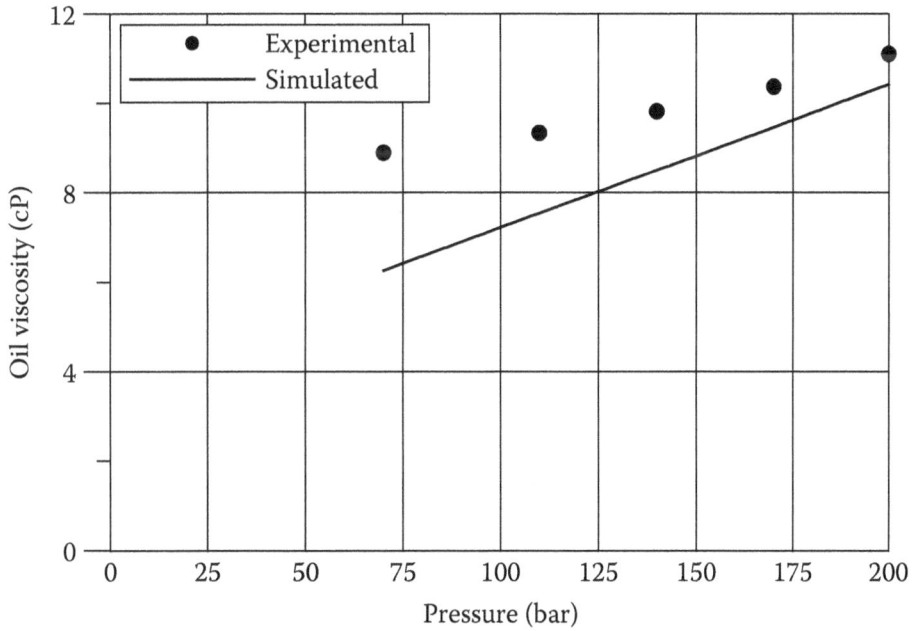

FIGURE 10.9 Measured and simulated viscosities for Oil 2 in Table 10.8 at 77°C. The viscosity data is shown in Table 10.9. The extended corresponding states model of Lindeloff et al. (2004) has been used in the simulations.

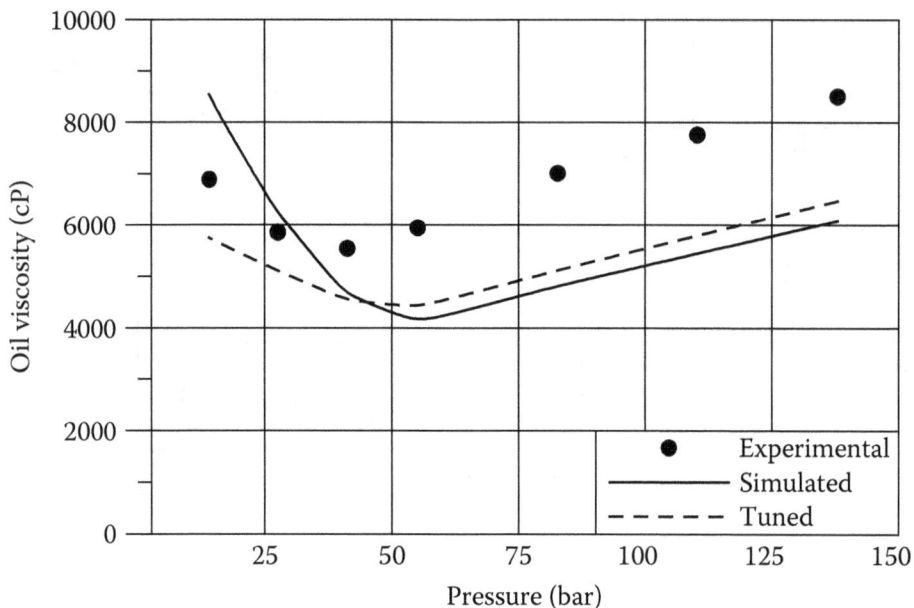

FIGURE 10.10 Measured and simulated viscosities for Oil 5 in Table 10.8 at 38°C. The viscosity data is shown in Table 10.9. The extended corresponding states model of Lindeloff et al. (2004) has been used in the simulations. The tuned viscosities are for viscosity data for three different temperatures tuned to at the same time.

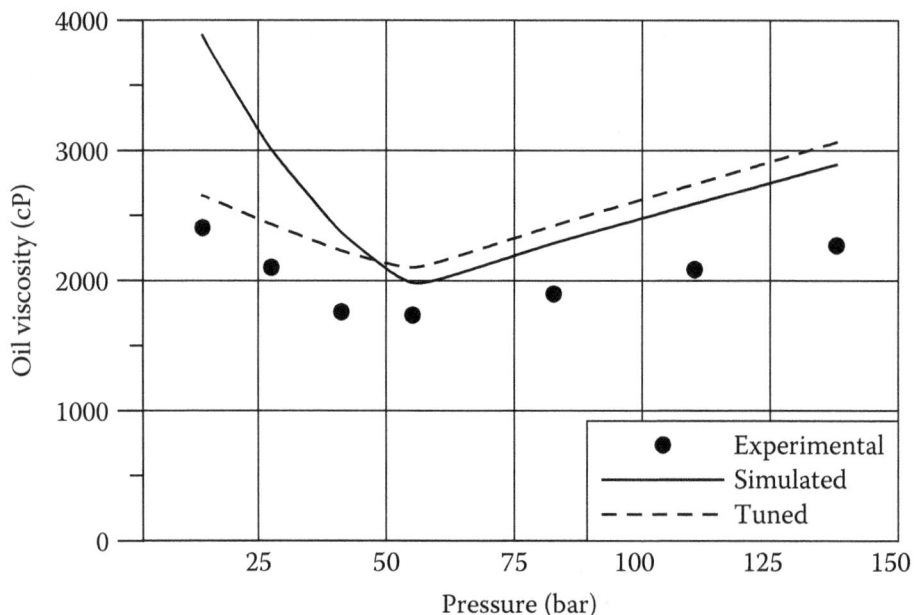

FIGURE 10.11 Measured and simulated viscosities for Oil 5 in Table 10.8 at 49°C. The viscosity data is shown in Table 10.9. The extended corresponding states model of Lindeloff et al. (2004) has been used in the simulations. The tuned viscosities are for viscosity data for three different temperatures tuned to at the same time.

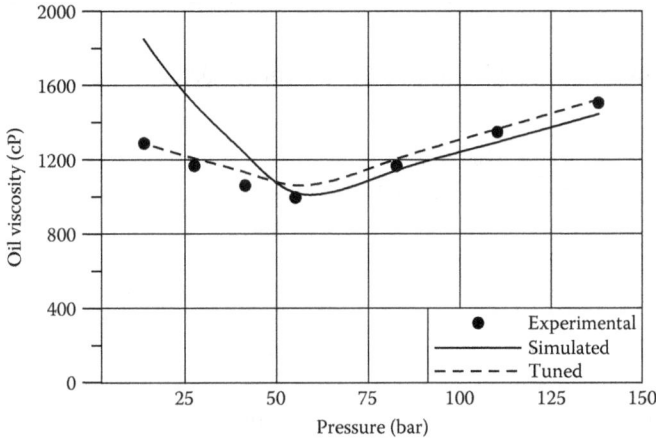

FIGURE 10.12 Measured and simulated viscosities for Oil 5 in Table 10.8 at 60°C. The viscosity data is shown in Table 10.9. The extended corresponding states model of Lindeloff et al. (2004) has been used in the simulations. The tuned results are for viscosity data for three different temperatures tuned to at the same time.

TABLE 10.10
Fluid Composition for Which Gas Viscosity Data Is Shown in Table 10.11

Component	Mol%	Molecular Weight	Density (g/cm³)
C_1	69.62		
C_2	13.14		
C_3	9.19		
iC_4	0.67		
nC_4	2.43		
iC_5	0.44		
nC_5	0.57		
C_6	0.56		
C_7	0.60	92	0.733
C_8	0.63	103	0.757
C_9	0.42	116	0.778
C_{10}	0.28	131	0.790
C_{11}	0.24	147	0.789
C_{12}	0.16	161	0.809
C_{13}	0.16	173	0.822
C_{14}	0.15	186	0.839
C_{15}	0.12	203	0.837
C_{16}	0.09	215	0.843
C_{17}	0.10	229	0.841
C_{18}	0.07	246	0.843
C_{19}	0.05	258	0.854
C_{20+}	0.31	384	0.880

presented some interesting gas viscosity data measured at various pressures and temperature, including pressures approaching 1400 bar and temperatures as high as 200°C. The fluid composition is shown in Table 10.10 and the viscosity data in Table 10.11. The data is plotted in Figure 10.13, which also shows the simulation results obtained with the corresponding states viscosity model.

TABLE 10.11

Gas Viscosity Data for the Fluid Composition in Table 10.10. The Data Is Plotted in Figure 10.13

50°C		100°C		150°C		200°C	
Pressure bar	Viscosity cP	Pressure bar	Viscosity cP	Pressure bar	Viscosity cP	Pressure bar	Viscosity cP
1381	0.119	1380	0.097	1380	0.085	1379	0.076
1211	0.108	1207	0.089	1207	0.078	1207	0.069
1037	0.099	1035	0.080	1035	0.070	1035	0.064
862	0.087	862	0.071	862	0.062	863	0.056
691	0.076	690	0.062	690	0.054	691	0.047
519	0.065	519	0.052	519	0.044	519	0.039
415	0.057	417	0.048	416	0.038	415	0.034

Source: Data from Kashefi, K. et al., *J. Petroleum Sci. Eng.* 112, 153–160, 2013.

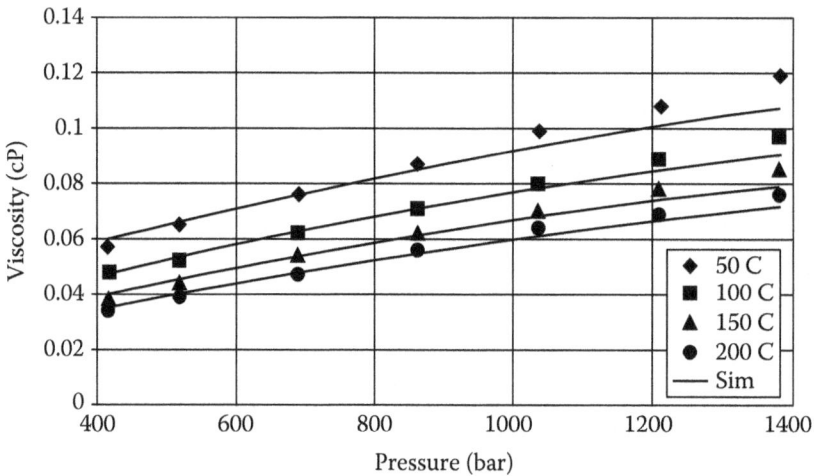

FIGURE 10.13 Gas viscosity data (Kashefi et al. 2013) for the fluid composition in Table 10.10. The full-drawn lines show the simulation results achieved with the corresponding states viscosity model.

The viscosity of the liquid phase condensing from a gas condensate fluid is important especially in flow line simulations, but it is rare to see viscosities reported for such liquids. One of the few data sources is Al-Meshari et al. (2007). Table 10.12 shows the molar compositions of a gas condensate mixture and a near critical reservoir fluid. Liquid viscosity data for the two fluids is shown in Table 10.13.

Table 10.14 shows the composition of a Gulf of Mexico (GoM) reservoir fluid for which viscosity data has been measured at high pressures (Hustad et al. 2014). Viscosity data is shown in Table 10.15 for the reservoir fluid with various concentrations of nitrogen. Also shown in Table 10.15 are viscosity simulation results obtained with the corresponding states viscosity model.

TABLE 10.12
Fluid Compositions for Which Viscosity Data Is Shown in Table 10.13

Component	Gas Condensate Mol%	Near Critical Fluid Mol%
N_2	7.08	4.39
CO_2	0.62	3.24
C_1	71.04	62.49
C_2	7.57	4.21
C_3	3.48	2.81
iC_4	0.64	1.00
nC_4	1.43	1.76
iC_5	0.50	0.92
nC_5	0.56	1.09
C_6	0.75	1.85
C_7	1.07	2.40
C_8	1.36	2.77
C_9	0.86	2.26
C_{10}	0.61	1.76
C_{11}	0.41	1.25
C_{12+}	2.02	5.80
Properties of C_{12+}		
Density (g/cm³)	0.8247	0.84
Molecular weight	232	240

TABLE 10.13
Liquid Viscosities for the Fluid Compositions in Table 10.12. The Constant Volume Depletion (CVD) and Differential Liberation Experiments (DL) Are Described in Chapter 3

Gas Condensate, CVD Experiment at 117°C		Near Critical Fluid, DL Experiment at 149°C	
Pressure bar	Viscosity cP	Pressure bar	Viscosity cP
395 (saturation point)	—	485	0.223
346	0.264	461	0.222
291	0.277	427	0.219
242	0.292	402	0.218
194	0.312	368	0.217
146	0.339	311	0.243
98	0.384	242	0.273
56	0.442	173	0.310
1	0.561	104	0.365
		35	0.508
		1	0.528

Source: Al-Meshari, A. et al., SPE 108434, Presented at *SPE ATCE*, Anaheim, USA, November 11–14, 2007.

TABLE 10.14
Composition of Gulf of Mexico Reservoir Fluid

Component	Mole%	Molecular Weight	Density (g/cm³)
N_2	0.123		
CO_2	0.066		
C_1	37.769		
C_2	5.435		
C_3	5.88		
iC_4	0.726		
nC_4	3.285		
iC_5	0.728		
nC_5	2.232		
C_6	2.669		
C_7	4.025	92.9	0.7188
C_8	4.029	106.6	0.7416
C_9	3.355	120.5	0.7619
$C_{10}-C_{11}$	5.438	139.2	0.7852
$C_{12}-C_{14}$	5.444	175.0	0.8154
$C_{15}-C_{17}$	4.589	212.2	0.8461
$C_{18}-C_{20}$	3.038	254.5	0.8645
$C_{21}-C_{25}$	2.838	313.7	0.8884
$C_{26}-C_{30}$	1.694	350.0	0.9035
$C_{31}-C_{35}$	1.763	434.2	0.9240
C_{36+}	4.872	808	1.0093

Source: Hustad. et al., *SPE Reservoir Evaluation & Engineering* 17, 384–395, 2014.

TABLE 10.15
Experimental and Simulated Viscosity Data at 94°C for the Gulf of Mexico Reservoir Fluid in Table 10.14 with Varying Concentrations of N_2. The Corresponding States Model Has Been Used in the Simulations

Added moles N_2 per Mole Reservoir Fluid	Viscosity (cP) at 1034 bar		Viscosity (cP) at 940 bar		Viscosity (cP) at 840 bar	
	Experimental	Simulated	Experimental	Simulated	Experimental	Simulated
0	2.026	2.179	1.865	2.053	1.706	1.916
0.09	1.712	1.844	1.573	1.734	1.451	1.614
0.18	1.407	1.461	1.295	1.369	1.194	1.270
0.27	1.101	1.130	1.025	1.069	0.960	1.004

Source: Hustad, O.S. et al., *SPE Reservoir Evaluation & Engineering* 17, 384–395, 2014.

10.2 THERMAL CONDUCTIVITY

Thermal conductivity is defined as the proportionality constant, λ, in the following relation (Fourier's law):

$$q = -\lambda \left[\frac{dT}{dx} \right] \tag{10.58}$$

where q is the heat flow per unit area, and (dT/dx) is the temperature gradient in the direction of the heat flow. The terms entering into the definition are illustrated in Figure 10.14.

Thermal conductivity is an important property, especially in flow studies, because it influences heat transfer and thereby the temperature profile in the pipeline. A correct representation of the temperature profile is needed, for example, to address the potential risk of solids precipitation (wax, hydrates, and scale). The thermal conductivity may be calculated using a corresponding states principle (Christensen and Fredenslund 1980; Pedersen and Fredenslund 1987).

According to the corresponding states theory, thermal conductivity can be found from the expression

$$\lambda_r = f(P_r, T_r) \tag{10.59}$$

where f is the same function for a group of substances obeying the corresponding states principle. For the reduced thermal conductivity, λ_r, the following equation is used:

$$\lambda_r(P,T) = \frac{\lambda(P,T)\, T_c^{1/6}\, M^{1/2}}{P_c^{2/3}} \tag{10.60}$$

Using simple corresponding states theory, the thermal conductivity of component x at temperature T and pressure P may be found from the equation

$$\lambda_x(P,T) = \frac{\left(\frac{P_{cx}}{P_{co}}\right)^{2/3}}{\left(\frac{T_{cx}}{T_{co}}\right)^{1/6}\left(\frac{M_x}{M_o}\right)^{1/2}}\lambda_o(P_o, T_o) \tag{10.61}$$

where $P_o = PP_{co}/P_{cx}$, $T_o = TT_{co}/T_{cx}$, and λ_o is the thermal conductivity of the reference substance at temperature T_o and pressure P_o. As is the case for viscosity, methane is a convenient choice as reference substance. However, some corrections must be introduced as compared with the simple corresponding states principle. The thermal conductivity of polyatomic substances (Hanley 1976) can be separated into two contributions, one for transport of translational energy and one for transport of internal energy.

$$\lambda = \lambda_{tr} + \lambda_{int} \tag{10.62}$$

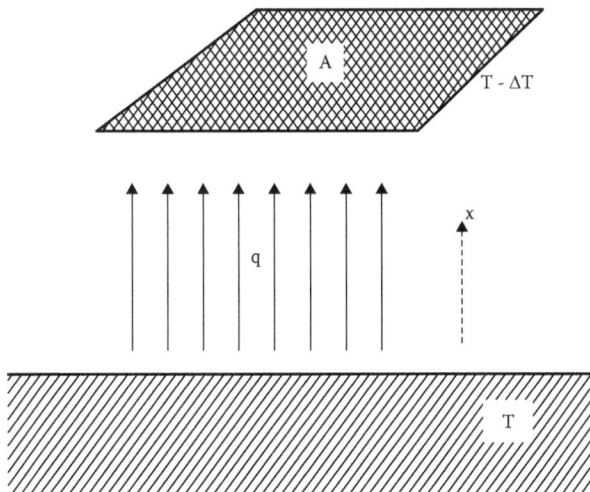

FIGURE 10.14 The terms used in the definition of thermal conductivity (Equation 10.58).

Christensen and Fredenslund (1980) have suggested that the corresponding states theory should only be applied to the translational term. A term $\lambda_{int,mix}$ is used to correct the deviations from the simple corresponding states model. The final expression for the calculation of thermal conductivity of a mixture at the temperature T and the pressure P is

$$\lambda_{mix}(P,T) = \frac{\left(\frac{P_{cx}}{P_{co}}\right)^{2/3}}{\left(\frac{T_{cx}}{T_{co}}\right)^{1/6}\left(\frac{M_x}{M_o}\right)^{1/2}}\left(\frac{\alpha_{mix}}{\alpha_o}\right)(\lambda_o(T_o,P_o) - \lambda_{int,o}(T_o)) + \lambda_{int,mix}(T) \qquad (10.63)$$

The mixture molecular weight M_{mix} of an N-component mixture is found from the Chapman–Enskog theory as described by Murad and Gubbins (1977):

$$M_{mix} = \frac{1}{16}\left[\frac{\sum_{i=1}^{N}\sum_{j=1}^{N}\left[z_i z_j \sqrt{\frac{1}{M_i} + \frac{1}{M_j}}\left(T_{ci}T_{cj}\right)^{1/4}\right]}{\left[\left(\frac{T_{ci}}{P_{ci}}\right)^{1/3} + \left(\frac{T_{cj}}{P_{cj}}\right)^{1/3}\right]^2}\right]^{-2} T_{c,mix}^{-1/3}P_{c,mix}^{4/3} \qquad (10.64)$$

where z is the mole fraction, i and j component indices, and $T_{c,mix}$ and $P_{c,mix}$ are given in Equations 10.19 and 10.22, respectively. The internal energy contributions to the thermal conductivity $\lambda_{int,o}$ of the reference substance and $\lambda_{int,mix}$ of the mixture are both found from

$$\lambda_{int} = \frac{1.18653\,\eta^*\left(C_P^{id} - 2.5R\right)f(\rho_r)}{M} \qquad (10.65)$$

$$f(\rho_r) = 1 + 0.053432\,\rho_r - 0.030182\,\rho_r^2 - 0.029725\,\rho_r^3 \qquad (10.66)$$

η^* is the gas viscosity at the actual temperature and a pressure of 1 atm, C_P^{id} is the ideal gas heat capacity at the actual temperature, and R is the gas constant. The α parameter of component i is found from (Pedersen and Fredenslund 1987):

$$\alpha_i = 1 + 0.0006004\,\rho_{ri}^{2.043}\,M_i^{1.086} \qquad (10.67)$$

where the reduced density is given by the expression in Equation 10.27. Equation 10.67 is also used to find α_0 of the reference component.

α_{mix} is found using the mixing rule

$$\alpha_{mix} = \sum_{i=1}^{N}\sum_{j=1}^{N} z_i z_j \sqrt{\alpha_i \alpha_j} \qquad (10.68)$$

This mixing rule will ensure that small molecules with small α values are given more importance than larger molecules with larger α values. The smaller molecules are more mobile than the larger ones and, therefore, contribute relatively more to the transfer of energy than do the larger ones.

The calculation of the thermal conductivity of the reference substance, methane, is based on a model of Hanley et al. (1975) of the following form:

$$\lambda(\rho,T) = \lambda_0(T) + \lambda_1(T)\rho + \Delta\lambda'(\rho,T) + \Delta\lambda_c(\rho,T) \qquad (10.69)$$

where the dilute gas thermal conductivity, λ_0, is calculated from

$$\lambda_0 = \frac{GT(1)}{T} + \frac{GT(2)}{T^{2/3}} + \frac{GT(3)}{T^{1/3}} + GT(4) + GT(5)T^{1/3} + GT(6)T^{2/3} + GT(7)T$$
$$GT(8)T^{4/3} + GT(9)T^{5/3} \tag{10.70}$$

The coefficients $GT(1)$ to $GT(9)$ can be seen in Table 10.16. The following empirical relation is used for λ_1:

$$\lambda_1(T) = A + B\left(C - \ln\frac{T}{F}\right)^2 \tag{10.71}$$

TABLE 10.16

Constants in Equations 10.70 through 10.74 for Thermal Conductivities in mW m⁻¹K⁻¹

Equation	Constant	Value
10.70	$GT(1)$	-2.147621×10^5
	$GT(2)$	2.190461×10^5
	$GT(3)$	-8.618097×10^4
	$GT(4)$	1.496099×10^4
	$GT(5)$	-4.730660×10^2
	$GT(6)$	-2.331178×10^2
	$GT(7)$	3.778439×10^1
	$GT(8)$	-2.320481
	$GT(9)$	5.311764×10^2
10.71	A	-0.25276292
	B	0.33432859
	C	1.12
	F	168.0
10.72	l_1	-7.0403639907
	l_2	1.2319512908×10
	l_3	$-8.8525979933 \times 10^2$
	l_4	7.2835897919×10
	l_5	0.74421462902
	l_6	-2.9706914540
	l_7	2.2209758501×10^3
10.74	m_1	-8.55109
	m_2	1.25539×10
	m_3	-1.02085×10^3
	m_4	2.38394×10^2
	m_5	1.31563
	m_6	-7.25759×10
	m_7	1.41160×10^3

where the constants A, B, C, and F are given in Table 10.16. The term $\Delta \lambda'$ is given by

$$\Delta\lambda'(\rho,T) = \exp\left(l_1 + \frac{l_4}{T}\right)\left[\exp\left(\rho^{0.1}\left(l_2 + \frac{l_3}{T^{3/2}}\right) + \theta\sqrt{\rho}\left(l_5 + \frac{l_6}{T} + \frac{l_7}{T_2}\right)\right) - 1.0\right] \qquad (10.72)$$

and dominates in the dense liquid region. The term $\Delta\lambda_c$ (ρ,T) is added to account for an increased thermal conductivity of pure components near the critical point. For mixtures, this term can be ignored. The values of the coefficients l_1–l_7 are given in Table 10.16.

As for viscosities, a "low-temperature term" (Pedersen and Fredenslund 1987) can be used to correct the failures of the methane model at reference temperatures with methane in solid form. Neglecting the critical term, the final expression for the thermal conductivity of methane is

$$\lambda(\rho,T) = \lambda_0(T) + \lambda_1(T)\rho + F_1'' \lambda'(\rho,T) + F_2\Delta\lambda''(\rho,T) \qquad (10.73)$$

F_1 and F_2 are defined in Equation 10.30. The following expression is used for $\Delta\lambda''(\rho,T)$:

$$\Delta\lambda''(\rho,T) = \exp\left(m_1 + \frac{m_4}{T}\right)\left[\exp\left(\rho^{0.1}\left(m_2 + \frac{m_3}{T^{3/2}}\right) + \theta\sqrt{\rho}\left(m_5 + \frac{m_6}{T} + \frac{m_7}{T^2}\right)\right) - 1.0\right] \qquad (10.74)$$

where the values of m_1–m_7 can be seen in Table 10.16.

10.2.1 DATA AND SIMULATION RESULTS FOR THERMAL CONDUCTIVITY

It has not been possible to find thermal conductivity data for petroleum reservoir fluids or other multicomponent hydrocarbon mixtures, but some data exists for binary gas mixtures and for narrow boiling petroleum fractions. Table 10.17 shows thermal conductivity data (Christensen and Fredenslund 1979) for a binary mixture of CO_2 and C_1, and Table 10.18 shows experimental thermal conductivities (Baltatu et al. 1985) for narrow boiling petroleum fractions. In both tables are also shown simulation results with the corresponding states model.

TABLE 10.17

Measured and Simulated Thermal Conductivities for Binary Mixture of 49.39 Mol% CO_2 and 50.61 Mol% C_1

Temperature (K)	Pressure (bar)	Measured Thermal Conductivity (mW m⁻¹K⁻¹)	Simulated Thermal Conductivity (mW m⁻¹K⁻¹)	% Dev.
267.12	17.91	24.45	24.32	−0.5
246.77	11.12	21.77	22.06	1.3
266.93	12.14	23.39	23.60	0.9
228.32	2.64	19.14	19.42	1.5
246.75	2.86	20.65	20.85	1.0
253.94	2.95	21.68	21.41	−1.3

Note: The simulation results are obtained with the model of Christensen and Fredenslund (1980).
Source: Data from Christensen, P.L. and Fredenslund, Aa., *J. Chem. Eng.* Data 24, 281–283, 1979.

TABLE 10.18

Measured and Simulated Thermal Conductivities for Narrow Boiling Petroleum Fractions at Atmospheric Pressure

Petroleum Product	Boiling Point (°C)	Density (g/cm³) at 15°C	Temperature (°C)	Experimental • (mW m⁻¹K⁻¹)	Simulated • (mW m⁻¹K⁻¹)	% Dev.
Petrol, lead-free	81.9	0.756	0.0	135.3	133.5	−1.3
Gasoline, B-70	111.8	0.750	110.2	89.6	88.2	−1.6
Kerosene, TS-1	196.8	0.789	23.2	116.9	119.5	2.2
Light kerosene (Kuwait crude)	231.3	0.799	60.2	119.7	109.9	−8.2
Gas oil	255.2	0.848	20.2	121.0	125.7	3.9
Kerosene 20% Max aromatic	258.1	0.786	60.2	121.6	100.0	−13.0
Diesel fuel (vacuum distilled)	313.3	0.851	200.2	92.4	83.9	−9.2
Aromatic heat transfer oil	394.6	0.948	125.2	110.9	111.2	−1.1

Note: The simulation results are obtained with the model of Pedersen and Fredenslund (1987).
Source: Data from Baltetu, M.E. et al., *Ind. Eng. Chem. Process Res. Dev.* 24, 325–332, 1985.

10.3 GAS/OIL INTERFACIAL TENSION

The molecules located in the surface of a liquid in equilibrium with a gas are exposed to forces differing from those acting in the bulk liquid or gas phases. The attraction from a neighboring gas molecule is less than the attraction from a neighboring liquid molecule. The surface layer is in tension and tends to contract to the smallest area compatible with the mass of material, container restraints, and external forces. The term surface tension, σ, is used for a liquid in contact with atmospheric air and defined as

$$\sigma = \left(\frac{\partial G}{\partial A}\right)_{T,V,N} \tag{10.75}$$

where

G = Gibbs free energy (as defined in Appendix A)
A = surface area
T = temperature
V = molar volume
N = number of molecules

The surface tension can be interpreted as the resistance from the liquid against an imposed increase in the surface area. The SI unit for surface tension is N/m. Another common unit is dyn/cm. These two units are related as $1 \text{ N/m} = 10^3 \text{ dyn/cm}$.

For the tension in the interface between gas and oil, it is more correct to use the term interfacial tension than surface tension, as the composition and properties of the gas also influence the tension in the interface. Knowledge of the oil/gas interfacial tension is needed to understand the pore-level flow processes in a reservoir and is also important for the slip between phases in a dynamic pipeline flow simulation.

10.3.1 MEASUREMENT OF INTERFACIAL TENSION

The interfacial tension between an oil and a gas can be determined using the pendant drop technique (Drelich and White 2002). As sketched in Figure 10.15, an oil drop hanging from a needle

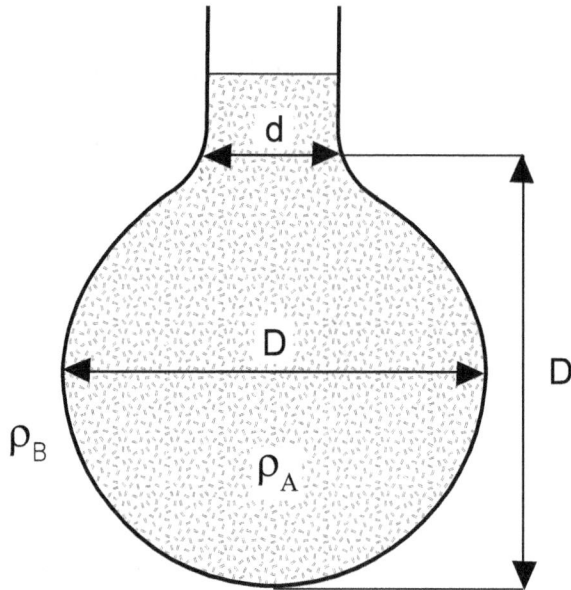

FIGURE 10.15 Pendant drop used to measure gas-oil interfacial tension.

made of stainless steel or glass is lowered into the gas phase for which the gas-oil interfacial tension is to be measured. The interfacial tension is calculated from the following equation:

$$\sigma = \frac{\Delta \rho\, g\, D^2}{H} \tag{10.76}$$

where $\Delta\rho$ is the difference between the liquid and gas density ((ρ_A–ρ_B) in Figure 10.15) in Equation 10.76, g is the gravitational acceleration, and D is the highest diameter of the pendant drop. H is a shape-dependent parameter, which is a function of the shape factor, d/D, where d is the diameter of the drop in a distance D from the bottom of the drop.

10.3.2 Models for Interfacial Tension

Brock and Bird (1955) have shown that the following relation can approximate the surface tension of a pure nonpolar component:

$$\sigma = A_c (1 - T_r)^\theta \tag{10.77}$$

where $\theta = 11/9$, T_r is the reduced temperature (T/T_c), and

$$A_c = P_c^{2/3} T_c^{1/3} (0.133 \alpha_c - 0.281) \tag{10.78}$$

α_c is the Riedel parameter (Riedel 1954)

$$\alpha_c = 0.9076 \left(1 + \frac{T_{Br} \ln P_c}{1 - T_{Br}} \right) \tag{10.79}$$

P_c is the critical pressure in atm, T_c the critical temperature in K, and T_{Br} the reduced boiling point (T_B/T_c).

It is not obvious what T_c and P_c to assign to mixtures, and furthermore it is questionable whether the surface tension of a mixture is related to the critical properties in the same way as is the case for a pure component. For a mixture, it is more convenient to express the surface tension in terms of phase properties, for example, densities that can either be measured or accurately calculated. To transform the temperature dependence to density dependence, the following approximation is used (Fisher 1967):

$$\rho_L - \rho_V = B'(T_c - T)^\beta \tag{10.80}$$

where ρ_L and ρ_V are the liquid and vapor phase densities, respectively. β is a constant, which for most pure components has a value between 0.3 and 0.5. The coefficient B' has the dimension of density/(temperature)$^\beta$. It can be transformed into a dimensionless constant B by rewriting Equation 10.80 to (Lee and Chien 1984)

$$\frac{\rho_L - \rho_V}{\rho_c} = B(1 - T_r)^\beta \tag{10.81}$$

where ρ_c is the critical density and T_r the reduced temperature (T/T_c). Substitution of Equation 10.81 into Equation 10.77 gives

$$\sigma^{\beta/\theta} = \frac{A_c^{\beta/\theta} V_c}{B}(\bar\rho_L - \bar\rho_V) \tag{10.82}$$

where V_c is the critical volume and $\bar\rho_L$ and $\bar\rho_V$ are the molar densities of the liquid and vapor phases, respectively. The molar densities are the densities divided by the molecular weight. The coefficient $\frac{A_c^{b/q} V_c}{B}$ is called the *Parachor* [P], which allows Equation 10.82 to be rewritten as

$$\sigma^{\beta/\theta} = [P](\bar\rho_L - \bar\rho_V) \tag{10.83}$$

This equation is often referred to as the Macleod–Sugden relation (Macleod 1923; Sugden 1924). The equation can be extended for application to mixtures as proposed by Weinaug and Katz (1943), who use $\beta/\theta = 1/4$

$$\sigma^{1/4} = \sum_{i=1}^{N}(\bar\rho_L [P]_i x_i - \bar\rho_V [P]_i y_i) \tag{10.84}$$

where i is a component index and x and y are mole fractions in the oil and gas phases, respectively. For densities in g/cm^3, the surface tension is in dyn/cm. For defined components, the Parachors have fixed values. Parachors of some common oil and gas constituents are listed in Table 10.19. The Parachor of a C_{7+} component may be calculated from

$$[P]_i = 59.3 + 2.34 M_i \tag{10.85}$$

where M_i is the molecular weight of the component. In a review article, Ali (1994) has presented some other correlations for Parachor.

Lee and Chien (1984) have proposed a second modification of Equation 10.83 with application to mixtures. The surface tension is found from

$$\sigma^{1/4} = \bar\rho_L [P]_L - \bar\rho_V [P]_V \tag{10.86}$$

TABLE 10.19

Parachors of Some Oil and Gas Constituents

Compound	Parachor
N_2	41.0
CO_2	78.0
H_2S	80.1
C_1	77.3
C_2	108.9
C_3	151.9
iC_4	181.5
nC_4	191.7
iC_5	225.0
nC_5	233.9
C_6	271.0

Source: Data from Poling, B.E. et al., *The Properties of Gases and Liquids*, McGraw-Hill, New York, 2000.

and the following expression is used for the liquid and vapor phase Parachors, $[P]_L$ and $[P]_V$.

$$[P]_L = \frac{A_{cL}^{1/4} V_{cL}}{B_L}; \quad [P]_V = \frac{A_{cV}^{1/4} V_{cV}}{B_V} \tag{10.87}$$

where

$$V_{cL} = \sum_{i=1}^{N} x_i V_{ci}; \quad V_{cV} = \sum_{i=1}^{N} y_i V_{ci} \tag{10.88}$$

$$B_L = \sum_{i=1}^{N} x_i B_i; \quad B_V = \sum_{i=1}^{N} y_i B_i \tag{10.89}$$

$$A_{cL} = P_{cL}^{2/3} T_{cL}^{1/3} (0.133\alpha_{cL} - 0.281); \quad A_{cV} = P_{cV}^{2/3} T_{cV}^{1/3} (0.133\alpha_{cV} - 0.281) \tag{10.90}$$

$$P_{cL} = \sum_{i=1}^{N} x_i P_{ci}; \quad P_{cV} = \sum_{i=1}^{N} y_i P_{ci} \tag{10.91}$$

$$T_{cL} = \sum_{i=1}^{N} x_i T_{ci}; \quad T_{cV} = \sum_{i=1}^{N} y_i T_{ci} \tag{10.92}$$

$$\alpha_{cL} = \sum_{i=1}^{N} x_i \alpha_{ci}; \quad \alpha_{cV} = \sum_{i=1}^{N} y_i \alpha_{ci} \tag{10.93}$$

x stands for liquid mole fraction, y for vapor mole fraction, and i is a component index. T_c is in K, P_c in atm, and V_c in cm³/mol. Values of B for a number of oil and gas constituents are given in Table 10.20. Pedersen et al. (1989) have suggested calculating the Parachor of the C_{7+} fraction as a whole using the following procedure. First, the critical molar volume of C_{7+} fraction as a whole is calculated from (Riedel 1954):

$$V_{c,C_{7+}} = \frac{RT_{c,C_{7+}}}{P_{c,C_{7+}} (3.72 + 0.26(\alpha_{c,C_{7+}} - 7.0))} \tag{10.94}$$

where $T_{c,C_{7+}}$ and $P_{c,C_{7+}}$ are weight averages of the T_cs and ρ_cs of all C_{7+} pseudocomponents. The term $\alpha_{c,C_{7+}}$ is found from Equation 10.79, using weight mean average values of T_{Br} and P_c for all C_{7+} pseudocomponents. A_c of the C_{7+} fraction is calculated from

$$A_{c,C_{7+}} = P_{c,C_{7+}}^{2/3} T_{c,C_{7+}}^{1/3} (0.133\alpha_{c,C_{7+}} - 0.281) \tag{10.95}$$

TABLE 10.20

Values of B Parameter (Lee and Chien 1984) Entered into Equation 10.86 for Determining the Parachor

Compound	B Parameter
Nitrogen	3.505
Carbon dioxide	3.414
Methane	3.403
Ethane	3.591
Propane	3.602
n-Butane	3.652
n-Pentane	3.690
n-Hexane	3.726

The Parachor of the total C_{7+} fraction is calculated from (Nokay 1959)

$$\log_{10}\left[\sigma_{C_{7+}}\right] = -8.93275 + 3.6884 \log_{10}\left(\frac{T_{c,C_{7+}}}{\rho_{C_{7+}}^{0.6676}}\right) \tag{10.96}$$

In this expression, the critical temperature is in R. ρc_{7+} is the density of the C_{7+} fraction at atmospheric conditions in g/cm^3. Knowing $[\rho_{C_{7+}}]$ from Equation 10.96, it is possible to determine B of the total C_{7+} fraction by rearranging Equation 10.87.

$$B_{C_{7+}} = \frac{A_{c,C_{7+}}^{1/4} V_{c,C_{7+}}}{[\rho_{C_{7+}}]} \tag{10.97}$$

B of the liquid and vapor phase may now be calculated from Equation 10.89 and the liquid and vapor phase Parachors from Equation 10.90. The surface tension can be calculated from Equation 10.86 with the phase densities calculated using a cubic equation of state with volume collection as outlined in Chapter 4.

Danesh et al. (1991) found, by analyzing data for interfacial tension of gas condensate systems, that improved accuracy could be obtained with the Weinaug–Katz correlation (Equation 10.84) by modifying the exponent β/θ from $1/4$ to

$$\frac{\beta}{\theta} = \frac{1}{3.583 + 0.16(\rho_L - \rho_V)} \tag{10.98}$$

where the liquid and vapor densities are in g/cm^3.

10.3.3 DATA AND SIMULATION RESULTS FOR INTERFACIAL TENSIONS

Simon et al. (1978) have presented high-pressure data for gas/oil interfacial tensions of a reservoir fluid to which considerable amounts of carbon dioxide were added. The composition of the reservoir fluid is given in Table 10.21, and the measured results for the surface tensions are shown in Table 10.22. Also shown are simulation results obtained with the Lee and Chien and Weinaug–Katz correlations. Table 10.23 shows gas/oil interfacial tension data (Firoozabadi et al. 1988) for another oil mixture, the composition of which is also shown in Table 10.21.

TABLE 10.21
Compositions (in Mol%) of Two Reservoir Oils. Measured Surface Tension Data for These Oils Is Shown in Table 10.22

Component	Simon et al. (1978)	Firoozabadi et al. (1988)
N_2	—	0.03
CO_2	0.01	2.02
C_1	31.00	51.53
C_2	10.41	8.07
C_3	11.87	5.04
iC_4	—	0.83
nC_4	7.32	2.04
iC_5	—	0.84
nC_5	4.41	1.05
C_6	2.55	1.38
C_{7+}	32.43	27.17
M_{C7+}	199	217
ρ_{C7+} (g/cm^3)	0.869	*0.891
at 1.01 bar and 15°C		

* Tuned to bubble point pressure of 82.3°C and 316.5 bar.

TABLE 10.22
Measured and Calculated Results for the Surface Tension of the Reservoir Oil in Table 10.21 Mixed with Carbon Dioxide

Molar Ratio (CO$_2$/Oil)	T (°C)	P (bar)	Surface Tension (mN/m)				
			Measured	L & C (Equation 10.86)	% Dev.	W–K (Equation 10.84)	% Dev.
55/45	54.4	137.9	0.434	0.428	−1.4	0.354	−18.4
55/45	54.4	156.6	0.0583	—	—	0.0548	−6.0
80/20	54.4	139.2	1.097	0.991	−9.7	0.679	−38.1
80/20	54.4	166.1	0.919	0.883	−3.9	0.437	−52.4
80/20	54.4	201.3	0.775	0.885	−14.2	0.315	−59.4

Note: L & C stands for Lee and Chien (1984) and W–K for Weinaug and Katz (1943).
Source: Data from Simon, R. et al., Soc. *Petroleum Eng. J.* 20–26, February 1978.

TABLE 10.23
Measured and Calculated Results for the Surface Tension of the Reservoir Oil in Table 10.21 at a Temperature of 82.2°C

P (bar)	Surface Tension (mN/m)				
	Measured	L & C (Equation 10.86)	% Dev.	W–K (Equation 10.84)	% Dev.
263.0	1.3	1.3	0.0	1.5	15.4
228.5	2.3	2.0	−13.0	2.4	4.3
194.0	3.3	3.1	−6.1	3.7	12.1
159.6	4.6	4.7	2.2	5.5	19.6

Note: L & C stands for Lee and Chien (1984) and W–K for Weinaug and Katz (1943).
Source: Data from Firoozabadi, A. et al., *SPE Reservoir Eng.* 3, 265–272, 1988.

10.4 DIFFUSION COEFFICIENTS

Consider a single-phase (gas or liquid) binary mixture of components A and B. Assume that a concentration gradient exists, meaning that the concentrations of A and B vary with position. Component fluxes $\vec{J}_A$ and $\vec{J}_B$ will in that case exist

$$\vec{J}_A = -D_{AB} \frac{dc_A}{dz}$$
$$\vec{J}_B = -D_{BA} \frac{dc_B}{dz}$$

(10.99)

where D_{AB} is the diffusion coefficient for component A in B, D_{BA} the diffusion coefficient for component B in A, and dc_A/dz is the concentration gradient of component A in direction z. Similarly, dc_B/dz is the concentration gradient of component B in direction z. Both diffusion coefficients are for diffusion in the z direction. For *n*-paraffins diffusing in *n*-paraffins, Hayduk and Minhas (1982) have proposed to express the diffusion coefficient (in m²/sec) as

$$D_i = 13.3 \times 10^{-12} \times T^{1.47} \frac{\eta^{\left(\frac{10.2}{V}\right)^{-0.791}}}{V^{0.791}}$$

(10.100)

where T is the temperature in Kelvin, η is the viscosity of the bulk phase in cP, and V is the molar volume of the bulk phase in cm³/g. Lindeloff and Krejbjerg (2002) have recommended the use of this expression in simulation of wax deposition at the inner side of a pipeline as a result of molecular diffusion. Wax formation is further dealt with in Chapter 11.

REFERENCES

Ali, J.K., Prediction of Parachors of petroleum cuts and pseudocomponents, *Fluid Phase Equilib.* 95, 383–398, 1994.

Al-Meshari, A., Kokal, S., and Sajjad, A., Measurement of gas condensates, near-critical and volatile oil densities, and viscosities at reservoir conditions, SPE 108434, presented at the *SPE ATCE*, Anaheim, November 11–14, 2007.

Baltatu, M.E., Ely, J.F., Hanley, H.J.M., Graboski, M.S., Perkins, R.A., and Sloan, E.D., Thermal conductivity of coal-derived liquids and petroleum fractions, *Ind. Eng. Chem. Process Res. Dev.* 24, 325–332, 1985.

Beggs, H.D. and Robinson, J.R., Estimating the viscosity of crude oil systems, *J. Pet. Technol.* 27, 1140–1141, 1975.

Brock, J.R. and Bird, R.B., Surface tension and the principle of corresponding states, *AICHE J.* 1, 174–177, 1955.

Christensen, P.L. and Fredenslund, Aa., Thermal conductivity of gaseous mixtures of methane with nitrogen and carbon dioxide, *J. Chem. Eng. Data.* 24, 281–283, 1979.

Christensen, P.L. and Fredenslund Aa., A corresponding states model for the thermal conductivity of gases and liquids, *Chem. Eng. Sci.* 35, 871–875, 1980.

Dandekar, A., Danesh, A., Tehrani, D.H., and Todd, A.C., A modified viscosity method for improved prediction of dense phase viscosities, presented at the *7th European IOR Symposium*, Moscow, Russia, October 27–29, 1993.

Danesh, A.S, Dandekar, A.Y., Todd, A.C., and Sarkar, R., A modified scaling law and parachor method for improved prediction of interfacial tension of gas–condensate systems, SPE 22710, presented at the *SPE ATCE*, Dallas, TX, October 6–9, 1991.

Drelich, J. and White, C.L., Measurement of interfacial tension in fluid-fluid systems, *Encyclopedia of Surface and Colloid Science*, Marcel Dekker Inc., New York, 2002.

Firoozabadi, A., Katz, D.L., Sonoosh, H., and Sajjadian, V.A., Surface tension of reservoir crude oil–gas systems recognizing the asphalt in the heavy fraction, *SPE Reserv. Eng.* 3, 265–272, 1988.

Fisher, M.E., The theory of equilibrium critical phenomena, *Rep. Prog. Phys.*, A.C. Stickland, Ed., 616, 1967.

Guo, X.Q., Wang, L.S., Rong, S.X., and Guo, T.M., Viscosity model based on equations of state for hydrocarbon liquids and gases, *Fluid Phase Equilib.* 139, 405–421, 1997.

Hanley, H.J.M., Prediction of the viscosity and thermal conductivity coefficients of mixtures, *Cryogenics* 16, 643–651, 1976.

Hanley, H.J.M., McCarty, R.D., and Haynes, W.M., Equation for the viscosity and thermal conductivity coefficients of methane, *Cryogenics.* 15, 413–417, 1975.

Hayduk, W. and Minhas, B.S., Correlations for predictions of molecular diffusivities in liquids, *Can. J. Chem. Eng.* 60, 295–299, 1982.

Herning, F. and Zippener, L., Calculation of the viscosity of technical gas mixtures from the viscosity of the individual gases, *Gas U. Wasserfach* 79, 69–73, 1936.

Hirschfelder, J.O., Curtiss, C.F., and Bird, R.B., *Molecular Theory of Gases and Liquids*, John Wiley & Sons, New York, 1954.

Hustad, O.S., Jia, N., Pedersen, K.S., Memon, A., and Lekumjorn, S., High pressure data and modeling results for phase behavior and asphaltene onsets of Gulf of Mexico oil mixed with nitrogen, *SPE Reserv. Evaluation Eng.* 17, 384–395, 2014.

Kashefi, K., Chapoy, A., Bell, K., and Tohidi Kalorazi, B., Viscosity of binary and multicomponent hydrocarbon fluids at high pressure and high temperature conditions: Measurements and predictions, *J. Pet. Sci. Eng.* 112, 153–160, 2013.

Lee, A., Gonzalez, M., and Eakin, B., The viscosity of natural gases, *J. Pet. Technol.* 18, 997–1000, 1966.

Lee, S.T. and Chien, M.C.H., A new multicomponent surface tension correlation based on scaling theory, *SPE/DOE Fourth Symposium on Enhanced Oil Recovery*, Tulsa, Oklahoma, April 14–16, 1984, pp. 147–158.

Lindeloff, N. and Krejbjerg, K., A compositional model simulating wax deposition in pipeline systems, *Energy Fuels.* 16, 887–891, 2002.

Lindeloff, N., Pedersen, K.S., Rønningsen, H.P., and Milter, J., The corresponding states viscosity model applied to heavy oil systems, *J. Can. Pet. Technol.* 43, 47–53, 2004.

Lohrenz, J., Bray, B.G., and Clark, C.R., Calculating viscosities of reservoir fluids from their compositions, *J. Pet. Technol.* 1171–1176, October 1964.

Macleod, D.B., On a relation between surface tension and density, *Trans. Faraday Soc.* 19, 38–42, 1923.

McCarty, R.D., A modified Benedict–Webb–Rubin equation of state for methane using recent experimental data, *Cryogenics.* 14, 276–280, 1974.

Murad, S. and Gubbins, K.E., Corresponding states correlation for thermal conductivity of dense fluids, *Chem. Eng. Sci.* 32, 499–505, 1977.

Nokay, R., Estimate petrochemical properties, *Chem. Eng.* 66, 147–148, 1959.

Pedersen, K.S. and Fredenslund, Aa., An improved corresponding states model for the prediction of oil and gas viscosities and thermal conductivities, *Chem. Eng. Sci.* 42, 182–186, 1987.

Pedersen, K.S., Fredenslund, Aa., Christensen, P.L., and Thomassen, P., Viscosity of crude oils, *Chem. Eng. Sci.* 39, 1011–1016, 1984.

Pedersen, K.S., Lund, T., and Fredenslund, Aa., Surface tension of petroleum mixtures, *J. Can. Pet. Technol.* 28, 118–123, 1989.

Poling, B.E., Prausnitz, J.M., and O'Connell, J.P. *The Properties of Gases and Liquids*, McGraw-Hill, New York, 2000.

Quinones-Cisneros, S.E., Zéberg–Mikkelsen, C.K., and Stenby, E.H., Friction theory prediction of crude oil viscosity at reservoir conditions based on dead oil properties, *Fluid Phase Equilib.* 212, 233–243, 2003.

Riedel, L., Eine neue universelle Dampfdruckformel. Untersuchungen über eine Erweiterung des Theorems der übereinstimmenden Zustände (in German), *Chem. Ing. Tech.* 26, 83–89, 1954.

Rønningsen, H.P., Prediction of viscosity and surface tension of North Sea petroleum fluids by using the average molecular weight, *Energy Fuels.* 7, 565–573, 1993.

Simon, R., Rosman, A., and Zana, E., Phase-behavior properties of CO_2-reservoir oil systems, *Soc. Pet. Eng. J.* 20–26, February 1978.

Standing, M.B. and Katz, D.L., Density of natural gases, *Trans. AIME.* 146, 140–149, 1942.

Stiel, L.I. and Thodos, G., The viscosity of nonpolar gases at normal pressures, *AICHE J.* 7, 611–615, 1961.

Sugden, S., The variation of surface tension with temperature and some related functions, *J. Chem. Soc.* 32–41, 1924.

Weinaug, C.F. and Katz, D.L., Surface tensions of methane–propane mixtures, *Ind. Eng. Chem.* 35, 239–246, 1943.

Westvik, K., Pseudocomponent characterization for the Lohrenz-Bray-Clark viscosity correlation, M.Sc. thesis, Stavanger University College, Norway, 1997.

11 Wax Formation

Most reservoir fluids contain heavy paraffinic compounds that may precipitate as a solid or solid-like waxy material if the fluid is cooled down. Wax precipitation may cause operational problems when unprocessed well streams are transported in undersea pipelines, in which the temperature may fall to that of the surrounding seawater. Wax may deposit as a solid layer on the inner side of the pipeline. With continued transport, this layer will build up and eventually plug the pipeline if not mechanically removed. Not all formed wax will deposit on the wall. Some will precipitate as solid particles in the bulk phase of the oil and will be transported in suspended form. The suspended wax particles will lead to an increase in the apparent viscosity of the oil and thereby affect the flow properties. This is further dealt with in Section 11.4.

The wax particles are essentially normal paraffins and slightly branched paraffins, but naphthenes with long paraffinic chains may also take part in wax formation. Typical wax-forming molecules are sketched in Figure 11.1. Experimental investigations (Bishop et al. 1995) reveal that compounds heavier than approximately C_{50} are rare in solid wax. This has to do with the paraffin concentration pattern. Normal paraffins and slightly branched isoparaffins are present in considerable concentrations in the lighter C_{7+} fractions. In the high-molecular-weight fractions, the degree of branching is higher and the molecules therefore less likely to enter into a solid structure. Figure 11.2 illustrates the qualitative development in molecular structure with increasing molecular weight. Paraffin components with no or little branching are found in high concentrations in the lighter C_{7+} fractions, but low melting temperatures limit the amount of lighter C_{7+} components in the formed wax. The solid wax phase is instead dominated by C_{20}–C_{50} paraffins.

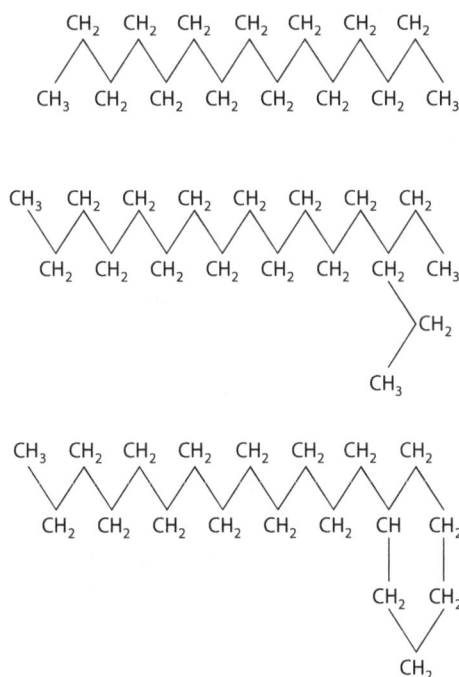

FIGURE 11.1 Typical wax-forming molecules.

DOI: 10.1201/9780429457418-11

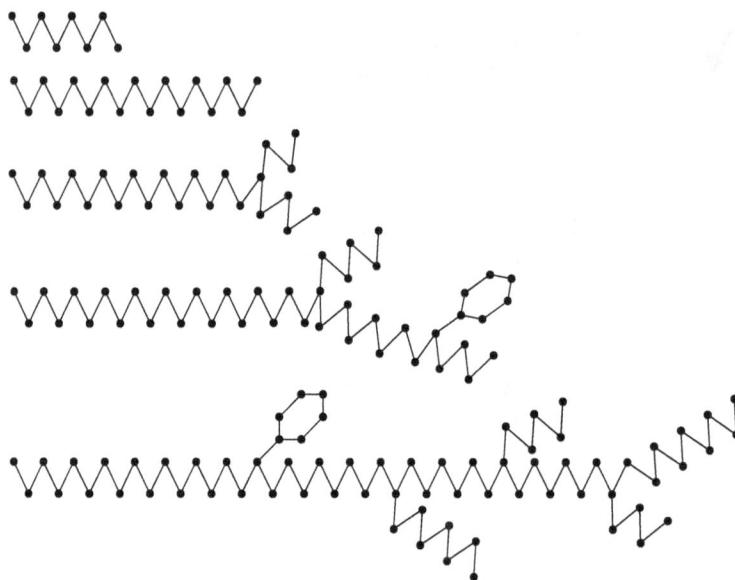

FIGURE 11.2 Development in molecular structure with increasing molecular weight. Branching and number of ring structures increase with molecular weight.

11.1 EXPERIMENTAL STUDIES OF WAX PRECIPITATION

11.1.1 ATMOSPHERIC PRESSURE

In 1991, a comprehensive experimental wax study was published in *Energy and Fuels* (Rønningsen et al. 1991; Pedersen, W.B. et al. 1991; Hansen et al. 1991; Pedersen K.S. et al. 1991). This study comprised experimental investigations of wax precipitation from 17 different stabilized North Sea oils. The composition of one of these oils is shown in Table 11.1. Wax appearance temperatures (WAT) measured using microscopy, differential scanning calorimetry (DSC), and viscometry are shown in Table 11.2. When using the microscopic technique (Rønningsen et al. 1991), the sample was heated from ambient temperature to 80°C, kept there for 10 minutes, and cooled at a rate of 0.5°C/min. The wax appearance temperature was reported as the highest temperature where crystals could be observed in a microscope. Rønningsen et al. found it necessary to preheat the samples to 80°C to get reproducible WAT results. Other WAT measurements using microscopy have been reported by Erickson et al. (1993) and Hammami and Raines (1997).

The crystallization taking place when wax is formed from an oil sample is associated with some release of heat (exothermic process). Likewise, it requires some heat to dissolve the wax particles again (endothermic process) later. These heat effects are used to detect the WAT in a DSC experiment. The DSC technique applied to wax measurements has been outlined in papers by Faust (1978), Hansen et al. (1991), and Létoffé et al. (1995). Starting above the WAT, typically at 80°C, the oil sample is cooled down at a constant rate. Above the WAT, the heat to be removed from the sample with time is almost constant and determined by the liquid heat capacity. At the WAT, the heat that has to be removed with time to keep a constant cooling rate increases steeply. This increase equals the heat released as a result of the solidification process. When no more wax is formed, the heat transfer with time returns to an almost constant level. The same curve is followed for the heat to be supplied to the sample to heat it up later at a constant heating rate. A typical DSC curve is sketched in Figure 11.3.

TABLE 11.1

Stable Oil Mixture (Oil 1 of Pedersen, K.S. et al. 1991). Measured WATs Are Given in Table 11.2

Component	Mole Percentage	Molecular Weight (g/mol)	Density (g/cm³)
C_1	1.139	—	—
C_2	0.507	—	—
C_3	0.481	—	—
iC_4	0.563	—	—
nC_4	0.634	—	—
iC_5	1.113	—	—
nC_5	0.515	—	—
C_6	2.003	—	—
C_7	5.478	91	0.749
C_8	8.756	115	0.768
C_9	7.222	117	0.793
C_{10}	5.414	132	0.808
C_{11}	5.323	148	0.815
C_{12}	4.571	159	0.836
C_{13}	5.289	172	0.850
C_{14}	4.720	185	0.861
C_{15}	4.445	197	0.873
C_{16}	3.559	209	0.882
C_{17}	3.642	227	0.873
C_{18}	3.104	243	0.875
C_{19}	2.717	254	0.885
C_{20}	2.597	262	0.903
C_{21}	1.936	281	0.898
C_{22}	2.039	293	0.898
C_{23}	1.661	307	0.899
C_{24}	1.616	320	0.900
C_{25}	1.421	333	0.905
C_{26}	1.233	346	0.907
C_{27}	1.426	361	0.911
C_{28}	1.343	374	0.915
C_{29}	1.300	381	0.920
C_{30+}	13.234	624	0.953

Note: The density is at 1.01 bar and 15°C.

TABLE 11.2

Wax Appearance Temperatures (WAT) in °C Determined by Microscopy, Differential Scanning Calorimetry (DSC), and Viscometry

Oil No.	Oil Characteristics	Microscopy	DSC	Viscometry
1	Biodegraded, aromatic oil	30.5	11.0	23
2	Paraffinic oil	38.5	17.0	28
3	Waxy oil	41.0	33.5	35

(Continued)

TABLE 11.2 (*Continued*)

Wax Appearance Temperatures (WAT) in °C Determined by Microscopy, Differential Scanning Calorimetry (DSC), and Viscometry

Oil No.	Oil Characteristics	Microscopy	DSC	Viscometry
4	Waxy condensate	48.0	32.5	31
5	Waxy oil	39.5	39.5	40
6	Waxy oil	39.0	39.5	39
7	Paraffinic oil	34.5	32.0	28
8	Paraffinic oil	38.0	32.0	31
9	Waxy oil	35.5	31.5	34
10	Light, paraffinic oil	41.0	31.5	29
11	Heavy, biodegraded, naphthenic oil	22.0	—	32
12	Paraffinic condensate	32.0	25.5	30
13	Very light, paraffinic condensate	<5	−26.0	<10
14	Waxy oil	33.5	23.0	30
15	Paraffinic oil	35.0	20.5	30
16	Paraffinic, asphaltenic oil	37.0	34.0	30
17	Paraffinic oil	39.0	24.0	34

Note: All the fluid compositions are from the North Sea. The composition of Oil No. 1 is shown in Table 11.1.

Source: From Rønningsen, H.P. et al., Wax precipitation from North Sea crude oils. 1. Crystallization and dissolution temperatures, and Newtonian and non-Newtonian flow properties, *Energy Fuels* 5, 895–908, 1991.

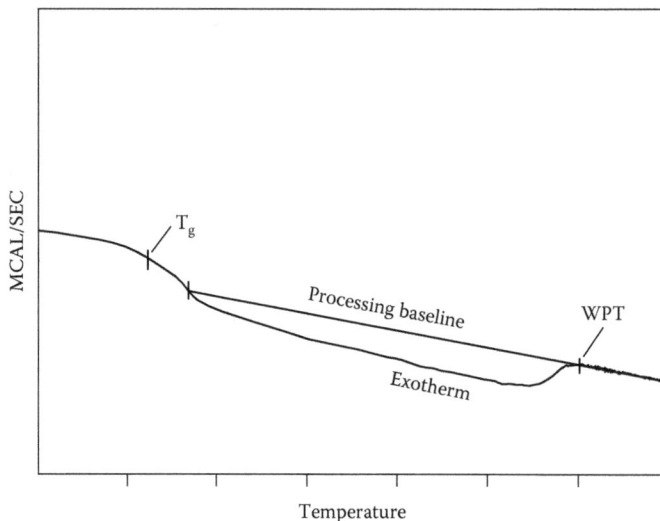

FIGURE 11.3 Typical differential scanning calorimetry (DSC) curve. WAP stands for wax precipitation temperature and T_g for glass transition temperature.

Because suspended wax makes the apparent viscosity of the oil increase, the WAT may also be determined from measurements of viscosity versus temperature. Starting at a temperature above the WAT, a sharp increase in viscosity with decreasing temperature indicates when the WAT is reached.

Ruffier et al. (1993) have reported WAT measurements carried out using ultrasonic equipment. The variation in the transit time and amplitude of an ultrasonic signal passing through the sample is used to detect the WAT.

The four mentioned articles published in *Energy & Fuels* in 1991 comprised 14 oils and 3 condensates (all stabilized). The measured WATs can be seen in Table 11.2. The highest observed WAT was for one of the condensates (Oil 4 in Table 11.2) and determined using microscopy. As already pointed out, C_{50+} compounds do not to any considerable extent take part in the wax formation. Because the molar concentration of C_{20}–C_{50} wax-forming compounds can be at least as high in a stable condensate as in a stabilized oil, condensates may well have WATs exceeding those of oil mixtures.

Pedersen, W.B. et al. (1991) measured the amount of wax formed from stable oils at different temperatures using a pulsed nuclear magnetic resonance (NMR) method. Protons present in the oils were excited by a pulse of radio frequency radiation. After the pulse, the decay in magnetization was characterized by its signal amplitude obtained 10 and 70 μs after the pulse. The first signal was proportional to the number of protons in the solid and liquid phases, and the latter signal was proportional to the number of protons in the liquid phase. The NMR signals were calibrated against samples of polyethylene dispersed in wax-free oil. Typical data for the amount of wax precipitated with temperature is shown in Figure 11.4. The highest wax content found in the 17 stable oils of that study was on the order of 15 weight percentage.

11.1.2 ELEVATED PRESSURES

The wax study published in *Energy & Fuels* in 1991 covered in the preceding section did not comprise experimental data at elevated pressure, but Rønningsen et al. (1997) later presented high-pressure data for 65 live oils. The measurements were undertaken using a high-pressure filter apparatus, as sketched in Figure 11.5. The oils were pumped through a tubing from a high-pressure cell (bottle 1), kept at 80°C through a temperature equilibration coil in series with a bath with a temperature control. The temperature of the bath was lowered at a rate of 6°C/hour. The differential pressure across the filter was recorded continuously. The precipitation of wax caused the differential pressure to increase rapidly, and the WAT was interpreted as the temperature at which the rise in pressure drop deviated from the normal increase owing to an increased viscosity of the oil. The onset temperature is easily detected by plotting the logarithm of the differential pressure versus inverse temperature. An example of such plot is shown in Figure 11.6. For each oil, the

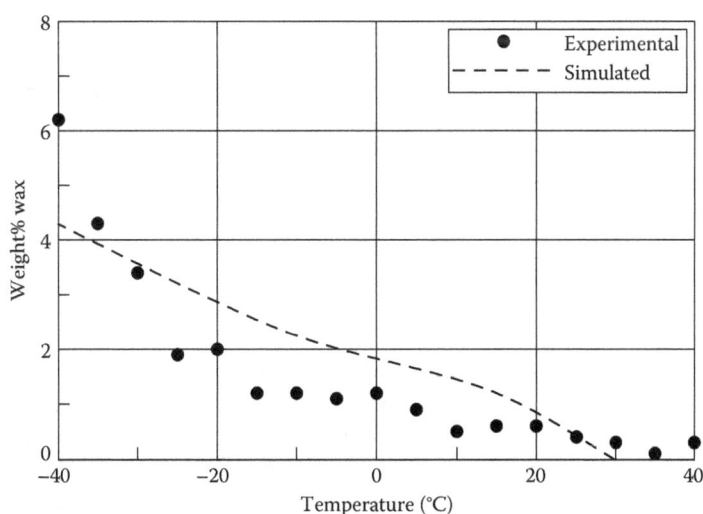

FIGURE 11.4   Measured (Pedersen, W.B. et al. 1991) and calculated (Rønningsen et al. 1997) amounts of wax precipitated at atmospheric pressure from the oil given in Table 11.1. The amounts of wax are in weight percentage of the total oil.

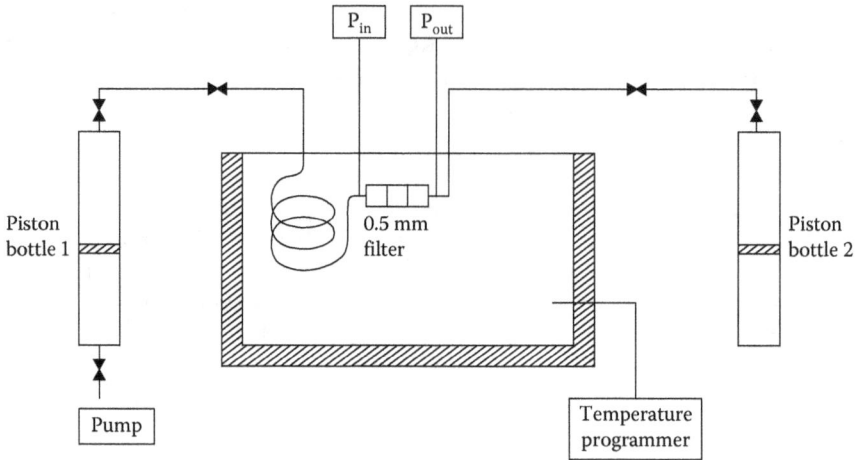

FIGURE 11.5   High-pressure technique for measuring wax appearance temperatures (WAT).

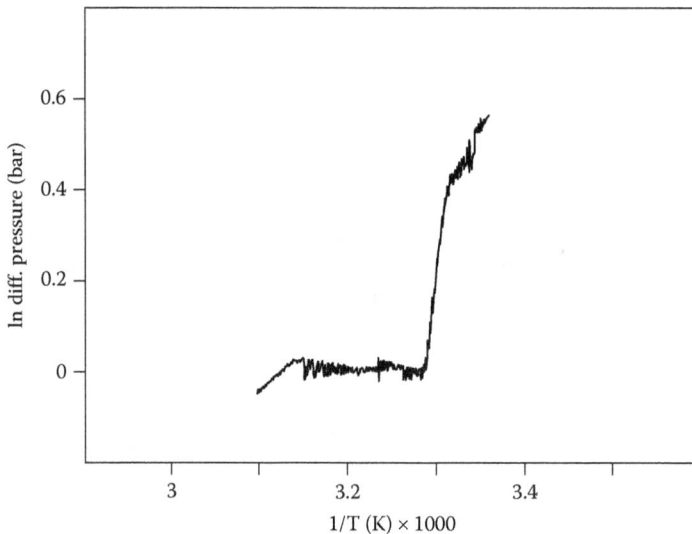

FIGURE 11.6   Detection of wax appearance temperature (WAT). At the WAT, the differential pressure increases sharply.

experiments were carried out as a series of experiments in which the gas content was decreased stepwise. All the experiments carried out for a particular oil were conducted at the same pressure (saturation point of oil sample with the highest gas content). Molar compositions at each stage are shown in Tables 11.3 and 11.4 for two of the investigated oils. Experimental WATs for the same oils are shown in Table 11.5 and plotted in Figures 11.7 and 11.8. It is seen that the WATs decrease with the content of dissolved gas. This decrease is approximately 0.15–0.20°C per mole percentage dissolved C_1–C_5.

The study of Rønningsen et al. (1997) further comprised a study of the effect of pressure on the WAT for (single phase) Oil 10c in Table 11.3. The results are shown in Table 11.6. The WAT

TABLE 11.3
Molar Composition of Oil 10 with Varying Gas Content Investigated by Rønningsen et al. (1997)

Component	Oil 10a	Oil 10b	Oil 10c	Oil 10d	Oil 10e	Mol Weight (g/mol)	Density (g/cm³)
N_2	0.48	0.25	0.08	0.01	0.00	—	—
CO_2	4.04	3.65	2.87	1.28	0.03	—	—
C_1	57.41	41.76	23.88	5.21	0.05	—	—
C_2	9.28	9.34	8.75	5.89	0.24	—	—
C_3	5.62	6.65	7.68	7.75	0.82	—	—
iC_4	1.00	1.29	1.63	1.91	0.38	—	—
nC_4	2.22	2.97	3.89	4.78	1.22	—	—
iC_5	0.83	1.19	1.64	2.17	0.98	—	—
nC_5	1.05	1.53	2.15	2.89	1.50	—	—
C_6	1.35	2.08	3.02	4.22	3.62	—	—
C_7	2.21	3.62	5.38	7.64	8.52	91.4	0.739
C_8	2.59	4.35	6.54	9.36	12.11	103.0	0.771
C_9	1.49	2.55	3.86	5.55	7.88	118.5	0.785
C_{10+}	10.43	18.78	28.63	41.34	62.64	252.0	0.860

TABLE 11.4
Molar Composition of Oil 11 with Varying Gas Content

Component	Oil 11a	Oil 11b	Oil 11c	Oil 11d	Oil 11e	Oil 11f	Mol Weight (g/mol)	Density (g/cm³)
N_2	0.29	0.13	0.04	0.01	0.02	0.00	—	—
CO_2	5.57	4.92	3.89	2.76	1.77	0.04	—	—
C_1	55.62	38.23	21.97	10.99	4.81	0.06	—	—
C_2	9.06	9.00	8.38	7.17	5.65	0.26	—	—
C_3	5.08	6.03	6.82	7.06	6.81	0.81	—	—
iC_4	0.91	1.18	1.45	1.61	1.67	0.37	—	—
nC_4	1.87	2.53	3.21	3.65	3.87	1.09	—	—
iC_5	0.70	1.02	1.35	1.59	1.74	0.84	—	—
nC_5	0.80	1.18	1.58	1.88	2.07	1.14	—	—
C_6	1.07	1.68	2.33	2.83	3.16	2.74	—	—
C_7	1.95	3.26	4.60	5.64	6.35	6.90	90.5	0.746
C_8	2.27	3.89	5.55	6.82	7.71	9.52	102.6	0.773
C_9	1.39	2.42	3.47	4.28	4.84	6.43	116.7	0.793
C_{10+}	13.42	24.53	35.38	43.72	49.56	69.80	290.0	0.876

Note: The densities are at 15°C at atmospheric pressure. The WATs are tabulated in Table 11.5 and plotted in Figure 11.7.

Source: From Rønningsen, H.P., Sømme, B.F., and Pedersen, K.S., An improved thermodynamic model for wax precipitation: experimental foundation and application, paper presented at *8th International Conference on Multiphase 97*, Cannes, France, June 18–20, 1997.

increases by approximately 0.02°C per bar, which is in good agreement with data for pure *n*-alkanes (Brockman 1992) and other live oil data (Brown et al. 1994).

Pan et al. (1997) used measurements of the volume as a function of temperature at constant pressure to detect the WAT of live oil samples. For an oil above the WAT, the thermal expansion coefficient is almost constant with temperature. Below the WAT, the expansion coefficient increases with decreasing temperature. This change in expansion coefficient is used to detect the WAT.

TABLE 11.5

Measured and Calculated Wax Appearance Temperatures (WAT) for Oils 10a–10e and Oils 11a–11f

Oil	Pressure (bar)	Exp WAT (°C)	Calculated WAT (°C)	ΔT (°C)
10a	420	16	18	2
10b	420	20	23	3
10c	420	25	29	4
10d	420	26	33	7
10e	420	28	38	10
11a	420	24	25	1
11b	420	27	31	4
11c	420	33	36	3
11d	420	39	38	−1
11e	420	40	40	0
11f	420	42	44	2

Note: The molar compositions for these oils are shown in Tables 11.3 and 11.4, and the WATs are plotted in Figures 11.6 and 11.7. The calculation results are for the model of Rønningsen et al. (1997).

Source: Data from Rønningsen, H.P., Sømme, B.F., and Pedersen, K.S., An improved thermodynamic model for wax precipitation: experimental foundation and application, paper presented at *8th International Conference on Multiphase 97*, Cannes, France, June 18–20, 1997.

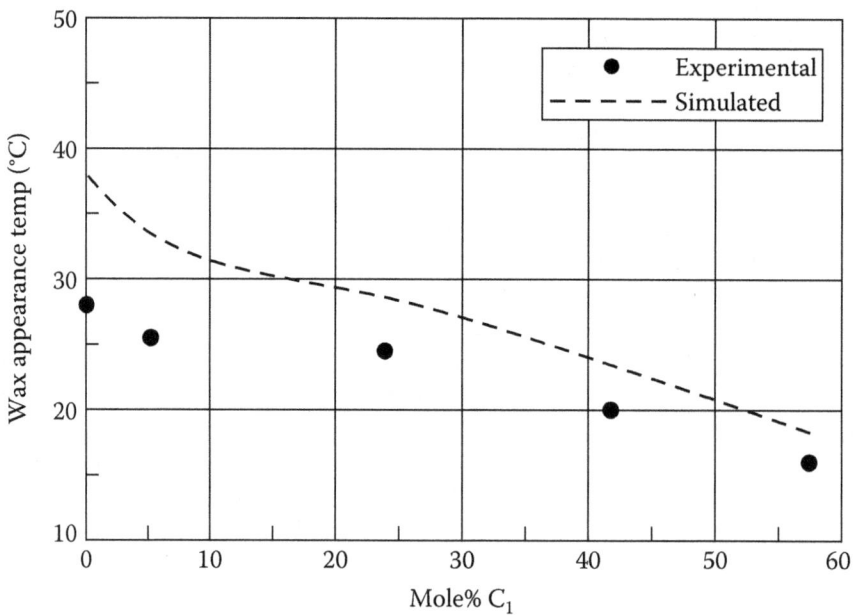

FIGURE 11.7 Measured and calculated WATs for the Oils 10a–10e in Table 11.3, as shown in Table 11.5. The calculation results are obtained using the model of Rønningsen et al. (1997).

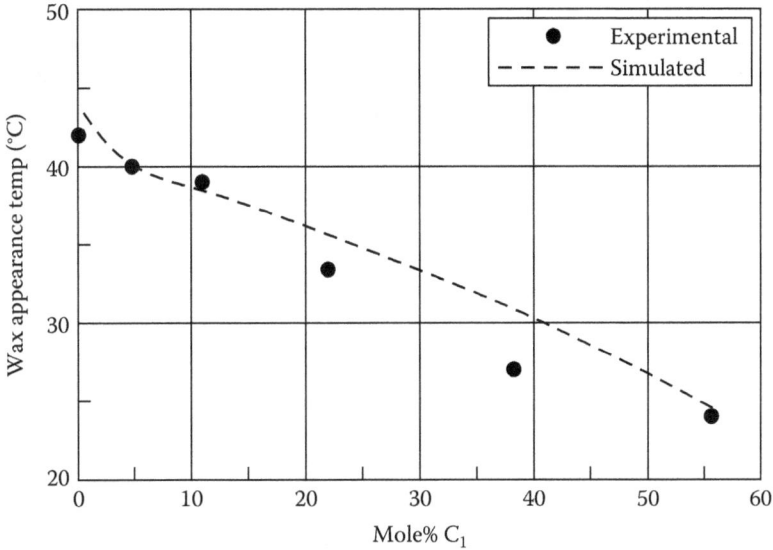

FIGURE 11.8 Measured and calculated WATs for the Oils 11a–11f in Table 11.4, as shown in Table 11.5. The calculation results are obtained using the model of Rønningsen et al. (1997).

TABLE 11.6

Measured and Calculated Wax Appearance Temperatures (WAT) for Oil 10c in Table 11.3

Pressure (bar)	Measured WAT (°C)	Calculated WAT (°C)	ΔT (°C)
400	25	27.7	+2.7
200	20	23.8	+3.8
100	18	20.6	+2.6

Note: Pressures above the saturation point. The calculation results are for the model of Rønningsen et al. (1997).

Table 11.7 shows the compositions and other data for two North Sea oils. For these two oils, Equinor (Rønningsen and Erstad 2023) has kindly made experimental data available for the amount of wax formed below the WAT at pressures ranging from atmospheric pressure to 200 bar. The two oils were flashed to standard conditions. For each of the stable oils, the n-paraffin concentrations (weight percents of stable oil) were measured for C_{10} and heavier fractions using high temperature gas chromatography. The results of the n-paraffin analyses are plotted in Figure 11.9. Most n-paraffins are found in the interval from C_{10}–C_{30}, and consistent with the observations of Bishop et al. (1995), the fractions heavier than C_{50} only contain traces of n-paraffins. DSC measurements were conducted to determine the weight percent of solid wax precipitated from the stable oil as a function of temperature at atmospheric pressure. The study further covered live oil mixtures prepared by adding gas of the composition in Table 11.7 to each of the stable oils at room temperature (20°C) until the saturation pressure was between 50 and 200 bar. While keeping a constant pressure, DSC measurements were made of the amount of precipitated wax while cooling the fluids to −20°. The weight percents of wax formed as a function of temperature were reported and are shown in Table 11.8 and plotted in Figure 11.10. The weight percents are of the stable oil.

TABLE 11.7

Molar Compositions of Two North Sea Oils for Which Wax Data Is Presented in Table 11.8. The Gas Composition Is the One Added to the Stable Oil to Perform Live Oil Wax Precipitation Measurements

Component	Oil A Mol%	Oil B Mol%	Gas Added Mol%
N_2	1.269	0.251	3.058
CO_2	2.319	0.002	1.424
C_1	21.367	54.39	87.390
C_2	4.384	3.793	5.792
C_3	4.697	2.092	1.728
iC_4	1.043	0.437	
nC_4	2.018	0.999	0.610
$iC5$	0.828	0.444	
nC_5	0.900	0.627	
C_6	1.251	1.089	
C_7	2.653	2.809	
C_8	3.239	2.589	
C_9	2.094	2.387	
C_{10+}	51.934	28.09	
C_{10+} molecular weight	322	288	
C_{10+} density (g/cm³)	0.892	0.854	
Key PVT data			
Reservoir pressure, P^{res} (bar)	629	287	
Reservoir temperature, T^{res} (°C)	133	84.6	
STO oil density (g/cm³)	0.880	0.851	
Single flash GOR (Sm³/Sm³)	43	132	
Saturation pressure at T^{res} (bar)	105	281	

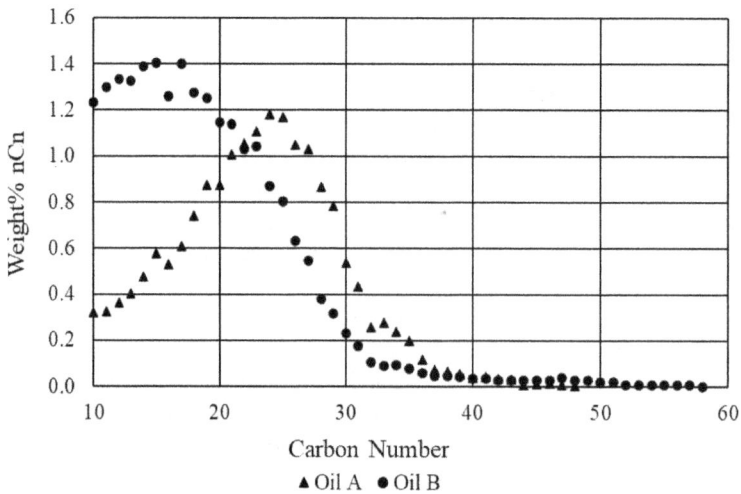

FIGURE 11.9 C_{10+} n-paraffin analyses for the oils from flashing Oil A and Oil B in Table 11.7 to standard conditions. The weight percentages are of the stable oils.

TABLE 11.8

Wax Precipitation Data for the Oils from Flashing Oil A and Oil B in Table 11.7 to Standard Conditions (Stable Oil). The Data at Elevated Pressures Are for the Stable Oils Saturated with the Gas Composition (NG) in Table 11.7 at the Mentioned Pressures and 20°C. The Data Is Plotted in Figure 11.10

Oil A (weight% of stable oil)				Oil B (weigh% of stable oil)			
Temp °C	1 bar Stable Oil	100 bar Stable oil +NG	200 bar Stable oil +NG	Temp °C	1 bar Stable Oil	50 bar Stable oil +NG	200 bar Stable oil +NG
40	0.00	0.00	0.00	47	0.00	0.01	0.00
35	0.02	0.02	0.01	40	0.21	0.18	0.08
30	0.43	0.21	0.09	35	0.33	0.25	0.15
25	2.60	2.02	1.43	30	0.54	0.36	0.18
20	4.97	4.44	3.80	25	0.93	0.53	0.21
15	7.02	6.63	6.06	20	2.03	1.22	0.55
10	8.75	8.49	7.97	15	3.67	2.56	1.59
5	10.14	10.02	9.54	10	5.40	4.09	2.98
0	11.20	11.18	10.73	5	7.06	5.58	4.35
−5	11.97	12.02	11.57	0	8.56	6.89	5.59
−10	12.46	12.57	12.12	−5	9.84	7.99	6.64
−15	12.70	12.88	12.41	−10	10.87	8.86	7.49
−20	12.74	12.97	12.45	−15	11.66	9.50	8.15
				−20	12.22	9.95	8.60

Data Source: Rønningsen, H. P. and Erstad, K., unpublished data kindly made available by Equinor, 2023.

Other live oil wax appearance temperature measurements have been presented by Hammami and Raines (1997), who used a laser-based solid detection system. The authors found that WATs measured using the solid detection technique were systematically lower than WATs measured using microscopy, a difference that was attributed to the fact that particles of some size are needed to detect WATs using the solid detection technique. This study confirms observations by Rønningsen et al. (1991) that different WAT measurement techniques may give slightly different results. Monger-McClure et al. (1997) have presented an extensive experimental study comparing WATs measured by four different experimental techniques, these being DSC, microscopy, filter plugging, and Fourier transform infrared (FTIR) energy scattering. The FTIR technique detects cloud points by measuring the increase in energy scattering associated with wax solidification. The samples investigated were from the Gulf of Mexico, Trinidad, and Oklahoma. The WATs measured were in the interval from 17°C to 56°C. The cloud points determined by each of the four methods agreed with the average value of all methods within 1.7°C.

Table 11.9 shows the composition of a gas condensate (Daridon et al. 2001) for which wax formation temperatures are shown in Table 11.10 together with dew point data. The data points are plotted in Figure 11.11.

Pan et al. (1997) used a temperature-controlled ultracentrifuge to separate precipitated wax from the surrounding live oil. This enabled direct measurements of the amount of precipitated wax.

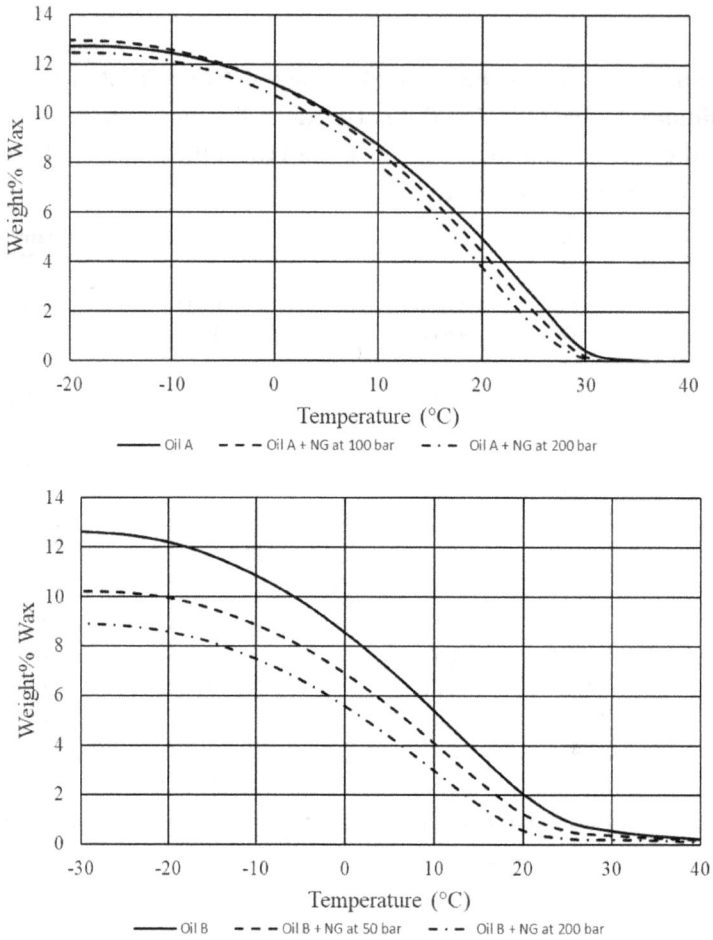

FIGURE 11.10 Wax precipitation curves for Oil A (upper) and Oil B (lower) in Table 11.7. The fully drawn lines are for the stable oils from flashing the oils to standard conditions. The dashed and dashed-dotted lines are for the stable oils saturated with the gas composition (NG) in Table 11.7 at 20°C and the shown pressures. The weigh%'s wax are of the stable oil.

TABLE 11.9

Molar Composition of Gas Condensate for which Measured Wax Appearance Temperatures (WAT) and Dew Points Are Shown in Table 11.10

Component	Mole Percentage
N_2	0.56
CO_2	2.90
C_1	69.13
C_2	8.16
C_3	4.01
iC_4	0.87
nC_4	1.60
iC_5	0.83
nC_5	0.74

Component	Mole Percentage
C_6	1.52
C_7	1.76
C_8	1.56
C_9	1.13
C_{10}	0.91
C_{11+}	4.30

Note: The C_{11+} molecular weight is measured to 237.5. The C_{11+} density is estimated to 0.846 g/cm³ based on the dew point data.

Source: From Daridon, J.-L., et al., Solid–wax–vapor phase boundary of a North Sea waxy crude: Measurement and modeling, *Energy Fuels* 15, 730–735, 2001.

TABLE 11.10

Wax Appearance Temperatures (WAT) and Dew Points Measured for the Gas Condensate Fluid in Table 11.9

Wax Appearance Temperatures		Dew Points	
Temperature (K)	Pressure (bar)	Temperature (K)	Pressure (bar)
295.15	400	293.15	312.6
294.45	350	303.25	317.5
293.55	300	—	—
292.75	250	—	—
292.95	200	—	—
291.75	150	—	—
290.15	100	—	—

Note: The data is plotted in Figure 11.11.

Source: From Daridon, J.-L., et al., Solid–wax–vapor phase boundary of a North Sea waxy crude: Measurement and modeling, *Energy Fuels* 15, 730–735, 2001.

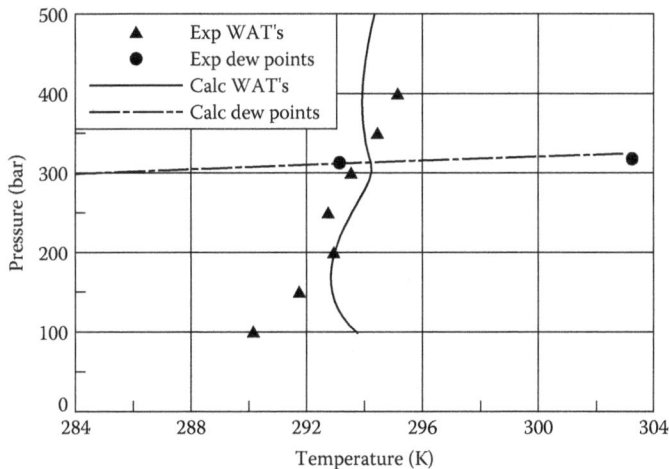

FIGURE 11.11 Measured (Daridon et al. 2001) and calculated WATs and dew points of gas condensate mixture in Table 11.9. The measured data can be seen in Table 11.10.

11.2 MODELING OF WAX PRECIPITATION

When wax is formed, some components undergo a solidification process, and when wax disappears, the components in the wax phase undergo a melting (or fusion) process. The thermodynamics of solidification and melting of a pure component is dealt with in Section 6.3.5. Figure 11.12 schematically shows a cell with a gas, an oil, and a wax phase in thermodynamic equilibrium. At equilibrium, the fugacity of any component i will be the same in each phase:

$$f_i^V = f_i^L = f_i^S \tag{11.1}$$

The fugacity of a component i in the gas and liquid phases in equilibrium with a wax phase may be calculated from the applied equation of state (Chapter 4). The challenging part is how to describe the wax phase.

Erickson et al. (1993) stressed the importance of distinguishing between *n*-paraffins and isoparaffins. Isoparaffins will have a depressing effect on the WATs because the melting enthalpies and melting temperatures of these compounds are lower than those of *n*-paraffins. Table 11.11 shows data (Loebel 1981–1982) for the melting temperature and the heat of fusion for some hydrocarbons with seven and eight carbon atoms. It is seen that even for hydrocarbons of the same carbon number, these properties may vary considerably. For a given carbon number, the *n*-paraffin component will have the highest melting enthalpy.

The experimental studies presented in Section 11.1 support the idea of Erickson et al. that it is necessary to distinguish between potentially wax-forming compounds and compounds that are unlikely to take part in a solid phase formation. A stable oil will be dominated by C_{10+}-components. If all the hydrocarbons in a stable oil are assigned melting properties similar to those of *n*-paraffins, the model will predict that almost all the oil constituents will be converted into wax if the fluid is cooled down to below the melting temperature of nC_{10}, which is 243.5 K. This does not correspond with experimental observations (e.g., those of Pedersen, W.B. et al. 1991), which show that the total content of potentially wax-forming constituents will seldom exceed 15 weight percentage of a stable oil. The considerations of Erickson et al. are further supported by the data presented in Table 11.11 for the melting temperature and the heat of fusion of some hydrocarbons of similar molecular weight. The heat of fusion is lower for the isoparaffins than for the *n*-paraffins. It can be seen from Equation 6.48

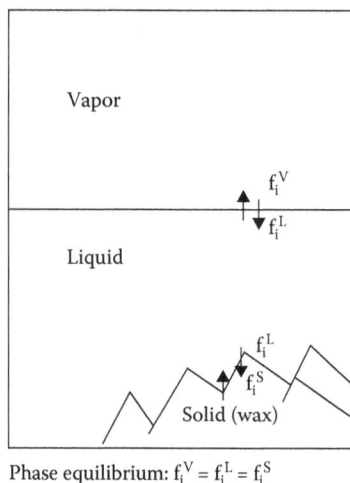

Phase equilibrium: $f_i^V = f_i^L = f_i^S$

FIGURE 11.12 Phase equilibrium between gas, oil, and wax. At equilibrium, the fugacity of any component i will be the same in each phase.

TABLE 11.11
Melting Points and Heat of Fusion of Some Hydrocarbons with Seven and Eight Carbon Atoms

Formula	Compound	Melting Point (°C)	Heat of Fusion (cal/g)
C_7H_{16}	n-heptane	−90.6	33.78
C_7H_{16}	2-methylhexane	−118.2	21.16
C_7H_{16}	2-methylpentane	−123.8	13.98
C_7H_{16}	2,2,3-trimethylbutane	−25.0	5.25
C_7H_{14}	Methylcyclohexane	−126.6	16.43
C_7H_8	Toluene	−94.99	17.17
C_8H_{18}	n-octane	−56.8	43.21
C_8H_{18}	3-methylheptane	−120.5	23.81
C_8H_{18}	4-methylheptane	−121.0	22.68
C_8H_{18}	2,2,4-trimethylpentane	−107.3	18.92
C_8H_{16}	Ethylcyclohexane	−11.3	17.75
C_8H_{16}	Trans-1,1-dimethylcyclohexane	−33.3	4.38
C_8H_{16}	Cis-1,2-dimethylcyclohexane	−49.9	3.50
C_8H_{16}	Trans-1,2-dimethylcyclohexane	−88.2	22.35
C_8H_{16}	Cis-1,3-dimethylcyclohexane	−75.6	23.05
C_8H_{16}	Trans-1,3-dimethylcyclohexane	−90.1	21.01
C_8H_{16}	Cis-1,4-dimethylcyclohexane	−87.4	19.82
C_8H_{16}	Trans-1,4-dimethylcyclohexane	−36.9	26.27
C_8H_{10}	o-xylene	−25.2	30.64
C_8H_{10}	m-xylene	−47.8	26.01
C_8H_{10}	p-xylene	13.2	37.83

Source: Data from Loebel, R., *Handbook of Chemistry and Physics*, CRC Press, Boca Raton, FL, 1981–1982.

that the heat of fusion (ΔH^f) influences the solid state fugacity through an exponential term. The tendency of a component to solidify below its melting temperature will, in other words, increase exponentially with its ΔH^f and be much higher for an n-paraffin than for an iso-paraffin. Erickson et al. measured the content of n-paraffins in the lighter C_{7+} fractions. They estimated the n-paraffin content in heavier fractions by assuming that the logarithm of the mole fraction of n-paraffins decreases linearly with carbon number.

The wax model of Pedersen (1995) as modified by Rønningsen et al. (1997) is based on the assumptions that only C_{7+} components can form wax, and only part of each C_{7+} carbon number fraction contributes to the wax formation. The wax-forming fraction essentially corresponds to the n-paraffin fraction. When an n-paraffin analysis does not exist as that plotted in Figure 11.9, the mole fraction of wax-forming components contained in a given carbon number fraction having the average molecular weight M_i and the average density ρ_i can be estimated from

$$z_i^S = z_i^{tot}\left[1-(A+B\times M_i)\left(\frac{\rho_i-\rho_i^P}{\rho_i^P}\right)^C\right] \qquad (11.2)$$

where z_i^{tot} is the total mole fraction of carbon number fraction i. A, B, and C are empirical constants determined from experimental wax precipitation data. The values used for A, B, and C are given in

---

TABLE 11.12

Constants in Expression Used to Split Pseudocomponents into Wax and Non–Wax-Forming Parts (Equation 11.2)

A	1.074
B	6.584×10^{-4}
C	0.1915

Table 11.12. ρ_i^P is the density of a normal paraffin of the same molecular weight as carbon number fraction i. The following expression is used for the n-paraffin density in g/cm^3:

$$\rho_i^P = 0.3915 + 0.0675 \ln M_i \tag{11.3}$$

The density term in Equation 11.2 serves to distinguish carbon number fractions of different wax contents. If the density of a carbon number fraction is close to that of the corresponding n-paraffin, the density term in Equation 11.2 will approach zero and the wax fraction will approach 1. For an aromatic oil, the densities of the carbon number fractions will be higher, and this will reduce the estimated wax content. When Equation 11.2 gives a negative value of z_i^S, the wax content is assumed to be equal to zero.

The model assumes that the wax phase can be described as an ideal solid solution, meaning that the fugacity of component i in the wax phase can be expressed as

$$f_i^S = x_i^S f_i^{\circ S} \tag{11.4}$$

where $f_i^{\circ S}$ can be found from Equation 6.48.

The melting enthalpy and the melting temperature of wax component i may, as suggested by Won (1986), be found from

$$\Delta H_i^f = 0.1426 M_i T_i^f \tag{11.5}$$

M stands for molecular weight and T^f for melting temperature in K. The following expression is by default used to determine the melting temperature of wax component i

$$T_i^f = 374.500.02617 M_i - \frac{20172}{M_i} \tag{11.6}$$

The gas and liquid fugacities are derived from the applied equation of state, and so is the fugacity of the pure wax-forming component i in liquid form:

$$f_i^{\circ L} = \varphi_i^{\circ L}(P) P \tag{11.7}$$

where $\varphi_i^{\circ L}(P)$ is the fugacity coefficient of pure liquid i at pressure P derived from the applied equation of state. Combining Equations 6.48, 11.4, and 11.7, and assuming $\Delta C_p = 0$, the following expression may be derived for the fugacity of component i in the wax phase:

$$f_i^S = x_i^S \varphi_i^{\circ L}(P) P \exp\left(-\frac{\Delta H_i^f}{RT}\left(1 - \frac{T}{T_i^f}\right) + \frac{\Delta V_i (P - P_{ref})}{RT}\right) \tag{11.8}$$

Experimental observations by Templin (1956) show that the solidification process of paraffinic hydrocarbons is associated with a 10% volume decrease. ΔV_i in Equation 11.8 is therefore assumed to be -10% of the molar volume of hydrocarbon i in liquid form at the same conditions.

The equation of state parameters (T_c, P_c, and acentric factor) of the C_{7+} components can be estimated using the C_{7+} characterization procedure of Pedersen et al., as outlined in Chapter 5. As an extension to the characterization procedure, the wax-forming and the non–wax-forming fractions of the C_{20+} pseudocomponents are assigned different critical pressures. The critical pressure of the wax-forming fraction of each C_{20+} pseudocomponent is found from

$$P_{ci}^{S} = P_{ci} \left(\frac{\rho_i^{P}}{\rho_i} \right)^{3.46} \tag{11.9}$$

P_{ci} equals the critical pressure of pseudocomponent i determined using the characterization procedure of Pedersen et al. ρ_i^{P} is the density of the wax-forming fraction of pseudocomponent i determined from Equation 11.3 and ρ_i the average density of the wax-forming and non–wax-forming fractions of pseudocomponent i. The critical pressure, P_{ci}^{no-S}, of the non–wax-forming fraction of pseudocomponent i

$$\left(\frac{z_i^{tot} - z_i^{S}}{z_i^{tot}} \right)$$

is found from

$$\frac{1}{P_{ci}} = \frac{\left(\dfrac{z_i^{tot} - z_i^{S}}{z_i^{tot}} \right)^2}{P_{ci}^{no-S}} + \frac{\left(\dfrac{z_i^{S}}{z_i^{tot}} \right)^2}{P_{ci}^{S}} + \frac{2 \dfrac{z_i^{tot} - z_i^{S}}{z_i^{tot}} \times \dfrac{z_i^{S}}{z_i^{tot}}}{\sqrt{P_{ci}^{no-S}} \sqrt{P_{ci}^{S}}} \tag{11.10}$$

where P_{ci} is the critical pressure of pseudocomponent i before being split into a wax-forming and a non–wax-forming fraction.

By using Equation 11.10, the contribution to the a parameter in the cubic equation of pseudocomponent i divided into a wax-forming and non–wax-forming component will be the same as that of the undivided pseudocomponent. This can be seen from Equation 4.33, showing the classical mixing rule used for the a parameter in a cubic equation. Maintaining a constant contribution to the a parameter minimizes the effect of the pseudocomponent division on other phase properties, for example, saturation points.

Using Equations 11.9 and 11.10, the wax-forming fraction of a given C_{20+} pseudocomponent will be assigned a lower critical pressure than the non–wax-forming fraction. This is consistent with the fact that the critical pressure of an n-paraffin is lower than that of an aromatic or a naphthenic compound of the same molecular weight. As exemplified by Rønningsen et al. (1997), neglecting this difference will make it difficult to match the experimentally observed depressing effect of dissolved gas on the WAT. Simulated WATs obtained for the compositions in Tables 11.3 and 11.4 are shown in Table 11.5 and plotted in Figures 11.7 and 11.8 in comparison with experimental observations. Table 11.6 shows a comparison of the measured and calculated pressure effect on the WAT of Oil 10c, the composition of which is given in Table 11.3. Simulation results for the weight percentage of wax precipitated with decreasing temperature for the fluid composition in Table 11.1 at atmospheric pressure are shown in Figure 11.4 (dashed line) in comparison with experimental results obtained using pulsed NMR spectroscopy.

Coutinho et al. (2006) assign the components in the wax phase an activity coefficient differing from 1.0. That model, thus, differs from that of Rønningsen et al. by assuming that the wax phase is a non-ideal mixture. The term activity coefficient is explained in Equation A.34 in Appendix A.

Table 11.13 shows the gas condensate mixture in Table 11.9, characterized for use with the model of Rønningsen et al. (1997). The component fractions from C_7 and following are split into a wax-forming (subscript S) and a non–wax-forming component (subscript no-S). From C_7 to C_{20}, T_c, P_c, and acentric factor of the wax-forming and non–wax-forming pseudocomponents are identical for the same carbon number range. The wax-forming pseudocomponents are assigned a melting temperature (T^f) and a heat (or enthalpy) of melting (ΔH^f), which, unlike the non-wax-forming components, enable them to enter a solid (wax) phase. From C_{20} and following, the P_c of the wax-forming fraction is lower than the P_c of the corresponding non–wax-forming pseudocomponent. The P_c of the wax-forming pseudocomponent is found from Equation 11.9 and the P_c of the corresponding non–wax-forming pseudocomponent from Equation 11.10. The binary interaction coefficients can be seen in Table 11.14. Simulation results for the variation in WAT with pressure for this gas condensate can be seen from Figure 11.11, which figure also shows the experimental data.

Table 11.15 shows the stable oil from flashing Oil B in Table 11.7 to standard conditions that has been characterized for the wax model of Rønningsen et al. (1997). Also shown is the simulated

TABLE 11.13
Gas Condensate Mixture in Table 11.9 Characterized for Wax Model of Rønningsen et al. (1997)

	Mole Percentage	T_c (K)	P_c (bar)	Acentric Factor	T_f (K)	ΔH_f (J/mol)
N_2	0.56	126.2	34.39	0.040	—	—
CO_2	2.90	304.2	74.74	0.225	—	—
C_1	69.13	190.6	46.61	0.008	—	—
C_2	8.16	305.4	49.49	0.098	—	—
C_3	4.01	369.8	43.02	0.152	—	—
iC_4	0.87	408.1	36.96	0.176	—	—
nC_4	1.60	425.2	38.50	0.193	—	—
iC_5	0.83	460.4	34.29	0.227	—	—
nC_5	0.74	469.6	34.19	0.251	—	—
C_6	1.52	507.4	30.08	0.296	—	—
$C_{7\,no\text{-}S}$	0.87	535.3	32.38	0.468	—	—
$C_{7\text{-}S}$	0.89	535.3	32.38	0.468	166.9	9564
$C_{8\text{-}no\text{-}S}$	0.89	555.9	30.15	0.500	—	—
$C_{8\text{-}S}$	0.67	555.9	30.15	0.500	188.8	12058
$C_{9\text{-}no\text{-}S}$	0.68	577.2	27.02	0.540	—	—
$C_{9\text{-}S}$	0.45	577.2	27.02	0.540	211.0	15238
$C_{10\text{-}11\text{-}no\text{-}S}$	0.93	601.6	23.93	0.591	—	—
$C_{10\text{-}11\text{-}S}$	0.57	601.6	23.95	0.591	233.0	19401
$C_{12\text{-}no\text{-}S}$	0.32	627.5	21.23	0.650	—	—
$C_{12\text{-}S}$	0.19	627.5	21.23	0.650	253.4	24357
$C_{13\text{-}14\text{-}no\text{-}S}$	0.53	650.3	19.42	0.707	—	—
$C_{13\text{-}14\text{-}S}$	0.29	650.3	19.42	0.707	268.3	29196
$C_{15\text{-}16\text{-}no\text{-}S}$	0.40	681.3	17.48	0.787	—	—
$C_{15\text{-}16\text{-}S}$	0.21	681.3	17.48	0.787	285.5	36420
$C_{17\text{-}18\text{-}no\text{-}S}$	0.30	708.5	16.25	0.860	—	—
$C_{17\text{-}18\text{-}S}$	0.15	708.5	16.25	0.860	298.0	43335
$C_{19\text{-}20\text{-}no\text{-}S}$	0.23	730.1	15.55	0.917	—	—
$C_{19\text{-}20\text{-}S}$	0.11	730.1	15.55	0.917	306.4	49128
$C_{21\text{-}24\text{-}no\text{-}S}$	0.31	762.8	17.44	1.005	—	—
$C_{21\text{-}24\text{-}S}$	0.13	762.8	10.23	1.005	317.1	58487
$C_{25\text{-}30\text{-}no\text{-}S}$	0.24	812.4	16.07	1.130	—	—
$C_{25\text{-}30\text{-}S}$	0.08	812.4	9.52	1.130	330.1	73756
$C_{31\text{-}80\text{-}no\text{-}S}$	0.17	916.5	14.86	1.293	—	—
$C_{31\text{-}80\text{-}S}$	0.06	916.5	8.90	1.293	344.3	98083

Note: The subscript S stands for a component that may potentially solidify (enter into wax phase), and the subscript no-S stands for a component that cannot enter into a wax phase. SRK binary interaction parameters can be seen in Table 11.14.

TABLE 11.14

Nonzero Binary Interaction Coefficients for Use with the Mixture in Table 11.13

	N_2	CO_2
CO_2	−0.032	—
C_1	0.028	0.120
C_2	0.041	0.120
C_3	0.076	0.120
iC_4	0.094	0.120
nC_4	0.070	0.120
iC_5	0.087	0.120
nC_5	0.088	0.120
C_6	0.080	0.120
C_{7+}	0.080	0.100

TABLE 11.15

Stable Oil from Flashing Oil B in Table 11.7 to Standard Conditions and for the Stable Oil Saturated with the Gas Composition (NG) in Table 11.7 at 20°C and 200 bar. The Fluid Compositions Are Characterized for the Wax Model of Rønningsen et al. (1997)

Component	Stable Oil Mol%	Stable Oil Saturated with NG at 200 bar Mol%	T_c °C	P_c bar	Acentric Factor	T_f °C	ΔH_f kJ/mol
N_2	0.001	1.699	−146.95	33.94	0.040		
CO_2	0.000	0.791	31.05	73.76	0.225		
C_1	0.452	48.751	−82.55	46.00	0.008		
C_2	0.215	3.313	32.25	48.84	0.098		
C_3	0.444	1.157	96.65	42.46	0.152		
iC_4	0.212	0.094	134.95	36.48	0.176		
nC_4	0.671	0.637	152.05	38.00	0.193		
iC_5	0.565	0.251	187.25	33.84	0.227		
nC_5	0.931	0.414	196.45	33.74	0.251		
C_6	2.352	1.045	234.25	29.69	0.296		
C_7–C_{10}	26.397	11.732	293.77	27.99	0.523		
C_{11}–C_{15}	23.878	10.612	368.55	19.60	0.689		
C_{16}–$C_{20\text{-no-S}}$	9.375	4.167	437.89	15.55	0.871		
C_{16}–$C_{20\text{-S}}$	6.109	2.715	437.89	15.55	0.871	16.35	44.54
C_{21}–$C_{25\text{-no-S}}$	6.333	2.815	493.15	17.13	1.019		
C_{21}–$C_{25\text{-S}}$	3.707	1.648	493.15	10.21	1.019	35.50	60.07
C_{26}–$C_{30\text{-no-S}}$	5.185	2.305	544.08	14.20	1.148		
C_{26}–$C_{30\text{-S}}$	1.325	0.589	544.08	9.54	1.148	48.22	75.33
C_{31}–$C_{35\text{-no-S}}$	3.932	1.747	591.74	12.77	1.250		
C_{31}–$C_{35\text{-S}}$	0.289	0.129	591.74	9.10	1.250	58.55	92.32
C_{36}–$C_{40\text{-no-S}}$	2.630	1.169	636.73	12.27	1.322		
C_{36}–$C_{40\text{-S}}$	0.107	0.048	636.73	8.81	1.322	76.61	109.57
C_{41}–$C_{50\text{-no-S}}$	2.813	1.250	697.37	11.96	1.366		
C_{41}–$C_{50\text{-S}}$	0.113	0.050	697.37	8.57	1.366	85.70	134.96
C_{51}–$C_{80\text{-no-S}}$	1.942	0.863	821.97	11.58	1.209		
C_{51}–$C_{80\text{-S}}$	0.023	0.010	821.97	8.41	1.209	93.69	162.95

Note: The subscript S stands for a component that may possibly solidify (enter into wax phase), and the subscript no-S stands for a component that cannot enter into a wax phase. SRK binary interaction parameters can be seen in Table 11.14.

composition of the oil after saturating the stable oil with gas of the composition in Table 11.7 at 200 bar at 20°C. The wax precipitation data for Oil B in Table 11.8 has been simulated using for the fluid composition in Table 11.15. The simulation results are plotted in Figure 11.13 together with the measured wax precipitation curves.

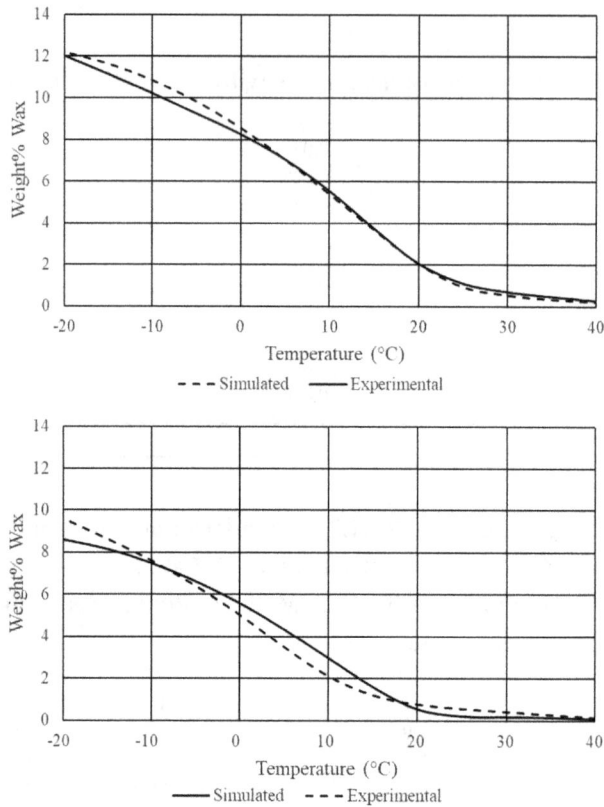

FIGURE 11.13 Measured (Rønningsen and Erstad 2023) and simulated wax precipitation curves for the stable oil from a flash of Oil B to standard conditions (upper) and for the stable oils saturated with the gas composition (NG) in Table 11.7 at 20°C at 200 bar (lower). The weigh%'s wax are of the stable oil.

11.3 WAX PT FLASH CALCULATIONS

A PT flash calculation considering a wax phase in addition to gas and oil may be accomplished by initially performing a usual vapor–liquid PT flash calculation as outlined in Chapter 6. It is afterwards to be checked whether the liquid phase from this flash will separate into a liquid and a wax phase. If the wax phase is treated as an ideal solution, as suggested by Rønningsen et al. (1997), the solid phase fugacity coefficient of component i is independent of composition and can be found from

$$\varphi_i^s = \frac{f_i^{\circ S}}{P} \tag{11.11}$$

This gives the following K-factor estimates for the solid/liquid phases (the K-factor as defined in Equation 6.4):

$$K_i^{SL} = \frac{\varphi_i^L}{\varphi_i^S}; \quad i = 1, 2, ..., N \tag{11.12}$$

where N is the number of components. The K-factors estimated in this manner are inserted into the Rachford–Rice equation (Equation 6.22), and the value of β found by solving this equation provides an initial estimate of the solid (wax) mole fraction of the total liquid + solid system.

New estimates of the wax and liquid phase compositions are found from (similar to Equations 6.5 and 6.6)

$$x_i^S = \frac{x_i K_i^{SL}}{1 + \beta \left(K_i^{SL} - 1 \right)} \tag{11.13}$$

$$x_i^L = \frac{x_i}{1 + \beta \left(K_i^{SL} - 1 \right)} \tag{11.14}$$

where x_i is the mole fraction of component i in the liquid phase from the initial vapor–liquid PT flash calculation. New estimates of the liquid and wax phase fugacity coefficients may be obtained from the applied liquid and wax phase models. A new estimate of the K-factors can be obtained from Equation 11.12, and a new wax phase mole fraction may be determined by solving Equation 6.22. This successive substitution PT flash calculation is continued until convergence. A new PT flash calculation is performed for the mixture consisting of the gas phase from the initial vapor–liquid PT flash combined with the liquid phase from the liquid–wax PT flash. The liquid phase from the new vapor–liquid flash is combined with the wax phase from the liquid–wax PT flash calculation and used as the feed in a new liquid–wax PT flash calculation. This calculation scheme is continued until convergence. More advanced flash calculation techniques may be applied, but the described calculation scheme will usually converge fairly rapidly because the wax phase will have a composition much different from the gas and liquid phases.

11.4   VISCOSITY OF OIL-WAX SUSPENSIONS

Oil containing solid wax particles may exhibit non-Newtonian flow behavior. This means that the viscosity varies with shear rate (dv_x/dy). Table 11.16 shows the composition of stable oil having a wax appearance temperature (WAT) of 38°C. Table 11.17 shows viscosity data for this oil measured at atmospheric pressure (Pedersen and Rønningsen 2000). At temperatures above the WAT, the oil behaves in a Newtonian manner (viscosity independent of shear rate). Below the WAT, the viscosity varies depending on shear rate. The viscosity data in Table 11.17 are plotted in Figure 11.14. This

TABLE 11.16
Molar Composition of Oil for Which Viscosity Data Is Shown in Table 11.17

Component	Mol%	Molecular Weight	Liquid Density (g/cm³)
C_1	0.281	—	—
C_2	0.448	—	—
C_3	1.171	—	—
iC_4	0.503	—	—
nC_4	1.198	—	—
iC_5	0.809	—	—
nC_5	0.933	—	—
C_6	1.989	—	—
C_7	5.981	89.0	0.758
C_8	9.859	100.7	0.789
C_9	7.160	115.5	0.808
C_{10+}	69.668	282.1	0.875

Source: From Pedersen, K.S. and Rønningsen, H.P., Effect of precipitated wax on viscosity—a model for predicting non-Newtonian viscosity of crude oils, *Energy Fuels* 14, 43–51, 2000.

TABLE 11.17

Viscosity Data at Atmospheric Pressure for Oil Composition Shown in Table 11.16

Temperature (°C)			Viscosity (cP)	
Above WAT				
80	—	—	2.33	—
70	—	—	2.60	—
60	—	—	3.11	—
50	—	—	3.88	—
40	—	—	4.94	—
Below WAT				
Shear Rate s⁻¹				
	30	100	300	500
34	6.44	6.2	5.9	6.7
32	13.0	9.2	6.8	7.9
30	20.3	11.7	8.0	9.0
28	27.5	14.1	9.4	10.9
26	31.6	16.4	10.9	12.9
24	36.0	19.4	13.4	16.1
22	43.0	23.9	19.2	22.8
20	53.9	32.4	29.5	33.0
18	73.2	48.2	42.7	43.3
16	104	70.5	57.7	55.3
14	152	100	75.0	69.9
12	212	134	89.8	78.6
10	283	172	105	89.0
8	369	210	126	104
6	470	267	150	114
4	575	326	177	128
2	725	395	200	148

Note: The WAT of the oil is 38°C.

Source: From Pedersen, K.S. and Rønningsen, H.P., Effect of precipitated wax on viscosity—a model for predicting non-Newtonian viscosity of crude oils, *Energy Fuels* 14, 43–51, 2000.

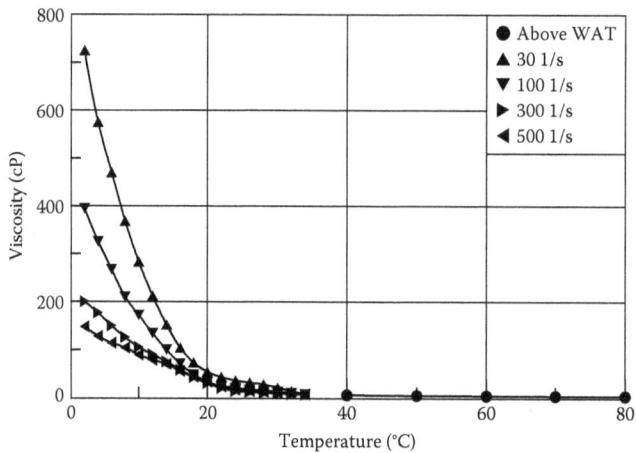

FIGURE 11.14 Viscosities of stable oil in Table 11.16 above and below WAT (Pedersen and Rønningsen 2000). Below the WAT, the viscosity is influenced by solid wax and behaves in a non-Newtonian manner. The actual data points can be seen in Table 11.17.

TABLE 11.18
Constants in Expression for Non-Newtonian
Viscosity (Equation 11.15)

D	37.82
E	83.96
F	8.559×10^6

plot illustrates the very pronounced influence that suspended solid wax may have on the apparent oil viscosity. It is also seen from Figure 11.14 that the viscosity below the WAT is much dependent on shear rate. The viscosity decreases with increasing shear rate. This suggests a Bingham plastic flow behavior, which is dealt with in Chapter 10. The impact of solid wax on viscosity is particularly pronounced for low shear rates. The restart of a pipeline with a waxy crude oil after shutdown can, therefore, be very difficult indeed. In a paper on the rheological behavior of gelled, waxy crude oils, Rønningsen (1992) has presented experimental data that may possibly help quantify the problem. A model for simulating the restart of pipelines containing non-Newtonian oil with solid wax particles has been presented by Davidson et al. (2004).

The apparent oil viscosity below the WAT may be calculated from (Pedersen and Rønningsen 2000)

$$\eta = \eta_{liq}\left[\exp(D\phi_{wax}) + \frac{E\phi_{wax}}{\sqrt{\frac{dv_x}{dy}}} + \frac{F\phi_{wax}^4}{\frac{dv_x}{dy}}\right] \tag{11.15}$$

where η_{liq} is the viscosity of the oil (not considering the influence of solid wax) and ϕ_{wax} the volume fraction of precipitated wax in the oil–wax suspension. The parameters D, E, and F can be seen in Table 11.18.

11.5 WAX INHIBITORS

Wax inhibitors may be added to waxy crude oils to facilitate transport in undersea pipelines. The most commonly used inhibitors lower the apparent viscosity and the pour point. The pour point is the lowest temperature at which oil will flow freely under its own weight under specific test conditions (ASTM D-92). During normal operation, the frictional pressure drop in the pipeline increases with viscosity. During shutdown, the temperature may drop to below the pour point, and it may be difficult to restart the pipeline. Rønningsen et al. (1991) distinguish between the maximum and minimum pour points. The latter is the pour point of a sample that has been thermally preconditioned at a high temperature (at least 80°C), and the former is the pour point of an unconditioned sample.

Basically, three groups of wax inhibitor chemicals are used:

Wax crystal modifiers
Detergents
Dispersants

The last two groups are surface-active agents, for example, polyesters and amine ethoxylates. These act by keeping the crystals dispersed as separate particles, thereby reducing their tendency to interact and adhere to solid surfaces.

Crystal modifiers are substances capable of building into wax crystals and altering the growth and surface characteristics of the crystal. The crystal modifiers will lower the pour point as well as the viscosity. The name *pour point depressant* is also used for this class of chemicals. Figure 11.15 schematically shows the wax inhibitor (crystal modifier) mechanism. The acetate group (CH_3COO^-) contained in the inhibitor is very unlike the paraffinic branches and will disturb further structuring of the paraffinic molecules.

Pedersen and Rønningsen (2003) have presented data for 12 different wax crystal modifiers. The data comprise WATs, pour point data, and viscosity data for a waxy North Sea crude oil (wax content of 15 weight percentage) treated with each of the 12 chemicals in three different concentrations (100, 500, and 1000 ppm weight). Table 11.19 shows the types of wax inhibitor chemicals tested. The term *wax inhibitor* is not to be understood as chemicals preventing wax precipitation from oil. On average, the WAT of the oil was only decreased by 3.3°C by addition of 1000-ppm wax inhibitor, but all 12 inhibitors showed some lowering of the viscosity below the WAT. In general, increasing the amount of inhibitor to above 500 ppm had little effect. The most efficient inhibitor was found to be No. 12 (mixture of EVA and copolymers of maleic acid anhydride and α-olefin). Figure 11.16

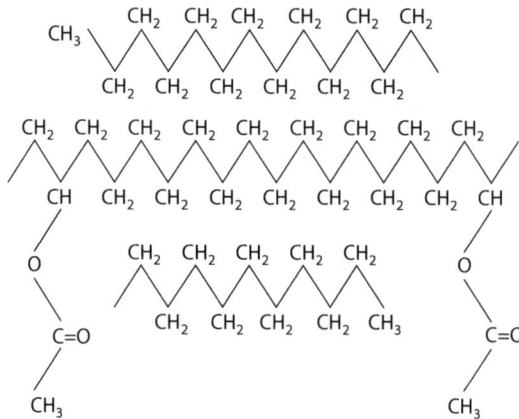

FIGURE 11.15   Wax inhibitor (crystal modifier) mechanism.

TABLE 11.19
Wax Inhibitors Studied by Pedersen and Rønningsen (2003)

Inhibitor No.	Type of Chemical
1	Polyalkyl methacrylate
2	Polyalkyl methacrylate
3	Polyalkyl methacrylate
4	Copolymer of ethylene and vinyl acetate (EVA)
5	Unknown
6	Copolymer of ethylene and vinyl acetate
7	Polymeric fatty ester
8	Polymeric fatty ester
9	Copolymer of polyalkyl acrylate and vinylpyridine
10	Methacrylic acid ester
11	EVA
12	Mixture of EVA and copolymers of maleic acid anhydride and α-olefin

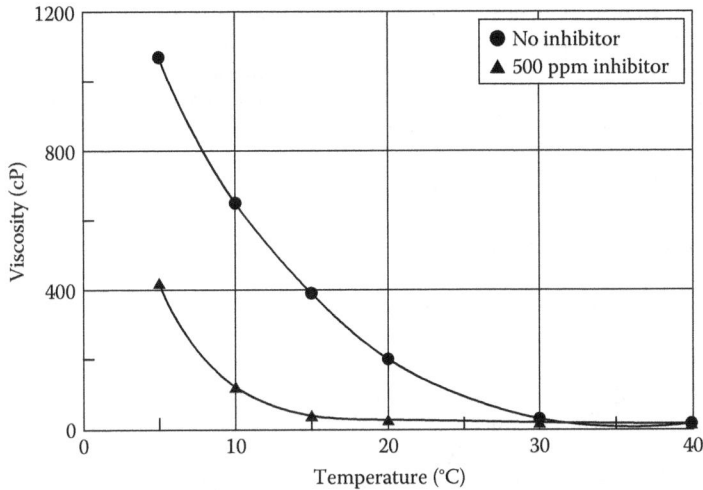

FIGURE 11.16   Viscosity of North Sea oil mixture untreated and with 500 ppm inhibitor (No. 12 in Table 11.1) added (Pedersen and Rønningsen 2003). The WAT of the untreated oil is 42°C.

shows a plot of viscosity versus temperature for the untreated oil and the oil treated with 500 ppm of inhibitor No. 12. On average, the pour point of the oil was lowered by 31°C as the result of adding 500 ppm of inhibitor (average of all 12 inhibitors).

Pedersen and Rønningsen (2003) have shown that the wax inhibitor effect can be modeled as a depression of the melting temperature of wax components within a given range of molecular weights. The inhibitor effect of inhibitor No. 12 shown in Figure 11.16 can be modeled by assuming that the wax inhibitor depresses the melting temperature of the C_{21}–C_{40} wax components by 15°C.

REFERENCES

Bishop, A.N., Philip, R.P., Allen, J., and Ruble, T.E., High molecular weight hydrocarbons and the precipitation of petroleum-derived waxes, *Organic Geochemistry: Development and Applications to Energy, Climate, Environment and Human History*, Grimalt, J.O. and Dorronsoro, C., Eds., AIGOA, Donoslia-San Sebastian, Spain, 1995.

Brockman, R., *Cloud Points of Hydrocarbon System, Influence of Pressure and Methane Addition*, Ph.D. thesis, University of Trondheim, Norway, 1992.

Brown, T.S., Niesen, V.G., and Erickson, D.D., The effects of light ends and high pressure on paraffin formation, SPE 28505, presented at the *SPE ATCE*, New Orleans, September 25–28, 1994.

Coutinho, J.A.P., Mirante, F., and Pauly, J., A new predictive UNIQUAC for modeling of wax formation in hydrocarbon fluids, *Fluid Phase Equilib*. 247, 8–17, 2006.

Daridon, J.-L., Pauly, J., Coutinho, J.A.P., and Montel, F., Solid–wax–vapor phase boundary of a North Sea waxy crude: Measurement and modeling, *Energy Fuels*. 15, 730–735, 2001.

Davidson, M.R., Nguyen, Q.D., Chang, C., and Rønningsen, H.P., A model for restart of a pipeline with compressible gelled waxy crude oil, *J. Non-Newtonian Fluid Mech*. 123, 269–280, 2004.

Erickson, D.D., Niesen, V.G., and Brown, T.S., Thermodynamic measurement and prediction of paraffin precipitated in crude oil, SPE 26604, presented at the *SPE ATCE*, Houston, TX, October 3–6, 1993.

Faust, H.R., The thermal analysis of waxes and petrolatums, *Thermochim. Acta* 26, 383–398, 1978.

Hammami, A. and Raines, M.A., Paraffin deposition from crude oils: Comparison of laboratory results to field data, SPE 38776, presented at the *SPE ATCE*, San Antonio, TX, October 5–8, 1997.

Hansen, A.B., Larsen, E., Pedersen, W.B., Nielsen, A.B., and Rønningsen, H.P., Wax precipitation from North Sea crude oils. 3. Precipitation and dissolution of wax studied by differential scanning calorimetry, *Energy Fuels*. 5, 914–923, 1991.

Létoffé, J.M., Claudy, P., Kok, M.V., Garcin, M., and Volle, J.L., Crude oils: Characterization of waxes precipitated on cooling by D.S.C. and thermomicroscopy, *Fuel.* 74, 810–817, 1995.

Loebel, R., *Handbook of Chemistry and Physics*, CRC Press, Boca Raton, FL, 1981–1982.

Monger-McClure, T.G., Tackett, J.E., and Merrill, L.S., DeepStar comparisons of cloud point measurements and paraffin prediction methods, SPE 38774, presented at the *SPE ATCE*, San Antonio, TX, October 5–8, 1997.

Pan, H., Firoozabadi, A., and Fotland, P., Pressure and composition effect on wax precipitation: Experimental data and model results, *SPE Prod. Facil.* 12, 250–258, 1997.

Pedersen, K.S., Prediction of cloud point temperatures and amount of wax precipitation, *SPE Prod. Facil.* 46–49, February 1995.

Pedersen, K.S. and Rønningsen, H.P., Effect of precipitated wax on viscosity—a model for predicting non-Newtonian viscosity of crude oils, *Energy Fuels.* 14, 43–51, 2000.

Pedersen, K.S. and Rønningsen, H.P., Influence of wax inhibitors on wax appearance temperature, pour point, and viscosity of waxy crude oils, *Energy Fuels.* 17, 321–328, 2003.

Pedersen, K.S., Skovborg, P., and Rønningsen, H.P., Wax precipitation from North Sea crude oils. 4. Thermodynamic modeling, *Energy Fuels.* 5, 924–932, 1991.

Pedersen, W.B., Hansen, A.B., Larsen, E., Nielsen, A.B., and Rønningsen, H.P., Wax precipitation from North Sea crude oils. 2. Solid-phase content as function of temperature determined by pulsed NMR, *Energy Fuels.* 5, 908–913, 1991.

Rønningsen, H.P., Rheological behavior of gelled, waxy North Sea crude oils, *J. Pet. Sci. Eng.* 7, 177–213, 1992.

Rønningsen, H.P., Bjørndal, B., Hansen, A.B., and Pedersen, W.B., Wax precipitation from North Sea crude oils. 1. Crystallization and dissolution temperatures, and Newtonian and non-Newtonian flow properties, *Energy Fuels.* 5, 895–908, 1991.

Rønningsen, H.P. and Erstad, K., unpublished data made available by Equinor, 2023.

Rønningsen, H.P., Sømme, B.F., and Pedersen, K.S., An improved thermodynamic model for wax precipitation: Experimental foundation and application, paper presented at the *8th International Conference on Multiphase 97*, Cannes, France, June 18–20, 1997.

Ruffier-Méray, V., Volle, J.L., Scanz, C., Le Maréchal, P., and Béhar, E., Influence of light ends on the onset crystallization temperature of waxy crudes within the frame of multiphase transport, SPE 26549, presented at the *SPE ATCE*, Houston, TX, October 3–6, 1993.

Templin, R.D., Coefficient of volume expansion for petroleum waxes and pure n-paraffins, *Ind. Eng. Chem.* 48, 154–161, 1956.

Won, K.W., Continuous thermodynamics for solid–liquid equilibria: Wax formation from heavy hydrocarbon mixtures, *Fluid Phase Equilib.* 30, 265–279, 1986.

12 Asphaltenes

Asphaltenes is a component class that may precipitate from petroleum reservoir fluids as a highly viscous and sticky material that may cause deposition problems in production wells and pipelines. Asphaltenes are defined as the constituents of an oil mixture that, at room temperature, are practically insoluble in *n*-pentane and *n*-heptane but soluble in benzene and toluene. The definition of asphaltene can be seen from Figure 12.1. The solubility of asphaltenes is also low in other paraffins. Because a major part of reservoir fluids consists of paraffins, asphaltene precipitation problems are quite frequent. Unlike wax precipitation as dealt with in Chapter 11, asphaltene precipitation is not limited to low temperatures. Precipitation may occur in the reservoir, in the production well, during pipeline transportation, and in process plants. Gas is often injected into an oil reservoir to obtain an enhanced recovery. Because natural gas essentially consists of paraffins, gas injection will tend to worsen asphaltene precipitation problems.

Being defined as a solubility class, the term *asphaltenes* may cover a range of different components, and that may be the reason the literature has different suggestions about the nature of the asphaltenes precipitating from oil mixtures. Some have in the past considered the asphaltene phase to be solid (Kawanaka et al. 1991; Chung 1992; MacMillan et al. 1995), but today the general perception is that an asphaltene phase precipitating during oil production is a highly viscous liquid (Burke et al. 1990; Godbole et al. 1995; Ting et al. 2003).

Also, there are different opinions about the solubility properties of already precipitated asphaltenes. If asphaltenes are precipitated from stabilized oils by the addition of large quantities of either *n*-pentane or *n*-heptane, it will give asphaltenes in almost pure form, and the cohesion between the individual asphaltene molecules may be so high that it becomes difficult to dissolve the asphaltenes again. As opposed to these types of experiments, experimental studies of oils precipitating asphaltenes at reservoir conditions (e.g., Angulo et al. 1995; Jamaluddin et al. 2000, 2002; Hustad et al. 2014) show that asphaltenes at those conditions may precipitate and dissolve again as an ordinary equilibrium phase.

For a constant composition, the solubility of asphaltenes decreases with decreasing pressure. For an oil of a fixed composition, the highest asphaltene precipitation is right at the bubble point. This is schematically illustrated in Figure 12.2. If the pressure is lowered to below the bubble point pressure, some gas will evaporate, and the gas concentration in the liquid phase will decrease. Gas components (N_2, CO_2, C_1, C_2, etc.) are bad solvents for the asphaltenes. When the concentration of gas components in the oil phase decreases, it makes the asphaltene components more soluble in the remaining liquid. The asphaltene phase will slowly dissolve and eventually disappear. The pressure at which the last asphaltenes go into solution is called the lower asphaltene onset pressure (lower AOP). Increasing the pressure from the bubble point will also make the asphaltene phase dissolve. Though paraffins are generally poor solvents for the asphaltenes, the solubility of asphaltenes in paraffins increases with pressure, and at a sufficiently high pressure, the upper asphaltene onset pressure (upper AOP), the asphaltene phase will disappear.

Figure 12.3 shows a typical phase diagram for an oil mixture splitting out an asphaltene phase in a pressure band around the bubble point. Not all oil mixtures containing asphaltenes will show asphaltene precipitation when pressure is reduced. However, for such oils, asphaltene precipitation may occur if gas is added to the oil. Figure 12.4 shows an example in which asphaltene formation is not seen until the gas content reaches ~8 mole%. The pressure span with asphaltene precipitation widens with an increasing amount of gas added.

Asphaltene components are heavy aromatics like the one sketched in Figure 12.5 with a molecular weight of 500–1000 g/mol (e.g., Koots and Speight 1975; Mullins et al. 2013).

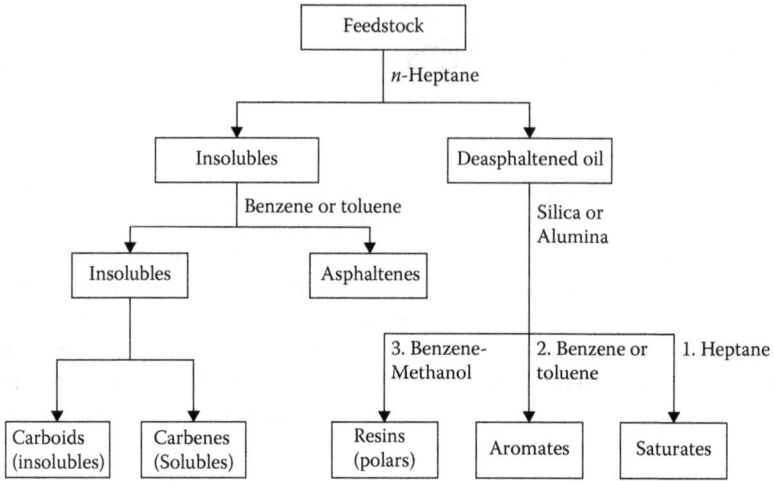

FIGURE 12.1   Asphaltenes and resins defined as solubility classes.

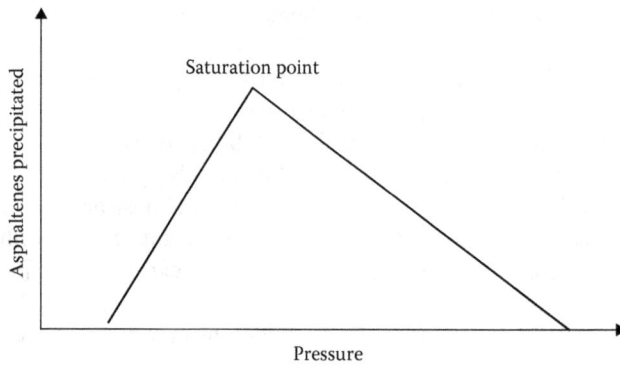

FIGURE 12.2   Asphaltene precipitation versus pressure. The highest amount of asphaltene precipitate is seen right at the saturation point.

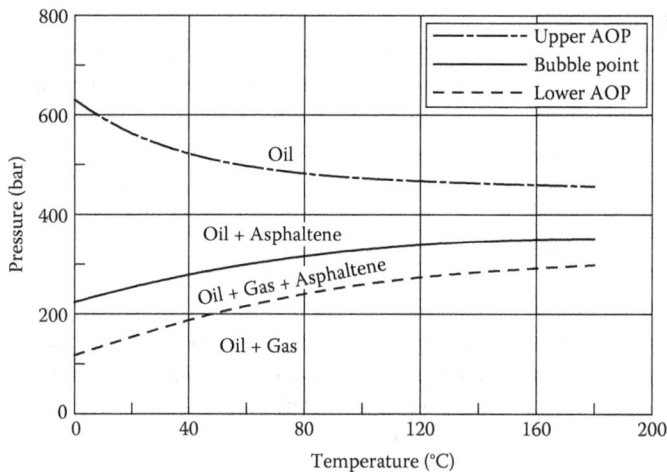

FIGURE 12.3   Phase diagram of an asphaltene-forming oil.

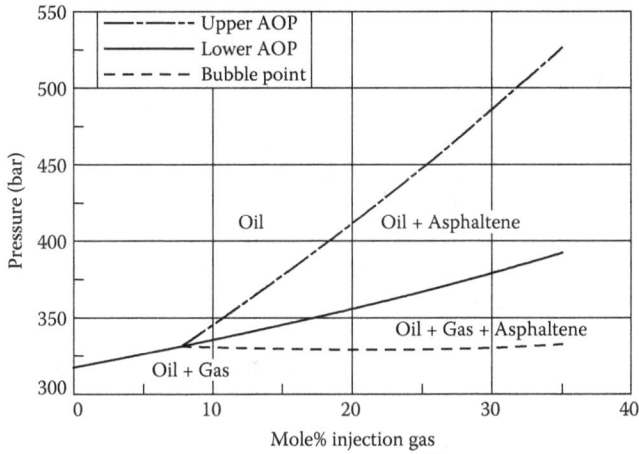

FIGURE 12.4 Asphaltene onset pressures and bubble point pressure for an oil as a function of mole percentage gas added.

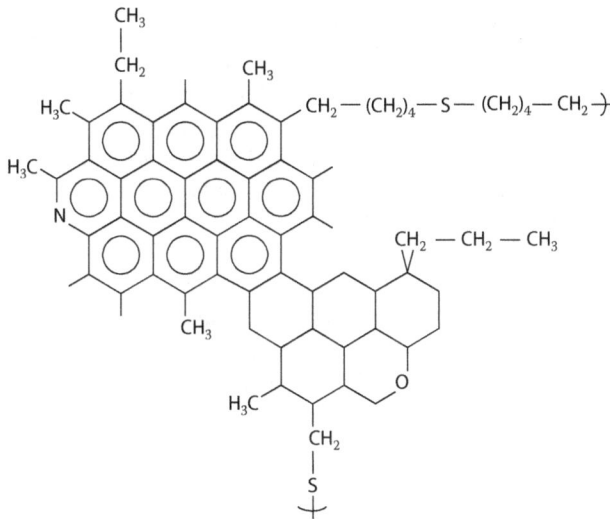

FIGURE 12.5 Typical structure of asphaltene molecule.

12.1 EXPERIMENTAL TECHNIQUES FOR STUDYING ASPHALTENE PRECIPITATION

Experimental studies are the key to a better understanding of asphaltenes as a component class and of the mechanisms behind asphaltene precipitation. Much literature on experimental asphaltene studies deals with asphaltene precipitation from stable oil mixtures, but techniques have been developed permitting asphaltene precipitation measurements on live oil mixtures. Tavakkoli et al. (2014) have analyzed the various experimental techniques.

12.1.1 QUANTIFICATION OF AMOUNT OF ASPHALTENES

The nC_5 or nC_7 precipitation technique (e.g., Burke et al. 1990) is used to determine the asphaltene content in a stabilized oil mixture. The *n*-paraffin is injected in large quantities (e.g., 40/1 n-C_5/

oil on a volume basis), which forces the asphaltenes to precipitate. The precipitate is filtered and washed to purify the asphaltenes. The information recorded is the amount of stabilized oil, amount of injected pentane, and amount of isolated asphaltene. The amount of asphaltene is sensitive to solvent (nC_5 or nC_7), and because the precipitation is further carried out at standard conditions, there is no guarantee that the precipitate is representative of the amount of asphaltenes that potentially may precipitate from the live oil at reservoir conditions.

An experimental determination of the asphaltene content is often carried out as part of a SARA analysis. S stands for saturates and covers paraffins and naphthenes (see Section 1.1). The first A stands for aromatics and covers aromatics with molecular weight below ~300. R stands for resins. It is a solubility class defined in Figure 12.1, which is dominated by aromatics with molecular weights in the range 300–500. Most of the heavier aromatics will be classified as asphaltenes, but dependent on molecular structure (solubility properties), aromatics with molecular weights in the range 500–700 could also be classified as resins.

12.1.2 DETECTION OF ASPHALTENE ONSET POINTS

Jamaluddin et al. (2000) have used four different techniques to detect asphaltene onset pressures for reservoir oil.

12.1.2.1 Gravimetric Technique

The gravimetric method is conducted in a conventional PVT cell. The temperature is kept fixed, and the pressure is reduced stepwise (intervals of ~50 bar). Precipitated asphaltene is allowed to segregate at the bottom of the cell. Samples are taken from the upper part of the cell where there is no asphaltene precipitate. The sampled oil is flashed to standard conditions, and the amount of asphaltenes contained in the oil is determined by n-paraffin precipitation. The amount of asphaltenes contained in the oil versus pressure is schematically shown in Figure 12.6. Above the upper AOP, the amount of asphaltenes contained in the oil phase is constant. At the upper AOP, the asphaltene content in the oil starts to decrease because some of the asphaltenes have precipitated and segregated at the bottom of the cell. At the bubble point, the amount of asphaltenes out of solution is at a maximum. This is in a gravimetric experiment seen as a minimum in the amount of asphaltenes in solution in the oil (corresponding to the maximum in Figure 12.2). Below the bubble point and until the pressure reaches the

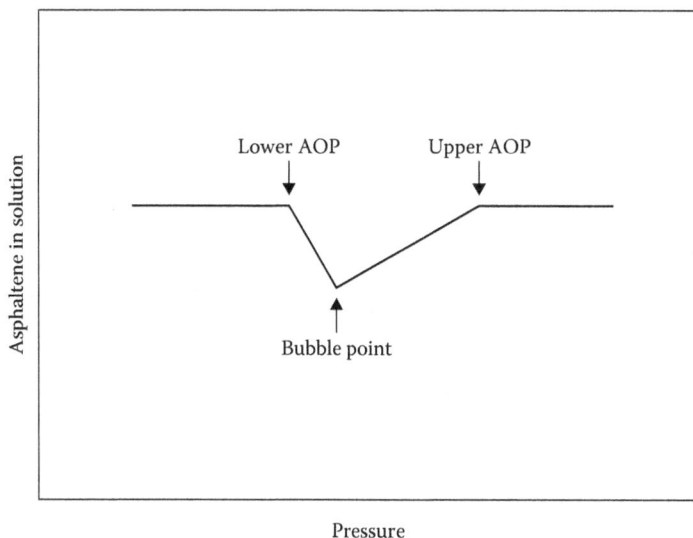

FIGURE 12.6 Amount of asphaltenes in solution in oil versus pressure in gravimetric experiment.

lower asphaltene AOP, the amount of asphaltenes in the oil will increase. At the lower AOP, it becomes constant and equal to the level above the upper AOP. All asphaltenes are now back in solution. The difference between the amount of asphaltenes in solution outside the asphaltene precipitation region and at the bubble point gives a measure of the amount of asphaltene that will precipitate from the reservoir fluid as a result of a depressurization. The gravimetric technique is therefore applicable for measuring asphaltene onset pressures as well as for quantifying the amount of asphaltene precipitate. However, the technique has the drawback that it is very time consuming (~24 h per data point).

12.1.2.2   Acoustic Resonance Technique

This technique allows the formation of a new phase to be determined acoustically. The new phase can be an asphaltene phase (Sivaraman et al. 1997), or it can be a gas phase (at the bubble point). The acoustic technique is unsuited for determining the lower AOP.

12.1.2.3   Light-Scattering Technique

The light-scattering technique is illustrated in Figure 12.7. The transmittance of near-infrared light through an oil phase is measured. In a homogeneous fluid with no suspended asphaltenes (P > upper AOP), light travels through the fluid with minimum scattering. Below the upper AOP, asphaltene particles appear and cause partial light scattering. A gradual reduction in light transmittance is seen until the bubble point at which total scattering takes place. Approaching the lower AOP, the light transmittance starts increasing again, thanks to redissolution of dispersed asphaltene particles.

12.1.2.4   Filtration and Other Experimental Techniques

In a filtration experiment, oil is circulated through a filter at a constant rate. If an asphaltene phase precipitates, as a result of either lowering the pressure or injecting gas, the upper AOP can be detected as an increased pressure drop over the filter. The technique is similar to that presented for wax in Chapter 11. The material retained on the filter is analyzed for saturates (alkanes and naphthenes), aromatics, resins, and asphaltenes (SARA) content.

Another method for upper AOP determination is the electrical conductivity measurement technique (Fotland et al. 1993). The electrical conductivity of an oil sample is measured continuously as the conditions are changed. For a constant composition measurement, the initial pressure is kept above the upper AOP and the pressure is then lowered until a marked shift is seen in electrical conductivity indicating that the upper AOP or the bubble point is reached. Below the bubble point, no further recording is possible.

A relatively simple technique for the determination of the asphaltene onset is through viscometric determination (Escobedo and Mansoori 1995). This detection method is based on experimental observations of an increase in the viscosity of a crude oil in which asphaltene precipitation occurs.

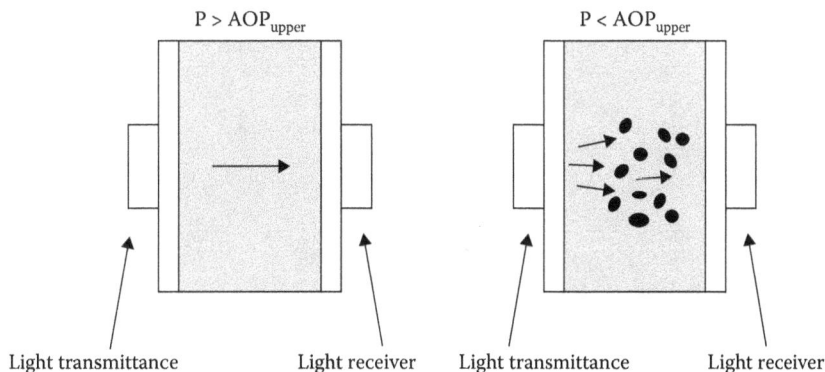

FIGURE 12.7   Principle of light-scattering technique for detection of asphaltene precipitation.

12.1.3 EXPERIMENTAL DATA FOR ASPHALTENE ONSET PRESSURES

Table 12.1 shows five oil compositions for which asphaltene onset pressures have been determined by depressurization experiments. The onset precipitation pressures can be seen in Tables 12.2 and 12.3.

Jamaluddin et al. (2000, 2002) have for two oils measured upper AOPs at four different temperatures (see Table 12.2). These onset pressures may give an idea about the development in onset pressure with temperature. One of the oils (Jamaluddin et al. 2002) has a minimum upper onset pressure at a temperature of around 120°C. The data in Table 12.2 also includes two lower AOP data points.

Jamaluddin et al. (2002) have also measured onset pressure data for an oil mixed with nitrogen in various ratios. This data is shown in Table 12.3. Gonzalez et al. (2012) have studied the effect of N_2, CO_2, and C_1 addition to a Gulf of Mexico reservoir oil and have found that the asphaltene onset pressure increases significantly when gas is added.

Table 12.4 shows two oil mixtures (Rydahl et al. 1997), which do not precipitate an asphaltene phase at any pressure unless gas in injected. It was investigated how much natural gas could be injected into the two oils at reservoir conditions before asphaltenes start to precipitate. The results are shown in Table 12.4, which also shows the composition of the natural gas injected and the reservoir conditions.

Hammami et al. (1995) have measured the asphaltene onset concentration for a typical live reservoir oil titrated with nC_6, nC_5, nC_4, propane, and ethane. The molar concentration of titrant at the asphaltene onset point was observed to decrease approximately in a linear fashion with decreasing molecular weight of the paraffinic solvents. This suggests that one hydrocarbon segment has approximately the same influence on asphaltene onset, independent of the size of the molecule it belongs to.

Hustad et al. (2014) have reported asphaltene onset and saturation point data for the oil mixture in Table 10.14 when adding N_2 gas. The data is shown in Table 12.5 and plotted in Figure 12.8.

TABLE 12.1

Oil Compositions (Mole Percentage) for Which Asphaltene Onset Data Is Shown in Tables 12.2 and 12.3

Component	Rydahl et al. (1997) Oil 2	Jamaluddin et al. (2000) Oil A	Jamaluddin et al. (2000) Oil B	Jamaluddin et al. (2002)	Kokal et al. (2003)
N_2	0.97	0.48	0.80	0.49	0.18
CO_2	0.20	0.92	0.05	11.37	5.21
H_2S	0.00	0.00	0.00	3.22	1.35
C_1	27.55	43.43	51.02	27.36	25.67
C_2	7.43	11.02	8.09	9.41	9.19
C_3	9.02	6.55	6.02	6.70	—
iC_4	1.29	0.79	1.14	0.81	7.29
nC_4	4.85	3.70	2.83	3.17	—
iC_5	1.67	1.28	1.58	1.22	4.99
nC_5	2.49	2.25	1.63	1.98	3.81
C_6	3.16	2.70	2.67	2.49	3.59
C_{7+}	41.39	26.88	24.17	31.79	38.72
C_{7+} M	217.7	228.1	368.9	248.3	204.9[a]
C_{7+} density (g/cm³) at 1.01 bar and 15°C	0.854	0.865	0.875	0.877	0.873

[a] Estimated from reported bubble point pressure at 99°C.

TABLE 12.2
Asphaltene Onset Pressures for the Oil Compositions in Table 12.1

Reference	Temperature (°C)	Upper AOP (bar)	Bubble Point (bar)	Lower AOP (bar)
Rydahl et al. (1997) Oil 2	90	200	—	—
Jamaluddin et al. (2000) Oil A	99	472.6	222.1	—
Jamaluddin et al. (2000) Oil A	104	454.2	226.4	—
Jamaluddin et al. (2000) Oil A	110	442.6	225.9	—
Jamaluddin et al. (2000) Oil A	116	429.2	226.8	135.1
Jamaluddin et al. (2000) Oil B	88	365.4	293.7	264.3
Jamaluddin et al. (2002)	83.2	372.3	172.4	—
Jamaluddin et al. (2002)	104.2	279.2	186.2	—
Jamaluddin et al. (2002)	120.0	251.7	199.9	—
Jamaluddin et al. (2002)	141.1	262.0	211.0	—
Kokal et al. (2003)	99	172.4	131.0	—

TABLE 12.3
Asphaltene Onset Pressures at 147°C in Nitrogen Injection Experiments for the Oil Composition of Jamaluddin et al. (2002) in Table 12.1

Mole Percentage N_2 Added Per Mole Initial Oil	Upper AOP (bar)	Bubble Point (bar)
0	267.0	213.9
5	386.4	281.0
10	541.0	337.2
20	822.0	491.8

TABLE 12.4
Molar Compositions of Two North Sea Oil Mixtures, an Injection Gas, and (Bottom Row) Gas Amounts Needed to Initiate Asphaltene Precipitation at Reservoir Conditions

Component	Rydahl et al. (1997) Oil 1	Rydahl et al. (1997) Oil 3	Injection Gas
N_2	1.50	1.63	1.88
CO_2	0.22	0.17	0.41
C_1	23.11	31.40	70.69
C_2	6.92	7.89	13.33
C_3	8.63	8.62	9.05
iC_4	1.30	1.25	1.08
nC_4	5.13	4.74	2.33
iC_5	1.78	1.62	0.45
nC_5	2.71	2.50	0.58
C_6	3.64	3.18	0.13
C_7	—	—	0.07
C_{7+}	45.08	36.98	—
C_{7+} M	218.2	212.0	—
C_{7+} density (g/cm³) at 1.01 bar and 15°C	0.8547	0.8519	—
Reservoir temperature (°C)	90	90	—
Reservoir pressure (bar)	350	320	—
Mol gas/mol oil at AOP	0.32	0.19	—

TABLE 12.5

Experimental and Simulated Asphaltene Onset Pressures at 94°C for the Reservoir Fluid Composition in Table 10.14 Swelled with Nitrogen

Mole% N_2 per Mole Reservoir Fluid	Asphaltene Onset Pressure bar	Saturation Pressure bar
0	—	176
9	414	300
18	758	362
27	940	680

Source: Hustad, O.S., et al., High pressure data and modeling results for phase behavior onsets of GoM oil mixed with nitrogen, *SPE Reservoir Eval. Eng.* 3, 384–395, 2014.

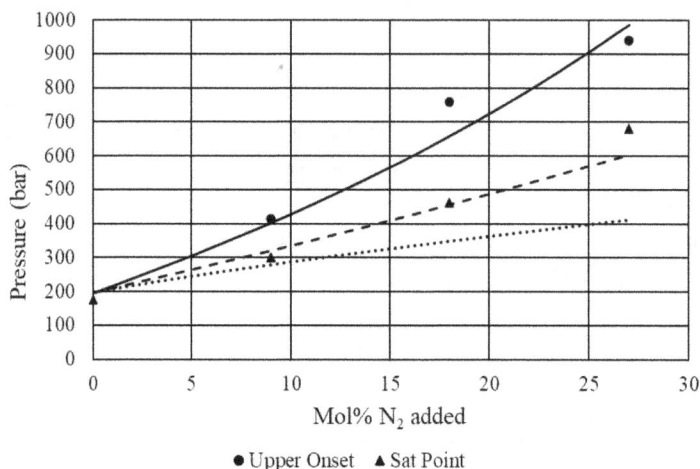

● Upper Onset ▲ Sat Point

FIGURE 12.8 Experimental and simulated asphaltene onset pressures and oil saturation pressures for the reservoir fluid in Table 10.14 with N_2 added at a temperature of 94°C. The experimental data can be seen in Table 12.5. The simulation results are for the SRK equation of state with the fluid characterized as shown in Table 12.11.

12.2 ASPHALTENE SCREENING METHODS

Since experimental asphaltene measurements are quite expensive and of limited use for fields without asphaltene problems, screening methods are often used to get a first impression of whether production from a certain field can be expected to give rise to asphaltene problems.

De Boer et al. (1995) have made a generalized plot, which is useful to give a first idea about the asphaltene precipitation potential of a petroleum reservoir fluid. This plot is shown schematically in Figure 12.9. For a given reservoir oil, the asphaltene precipitation potential increases with increasing pressure above the bubble point pressure. The higher the pressure, the more asphaltenes can be kept in solution in the oil. The second parameter in the de Boer plot is the density of the reservoir fluid (x-axis in Figure 12.9). As mentioned in Chapter 1, the density of hydrocarbon constituents of the same molecular weight increases in the order paraffins → naphthenes → aromatics. Paraffins are bad solvents for asphaltenes, whereas asphaltenes are soluble in aromatics (see Figure 12.1). For this reason, asphaltene precipitation is more likely to take place from a reservoir fluid of low density (dominated by paraffins) than from a reservoir fluid of high density (dominated by aromatics). This is exactly what is reflected in the de Boer plot. It is noteworthy that the asphaltene content in the reservoir fluid is not a parameter in the de Boer plot. A reservoir fluid with a high asphaltene content is also likely to have a high content of lower molecular weight aromatics. Aromatics are good solvents for asphaltenes, and no precipitation will take place even when the asphaltene content is quite high. This is consistent with the de Boer plot saying that asphaltene precipitation is unlikely to take place from high-density (highly aromatic) reservoir fluids.

The Asphaltene Stability Index method is sketched in Figure 12.10. The Stable region with black dots represents fluids unlikely to cause asphaltene problems, either because the content of saturates (paraffins and naphthenes) is low or because the content of asphaltenes is low as compared to the content of resins. The Unstable region with circles represents fluids that are likely to cause asphaltene problems.

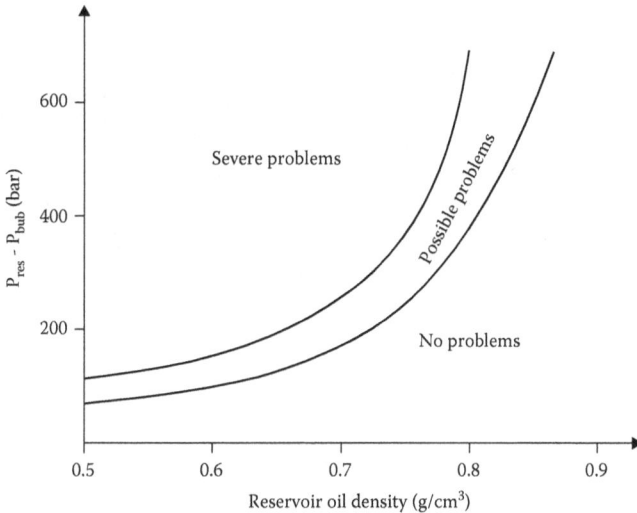

FIGURE 12.9 De Boer plot showing asphaltene precipitation potential as a function of undersaturation (reservoir pressure–saturation pressure) and reservoir fluid density.

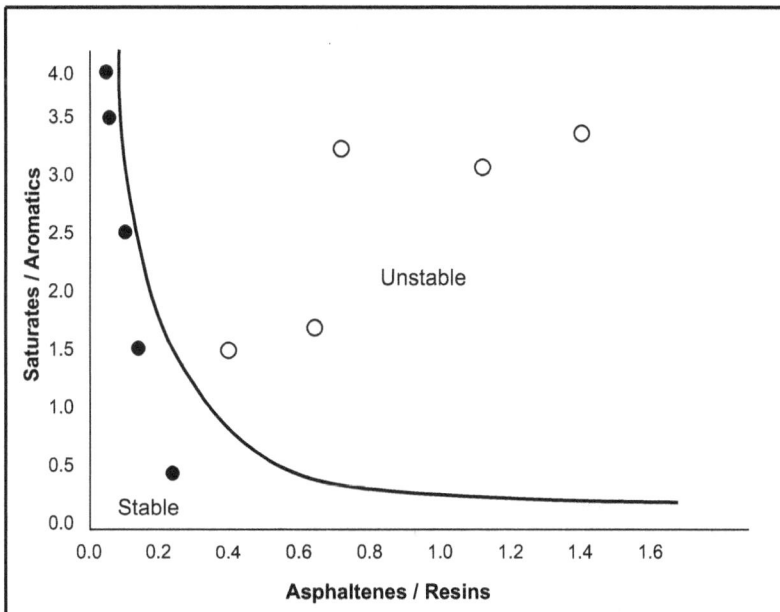

FIGURE 12.10 Asphaltene Stability Index screening method.

Boesen et al. (2018) have evaluated four commonly used asphaltene screening methods on 36 reservoir fluids with published asphaltene data. None of the screening methods could identify all problematic reservoir fluids, but the SARA-based Asphaltene Stability Index method sketched in Figure 12.10 was found to be the most reliable.

12.3   CONSTRUCTION OF ASPHALTENE PHASE DIAGRAMS

Agger and Sørensen (2018) have presented an algorithm to construct complete asphaltene PT and Px phase diagrams when an equation of state model is used. Figure 12.11 shows an example of a PT diagram generated using this algorithm. The calculation starts by searching for an "asphaltene dew point" at a high temperature and a low pressure (~1 bar), that is, searching for a PT point where a gas phase is in equilibrium with an incipient asphaltene phase. This is point A in Figure 12.11. The asphaltene saturation point line is followed to point B, where the gas phase has changed character to a liquid phase. Here a new incipient gas phase appears. Point B is therefore a bubble point, but a special one since the liquid phase is not only in equilibrium with an incipient gas phase but also with an incipient asphaltene phase. The bubble point line is followed to point C where the gas and liquid compositions become identical, which is characteristic of a critical point but a special one since the gas and liquid compositions are in equilibrium with an asphaltene phase. From point C to D, a gas phase is in equilibrium with an asphaltene phase and an incipient liquid phase. At point D, the liquid phase and the asphaltene phase become identical, forming a critical composition in equilibrium with a gas phase. From point D to B, a gas and a liquid phase are in equilibrium with an incipient asphaltene phase (lower asphaltene onset pressure line). The section from point B to E is a normal bubble point line with a liquid in equilibrium with an incipient gas phase. On the line from point E to F, a liquid phase is in equilibrium with an incipient asphaltene phase (upper asphaltene onset pressure line). From point E to G, a liquid and asphaltene phase are in equilibrium with an incipient gas phase (bubble point line at which the liquid phase coexists with the asphaltene phase). Finally, on the line from point E to H, gas and liquid are in equilibrium with an incipient asphaltene phase (lower asphaltene onset pressure line).

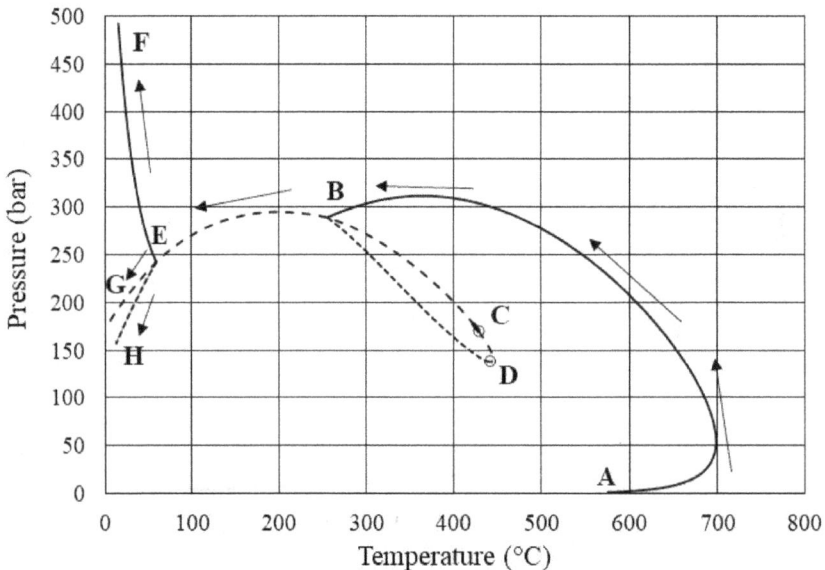

FIGURE 12.11   Asphaltene phase diagram showing the order in which the boundary lines are generated. The meaning of the phase boundary lines is explained in the text.

12.4   ASPHALTENE MODELING

12.4.1   FUNDAMENTALS

Whether asphaltene precipitation takes place from a constant composition fluid at a particular temperature is determined by the pressure. If a pressure depletion experiment is carried out at a constant temperature starting at a high pressure with all asphaltenes in solution in the oil, the upper AOP is the pressure at which it becomes thermodynamically favorable for the oil to split into two phases. As explained in Section 6.3.1, this means that the system can lower its Gibbs free energy by splitting out a second phase (here an asphaltene phase) rather than remaining single-phase oil. Recalling the fundamental thermodynamic relation (Equation A.24 in Appendix A)

$$dG = \Delta V \, dP - \Delta S \, dT \qquad (12.1)$$

and considering the system at a constant temperature, it is possible to conclude that the total volume of the oil and asphaltene phases below the upper AOP is higher than the volume of the oil had it remained single phase. A successful asphaltene model must account for this volumetric behavior, which is schematically illustrated in Figure 12.12. The mechanism behind asphaltene precipitation is therefore not primarily to be sought in interactions between asphaltene molecules, for example, asphaltene–asphaltene association. Asphaltene precipitation is essentially determined by the reaction of the surrounding medium (the oil) to the presence of asphaltene components. If the fluid can increase its molar volume by splitting off an asphaltene phase, it will do so. Otherwise, the asphaltenes will remain in solution in the oil. At pressure above the asphaltene onset point, the molecules in the oil phase, even though they are tightly packed, must make room for the large asphaltene molecules. Had the asphaltenes been contained in a separate phase, the oil molecules would have been more closely packed and the molar volume of the oil phase plus the asphaltene phase would have been smaller than with asphaltenes in solution in the oil. At lower pressures, the distance between the molecules in the oil phase increases, and the asphaltenes have less influence on the oil's molar volume. At the asphaltene precipitation point, the total molar volume would be the same with all asphaltenes in solution in the oil and with an infinitesimally small asphaltene phase split off. At pressures between the onset pressure and the bubble point of the oil, the total molar volume of the oil phase plus a separate asphaltene rich phase is higher than the molar volume would be if all the asphaltenes were in solution in the oil.

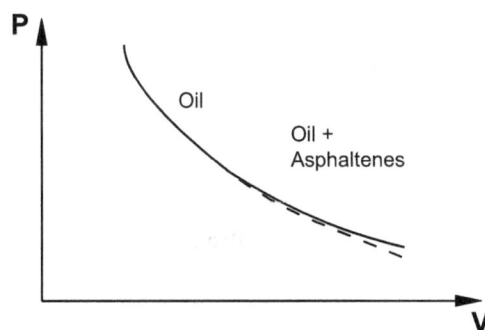

FIGURE 12.12 Pressure–volume relation for single-phase oil and for an oil splitting out an asphaltene phase. The full drawn line is the thermodynamically favorable one. The dashed line is an imaginary one, which the oil would have followed had it remained single phase below the upper asphaltene onset pressure.

12.4.2 Models Based on Cubic Equation of State

Rydahl et al. (1997) have presented asphaltene simulations using the SRK equation of state. They consider the aromatic fraction of C_{50+} to be asphaltenes. Having characterized the oil composition as described in Chapter 5, each pseudocomponent heavier than C_{49} is split into an asphaltene and a nonasphaltene component. The asphaltene components are by default assigned the properties in Table 12.6.

The critical temperature T_{ci}^{no-A} of the nonasphaltene fraction ($Frac_i^{no-A}$) of pseudocomponent i is found from the relation

$$T_{ci} = Frac_i^{no-A} T_{ci}^{no-A} + Frac_i^A T_{ci}^A \qquad (12.2)$$

where T_{ci} is the critical temperature of pseudocomponent i before being split into an asphaltene and a nonasphaltene component. $Frac_i^A$ is the asphaltene forming fraction of pseudocomponent i. The critical pressure P_{ci}^{no-A} of the non-asphaltene-forming fraction of pseudocomponent i is found from the equation

$$\frac{1}{P_{ci}} = \frac{\left(Frac_i^{no-A}\right)^2}{P_{ci}^{no-A}} + \frac{\left(Frac_i^A\right)^2}{P_{ci}^A} + \frac{2Frac_i^{no-A} \times Frac_i^A}{\sqrt{P_{ci}^{no-A}}\sqrt{P_{ci}^A}} \qquad (12.3)$$

whereas the acentric factor of the non-asphaltene-forming fraction of pseudocomponent i is found from

$$\omega_i = Frac_i^{no-A} \omega_i^{no-A} + Frac_i^A \omega_i^A \qquad (12.4)$$

Equations 12.2 through 12.4 ensure that the equation of state mixture a and b parameters are only marginally influenced from splitting up the C_{50+} components in an asphaltene and a nonasphaltene component. This has the advantage that approximately the same bubble point and gas/liquid ratios will be simulated before and after the split.

The binary interaction parameters between asphaltene components and C_1–C_9 hydrocarbons are by default 0.017. Binary interaction parameters of zero are default used for all other hydrocarbon–hydrocarbon interactions. Tuning the model to an experimental asphaltene onset pressure may either be accomplished by tuning the asphaltene T_c and P_c or by tuning the asphaltene content in the oil.

An example of a fluid composition characterized for a cubic equation of state using this procedure is shown in Table 12.7 [oil of Jamaluddin et al. (2002) characterized for the SRK equation]. Binary interaction coefficients are shown in Table 12.8. Table 12.9 shows Oil A of Jamaluddin et al. (2000) characterized for the PR equation and Table 12.10 Oil 1 of Rydahl et al. (1997) characterized

---

TABLE 12.6

Default Asphaltene Component Properties

T_c^A	1398.5 K
P_c^A	14.95 bar
ω^A	1.274

Source: Rydahl, A.K., Pedersen, K.S., and Hjermstad, H.P., Modeling of live oil asphaltene precipitation, presented at *AIChE Spring Meeting,* Houston, March 9–13, 1997.

TABLE 12.7

Oil Composition Characterized for the SRK Equation of State

Component	Mole Percentage	Molecular Weight	T_c (°C)	P_c (bar)	Acentric Factor
N_2	0.490	28.0	−147.0	33.94	0.040
CO_2	11.369	44.0	31.1	73.76	0.225
H_2S	3.220	34.1	100.1	89.37	0.100
C_1	27.357	16.0	−82.6	46.00	0.008
C_2	9.409	30.1	32.3	48.84	0.098
C_3	6.699	44.1	96.7	42.46	0.152
iC_4	0.810	58.1	135.0	36.48	0.176
nC_4	3.170	58.1	152.1	38.00	0.193
iC_5	1.220	72.2	187.3	33.84	0.227
nC_5	1.980	72.2	196.5	33.74	0.251
C_6	2.490	86.2	234.3	29.69	0.296
C_7–C_{25}	25.278	176.6	393.1	20.39	0.756
C_{26}–C_{49}	5.524	473.1	620.8	14.19	1.262
C_{50}–C_{64}–PN	0.545	774.8	683.8	13.72	1.313
C_{50}–C_{64}–A	0.192	774.8	1013.6	18.11	1.274
C_{65}–C_{80}–PN	0.183	989.2	845.3	14.10	0.876
C_{65}–C_{80}–A	0.064	989.2	1013.6	18.11	1.274

Source: Jamaluddin, A.K.M., et al., An investigation of asphaltene instability under nitrogen injection, SPE 74393 presented at *SPE International Petroleum Conference and Exhibition in Villahermosa*, Mexico, February 10–12, 2002.

Note: PN stands for paraffins and naphthenes and A for asphaltenes. Binary interaction parameters can be seen in Table 12.8. The plus composition is shown in Table 12.1. Simulated and experimental asphaltene onset pressures and bubble point pressures are plotted in Figure 12.13.

TABLE 12.8

Nonzero Binary Interaction Coefficients (k_{ij}) Used with the Oil Compositions in Tables 12.7, 12.9, and 12.10

	N_2	CO_2	H_2S	C_1–C_9
CO_2	−0.032	—	—	—
H_2S	0.170	0.099	—	—
C_1	0.028	0.120	0.080	—
C_2	0.041	0.120	0.085	—
C_3	0.076	0.120	0.089	—
iC_4	0.094	0.120	0.051	—
nC_4	0.070	0.120	0.060	—
iC_5	0.087	0.120	0.060	—
nC_5	0.088	0.120	0.069	—
C_6	0.080	0.120	0.050	—
C_7_PN	0.080	0.100	—	—
C_{7+}–A	0.080	0.100	—	0.017

TABLE 12.9
Oil Mixture A of Jamaluddin et al. (2000) Characterized for the PR Equation of State. Experimental AOP Data Is Given in Table 12.2

Component	Mole Percentage	Molecular Weight	T_c (°C)	P_c (bar)	Acentric Factor
N_2	0.480	28.0	−146.95	33.940	0.04
CO_2	0.919	44.0	31.05	73.760	0.225
C_1	43.391	16.0	−82.55	46.000	0.008
C_2	11.010	30.1	32.25	48.840	0.098
C_3	6.544	44.1	96.65	42.460	0.152
iC_4	0.789	58.1	134.95	36.480	0.176
nC_4	3.787	58.1	152.05	38.000	0.193
iC_5	1.279	72.2	187.25	33.840	0.227
nC_5	2.248	72.2	196.45	33.740	0.251
C_6	2.698	86.2	234.25	29.690	0.296
C_7–C_{25}	22.738	180.4	418.66	19.460	0.6803
C_{26}–C_{49}	3.747	460.3	683.62	13.080	1.2077
C_{50}–C_{64}–PN	0.230	769.2	845.45	10.660	0.9492
C_{50}–C_{64}–A	0.070	769.2	1172.58	17.300	1.274
C_{65}–C_{80}–PN	0.054	982.8	1051.88	10.280	0.1823
C_{65}–C_{80}–A	0.016	982.8	1172.58	17.300	1.274

Note: PN stands for paraffins and naphthenes and A for asphaltenes. Binary interaction coefficients (k_{ij}) can be seen in Table 12.8. The plus composition oil is shown in Table 12.1. Simulated and experimental asphaltene onset pressure and bubble point pressures are plotted in Figure 12.14.

TABLE 12.10
Oil Mixture 1 of Rydahl et al. (1997) Characterized for the SRK Equation of State

Component	Mole Percentage Oil	Mole Percentage Gas	Molecular Weight	T_c (°C)	P_c (bar)	Acentric Factor
N_2	1.497	1.88	28.0	−147.0	33.94	0.040
CO_2	0.220	0.41	44.0	31.1	73.76	0.225
C_1	23.066	70.69	16.0	−82.6	46.00	0.008
C_2	6.907	13.33	30.1	32.3	48.84	0.098
C_3	8.614	9.05	44.1	96.7	42.46	0.152
iC_4	1.298	1.08	58.1	135.0	36.48	0.176
nC_4	5.290	2.33	58.1	152.1	38.00	0.193
iC_5	1.777	0.45	72.2	187.3	33.84	0.227
nC_5	2.705	0.58	72.2	196.5	33.74	0.251
C_6	3.633	0.13	86.2	234.3	29.69	0.296
C_7	—	0.07	92.8	253.9	31.78	0.458
C_7–C_{25}	38.340	—	169.4	386.0	20.87	0.741
C_{26}–C_{49}	5.972	—	463.3	611.9	13.81	1.252
C_{50}–C_{80}–PN	0.528	—	814.2	714.8	12.80	1.216
C_{50}–C_{80}–A	0.155	—	814.2	1184.1	15.57	1.274

Note: Figure 12.15 shows the simulated pressure range with asphaltene precipitation when gas is injected into the oil at 90°C. Binary interaction coefficients are given in Table 12.8. The plus composition of the oil is shown in Table 12.4, which also shows the composition of the injection gas.

for the SRK equation. The asphaltene T_c and P_c have been found by tuning to the experimental onset pressures and deviate from the default values in Table 12.6. The nonzero binary interaction coefficients can be seen in Table 12.8. Experimental and simulated asphaltene onset pressures and bubble point pressures are plotted in Figures 12.13 through 12.15.

Table 12.11 shows the fluid composition in Table 10.14 characterized for the SRK equation of state and the asphaltenes split out as a separate component (A). Figure 12.8 shows the experimental data and simulation results for the asphaltene onset pressure and the saturation (bubble point) pressures when the fluid in Table 12.11 is contacted by nitrogen. The experimental data can be seen in Table 12.5.

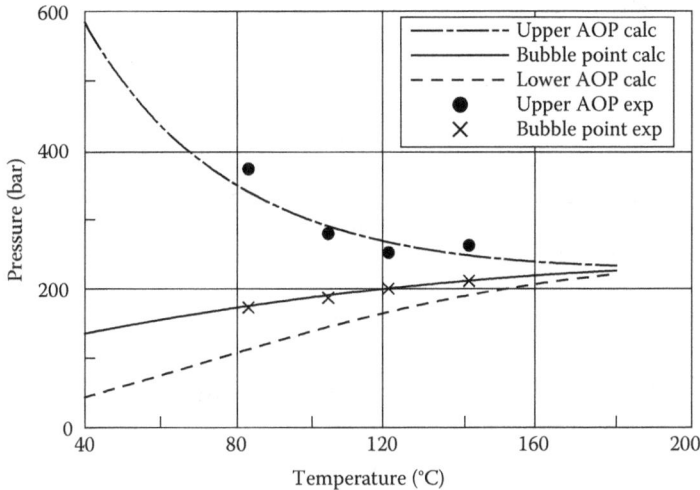

FIGURE 12.13 Experimental and simulated asphaltene onset pressures and bubble point pressures for fluid composition of Jamaluddin et al. (2002). Experimental onset pressures and bubble point pressures are given in Table 12.2. The oil composition can be seen in Table 12.1 and the characterized fluid in Table 12.7

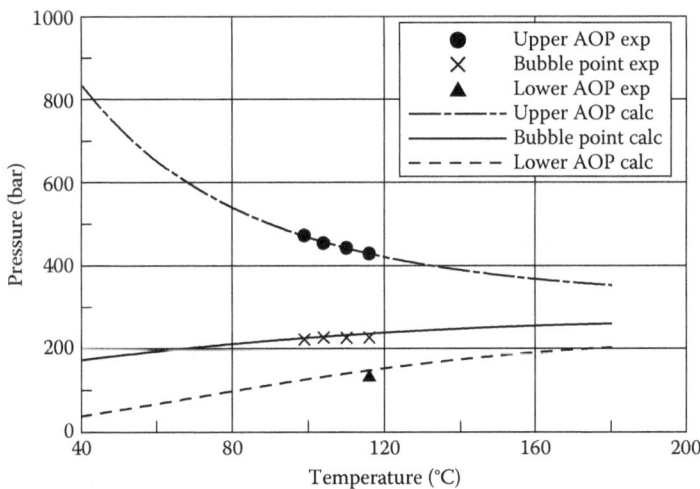

FIGURE 12.14 Experimental and simulated asphaltene onset pressures and bubble point pressures for Oil A of Jamaluddin et al. (2000). Experimental onset pressures and bubble point pressures are given in Table 12.2. The oil composition can be seen in Table 12.1 and the characterized fluid in Table 12.9.

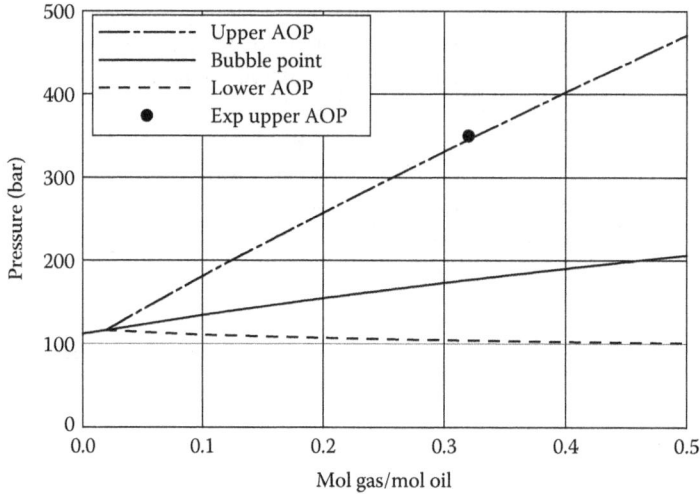

FIGURE 12.15 Simulated pressure range with asphaltene precipitation for oil mixture 1 of Rydahl et al. (1997) as a function of amount of gas added. Also shown is an experimental upper AOP. The oil and gas compositions can be seen in Table 12.4 and the characterized fluid in Table 12.9.

TABLE 12.11

Reservoir Fluid in Table 10.14 Characterized for the SRK Equation of State and with the Asphaltenes Split Out as a Separate Component (A)

		Component Properties			Nonzero k_{ij}s		
Component	Mol %	Critical Temperature °C	Critical Pressure bar	Acentric Factor	N_2	CO_2	C_{44}–C_{80}-A
N_2	0.123	−146.9	33.94	0.0400			0.250
CO_2	0.066	31.1	73.76	0.2250	−0.032		0.170
C_1	37.770	−82.5	46.00	0.0080	0.028	0.120	0.120
C_2	5.435	32.2	48.84	0.0980	0.041	0.120	0.110
C_3	5.880	96.6	42.46	0.1520	0.076	0.120	0.100
iC_4	0.726	134.9	36.48	0.1760	0.094	0.120	0.080
nC_4	3.285	152.1	38.00	0.1930	0.070	0.120	0.080
iC_5	0.728	187.2	33.84	0.2270	0.087	0.120	0.070
nC_5	2.232	196.4	33.74	0.2510	0.088	0.120	0.070
C_6	2.669	234.2	29.69	0.2960	0.080	0.120	0.060
C_7	4.025	254.2	31.80	0.4587	0.120	0.100	0.040
C_8	4.029	278.3	28.20	0.4986	0.120	0.100	0.030
C_9	3.355	303.5	26.86	0.5385	0.120	0.100	0.020
C_{10}–C_{13}	7.181	347.6	22.25	0.6327	0.120	0.100	
C_{14}–C_{17}	5.457	410.8	18.09	0.7869	0.120	0.100	
C_{18}–C_{21}	4.146	461.6	16.32	0.9193	0.120	0.100	
C_{22}–C_{25}	3.151	506.2	15.42	1.0351	0.120	0.100	
C_{26}–C_{30}	2.897	553.0	14.83	1.1488	0.120	0.100	
C_{31}–C_{35}	2.055	601.8	14.46	1.2508	0.120	0.100	
C_{36}–C_{43}	2.121	659.9	14.26	1.3349	0.120	0.100	
C_{44}–C_{54}	1.380	709.4	14.53	1.3741	0.120	0.100	
C_{55}–C_{80}	0.817	814.6	15.72	1.0813	0.120	0.100	
C_{44}–C_{80}-A	1.696	738.2	24.84	1.2740			

12.4.3 CPA MODELS

The CPA model is presented in Section 16.1.3. Arya et al. (2015, 2016) have used the CPA model to simulate asphaltene precipitation onset conditions for various reservoir fluids with and without gas injection. Self-association between asphaltene molecules and cross-association between asphaltene and heavy component molecules are considered.

12.4.4 PC-SAFT MODELS

The PC-SAFT equation of state presented in Section 4.9 is generally superior to a cubic equation of state for the simulation of liquid phase compressibilities. As this is a key requirement to an asphaltene model, the PC-SAFT equation might be a good candidate for the modeling of asphaltene precipitation. This is confirmed by the work of Ting et al. (2007). Figure 12.16 shows an asphaltene phase diagram using the characterization in Table 12.12. The match is about as good as that seen with the SRK equation of state (Figure 12.13).

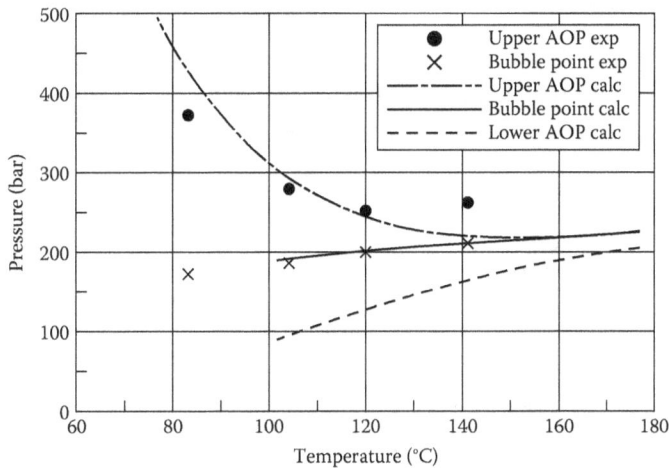

FIGURE 12.16 Simulated asphaltene phase diagram for oil composition in Table 12.1 of Jamaluddin et al. (2002) using the PC-SAFT equation and the characterization in Table 12.13. Experimental data can be seen in Table 12.2 and SRK simulation results from Figure 12.13.

TABLE 12.12
PC-SAFT Parameters Used to Represent Oil Composition of Jamaluddin et al. (2002)

Component	Molecular Weight	Mole Percentage	σ (Å)	m	e/ k (K)
CO_2/H_2S	41.80	14.583	2.7852	2.0729	169.21
N_2	28.01	0.495	3.3130	1.2053	90.96
C_1	16.04	27.334	3.7039	1.0000	150.03
Light gases	44.60	21.917	3.6130	2.0546	204.96
Paraffins and naphthenes	207.63	23.853	3.9320	5.9670	254.05
Aromatics (including resins)	270.5	11.750	3.8160	6.4730	342.08
Asphaltenes	1700	0.0676	4.3000	29.5000	380.00 [*1]

[*1] Modified from 420 K in Gonzalez et al. to 380 K.

Source: Gonzalez, D.L., et al., Prediction of asphaltene instability under gas injection with the PC-SAFT equation of state, *Energy Fuels* 19, 1230–1234, 2005.

Note: The raw composition is shown in Table 12.1.

Table 12.5 shows experimental and simulated asphaltene onset pressures and bubble point pressures for the Gulf of Mexico (GoM) reservoir fluid in Table 10.14 with N_2 in various concentrations. The PC-SAFT equation has been used in the simulations with the parameters in Table 12.13. The simulation results were comparable to those shown in Figure 12.8. obtained with the SRK-Peneloux model in Table 12.11.

Other papers on asphaltene modeling using PC-SAFT are presented by Gonzalez et al. (2005), Panuganti et al. (2012), and Punnapala et al. (2013).

12.4.5 Assessment of Asphaltene Models

Lekumjorn et al. (2020) have simulated asphaltene onset data for 14 reservoir fluids with and without gas injection. For all the fluids, the asphaltene onset conditions could be modeled with the SRK equation of state. An equally good match could be obtained with the CPA and PC-SAFT models,

TABLE 12.13

PC-SAFT Model Parameters Used to Simulate the Asphaltene Onset Pressures and Saturation Pressures in Table 12.5 for the Fluid Composition in Table 10.14

Component	Mol%	m	σ(Å)	e/k (K)
N_2	0.12	1.205	3.313	90.960
CO_2	0.07	2.073	2.785	169.210
C_1	37.77	1.000	3.704	150.030
C_2	5.44	1.607	3.521	191.420
C_3	5.88	2.002	3.618	208.110
iC_4	0.73	2.262	3.757	216.530
nC_4	3.28	2.332	3.709	222.880
iC_5	0.73	2.562	3.830	230.750
nC_5	2.23	2.690	3.773	231.200
C_6	2.67	3.058	3.798	236.770
C_7	4.02	3.114	3.790	249.483
C_8	4.03	3.516	3.786	251.013
C_9	3.36	3.929	3.783	252.492
C_{10}–C_{11}	5.44	4.480	3.778	254.375
C_{12}–C_{14}	5.44	5.544	3.772	256.256
C_{15}–C_{17}	4.59	6.593	3.769	260.434
C_{18}–C_{20}	3.04	7.821	3.768	262.374
C_{21}–C_{25}	2.84	9.484	3.770	266.129
C_{26}–C_{30}	1.69	10.457	3.771	269.291
C_{31}–C_{49}	3.31	14.767	3.775	275.595
C_{50}–C_{80}	2.56	28.648	3.597	247.386
C_{50}–C_{80}–A	0.76	11.000	4.530	500.000

Nonzero Binary Interaction Coefficients

k_{ij}	N_2	CO_2	C_1–C_9
CO_2	−0.0315		
C_1	0.0278	0.12	
C_2	0.0407	0.12	
C_3	0.0763	0.12	

Nonzero Binary Interaction Coefficients

k_{ij}	N_2	CO_2	C_1–C_9
iC_4	0.0944	0.12	
nC_4	0.07	0.12	
iC_5	0.0867	0.12	
nC_5	0.0878	0.12	
C_6	0.08	0.12	
C_7	0.13	0.1	
C_8	0.13	0.1	
C_9	0.13	0.1	
C_{10}–C_{11}	0.13	0.1	
C_{12}–C_{14}	0.13	0.1	
C_{15}–C_{17}	0.13	0.1	
C_{18}–C_{20}	0.13	0.1	
C_{21}–C_{25}	0.13	0.1	
C_{26}–C_{30}	0.13	0.1	
C_{31}–C_{49}	0.13	0.1	
C_{50}–C_{80}	0.13	0.1	
C_{50}–C_{80}–A	0.17	0.1	0.017

but the two latter more complex models did not present any advantages as compared to the classical SRK equation.

Abutaqiya et al. (2020) and Mogensen et al. (2023) have successfully modeled asphaltene precipitation using the PR equation of state. This further supports that modeling of asphaltene precipitation can be done with a classical cubic equation of state and does not require a more complex equation.

12.5 PRACTICAL SOLUTION TO ASPHALTENE PROBLEM

Hasket and Tarteral (1965) have presented a practical solution to an asphaltene problem at the Hassi Messaud field in Algeria. The composition of the fluid produced from the field is shown in Table 12.14. The characterized fluid for use with the SRK equation of state is shown in Table 12.15 and the simulated asphaltene phase envelope in Figure 12.17. The asphaltene phase envelope shows that the asphaltene deposition happens in a narrow pressure interval around the bubble point pressure. Initially asphaltene deposition was seen over a relatively broad depth interval in the well as sketched in Figure 12.18. The asphaltene deposition made the pressure increase below the deposit and decrease below the deposit. This made the depth interval with asphaltene deposition decrease as can also be seen from Figure 12.18. Continued asphaltene deposition made the pressure drop over the deposited asphaltene layer increase and the velocity of the well stream increase when passing the interval with asphaltene deposit. The deposited asphaltene layer effectively acted as a choke with an inlet pressure equal to the upper asphaltene onset pressure and an outlet pressure equal to the lower asphaltene onset pressure. Because the fluid velocity when passing this natural choke is high, there is not time enough for the precipitated asphaltenes to deposit and the asphaltene deposition almost stops. This solution cannot be generally applied as it is dependent on the pressure interval with asphaltene precipitation being relatively narrow.

TABLE 12.14

Molar Composition of Hassi Messaoud Reservoir Fluid. The C_{7+} Molecular Weight Is 180 and the C_{7+} Density Is 0.835 g/cm³

Component	Mol %
N_2	1.80
CO_2	1.32
C_1	33.15
C_2	13.95
C_3	9.91
iC_4	1.29
nC_4	4.66
iC_5	1.40
nC_5	2.48
C_6	3.59
C_7	3.30
C_8	3.25
C_9	2.89
C_{10}	2.50
C_{11}	2.13
C_{12+}	12.38

TABLE 12.15

Reservoir Fluid Composition in Table 12.14 Characterized for the SRK Equation. The C_{50}–C_{80} Fraction Is Split into a Non-Asphaltene and an Asphaltene (A) Component. Figure 12.17 Shows the Simulated Asphaltene Phase Envelope

Component	Mol %	Critical Temperature °C	Critical Pressure bar	Acentric Factor
N_2	1.800	−146.95	33.94	0.040
CO_2	1.320	31.05	73.76	0.225
C_1	33.150	−82.55	46.00	0.008
C_2	13.950	32.25	48.84	0.098
C_3	9.910	96.65	42.46	0.152
iC_4	1.290	134.95	36.48	0.176
nC_4	4.660	152.05	38.00	0.193
iC_5	1.400	187.25	33.84	0.227
nC_5	2.480	196.45	33.74	0.251
C_6	3.590	234.25	29.69	0.296
C_7–C_{10}	11.940	292.29	27.80	0.522
C_{11}–C_{20}	11.289	397.61	18.98	0.757
C_{21}–C_{49}	3.180	552.69	15.22	1.130
C_{50}–C_{80}	0.038	763.48	15.14	1.284
C_{50}–C_{80}-A	0.0034	1300.00	11.70	1.274

	NonZero Binary Interaction Coefficients		
	N_2	CO_2	C_{50}–C_{80}-A
N_2			0.120
CO_2	−0.032		0.120
C_1	0.028	0.120	0.100
C_2	0.041	0.120	0.100
C_3	0.076	0.120	0.090
iC_4	0.094	0.120	0.080

nC$_4$	0.070	0.120	0.080
iC$_5$	0.087	0.120	0.070
nC$_5$	0.088	0.120	0.070
C$_6$	0.080	0.120	0.050
C$_7$–C$_{10}$	0.080	0.100	0.040
C$_{11}$–C$_{20}$	0.080	0.100	
C$_{21}$–C$_{49}$	0.080	0.100	
C$_{50}$–C$_{80}$	0.080	0.100	

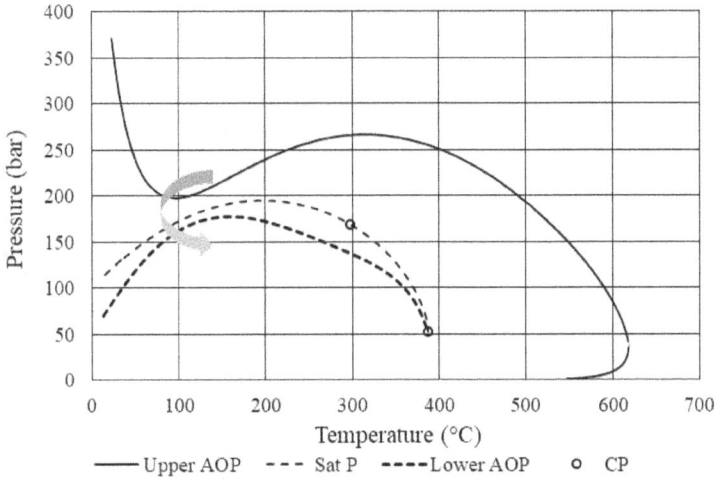

FIGURE 12.17   Simulated asphaltene phase envelope for the fluid composition in Table 12.14 characterized for the SRK equation, as shown in Table 12.15.

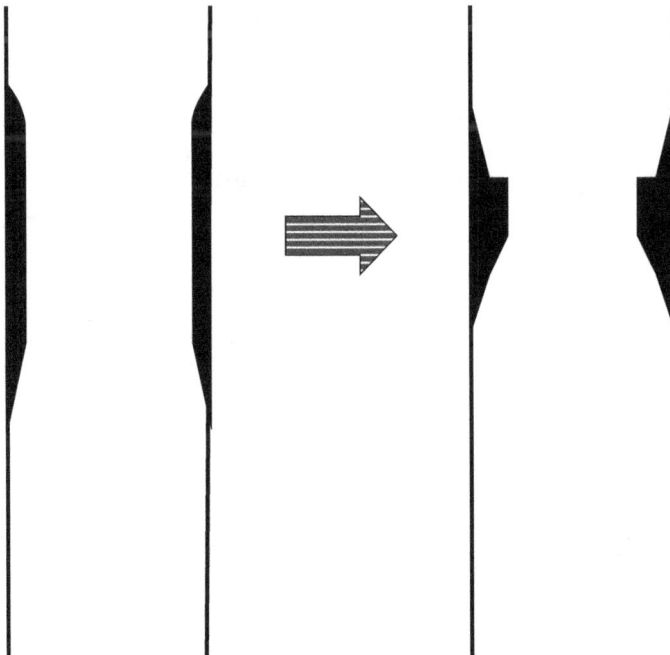

FIGURE 12.18   Asphaltene deposition of Hassi Messaoud well forming a natural choke.

12.6 ASPHALTENE TAR MAT CALCULATION

Chapter 14 deals with compositional variations with depth. The concentration of heavy molecular weight components increases with depth. This is a combined effect of gravity and the temperature increase with depth. Asphaltenes are high molecular weight components and the concentration of asphaltene components will therefore increase with depth. As is illustrated in Figure 12.19, a tar mat may be found beneath the oil zone. Figure 12.19 also sketches the mechanisms deciding whether a tar mat will be found or not and in which depth it is possibly found. The reservoir pressure increases with depth as is illustrated by the full-drawn line in Figure 12.19. The increasing concentration of asphaltenes with depth makes the asphaltene onset pressure increase as is illustrated by the dashed line in Figure 12.19. At the point where the asphaltene onset pressure and the reservoir pressure become equal, the oil zone will end, and a tar mat zone will start.

Simulation of a tar mat is illustrated in Figure 12.20 for the fluid composition in Table 12.16. At a temperature of 99°C, the fluid composition has an asphaltene onset pressure of 172 bar. The fluid was sampled at a reservoir pressure of 192 bar in a depth of 500 m. As can be seen from Figure 12.20, the increase in onset pressure with depth exceeds the increase in reservoir pressure with depth and in a

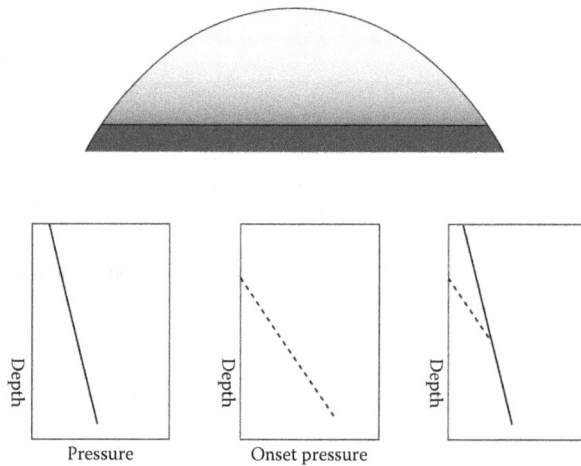

FIGURE 12.19 Schematic illustration of mechanisms behind a contact between an oil zone and a tar mat zone. The contact is at the depth at which the reservoir pressure and asphalt onset pressure coincide.

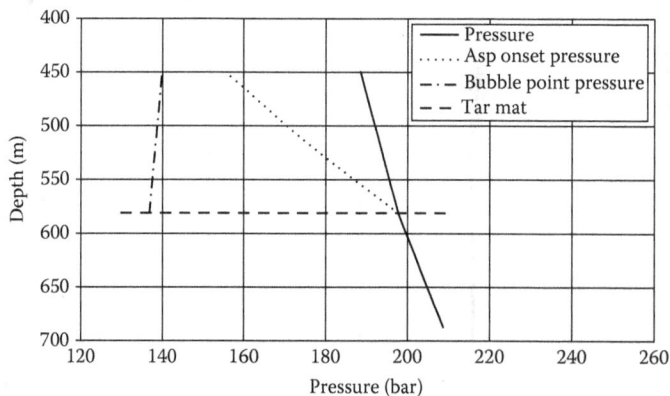

FIGURE 12.20 Tar mat simulation for the fluid composition in Table 12.16.

TABLE 12.16

SRK–Peneloux Model Parameters Used to Simulate the Asphaltene Tar Mat in Figure 12.20

Component	Mol%	Molecular Weight	T_c°C	P_c bar	Acentric Factor	Volume Correction cm³/mole
N_2	0.186	28.01	−146.95	33.94	0.040	0.92
CO_2	5.397	44.01	31.05	73.76	0.225	3.03
H_2S	1.398	34.08	100.05	89.37	0.100	1.78
C_1	26.590	16.04	−82.55	46.00	0.008	0.63
C_2	9.519	30.07	32.25	48.84	0.098	2.63
C_3	3.925	44.10	96.65	42.46	0.152	5.06
iC_4	1.101	58.12	134.95	36.48	0.176	7.29
nC_4	2.525	58.12	152.05	38.00	0.193	7.86
iC_5	5.169	72.15	187.25	33.84	0.227	10.93
nC_5	3.947	72.15	196.45	33.74	0.251	12.18
C_6	3.719	86.18	234.25	29.69	0.296	17.98
C_7	3.713	96.00	266.97	34.39	0.468	5.57
C_8	3.344	107.00	285.44	31.02	0.500	11.14
C_9	3.012	121.00	306.20	27.57	0.540	18.31
C_{10}–C_{12}	7.356	146.39	340.37	23.52	0.613	28.67
C_{13}–C_{14}	3.766	182.11	380.44	20.08	0.707	38.13
C_{15}–C_{16}	3.055	213.58	411.76	18.19	0.787	41.84
C_{17}–C_{19}	3.536	249.43	444.63	16.84	0.874	40.62
C_{20}–C_{23}	2.385	295.84	484.01	15.78	0.978	32.17
C_{24}–C_{29}	2.424	363.23	536.14	14.86	1.110	11.54
C_{30}–C_{35}	1.510	447.79	595.26	14.23	1.240	−24.08
C_{56}–C_{46}	1.447	559.09	668.53	13.87	1.342	−80.39
C_{47}–C_{80}	0.784	789.61	735.72	13.48	1.234	−262.88
C_{50}–C_{80}–A	0.192	789.61	1125.35	14.95	1.274	−15.56

Nonzero Binary Interaction Coefficients				
k_{ij}	N_2	CO_2	H_2S	C_1–C_9
CO_2	−0.0315			
H_2S	0.1696	0.0989		
C_1	0.0278	0.1200	0.0800	
C_2	0.0407	0.1200	0.0852	
C_3	0.0763	0.1200	0.0885	
iC_4	0.0944	0.1200	0.0511	
nC_4	0.07	0.1200	0.0600	
iC_5	0.0867	0.1200	0.0600	
nC_5	0.0878	0.1200	0.0689	
C_6	0.08	0.1200	0.0500	
C_7	0.08	0.1000		
C_8	0.08	0.1000		
C_9	0.08	0.1000		
C_{10}–C_{12}	0.08	0.1000		
C_{13}–C_{14}	0.08	0.1000		
C_{15}–C_{16}	0.08	0.1000		
C_{17}–C_{19}	0.08	0.1000		
C_{20}–C_{23}	0.08	0.1000		
C_{24}–C_{29}	0.08	0.1000		
C_{30}–C_{35}	0.08	0.1000		
C_{56}–C_{46}	0.08	0.1000		
C_{47}–C_{80}	0.08	0.1000		
C_{50}–C_{80}–A	0.08	0.1000		0.023

depth of 581 m the asphaltene onset pressure and the reservoir pressure coincide. Here the oil zone stops and beneath is an asphaltene tar mat.

Hirschberg (1988) has reported the presence of a tar mat at the bottom of a reservoir, which originates from compositional grading and consists essentially of asphaltenes.

REFERENCES

Abutaqiya, M.I.L., Sisco, C.J., Khemka, Y.C.J., Safa, M.A., Ghloum, E.F., Rashed, A.M., Gharbi, R., Santhanagopalan, S., Al-Qahtani, M., Al-Kandari, E., Vargas, F.M., Accurate modeling of asphaltene onset pressure in crude oils under gas injection using Peng–Robinson equation of state, *Energy Fuels.* 34, 4055–4070, 2020.

Agger, C.S. and Sørensen, H., Algorithm for constructing complete asphaltene PT and Px Phase diagrams, *Ind. Eng. Chem. Res.* 57, 392–400, 2018.

Angulo, R., Borges, A., Franseca, M., and Gil, C., Experimental asphaltene precipitation study. Phenomenological behavior of Venezuelan live crude oils, proceeding from *ISCOP' 95*, Rio de Janeiro, Brazil, November 26–29, 1995.

Arya, A., von Solms, N., and Kontogeorgis, G., Determination of asphaltene onset conditions using the cubic plus association equation of state, *Fluid Phase Equilib.* 400, 8–19, 2015.

Arya, A., von Solms, N., and Kontogeorgis, G., Investigation of the gas injection effect on asphaltene onset precipitation using Cubic-Plus-Association equation of state, *Energy Fuels.* 30, 3560–3574, 2016.

Boesen, R.R., Sørensen, H., and Pedersen, K.S., Asphaltene predictions using screening methods and equations of state, SPE-190401-MS, presented at the *SPE EOR Conference*, Muscat, Oman, March 26–28, 2018.

Burke, N.E., Hobbs, R.E., and Kashow, S.F., Measurement and modeling of asphaltene precipitation, *J. Petl. Technol.* 1440–1446, 1990.

Chung, T.-H., Thermodynamic modeling for organic solid precipitation, SPE 24851, presented at the *SPE ATCE*, Washington, DC, October 4–7, 1992.

De Boer, R.B., Leerlooyer, K., Eigner, M.R.P., and van Bergen, A.R.D., Screening of crude oils for asphaltene precipitation: Theory, practice, and the selection of inhibitors, *SPE Prod. Facil.* 55–61, 1995.

Escobedo, J. and Mansoori, G.A., Viscometric determination of the onset of asphaltene flocculation: A novel method, *SPE Prod. Facil.* 115–118, 1995.

Fotland, P., Anfindsen, H., and Fadnes, F.H., Detection of asphaltene precipitation and amounts precipitated by measurement of electrical conductivity, *Fluid Phase Equilib.* 82, 157–164, 1993.

Godbole, S.P., Thele, K.J., and Reinbold, E.W., EOS modeling and experimental observations of three-hydrocarbon phase equilibria, *SPE Reserv. Eng.* 10, 101–108, 1995.

Gonzalez, D.L., Mahmoodaghdam, E., Lim, F., and Joshi, N., Effects of gas additions to deepwater Gulf of Mexico reservoir oil: Experimental investigation of asphaltene precipitation and deposition, SPE 159098, presented at the *SPE ATCE*, San Antonio, TX, October 8–10, 2012.

Gonzalez, D.L., Ting, P.D., Hirazaki, G.J., and Chapman, W.G., Prediction of asphaltene instability under gas injection with the PC-SAFT equation of state, *Energy Fuels.* 19, 1230–1234, 2005.

Hammami, A., Chang-Yen, D., Nighswander, J.A., and Stange, E., An experimental study of the effect of paraffinic solvents on the onset and bulk precipitation of asphaltenes, *Fuel Sci. Technol. Int.* 13, 1167–1184, 1995.

Haskett, C.E. and Tarteral, M., A practical solution to the problem of asphaltene deposits—Hassi Messaoud field, Algeria, *J. Pet. Technol.* 17, 387–391, 1965.

Hirschberg, A., Role of asphaltenes in compositional grading of a reservoir's fluid column, *J. Petl. Technol.* 40, 89–94, 1988.

Hustad, O.S., Jia, N., Pedersen, K.S., Memon, A., and Lekumjorn, S., High pressure data and modeling results for phase behavior onsets of GoM oil mixed with nitrogen, *SPE Reserv. Eval. Eng.* 3, 384–395, 2014.

Jamaluddin, A.K.M., Joshi, N., Iwere, F., and Gurpinar, F., An investigation of asphaltene instability under nitrogen injection, SPE 74393, presented at the *SPE International Petroleum Conference and Exhibition in Villahermosa*, Mexico, February 10–12, 2002.

Jamaluddin, A.K.M., Joshi, N., Joseph, M.T., D'Cruz, D., Ross, B., Creek, J., Kabir, C.S., and McFadden, J.D., Laboratory techniques to defines the asphaltene precipitation envelope, presented at the *Petroleum Society's Canadian International Petroleum Conference in Calgary*, Canada, June 4–8, 2000.

Kawanaka, S., Park, S.J., and Monsoori, G.A., Organic deposition from reservoir fluids: A thermodynamic predictive technique, *SPE Reserv. Eng.* 185–192, 1991.

Kokal, S., Al-Dawood, N., Fontanilla, J., Al-Ghamdi, A., Nasr-El-Din, H. and Al-Rufaie, Y., Productivity Decline in Oil Wells Related to Asphaltene Precipitation and Emulsion Blocks, SPE Prod. Facilities 18, 247–256, 2003.

Koots, J.A. and Speight, J.G., Relation of petroleum resins to asphaltenes, *Fuel.* 54, 179–184, 1975.

Lekumjorn, S., Boesen, R.R., Sørensen, H., and Krejbjerg, K., Asphaltene modeling with cubic and more complex Equations of state, SPE paper 203491-MS, presented at the *Abu Dhabi International Petroleum Exhibition & Conference*, Abu Dhabi, November 9–12, 2020.

MacMillan, D.J., Tackeff, J.E. Jr., Jessee, M.A., and Monger-McClure, T.G., A unified approach to asphaltene precipitation: Laboratory measurement and modeling, *J. Pet. Technol.* 47, 788–793, 1995.

Mogensen, K., Grutters, M., and Merrill, R., Hybrid algorithm predicts asphaltene envelope for entire ADNOC fluid portfolio, SPE-212633-MS, presented at the S*PE Reservoir Characterisation and Simulation Conference and Exhibition*, Abu Dhabi, January 24–26, 2023.

Mullins, O.C., Andrew, E., Pomerantz, A.E., Zuo, J.Y., Andrews, A.B., Hammond, P., Dong, C., Elshahawi, H., Seifert, D.J., Jayant, P., Rane, J.P., Banerjee, S., and Pauchard, V., Asphaltene nanoscience and reservoir fluid gradients, tar mat formation, and the oil-water interface, SPE 166278, presented at the *SPE ATCE*, New Orleans, LA, September 30–October 2, 2013.

Panuganti, S., Vargas, F.M., Gonzalez, D.L., Kurup, A.S., and Chapman, W.G., PC-SAFT characterization of crude oils and modeling of asphaltene phase behavior, *Fuel.* 93, 658–669, 2012.

Punnapala, S. and Vargas, F.M., Revisiting the PC-SAFT characterization procedure for an improved asphaltene precipitation prediction, *Fuel.* 108, 417–429, 2013.

Rydahl, A.K., Pedersen, K.S., and Hjermstad, H.P., Modeling of live oil asphaltene precipitation, presented at the *AIChE Spring Meeting*, Houston, March 9–13, 1997.

Sivaraman, A., Hu, Y., Thomas, F.B., Bennion, D.B., and Jamaluddin, A.K.M., Acoustic resonance: An emerging technology to identify wax and asphaltene precipitation onset conditions in reservoir fluids, Paper presented at the *Annual Technical Meeting*, Calgary, Alberta, June 1997

Tavakkoli, M., Panuganti, S.R., Taghikhani, V., Pishvaie, M.R., and Chapman, W.G., Precipitated asphaltene amount at high-pressure and high-temperature conditions, *Energy Fuels.* 28, 1596–1610, 2014.

Ting, P.D., Gonzalez, D.L., Hirasaki, G.J., and Chapman, W.G., Application of the PC-SAFT equation of state to asphaltene phase behavior, *Asphaltenes, Heavy Oils, and Petroleomics*, Springer, New York, NY, 301–327, 2007.

Ting, P.D., Hirasaki, G.J., and Chapman, W.G., Modeling of asphaltene phase behavior with the SAFT equation of state, *Pet. Sci. Technol.* 21, 647–661, 2003.

13 Gas Hydrates

When water approaches its freezing point, water lattices with cavities inside may start to form. These lattices will not be stable unless some of the cavities are filled with gas molecules. A water structure stabilized by gas molecules is called a *gas hydrate*. It has an appearance similar to that of snow or ice but can survive at temperatures far above the freezing point of water. Methane (C_1) and carbon dioxide (CO_2) are examples of gas components of the right molecular size to stabilize hydrate lattices. Gas hydrates may be stable at temperatures as high as 35°C, and the hydrate formation temperature increases with pressure.

Multiphase mixtures of reservoir fluids and formation water from deep water production are frequently transported to onshore or offshore processing plants through undersea pipelines. The distance from a petroleum field to the processing plant can be long, and a considerable temperature drop may take place. During shutdown, the temperature may drop to that of the surrounding seawater (~4°C). As the pressures in the pipelines can be quite high, the pipeline conditions may promote the formation of gas hydrates. Because hydrate formation may lead to plugging of pipelines and ultimately shutdown of production, it is of utmost importance to assess the risk of hydrate formation when designing new multiphase transport pipelines.

Figure 13.1 illustrates the phase behavior of a pure hydrate-forming gas component. Line AB represents equilibria between hydrate, gas in gaseous form, and ice. Line BC represents equilibria between hydrate, gas in gaseous form, and liquid water. Line CD represents equilibria between hydrate, "gas" in liquid form, and water in liquid form.

For a mixed gas composition, the situation is slightly more complicated. The hydrate curve may possibly intersect the phase envelope of the gas mixture as is illustrated in Figure 13.2. Lines AB,

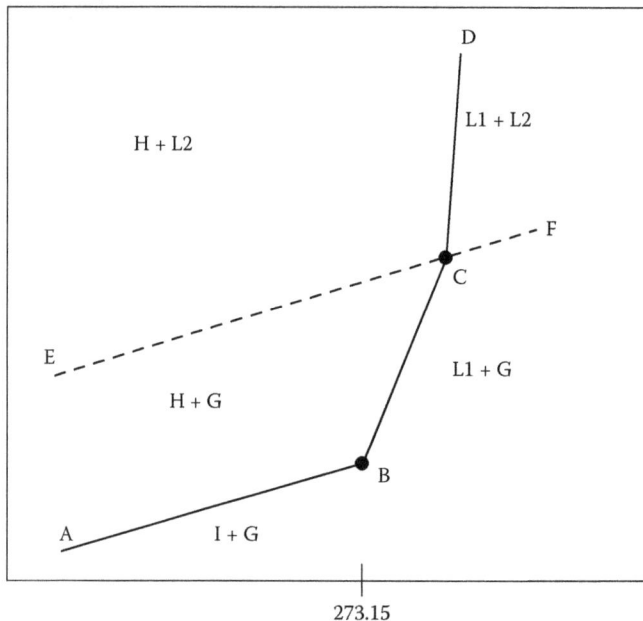

FIGURE 13.1 The full-drawn line shows the hydrate formation conditions of a pure gas component. The dashed line (EF) shows the vapor pressure curve of the hydrate-forming gas compound. G stands for gas, H for hydrate, I for ice, L1 for liquid water, and L2 for hydrate-forming compound in liquid form.

DOI: 10.1201/9780429457418-13

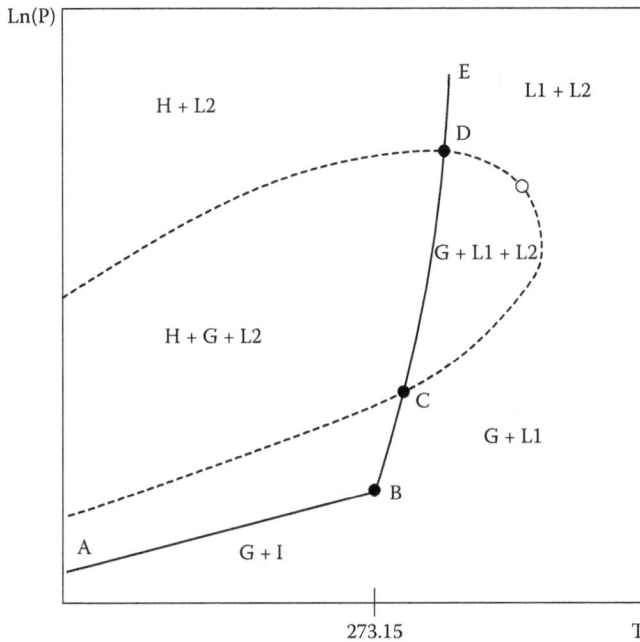

FIGURE 13.2 The full-drawn line shows the hydrate formation conditions of a gas mixture. The dashed line shows the phase envelope of the gas mixture. G stands for gas, H for hydrate, I for ice, L1 for liquid water, and L2 for hydrate-forming compounds in liquid form.

BC, and DE correspond to lines AB, BC, and CD in Figure 13.1. At line CD in Figure 13.2, hydrate, "gas" in gaseous form, "gas" in liquid form, and water in liquid form coexist and in equilibrium.

Makogan (1997), Holder and Bisnoi (2000), and Sloan and Koh (2008) have given detailed descriptions of gas hydrates including natural occurrences, experimental investigations, and gas hydrate kinetics.

13.1 TYPES OF HYDRATES

Three different types of hydrate lattices may form when gas and oil constituents are in contact with water—structure I, II, and H hydrates. Each type of lattice contains a number of cavities of differing sizes. Some physical data for each of the two types of hydrate structures are given in Table 13.1. In a stable hydrate, gaseous compounds (guest molecules) occupy some of the cavities. Structure I and II hydrates contain cavities of two different sizes, small ones and large ones. Structure H hydrates contain cavities of three different sizes, small, medium-sized, and huge ones. The small and the medium-sized structure H cavities are in fact of approximately the same size as the large cavities in structure I and structure II. The terms *small* and *medium sized* are used to distinguish them from the huge structure H cavities. The cavity structures are sketched in Figure 13.3. Some guest molecules may enter into more than one cavity size, but others are limited to just one cavity size. Methane (C_1) may, for example, enter both small and large cavities of structure I and structure II hydrates as well as the small and medium-sized cavities of structure H hydrates. Isobutane (iC_4) may, on the other hand, only enter into the large cavity of structure II hydrates. Table 13.2 shows what compounds of interest in gas and oil production may enter into the cavities of hydrate structures I, II, and H. Components like 2,2-dimethyl-propane (2,2-dim-C_3), cyclopentane (cC_5), cyclohexane (cC_6), and benzene are called *structure II heavy hydrate formers* and may significantly influence the hydrate formation conditions (Danesh et al. 1993; Tohidi et al. 1996, 1997).

TABLE 13.1

Physical Data for Each of the Three Hydrate Structures

		Structure I	Structure II	Structure H
Number of water molecules per unit cell		46	136	34
Number of small cavities per unit cell		2	16	3
Number of medium-sized cavities per unit cell		0	0	2
Number of large cavities per unit cell		6	8	0
Number of huge cavities per unit cell		0	0	1
Cavity diameter (Å)	Small	7.95	7.82	8.11
	Medium sized	—	—	8.66
	Large	8.60	9.46	—
	Huge	—	—	11.42

Note: The parameters for structures I and II are from Erickson (1983) and the parameters for structure H hydrate from Mehta and Sloan (1996).

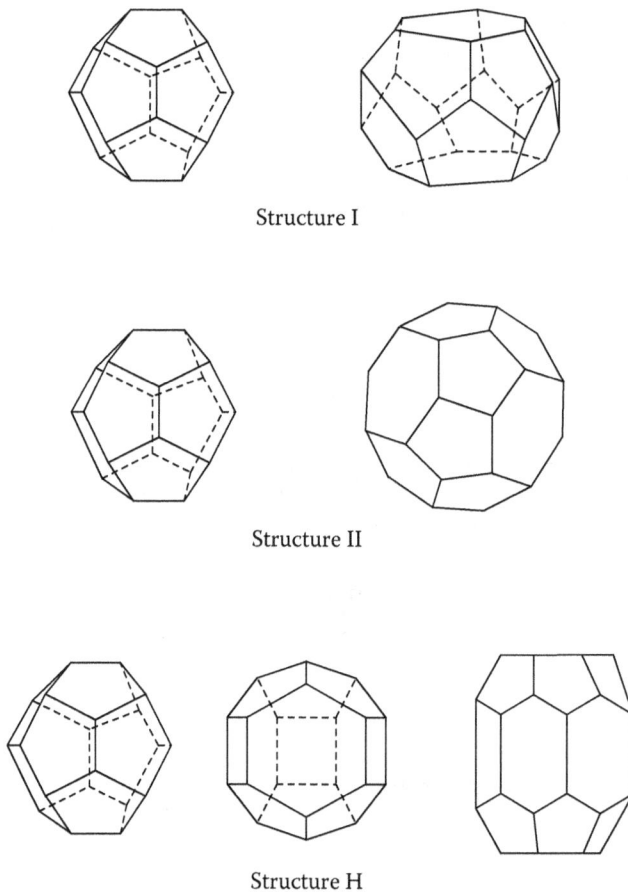

Structure I

Structure II

Structure H

FIGURE 13.3 Hydrate cavity structures.

TABLE 13.2
Guest Molecules of Structure I, II, and H Hydrates

Compound	Structure I		Structure II		Structure H	
	Small Cavities	Large Cavities	Small Cavities	Large Cavities	Small/Medium Cavities	Huge Cavities
N_2	+	+	+	+	+	—
CO_2	+	+	+	+	—	—
H_2S	+	+	+	+	—	—
C_1	+	+	+	+	+	—
C_2	—	+	—	+	—	—
C_3	—	—	—	+	—	—
iC_4	—	—	—	+	—	—
nC_4	—	—	—	+	—	—
2,2-dim-C_3	—	—	—	+	—	—
c-C_5	—	—	—	+	—	—
c-C_6	—	—	—	+	—	—
Benzene	—	—	—	+	—	—
iC_5	—	—	—	—	—	+
2,2-Dimethylbutane	—	—	—	—	—	+
2,3-Dimethylbutane	—	—	—	—	—	+
2,2,3-Trimethylbutane	—	—	—	—	—	+
3,3-Dimethylpentane	—	—	—	—	—	+
Methylcyclopentane	—	—	—	—	—	+
Methylcyclohexane	—	—	—	—	—	+
Cis-1,2-Dimethylcyclohexane	—	—	—	—	—	+
Ethylcyclopentane	—	—	—	—	—	+
Cyclooctane	—	—	—	—	—	+
1,1-Dimethylcyclohexane	—	—	—	—	—	+
Cycloheptane	—	—	—	—	—	+

Structure H consists of three different cavity sizes. These are conveniently modeled as just two cavity sizes, a small/medium-sized one and a huge one. The huge cavity can accommodate molecules containing five to eight carbon atoms. The small/medium-sized cavities will usually be accommodated with N_2 or C_1. Structure H hydrates were discovered by Ripmeester et al. (1987) and are not as commonly seen during oil and gas production as are structure I and II hydrates. As there may be mixture compositions and conditions favoring structure H formation, it is however good practice to also consider structure H hydrates when evaluating the risk of hydrate formation (Mehta and Sloan 1999).

That a component can enter into more than one hydrate structure (I, II, and H) does not necessarily mean that the pure component will form more than one structure if mixed with water. Methane (C_1) can, for example, enter into the small and medium-sized cavities of all three structures, and it can enter into the large cavities of structure I and structure II hydrates. In pure form, methane will form structure I hydrates. Only when mixed with other components can methane enter into structure II and structure H hydrates. Table 13.3 shows the hydrate structure formed by pure component hydrate formers. Normal butane (nC_4) and heavier can form hydrates only if smaller components are also present. For example, structure H hydrates cannot form from a binary mixture of isopentane (iC_5) and water. Occupancy of the huge cavities is not sufficient to stabilize the hydrate structure.

TABLE 13.3
Hydrate Structures Formed by Pure Components

Component	Structure I	Structure II
N_2	x	—
CO_2	x	—
H_2S	x	—
C_1	x	—
C_2	x	—
C_3	—	x
iC_4	—	x

Some of the small or medium-sized cavities must also be occupied. This requires the presence of small molecules like C_1 or N_2.

13.2   MODELING OF HYDRATE FORMATION

Hydrates may form when the hydrate state is energetically favorable as compared to a nonhydrate state (liquid water or ice). The transformation of water from a nonhydrate state to a hydrate state can be regarded as consisting of two steps

1. Liquid water or ice (α) $\rightarrow$ Empty hydrate lattice (β)
2. Empty hydrate lattice (β) $\rightarrow$ Filled hydrate lattice (H)

where α, β, and H are used to identify each of the three states considered. The β state is purely hypothetical and is considered only to facilitate hydrate calculations. The energetically favorable state (H or α) is the one of lower chemical potential (the term *chemical potential* is further dealt with in Appendix A). The difference between the chemical potential of water in the hydrate state (H) and in a pure water state (α) can be expressed as

$$\mu_w^H - \mu_w^\alpha = \left(\mu_w^H - \mu_w^\beta\right) + \left(\mu_w^\beta - \mu_w^\alpha\right) \tag{13.1}$$

The first term on the right-hand side, $\left(\mu_w^H - \mu_w^\beta\right)$, can be regarded as the stabilizing effect on the hydrate lattice caused by adsorption of gas molecules. The difference between the chemical potential of water in the empty and in the filled hydrate lattice is calculated from

$$\left(\mu_w^H - \mu_w^\beta\right) = RT \sum_{i=1}^{NCAV} v_i \ln\left(1 - \sum_{k=1}^{N} Y_{ki}\right) \tag{13.2}$$

where v_i is the number of cavities of type i per water molecule, and Y_{ki} denotes the possibility that a cavity i is occupied by a gas molecule of type k. NCAV is the number of cavities per unit cell in the hydrate lattice, and N is the number of gaseous compounds, which may enter into a cavity in the hydrate lattice. The probability Y_{ki} is calculated using the Langmuir adsorption theory:

$$Y_{ki} = \frac{C_{ki} f_k}{1 + \sum_{j=1}^{N} C_{ji} f_j} \tag{13.3}$$

where f_k is the fugacity of component k. C_{ki} is the temperature-dependent adsorption constant, specific for the cavity of type i and for the gas component k. The adsorption constant accounts for the water–gas interactions in the hydrate lattice. In the work of van der Waals and Platteeuw (1959) and several later works (e.g., Parrish and Prausnitz 1972; Anderson and Prausnitz 1986), the adsorption constant C is calculated from a Kihara potential model (Kihara 1953):

$$C_{ki} = \frac{4\pi}{kT} \int_0^R \exp\left(\frac{-w(r)}{kT}\right) r^2 dr \tag{13.4}$$

where w(r) is the potential function of guest k in cavity i in the radial distance r from the center of the molecule. R is the radius of the cell, and k is the Boltzmann constant. The potential function is modeled using a Kihara spherical core pair potential. This is a function of r, R, and three Kihara parameters specific for each guest component. The Kihara parameters are estimated from experimental data for the hydrate formation conditions.

A somewhat simpler expression for C is (Parrish and Prausnitz 1972; Munck et al. 1988)

$$C_{ki} = \frac{A_{ki}}{T} \exp\left(\frac{B_{ki}}{T}\right) \tag{13.5}$$

For each compound (k) capable of entering into a cavity of type i, A_{ki} and B_{ki} must be determined from experimental data.

As can be seen from Equation 13.3, the fugacities of the components occupying the hydrate cavities enter into the expression used to calculate $(\mu_w^H - \mu_w^\beta)$ in Equation 13.1. At equilibrium, each component will have the same fugacity in all phases. The simulated hydrate formation conditions will therefore depend on the fugacities of the hydrate-forming components in the hydrocarbon gas and/or liquid phases in equilibrium with the hydrate phase. Component fugacities in hydrocarbon phases in equilibrium with a hydrate phase are usually calculated from a cubic equation of state (EOS), for example, the SRK (Equation 4.20) or the PR equation (Equation 4.36). The calculated component fugacities will differ slightly depending on which EOS is used. This may be accounted for by estimating hydrate parameters specific for the selected EOS. Table 13.4 shows hydrate parameters for use with the SRK and PR equations.

The chemical potential of water in the α-state may be expressed as

$$\mu_w^\alpha = \mu_w^0 + RT\ln\left(\frac{f_w^\alpha}{f_w^0}\right) \tag{13.6}$$

where μ_w^0 is the chemical potential of pure water as liquid or ice at temperature T. $f_w^\pm$ is the fugacity of water in the α-phase, and f_w^0 is the fugacity of pure ice or liquid water.

When a hydrate phase (H) exists in equilibrium with a fluid water phase (α), the following equilibrium criterion is fulfilled:

$$\mu_w^H = \mu_w^\alpha \tag{13.7}$$

This allows a combination of Equations 13.2 and 13.6 to

$$\frac{\mu_w^\beta - \mu_w^0}{RT} = \ln\left(\frac{f_w^\alpha}{f_w^0}\right) - \sum_{i=1}^{NCAV} v_i \ln\left(1 - \sum_{k=1}^{N} Y_{ki}\right) \tag{13.8}$$

expressing the difference in chemical potential between water in the state of an empty hydrate lattice and pure liquid or solid water.

TABLE 13.4

A and B Parameters for Calculating Langmuir Constants in Equation 13.5

Guest Component	Structure	Small Cavity		Large/Huge Cavity	
		A (K/bar)	B (K)	A (K/bar)	B (K)
Soave–Redlich–Kwong Equation (Equation 4.20)					
N_2	I	5.280×10^{-2}	932.3	3.415×10^{-2}	2240
N_2	II	7.507×10^{-3}	2004	9.477×10^{-2}	1596
N_2	H	1.318×10^{-5}	3795	—	—
CO_2	I	4.856×10^{-11}	7470	9.862×10^{-2}	2617
CO_2	II	6.082×10^{-5}	3691	1.683×10^{-1}	2591
H_2S	I	9.928×10^{-3}	2999	1.613×10^{-2}	3737
H_2S	II	2.684×10^{-4}	4242	8.553×10^{-1}	2325
C_1	I	4.792×10^{-2}	1594	1.244×10^{-2}	2952
C_1	II	2.317×10^{-3}	2777	1.076	1323
C_1	H	2.763×10^{-4}	3390	—	—
C_2	I	—	—	2.999×10^{-3}	3861
C_2	II	—	—	7.362×10^{-3}	4000
C_3	II	—	—	8.264×10^{-3}	4521
iC_4	II	—	—	8.189×10^{-2}	4013
nC_4	II	—	—	1.262×10^{-3}	4580
$c\text{-}C_5$	II	—	—	1.161×10^{-2}	5479
$c\text{-}C_6$	II	—	—	$4,365 \times 10^{-4}$	5951
Neo C_5	II	—	—	5.472×10^{-4}	5570
Benzene	II	—	—	2.628×10^{-4}	5951
iC_5	H	—	—	1.639×10^{4}	1699
2,2-Dimethylbutane	H	—	—	1.606×10^{3}	3175
2,3-Dimethylbutane	H	—	—	1.724×10^{2}	3608
2,2,3-Trimethylbutane	H	—	—	7.960×10^{8}	−39.00
3,3-Dimethylpentane	H	—	—	2.789×10^{3}	3183
Methylcyclopentane	H	—	—	6.336×10^{1}	4024
Methylcyclohexane	H	—	—	1.802×10^{3}	3604
Cis-1,2-Dimethylcyclohexane	H	—	—	6.873×10^{2}	4114
Ethylcyclopentane	H	—	—	1.315×10^{2}	4207
1,1-Dimethylcyclohexane	H	—	—	2.344×10^{3}	4089
Cycloheptane	H	—	—	$3,912 \times 10^{1}$	5050
Cyclooctane	H	—	—	1.625×10^{3}	4135
Peng–Robinson equation (Equation 4.36)					
N_2	I	6.915×10^{-2}	1740	3.342×10^{-2}	2028
N_2	II	6.558×10^{-2}	1444	1.530	229.0
N_2	H	4.836×10^{-5}	3555	—	—
CO_2	I	2.614×10^{1}	38.60	1.113×10^{-3}	3856
CO_2	II	3.071×10^{-3}	2652	4.824×10^{-3}	3183
H_2S	I	3.211×10^{-2}	3357	2.329×10^{-1}	2716
H_2S	II	7.187×10^{-2}	2548	9.357×10^{-4}	4221
C_1	I	8.287×10^{2}	−881.1	2.019×10^{-3}	3405
C_1	II	6.954×10^{-3}	1865	6.354×10^{-3}	2785
C_1	H	2.890×10^{-4}	3484	—	—
C_2	I	—	—	8.547×10^{-3}	3583
C_2	II	—	—	9.765×10^{-3}	3770
C_3	II	—	—	2.970×10^{-5}	6081

Guest Component	Structure	Small Cavity		Large/Huge Cavity	
		A (K/bar)	B (K)	A (K/bar)	B (K)
iC$_4$	II	—	—	2.372×10^{-3}	4988
nC$_4$	II	—	—	2.146×10^{-6}	6305
c-C$_5$	II	—	—	4.814×10^{-3}	5648
c-C$_6$	II	—	—	4.293×10^{-4}	5881
Neo C$_5$	II	—	—	6.684×10^{-2}	4099
Benzene	II	—	—	9.029×10^{-4}	5429
iC$_5$	H	—	—	4.248×10^{3}	1639
2,2-dimethylbutane	H	—	—	2.024×10^{3}	2685
2,3-dimethylbutane	H	—	—	1.062×10^{3}	2668
2,2,3-trimethylbutane	H	—	—	2.503×10^{8}	−177.3
3,3-dimethylpentane	H	—	—	1.223×10^{3}	2973
Methylcyclopentane	H	—	—	1.371×10^{2}	3413
Methylcyclohexane	H	—	—	1.127×10^{1}	4547
Cis-1,2-dimethylcyclohexane	H	—	—	1.364×10^{3}	3371
Ethylcyclopentane	H	—	—	4.550×10^{2}	3414
1,1-dimethylcyclohexane	H	—	—	2.243×10^{3}	3672
Cycloheptane	H	—	—	5.013	5036
Cyclooctane	H	—	—	1.278×10^{4}	3101

Source: From Munck, J., et al. Computations of the formation of gas hydrates, *Chem. Eng. Sci.* 43, 2661–2672, 1988; Madsen, J., et al. Modeling of structure H hydrates using a Langmuir adsorption model, *Ind. Eng. Chem. Res.* 39, 1111–1114, 2000; Rasmussen, C.P. and Pedersen, K.S., Challenges in modeling of gas hydrate phase equilibria, *4th International Conference on Gas Hydrates*, Yokohama Japan, May 19–23, 2002.

An alternative expression for the difference in chemical potential on the left-hand side of Equation 13.8 can be derived using the following general thermodynamic relation (can be derived from Equation A.5 in Appendix A):

$$d\left(\frac{\Delta\mu}{RT}\right) = -\frac{\Delta H}{RT^2}dT + \frac{\Delta V}{RT}dP \tag{13.9}$$

where R is the gas constant, and ΔH and ΔV are the changes in molar enthalpy and molar volume associated with the phase transition from fluid water or solid ice to empty hydrate. The left-hand side of Equation 13.8 may then be expressed as

$$\frac{\mu_w^{\beta} - \mu_w^0}{RT} = \frac{\Delta\mu_w^0}{RT_0} - \int_{T_0}^{T} \frac{\Delta H_0 + \Delta C_P(T - T_0)}{RT^2}dT + \int_{P_0}^{P} \frac{\Delta V}{R\overline{T}}dP \tag{13.10}$$

where T_0, P_0 indicate a reference state at which $\Delta\mu_w$ is known and equal to $\Delta\mu_w^0$. ΔH_0 is molar enthalpy difference between water in the empty hydrate lattice (β-state) and in the form of pure liquid or ice at temperature T_0. Similarly, ΔC_P is the difference in molar heat capacity of water in the β state and in the state of liquid water or ice.

In Equation 13.10, it has been assumed that ΔC_P is independent of T and ΔH independent of pressure. The temperature dependence of the second term has been approximated using the average temperature

$$\overline{T} = \frac{T + T_0}{2} \tag{13.11}$$

If the reference pressure P_0 is chosen to be equal to be zero, Equation 13.10 can be rewritten as

$$\frac{\mu_w^\beta - \mu_w^0}{RT} = \frac{\Delta\mu_w^0}{RT_0} - \int_{T_0}^{T} \frac{\Delta H_0 + \Delta C_P(T - T_0)}{RT^2} dT + \frac{P\Delta V}{R\bar{T}} \tag{13.12}$$

The constants needed to calculate the change in chemical potential for the $\beta \rightarrow \alpha$ transition are shown in Table 13.5 (from Erickson 1983; Mehta and Sloan 1996).

Equations 13.8 and 13.12 may be combined to

$$\frac{\Delta\mu_w^0}{RT_0} - \int_{T_0}^{T} \frac{\Delta H_0 + \Delta C_P(T - T_0)}{RT^2} dT + \frac{P\Delta V}{R\bar{T}} = \ln\left(\frac{f_w^\alpha}{f_w^0}\right) - \sum_{i=1}^{NCAV} v_i \ln\left(1 - \sum_{k=1}^{N} Y_{ki}\right) \tag{13.13}$$

which for a given P may be used to calculate the hydrate formation temperature, which is the highest T for which Equation 13.13 is fulfilled.

The fugacity of pure liquid water may be evaluated from a cubic equation of state. The fugacity, $f_w^\pm$, of water in a mixed liquid water phase may be derived using the methods in Chapter 16. When Equation 13.13 is fulfilled, the chemical potential of water in the α-phase equals the chemical potential of water in a hydrate phase. Taking Figures 13.1 and 13.2 as examples, hydrates may be present to the left of the curves, where

$$\mu_w^H - \mu_w^\alpha < 0 \tag{13.14}$$

In this equation, $\mu\alpha_w$ is the chemical potential of water in the water phase if no gas hydrates were formed and $\mu\eta_w$ is the chemical potential of water in the hydrate structure.

Whether this hydrate is structure I, structure II, or structure H depends on which of the three structures has the lowest chemical potential considering conditions (T and P) and mixture composition. To the right of the hydrate curves in Figures 13.1 and 13.2,

$$\mu_w^H - \mu_w^\alpha > 0 \tag{13.15}$$

At equilibrium, no hydrate can exist, and the water will be in the form of gas, liquid, or ice.

It is sometimes questioned whether gas hydrates may form from a gas phase, or a liquid or solid water phase is needed to initiate gas hydrate formation. From an equilibrium point of view, nothing prevents hydrates from forming a gas phase. What matters is the difference in chemical potential between water in the hydrate state and in the phase (gas, liquid, or ice) from which hydrates may

---

TABLE 13.5

Parameters for Phase Transition from Water in the Form of Liquid (Liq) of Ice (Ice) to Hydrate. ΔC_p Is Assumed to Be Independent of Temperature. The Other Quantities Are at 273.15 K

Property	Unit	Structure I	Structure II	Structure H
$\Delta\mu_w$ (liq)	J/mol	1264	883	1187.33
ΔH_0 (liq)	J/mol	−4858	−5201	−5162.43
ΔH_0 (ice)	J/mol	1151	808	846.57
ΔV(liq)	cm³/mol	4.6	5.0	5.45
ΔV(ice)	cm³/mol	3.0	3.4	3.85
ΔC_p (liq)	J/mol/K	−39.16	−39.16	−39.16

potentially form. However, it may take some time for the water molecules in a gas phase to get together and form a hydrate lattice. Hydrate formation may not start until after several hours, days, or even weeks. The time it takes from when the conditions are favorable for hydrate formation until hydrate formation actually starts is called the *induction period* (Skovborg et al. 1993), and the discipline dealing with hydrate growth from the time hydrate formation starts until the system reaches thermodynamic equilibrium is called *hydrate kinetics*.

13.3   HYDRATE INHIBITORS

The chemical potential of water in a liquid water phase, α, may be expressed as (Equations A.31 and A.33 in Appendix A)

$$\mu_w^\alpha = \mu_w^0 + RT \ln f_w^\alpha = \mu_w^0 + RT \ln \left(x_w^\alpha \, \phi_w^\alpha P \right) \tag{13.16}$$

where f stands for fugacity and ϕ for fugacity coefficient. If the water concentration is lowered maintaining a constant fugacity coefficient, the chemical potential of water will decrease. The water mole fraction (x_w) may be lowered by adding a substance soluble in water, for example, an alcohol or a glycol. By lowering the chemical potential of water in the aqueous phase, it becomes less favorable to form hydrates. For a fixed pressure, this means that the hydrate formation temperature becomes lower for a diluted water phase than for a pure water phase. Additives used to lower the hydrate formation temperature are called *hydrate inhibitors*. They act not only by diluting the water but the fugacity coefficient of water (ϕ_w) may also be lowered as a result of adding a water-soluble substance, which will further serve to decrease the hydrate formation temperature. The most commonly used hydrate inhibitors are methanol (MeOH) and mono, di, and tri ethylene glycol (MEG, DEG, and TEG). The salts naturally found in formation water also act as inhibitors. It has been suggested to add formate salts (Fadnes et al. 1998) to pipelines transporting hydrocarbons and water to suppress the hydrate formation temperature.

A number of empirical relations exist for calculating the effect of adding a hydrate inhibitor. The most well known is the Hammerschmidt (1969) correlation:

$$w_{inhibitor} = \frac{100 \, \Delta T}{\dfrac{K}{M_{Inhibitor}} + \Delta T} \tag{13.17}$$

ΔT is the temperature difference between the hydrate formation temperature if no inhibitor is present and the hydrate formation temperature if the water phase contains an inhibitor weight percentage of $w_{inhibitor}$. $M_{inhibitor}$ is the molecular weight of the inhibitor, and K is an inhibitor-dependent constant. Table 13.6 lists values of K for frequently used inhibitors. These values have been found from freezing point depression data for binary mixtures of water and the given inhibitor.

Instead of using a simplified expression to calculate the inhibitor effect, a rigorous thermodynamic model may be used for the water phase, as is further dealt with in Chapter 16.

TABLE 13.6
Values of K in Equation 13.17 for Frequently Used Inhibitors

Inhibitor	K (kg K/kmole)	Reference
MeOH	1623.96	Lide (1981) and Dean (1999)
NaCl	3695.32	Lide (1981)
KCl	3241.29	Lide (1981)
CaCl$_2$	9106.73	Lide (1981)

13.4 HYDRATE SIMULATION RESULTS

The simulation results in this chapter are based on the hydrate model of Munck et al. (1988). The SRK or the PR EOS is used for the fluid phases and the hydrate parameters entering into Equation 13.5 are taken from Table 13.4.

Figure 13.4 shows experimental and simulated data for the hydrate formation conditions of a mixture of pure water and C_1. The SRK EOS is used for the fluid phases.

As can be seen in Table 13.3, C_1 forms structure I hydrate in pure form. Even though C_1 is the dominant component in natural gas, most natural gases will form structure II hydrates. Natural gas also contains structure-II-forming components such as C_3, iC_4, and nC_4. As C_1 only has a slight preference for structure I, even small amounts of structure-II-forming components may be sufficient to shift the preference to structure II. Figure 13.5 shows experimental (Bisnoi and Dholabhai 1999)

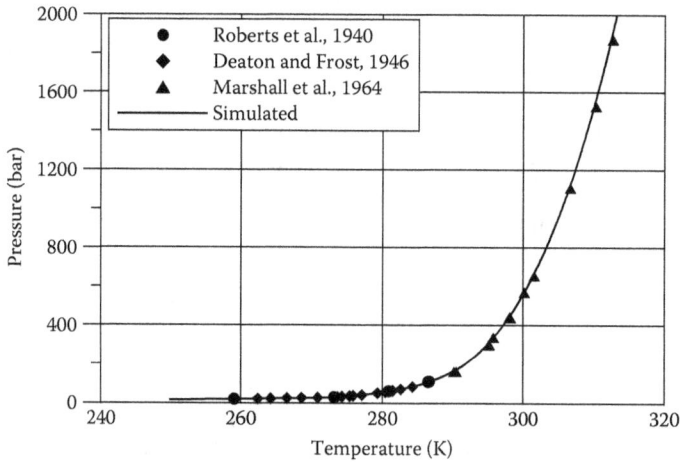

FIGURE 13.4 Experimental and simulated results for the hydrate formation conditions of methane (C_1). The SRK equation of state and Equation 13.5 with the SRK hydrate parameters in Table 13.4 have been used in the simulations.

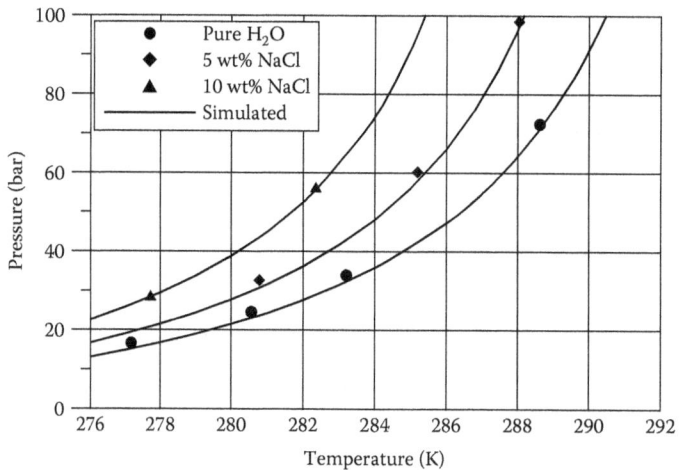

FIGURE 13.5 Experimental (Bisnoi and Dholabhai 1999) and simulated results for the hydrate formation conditions of the three-component gas mixture in Table 13.7 with varying concentration of NaCl in the aqueous phase. The PR equation of state with the Huron and Vidal mixing rule (Chapter 16) and Equation 13.5 the PR hydrate parameters in Table 13.4 have been used in the simulations.

and simulated results for the hydrate formation conditions of the three-component gas mixture in Table 13.7. The PR EOS with the Huron and Vidal mixing rule (Chapter 16) was used for the fluid phases. The figure illustrates that salts, in this case sodium chloride (NaCl), act as hydrate inhibitors.

Figure 13.6 shows experimental (Bisnoi and Dholabhai 1999) and simulated results for the hydrate formation conditions of the natural gas mixture in Table 13.7. Me15Na5 in Figure 13.6 means water containing 15 weight percentage methanol and 5 weight percentage NaCl. The PR EOS with the Huron and Vidal mixing rule (Chapter 16) has been used to simulate the fluid phases. The suppression of the hydrate formation temperature has been achieved by a combined influence of NaCl and methanol.

Figure 13.7 shows experimental (Ng et al. 1985) and simulated results for the hydrate formation conditions of the gas condensate mixture in Table 13.8. MEG stands for mono ethylene glycol and MeOH for methanol. The SRK EOS with the Huron and Vidal mixing rule (Chapter 16) was used.

TABLE 13.7

Gas Compositions (in Mol%) Whose Experimental (Bisnoi and Dholabhai 1999) and Simulated Hydrate Formation Conditions Can Be Seen from Figures 13.5 and 13.6

Component	Three-Component Mixture	Natural Gas
CO_2	20	0.5
C_1	78	82.0
C_2	—	11.3
C_3	2	4.2
iC_4	—	0.9
nC_4	—	0.6
iC_5	—	0.1
nC_5	—	0.2
nC_6	—	0.2

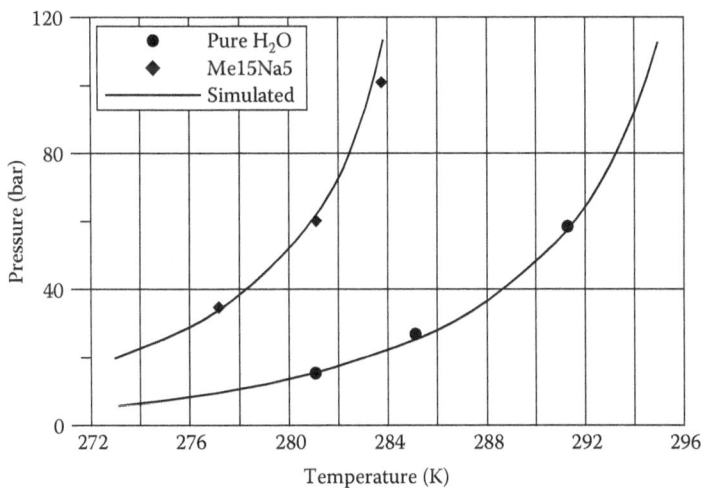

FIGURE 13.6 Experimental (Bisnoi and Dholabhai 1999) and simulated results for the hydrate formation conditions of the natural gas mixture in Table 13.7. Me15Na5 means water containing 15 weight percentage methanol and 5 weight percentage NaCl. The PR equation of state with the Huron and Vidal mixing rule (Chapter 16) and Equation 13.5 with the PR hydrate parameters in Table 13.4 have been used in the simulations.

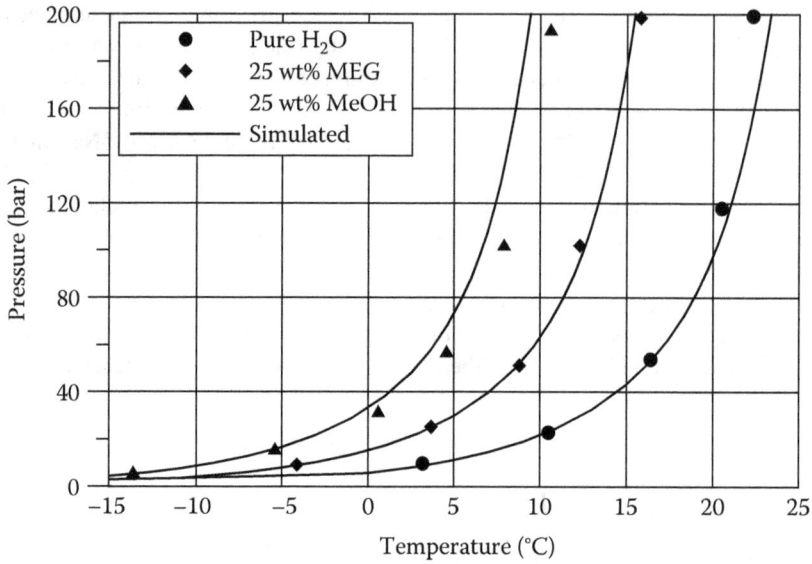

FIGURE 13.7 Experimental (Ng et al. 1985) and simulated results for the hydrate formation conditions of the gas condensate mixture in Table 13.8. MEG stands for mono ethylene glycol and MeOH for methanol. The SRK equation of state with the Huron and Vidal mixing rule (Chapter 16) and Equation 13.5 with the SRK hydrate parameters in Table 13.4 have been used in the simulations. The hydrocarbon fluid was characterized using the procedure of Pedersen et al. as described in Chapter 5.

TABLE 13.8

Composition of Gas Condensate Mixture Whose Experimental (Ng et al. 1985) and Simulated Hydrate Formation Data Can Be Seen from Figure 13.7

Component	Mole Percentage	Component	Mole Percentage
C_1	74.1333	C_{10}	0.6047
C_2	7.2086	C_{11}	0.3296
C_3	4.4999	C_{12}	0.1529
iC_4	0.8999	C_{13}	0.1012
nC_4	1.8088	C_{14}	0.0538
iC_5	0.8702	C_{15}	0.0208
nC_5	0.8889	C_{16}	0.0117
C_6	1.4582	C_{17}	0.0080
Methyl-cC_5	0.3635	C_{18}	0.0065
Benzene	0.0424	C_{19}	0.0021
c-C_6	0.7284	C_{20}	0.0014
C_7	1.5170	C_{21}	0.0008
Methyl-cC_6	1.1961	C_{22}	0.0007
Toluene	0.3874	C_{23}	0.0005
C_8	1.4400	C_{24}	0.0004
m-Xylene	0.3577	C_{25}	0.0004
o-Xylene	0.0654	C_{26}	0.0003
C_9	0.8364	C_{27}	0.0003

The hydrocarbon fluid was characterized using the procedure of Pedersen et al., as presented in Chapter 5. On a weight basis, MeOH is a more efficient hydrate inhibitor than MEG. The molecular weight of MeOH is 32.04 and is lower than that of MEG which is 62.07. A hydrate inhibitor primarily acts by diluting the water phase (see Equation 13.16). If the weight amount is the same, the number of MeOH molecules will be 62.07/32.04 times higher than the number of MEG molecules. This is the primary reason for MeOH being the better hydrate inhibitor of the two on a weight basis.

Figure 13.8 shows experimental (Ng et al. 1987) and simulated results for the hydrate formation conditions of the separator liquid in Table 13.9. Hydrate data is shown for a pure water aqueous phase and for aqueous phases with 13 and 25 weight% MeOH. The SRK EOS with the Huron and

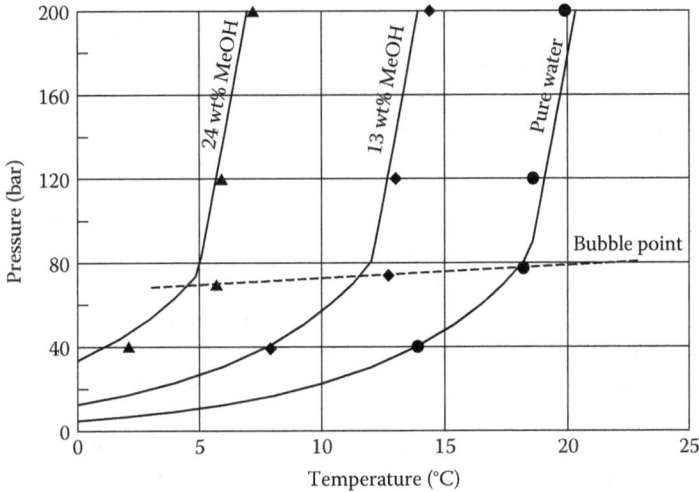

FIGURE 13.8 Experimental (Ng et al. 1987) and simulated results for the hydrate formation conditions of the separator liquid in Table 13.9. MeOH stands for methanol. The SRK equation of state with the Huron and Vidal mixing rule (Chapter 16) and Equation 13.5 with the SRK hydrate parameters in Table 13.4 have been used in the simulations. The separator liquid was characterized using the procedure of Pedersen et al. as described in Chapter 5.

TABLE 13.9

Composition of Separator Liquid Whose Experimental (Ng et al. 1987) and Simulated Hydrate Formation Data Can Be Seen from Figure 13.8

Component	Mole Percentage
N_2	0.16
CO_2	2.10
C_1	26.19
C_2	8.27
C_3	7.50
iC_4	1.83
nC_4	4.05
iC_5	1.85
nC_5	2.45
C_{6+}	45.60
Average M	90.2

Note: In the simulations, all C_{6+} is assumed to be C_{7+}, and a C_{7+} density of 0.84 g/cm³ is assumed.

Vidal mixing rule (Chapter 16) has been used to simulate the data. The hydrocarbon fluid was characterized using the procedure of Pedersen et al. as presented in Chapter 5. The hydrate curve is seen to be steeper above the bubble point of the separator liquid than at the lower pressures with both hydrocarbon gas and liquid present.

Figure 13.9 shows experimental (Deaton and Frost 1946) and simulated results for the hydrate formation conditions of a five-component gas mixture. The SRK and the PR equations of state with the Huron and Vidal mixing rule (Chapter 16) were used in the simulations. Different fugacities are obtained depending on which EOS is applied. Even though EOS-specific hydrate parameters are used to compensate for fugacity differences, there are still minor differences in the hydrate formation conditions depending on which EOS is used.

Despite MeOH being a more efficient inhibitor than MEG in the same weight concentration. MEG is often the preferred hydrate inhibitor. At 4°C the pure component vapor pressure of MEG is around 1.5×10^{-5} bar whereas MeOH at the same temperature has a pure component vapor pressure of around 5×10^{-2} bar. Because of this difference, the pollution of the gas phase will be higher with MeOH as inhibitor than with MEG.

In colder regions, large concentrations of MEG can be acquired to depress the hydrate-formation temperature enough to avoid hydrate formation. Hemmingsen et al. (2011) have published data for hydrate temperature depressions of a system consisting of 88.13 mol% methane and 11.87 mol% propane with MEG concentrations of up to 60 weight percent. The data is shown in Table 13.10. In general, the modeling of aqueous systems may require a non-classical mixing rule as described in Chapter 16, but a MEG-water mixture can be described well using a cubic EOS with classical mixing rules (Equations 4.33 to 4.35) with a H_2O-MEG k_{ij} of -0.063 for SRK and -0.065 for PR. PR simulation results for the data in Table 13.10 are shown in Figure 13.10.

Processing of gas mixtures may lead to severe cooling. The gas will often contain trace amounts of water and the cooling may potentially lead to the formation of gas hydrates. Løkken et al. (2008) have determined the water concentration in a natural gas in equilibrium with an incipient amount of gas hydrates at temperatures ranging from 20°C to −20°C. A gas in equilibrium

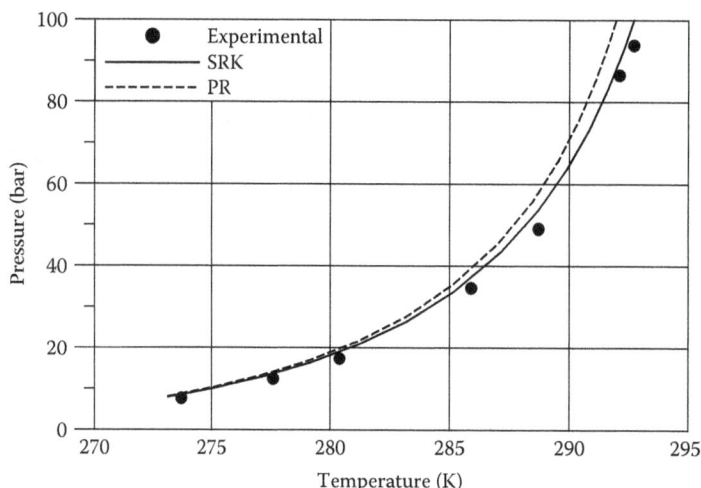

FIGURE 13.9 Experimental (Deaton and Frost 1946) and simulated results for the hydrate formation conditions of a gas consisting of (mole percentages) N_2: 0.3, CO_2: 0.4, C_1: 91.0, C_2: 3.2, C_3: 2.0, and C_4: 3.1. The SRK and the PR equations of state with the Huron and Vidal mixing rule (Chapter 16) and Equation 13.5 with the hydrate parameters in Table 13.4 have been used in the simulations.

TABLE 13.10

Experimental Hydrate Dissociation Temperatures for Hydrates Formed from a Gas Composed of 88.13 mol% Methane and 11.87 mol% Propane in the Presence of and Aqueous Mono Ethylene Glycol (MEG) Solutions Consisting of Distilled Water and MEG

Weight% MEG	Pressure bar	Temperature (°C) ±0.1
0 (distilled water)	11.2	6.8
	80.5	22.2
	133.6	24.4
40	12.2	−5.2
	84.8	8.8
	133.3	10.2
50	22.5	−5.2
	89.4	2.1
	174.9	4.3
60	23.6	−14.3
	93.2	−6.7
	172.7	−5.3

Source: Hemmingsen P.V. et al., Hydrate temperature depression of MEG solutions at concentration up to 60 wt%. Experimental data and simulation results. *Fluid Phase Equilib*. 307, 175–179, 2011.

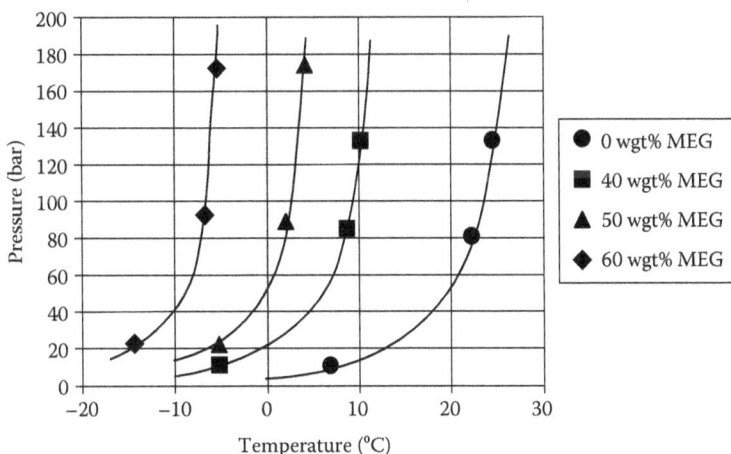

FIGURE 13.10 Hydrate formation temperatures for a mixture of 88.13 mol% methane and 11.87 mol% propane with mono ethylene glycol (MEG) concentrations from 0 to 60 weight percent (Hemmingsen et al. 2011). The fluid phases were modeled using the PR equation with classical mixing rules. A $k_{ij} = -0.065$ was used for H_2O-MEG.

with hydrates will be at its hydrate onset point and the data Løkken et al. of can be used to evaluate how high a water concentration is allowed in a gas before there is a risk of hydrate formation. The gas composition is shown in Table 13.11. Table 13.12 shows the water content in the gas in equilibrium with hydrate at 150 bar. The data is plotted in Figure 13.11, which also shows simulation results. The SRK equation with the Huron and Vidal mixing rule (Chapter 16) was applied for the gas phase.

TABLE 13.11
Gas Composition Used in Study of Water Concentration in Gas Phase in Equilibrium with Gas Hydrates (Data Is Shown in Table 13.12)

Component	Mol%
Nitrogen	0.6032
Carbon dioxide	2.6094
Methane	80.138
Ethane	9.4689
Propane	4.6227
i-Butane	0.6420
n-Butane	1.1427
2,2-dimethylpropane	0.0136
i-Pentane	0.2349
n-Pentane	0.2272
Cyclopentane	0.0121
2,2-dimethylbutane	0.0031
2,3-dimethylbutane	0.0068
2-methylpentane	0.0416
3-methylpentane	0.0216
C_6	0.0535
C_7	0.1056
C_8	0.0441
C_9	0.0074
C_{10}	0.0016
C_{11}	0.00011
C_{12}	0.00004
C_{13}	0.00004

TABLE 13.12
Gas Phase Water Concentration in Natural Gas in Table 13.11 in Equilibrium with Gas Hydrate at 150 bar. 1 ppm mol Equals 0.0001 mol%

Temperature (°C)	H_2O Concentration (ppm mol) in Gas Phase
−20	19
−10	37
0	72
10	150
20	295

Source: Løkken T.V. et al., Water content of high pressure natural gas: Data, prediction and experience from field, presented at *International Gas Union Research Conference*, Paris, October 8–10, 2008.

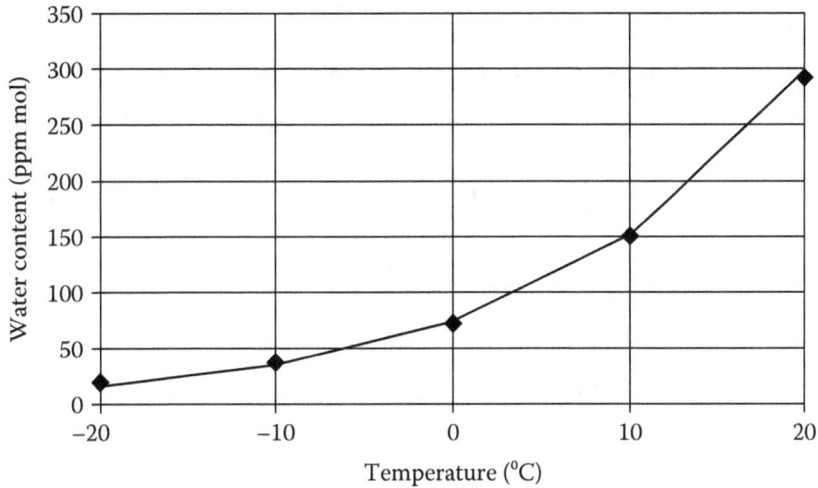

FIGURE 13.11   Measured (Løkken et al. 2008) and simulated water concentrations in gas phase in equilibrium with gas hydrate at a pressure of 150 bar. The gas composition is given in Table 13.11. The SRK equation with the Huron and Vidal mixing rule (Chapter 16) has been used to simulate the gas phase.

TABLE 13.13

Hydrate Formation Data for C_1 Mixed with a $CaBr_2$ Brine Solution and for Mixture of C_1 and C_2 Mixed with a KCl Brine Solution. The Data Is Plotted in Figure 13.12

74.7 mol% C_1 and 25.3 mol% C_2 + 32 weight% $CaBr_2$ Brine		C_1 + 20 weight% KCl Brine		C_1 + H_2O	
Pressure bar	Temperature °C	Pressure bar	Temperature °C	Pressure bar	Temperature °C
375	9.39	185.5	9.11	343.0	22.60
656	13.28	384.0	14.22	693.0	28.50
903	16.22	603.8	17.78	1037.2	32.60
1210	19.00	806.7	20.39		
1491	21.39	996.9	22.50		
1818	23.78				

Data Sources: Hu, Y., Makogan, T.Y., Karanjkar, P.U., Lee, K.-H., Lee, B.R. and Sum, R.K., Gas hydrate phase equilibrium with $CaBr_2$ and $CaBr_2$+MEG at ultra-high pressures, *The Journal of Natural Gas Engineering* 2, 42–49, 2017a. Hu, Y., Lee, K-.H., Lee, B.R., Sum, A.K., Gas hydrate formation from high concentration KCl brines at ultra-high pressures, *Journal of Industrial and Engineering Chemistry* 50, 142–146, 2017b.

Table 13.13 shows experimental hydrate formation data for C_1 contacted with a $CaBr_2$ brine solution and for a mixture of C_1 and C_2 contacted with a KCl brine solution (Hu et al. 2017a, 2017b). The experimental data is plotted in Figure 13.12 with simulation results with the SRK equation. The Huron and Vidal mixing rule is used for C_1-H_2O and C_2-H_2O, while H_2O-$CaBr_2$ and H_2O-KCl are represented using the classical mixing rule with k_{ij}s of respectively −0.2011 and −0.1585.

TABLE 13.14

Results of Flash for 100 bar and 4°C for a Natural Gas Mixed with Water and Methanol (MeOH)

	Feed Mol%	HC Vapor Mol%	HC Liquid Mol%	Aqueous Phase Mol%	Hydrate II Mol%
H_2O	45.00	0.01	0.01	79.67	85.64
MeOH	5.00	0.10	0.23	19.88	0.00
N_2	0.32	0.74	0.42	0.00	0.03
CO_2	0.41	0.82	0.84	0.05	0.07
C_1	35.72	74.85	59.29	0.36	9.44
C_2	6.19	12.01	15.26	0.04	1.36
C_3	5.01	7.76	13.95	0.01	3.09
iC_4	0.54	0.80	1.82	0.00	0.28
nC_4	1.33	2.20	5.55	0.00	0.08
iC_5	0.19	0.29	0.95	0.00	0.00
nC_5	0.21	0.31	1.11	0.00	0.00
nC_6	0.105	0.12	0.58	0.00	0.00
Total	100.00	36.87	8.86	24.87	29.40

Note: The SRK equation of state with the Huron and Vidal mixing rules has been used for the fluid phases. For the hydrate phase, the SRK parameters in Table 13.4 have been used. HC stands for hydrocarbon.

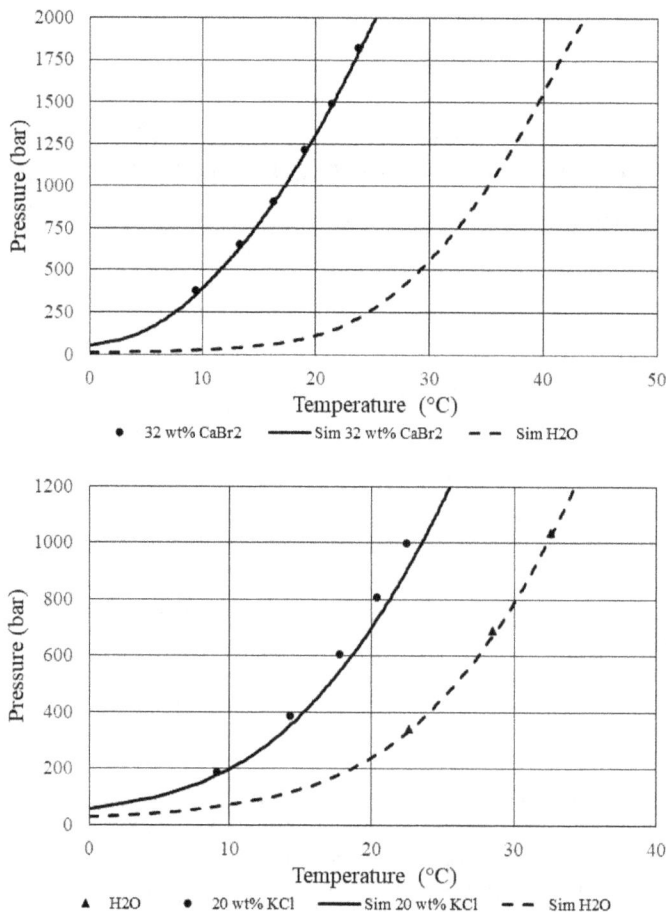

FIGURE 13.12 Experimental and simulated hydrate formation conditions for C_1+C_2 mixture with 32 weight% $CaBr_2$ brine (upper) and C_1 with 20 weight% KCl brine (lower). The experimental data is shown in Table 13.13.

13.5 HYDRATE P/T FLASH CALCULATIONS

Hydrate formation cannot always be prevented in pipelines transporting unprocessed well streams carrying formation water. It may not be economically feasible to insulate a pipeline sufficiently to stay out of the hydrate region, and there may be environmental or technical reasons for not using hydrate inhibitors. Especially in arctic regions and when producing from deepwater reservoirs, the use of hydrate inhibitors may not be an option, and such fields cannot be brought into production unless it is possible to find a way to transport solid hydrates as slurries. In these situations, it is essential to be able to quantitatively determine the amount of hydrates formed. It is, in other words, necessary to be able to carry out flash calculations (see Chapter 6) where at least one phase is a gas hydrate phase. This requires a method for evaluating component fugacities in a hydrate phase.

13.5.1 HYDRATE FUGACITIES

A hydrate structure contains water and guest molecules, the fugacities of which may be calculated as described in the following (Cole and Goodwin 1990; Michelsen 1991). Only structure I and structure II hydrates are considered.

Equations 13.2 and 13.3 may be combined to give

$$\left(\mu_w^H - \mu_w^\beta\right) = RT \sum_{i=1}^{NCAV} v_i \ln\left(\frac{1+\sum_{j=1}^{N}C_{ji}f_j}{1+\sum_{j=1}^{N}C_{ji}f_j} - \frac{\sum_{k=1}^{N}C_{ki}f_k}{1+\sum_{j=1}^{N}C_{ji}f_j}\right)$$

(13.18)

$$= RT \sum_{i=1}^{NCAV} v_i \ln\left(\frac{1}{1+\sum_{j=1}^{N}C_{ji}f_j}\right) = -RT \sum_{i=1}^{NCAV} v_i \ln\left(1+\sum_{j=1}^{N}C_{ji}f_j\right)$$

Expanding this equation into two cavities and making use of the relation between chemical potential and fugacity (Equation A.31 in Appendix A), the following expression may be derived for the fugacity of water in the hydrate phase

$$\ln f_w^H - \ln f_w^\beta = -v_1 \ln\left(1+\sum_{k=1}^{N}C_{k1}f_k\right) - v_2 \ln\left(1+\sum_{k=1}^{N}C_{k2}f_k\right)$$

(13.19)

where f_w^β is the fugacity of water in the empty hydrate lattice, which may be derived from Equation 13.12:

$$\ln\left(\frac{f_w^\beta}{f_w^0}\right) = \frac{\Delta\mu_w^0}{RT_0} - \int_{T_0}^{T}\frac{\Delta H_0 + \Delta C_P(T-T_0)}{RT^2}dT + \frac{P\Delta V}{R\overline{T}}$$

(13.20)

where f_w^0 is the fugacity of pure water. In Equation 13.3, Y_{ki} expressed the probability that cavity i is occupied by a molecule of type k. Y_{ki} may also be regarded as the fractional occupancy, n_{ki}, of molecules of type k in cavity i

$$n_{ki} = \frac{C_{ki}f_k}{1+\sum_{j=1}^{N}C_{ji}f_j}; k = 1, 2, \ldots, N$$

(13.21)

N is the number of components. Equation 13.21 enables calculation of the ratio N_k between the mole fraction of molecules of type k (x_k) and the mole fraction of water (x_w) in a hydrate structure:

$$N_k \frac{x_k}{x_w} = v_1 n_{k1} + v_2 n_{k2} \tag{13.22}$$

where v_1 is the number of small cavities per water molecule in the considered hydrate structure (I or II) and v_2 the number of large cavities per mole water in the same hydrate structure. The fraction of empty cavities of type 1 and 2 becomes

$$n_{01} = 1 - \sum_{k=1}^{N} n_{k1} = \frac{1}{1 + \sum\limits_{k=1}^{N} C_{k1} f_k}; \quad n_{02} = 1 - \sum_{k=1}^{N} n_{k2} = \frac{1}{1 + \sum\limits_{k=1}^{N} C_{k2} f_k} \tag{13.23}$$

Equation 13.23 may be combined with Equation 13.21 to give

$$n_{k1} = C_{k1} n_{01} f_k; \quad n_{k2} = C_{k2} n_{02} f_k \tag{13.24}$$

which may also be expressed as

$$\frac{n_{k1}}{n_{k2}} = \frac{C_{k1}}{C_{k2}} \frac{n_{01}}{n_{02}} = \alpha_k \frac{n_{01}}{n_{02}}, \quad \text{where } \alpha_k = \frac{C_{k1}}{C_{k2}} \tag{13.25}$$

The total number of empty cavities per mole of water (N_0) can be expressed as

$$N_0 = v_1 n_{01} + v_2 n_{02} = v_1 + v_2 - \sum_{k=1}^{N} N_k \tag{13.26}$$

The term $\theta = \dfrac{v_2 n_{02}}{N_0}$ is introduced and represents the number of empty large cavities divided by the total number of empty cavities. Substitution of θ into Equation 13.26 gives

$$v_1 n_{01} = (1 - \theta) N_0 \tag{13.27}$$

Equations 13.22 and 13.25 may be combined to give

$$v_1 n_{k1} = N_k \frac{\alpha_k (1 - \theta)}{\theta + \alpha_k (1 - \theta)}; \quad v_2 n_{k2} = N_k \frac{\theta}{\theta + \alpha_k (1 - \theta)} \tag{13.28}$$

The fraction of empty and filled cavities of type 2 must sum to 1, which can be expressed as

$$v_2 \left(\sum_{k=1}^{N} n_{k2} + n_{02} \right) = v_2 \tag{13.29}$$

This equation may, using the definition of θ and Equation 13.28, be rewritten to

$$F(\theta) = \sum_{k=1}^{N} N_k \frac{\theta}{\theta + \alpha_k (1 - \theta)} + \theta N_0 - v_2 = 0 \tag{13.30}$$

Whereas smaller molecules may enter both cavities, larger ones can only enter into the large (type 2) cavities. For the latter type of molecules, α_k becomes zero. If the first N_S components can enter both cavities and the last $N-N_S$ components can only enter into type 2 cavities, Equation 13.30 may be rewritten to

$$F(\theta) = \sum_{k=1}^{N_S} N_k \frac{\theta}{\theta + \alpha_k (1-\theta)} + \sum_{k=N_S+1}^{N} N_k + \theta N_0 - v_2 = 0 \qquad (13.31)$$

F is monotonically increasing with θ what makes it fairly straightforward to determine θ fulfilling Equation 13.31.

By substituting n_{k2} from Equation 13.24 and $v_2 = \dfrac{\theta N_0}{n_{02}}$ into Equation 13.28 (second equation), the following expression can be derived for the fugacities.

$$f_k = \frac{N_k}{N_0} \frac{1}{C_{k2}} \frac{1}{\theta + \alpha_k (1-\theta)} \qquad (13.32)$$

Equations 13.24, 13.27, 13.32, and 13.19 may be expressed as

$$\ln f_w^H - \ln f_w^\beta = v_1 \ln \left[\frac{N_0(1-\theta)}{v_1} \right] + v_2 \ln \left[\frac{N_0 \theta}{v_2} \right] \qquad (13.33)$$

13.5.2 Flash Simulation Technique

To make flash calculations considering hydrate phases, the procedures in Chapter 6 must be slightly modified. The techniques in Chapter 6 are based on the assumption that the component fugacities do not reach extreme values for any concentration. If all the cavities in a hydrate structure are filled with guest molecules, the addition of more guest molecules will make the fugacity of the guest molecules approach infinity. An overloaded hydrate lattice in one of the iterative steps of a hydrate flash calculation will make it almost impossible to converge a hydrate flash calculation.

For this reason, hydrate flash calculations are made using an "inverse" calculation procedure (Bisnoi et al. 1989).

1. Initial estimates are established of the fugacity coefficients of all the components in all phases except in the hydrate phases and in any pure solid phases. The hydrocarbon components are split between a gas and a liquid phase using the Wilson K-factor approximation (Equation 6.13). The K-factor of component i is defined as $y_i/x_i = \varphi_i^L/\varphi_i^V$, where y_i is the mole fraction of component i in the vapor phase, x_i the mole fraction of component i in the liquid phase, φ_i^V the fugacity coefficient of component i in the vapor phase, and φ_i^L the fugacity coefficient of component i in the liquid phase. The gas phase is assumed to be an ideal gas, that is, all component fugacity coefficients are assumed to be equal to 1.0. With this assumption, the liquid phase fugacity coefficients become equal to the K-factors. The hydrocarbon phases are assumed to be free of aqueous components, and the water phase free of hydrocarbons. The component fugacity coefficients in the water are assumed equal to those of the aqueous components in pure form at the actual conditions. This corresponds to assuming that the water phase is an ideal solution.
2. The phase amounts and compositions corresponding to the fugacity coefficients established in 1 are calculated (Michelsen 1988).

3. Mixture fugacities (f_k^{mix}, k = 1, 2, ..., N) are calculated. For a nonaqueous component, the mixture fugacity equals the molar average of the fugacities of the given component in the hydrocarbon phases. For the aqueous components, the mixture fugacity is set equal to the fugacity of the component in the water phase.
4. A correction term θ is calculated (not to be mistaken for θ in Subsection 13.5.1), which is based on Equations 13.2 and 13.3:

$$\theta = \ln f_w^H - \ln f_w^{mix} = \sum_{i=1}^{NCAV} v_i \ln\left(1 - \sum_{k=1}^{N} \frac{C_{ki}f_k}{1 + \sum_{j=1}^{N} C_{ji}f_j}\right) + \ln f_w^\beta - \ln f_w^{mix} \tag{13.34}$$

The superscript H means hydrate phase, and β refers to the empty hydrate lattice. New estimates of the component fugacities in the hydrate phase are obtained from

$$\ln f_k^H = \ln f_k^{mix} + \theta \quad k = 1, 2, ..., N \tag{13.35}$$

5. With NHYD hydrate forming components, hydrate-phase compositions are estimated from

$$\frac{x_k}{x_w} = \sum_{i=1}^{NCAV} v_i \frac{C_{ki}f_k}{1 + \sum_{j=1}^{NHYD} C_{ji}f_j} \quad k = 1, 2, ..., N \tag{13.36}$$

which is derived by combining Equations 13.21 and 13.22. It enables the fugacity coefficient of water and other hydrate-forming components to be calculated. To prevent nonhydrate formers from entering into the hydrate phases, nonhydrate formers are assigned large fugacity coefficients ($\ln \varphi_i = 50$) in the hydrate phase.

6. Based on the current estimate of the component fugacity coefficients in each phase, a new estimate of the phase amounts and phase compositions is established (Michelsen 1988).
7. If not converged, repeat from 3.

The preceding concept can easily be extended to also handle ice and other pure solid phases. All it requires is a fugacity expression for the solid phase as a function of temperature.

Boesen et al. (2014) have presented another approach for hydrate flash calculations. The flash algorithm considers the hydrocarbon fluid phases and the aqueous phases (incl. hydrate phases) as two sub-systems. Thermodynamic equilibrium exists within each sub-system whereas equilibrium between the two sub-systems will develop as a result of component exchange between the phases. The algorithm is well suited to study hydrate kinetics but can also be used in phase equilibrium calculations when good initial estimates exist. For example, in flow simulations a good estimate will often exist from the previous segment or the previous time step.

Table 13.14 shows the results of a flash calculation for a natural gas mixed with water and methanol. Four phases are present at equilibrium. As can be seen from the table, not all methanol ends up in the aqueous phase. Some is lost to the hydrocarbon phases. This must be taken into consideration when calculating the amount of methanol needed to inhibit hydrate formation in a hydrocarbon-water mixture.

REFERENCES

Anderson, F.E. and Prausnitz, J.M., Inhibition of gas hydrates by methanol, *AICHE J.* 32, 1329–1333, 1986.

Bisnoi, P.R. and Dholabhai, P.D., Equilibrium conditions for hydrate formation for a ternary mixture of methane, propane and carbon dioxide, and a natural gas mixture in the presence of electrolytes and methanol, *Fluid Phase Equilib.* 158–160, 821–827, 1999.

Bisnoi, P.R., Gupta, A.K., Englezos, P., and Kalogerakis, N., Multiphase equilibrium flash calculations for systems containing gas hydrates, *Fluid Phase Equilib.* 53, 97–104, 1989.

Boesen, R.B., Sørensen, H., and Pedersen, K.S., New approach for hydrate flash calculations, presented at the *8th International Conference on Gas Hydrates*, Beijing, China, July 29–August 1, 2014.

Cole, W.A. and Goodwin, S.P., Flash calculations for gas hydrates: A rigorous approach, *Chem. Eng. Sci.* 45, 569–573, 1990.

Danesh, A., Tohidi, B., Burgass, R.W., and Todd, A.C., Benzene can form gas hydrates, *Trans. ICHEME.* 71(Pt. A), 457–459, 1993.

Dean, J.A., *Lange's Handbook of Chemistry*, 15th ed., McGraw-Hill, New York, 1999.

Deaton, W.M. and Frost, E.M., *Gas Hydrates and Their Relation to the Operation of Natural-Gas Pipe Lines*, U.S. Bureau of Mines Monograph 8, Washington D.C. US, 1946.

Erickson, D.D., *Development of a Natural Gas Hydrate Prediction Computer Program*, M.Sc. thesis, Colorado School of Mines, 1983.

Fadnes, F.H., Jacobsen, T., Bylov, M., Holst, A., and Downs, J.D., Studies on the prevention of gas hydrates formation in pipelines using potassium formate as a thermodynamic inhibitor, SPE 50688, presented at the *SPE European Petroleum Conference in the Hague*, The Netherlands, October 20–22, 1998.

Hammerschmidt, E.G., Possible technical control of hydrate formation in natural gas pipelines, *Brennstoff-Chemie.* 50, 1969, 117–123.

Hemmingsen, P.V., Burgass, R., Pedersen, K.S., Kinnari, K., and Sørensen, H., Hydrate temperature depression of MEG solutions at concentration up to 60 wt%. Experimental data and simulation results. *Fluid Phase Equilib.* 307, 175–179, 2011.

Holder, G.D. and Bisnoi, P.R., *Gas Hydrates—Challenges for the Future*, Academy of Sciences, New York, 2000.

Hu, Y., Lee, K.-.H., Lee, B.R., and Sum, A.K., Gas hydrate formation from high concentration KCl brines at ultra-high pressures, *J. Ind. Eng. Chem.* 50, 142–146, 2017b.

Hu, Y., Makogan, T.Y., Karanjkar, P.U., Lee, K.-H., Lee, B.R., and Sum, R.K., Gas hydrate phase equilibrium with CaBr2 and CaBr$_2$+MEG at ultra-high pressures, *J. Nat. Gas Eng.* 2, 42–49, 2017a.

Kihara, T., Virial coefficients and models and molecules in gases, *Rev. Mod. Phys.* 25, 831–843, 1953.

Lide, D.R., *Handbook of Chemistry and Physics*, 62nd ed., CRC Press, Boca Raton, FL, 1981.

Løkken, T.V., Berskås, A., Christensen, K.O., Nygaard, C.F., and Solbraa, E., Water content of high pressure natural gas: Data, prediction and experience from field, presented at the *International Gas Union Research Conference*, Paris, October 8–10, 2008.

Madsen, J., Pedersen, K.S., and Michelsen, M.L., Modeling of structure H hydrates using a Langmuir adsorption model, *Ind. Eng. Chem. Res.* 39, 1111–1114, 2000.

Makogan, Y.F., *Hydrates of Hydrocarbons*, PennWell Publishing Company, Tulsa, OK, 1997.

Mehta, P.A. and Sloan, E.D., Improved thermodynamic parameters for prediction of structure H hydrate equilibria, *AICHE J.* 42, 2036–2046, 1996.

Mehta, P.A. and Sloan, E.D., Structure H hydrates: Implications for the petroleum industry, *SPE J.* 4, 3–8, 1999.

Michelsen, M.L., Calculation of multiphase equilibrium in ideal solutions, SEP 8802, *The Department of Chemical Engineering*, The Technical University of Denmark, 1988.

Michelsen, M.L., Calculation of hydrate fugacities, *Chem. Eng. Sci.* 46, 1192–1193, 1991.

Munck, J., Skjold-Jørgensen, S., and Rasmussen, P., Computations of the formation of gas hydrates, *Chem. Eng. Sci.* 43, 2661–2672, 1988.

Ng, H.-J., Chen, C.-J., and Robinson, D.B., The effect of ethylene glycol or methanol on hydrate formation in systems containing ethane, propane, carbon dioxide, hydrogen sulfide or a typical gas condensate, *Research Report RR-92*, Gas Processors Association, Tulsa, Oklahoma, 1985.

Ng, H.-J., Chen, C.-J., and Sæterstad, T., Hydrate formation and inhibition in gas condensate and hydrocarbon liquid system, *Fluid Phase Equilib.* 36, 99–106, 1987.

Parrish, W.R. and Prausnitz, J.M., Dissociation pressures of gas hydrates formed by gas mixtures, *Ind. Eng. Chem. Process Des. Dev.* 11, 26–35, 1972.

Rasmussen, C.P. and Pedersen, K.S., Challenges in modeling of gas hydrate phase equilibria, presented at the *4th International Conference on Gas Hydrates*, Yokohama, Japan, May 19–23, 2002.

Ripmeester, J.A., Tse, J.S., Ratcliffe, C.I., and Powell, B.M., A new clathrate hydrate structure, *Nature (London)* 325, 135–136, 1987.

Skovborg, P., Ng, H.J., Rasmussen, P., and Mohn, U., Measurement of induction times for the formation of methane and ethane gas hydrates, *Chem. Eng. Sci.* 48, 445–453, 1993.

Sloan, E.D. and Koh, C., *Clathrate Hydrates of Natural Gases*, 3rd ed., CRC Press, Boca Raton, 2008.

Tohidi, B., Danesh, A., Burgass, R.W., and Todd, A.C., Equilibrium data and thermodynamic modelling of cyclohexane gas hydrates, *Chem. Eng. Sci.* 51, 159–163, 1996.

Tohidi, B., Danesh, A., Todd, A.C., Burgass, R.W., and Østergaard, K.K., Equilibrium data and thermodynamic modelling of cyclopentane and neopentane hydrates, *Fluid Phase Equilib.* 138, 241–250, 1997.

van der Waals, J.H. and Platteeuw, J.C., Clathrate solutions, *Adv. Chem. Phys.* 2, 1–57, 1959.

14 Compositional Variations with Depth

Petroleum reservoirs show variations in pressure, temperature, and composition with depth. This is illustrated in Figure 14.1, which schematically shows pressure, temperature, and C_1 mole fraction at three different depths. Pressure and temperature increase with depth. The concentration of lighter components decreases with depth, whereas the concentration of heavier components increases.

Table 14.1 shows typical changes in pressure, temperature, and composition per 100-m vertical depth in a petroleum reservoir. Gravity, temperature gradient and viscosity influence the compositional gradient with depth. Capillary forces, convection, and secondary migration of hydrocarbons into the reservoir may also influence the gradients, but only gravity, temperature gradient and viscosity will be considered here.

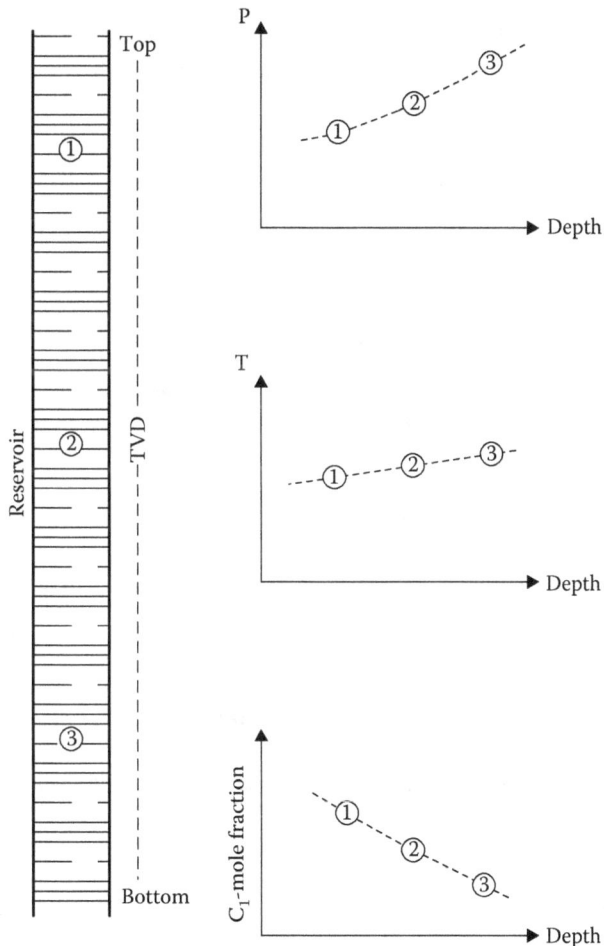

FIGURE 14.1 Variation in pressure, temperature, and C_1 concentration with depth. TVD stands for true vertical depth.

DOI: 10.1201/9780429457418-14

TABLE 14.1

Typical Changes with Depth in a Petroleum Reservoir

Effect per 100 m Downwards	Typical Oil	Typical Gas
Pressure (bar)	4–6	2
Temperature (°C)	2	2
Saturation point (bar)	–8	8
Molecular weight	8–21	0.3–0.4
Density (g/cm³)	0.02–0.15	0.005–0.007
C_1 mole percentage	–1.6	–0.5
C_2–C_6 mole percentage	–0.2	–0.2
C_{7+} mole percentage	1.8–1.9	0.2–0.3

14.1   ISOTHERMAL RESERVOIRS

At equilibrium in a closed system with negligible height differences, the chemical potential (and fugacity) of a component i is the same at all positions in the system. The criterion of equal chemical potentials is, however, not valid for a system with considerable height differences. For such a system, height differences must also be considered.

14.1.1   THEORY OF ISOTHERMAL RESERVOIRS

For an isothermal system, the equilibrium relation for component i in two different depths, a reference depth, h_0, and another depth, h, becomes

$$\mu_i(h) - \mu_i(h^\circ) = M_i\, g\,(h - h^\circ) \tag{14.1}$$

where μ is the chemical potential, h stands for depth, M for molecular weight, and g for gravitational acceleration. h° is a reference depth. The chemical potential is related to the fugacity through the following relation (Appendix A):

$$d\mu_i = RT\, d\ln f_i = RT\, d\ln(\varphi_i x_i P) \tag{14.2}$$

where T is the temperature. For an isothermal reservoir, Equation 14.1 may, using this equation, be rewritten to

$$\ln f_i^h - \ln f_i^{h^\circ} = \frac{M_i\, g\,(h - h^\circ)}{RT} \tag{14.3}$$

The fugacity and the fugacity coefficient of component i are related through (Equation A.33)

$$f_i = \varphi_i\, z_i\, P \tag{14.4}$$

and Equation 14.3 may for an N component system be written as

$$\ln(\varphi_i^h z_i^h P^h) - \ln(\varphi_i^{h0} z_i^{h0} P^{h0}) = \frac{M_i g(h - h^\circ)}{RT}; i = 1,\ 2,...,\ N \tag{14.5}$$

The mole fractions of the components must sum to 1.0 giving one additional equation:

$$\sum_{i=1}^{N} z_i = 1 \qquad (14.6)$$

If the pressure P^{h° and the composition ($z_i^{h^\circ}$, $i = 1, 2, ..., N$) are known in the reference depth h°, there are $N + 1$ variables for a given depth h, namely, (z_i^h, $i = 1, 2, ..., N$) and P^h.

A set of $N + 1$ equations with $N + 1$ variables may be solved to give the molar composition and pressure as a function of height. Schulte (1980) has outlined how to solve these equations and has also given examples of compositional gradients calculated using Equations 14.5 and 14.6 with fugacity coefficients derived from an equation of state.

It is stated in Chapter 4 that the Peneloux volume correction (parameter c in Equations 4.43 and 4.48) does not influence equilibrium phase compositions. This is true for a separator, a PVT cell, and other systems at a constant pressure, but not for a reservoir with pressure increasing with depth. The fugacity coefficient of component i, derived from the original SRK or PR equations (Soave 1972; Peng and Robinson 1976) are related to the Peneloux corrected (Peneloux et al. 1982) fugacity coefficient as follows:

$$\ln\varphi_{i,SRK} - \ln\varphi_{i,SRK\text{-}Pen} = \frac{c_i P}{RT} \qquad (14.7)$$

$$\ln\varphi_{i,PR} - \ln\varphi_{i,PR\text{-}Pen} = \frac{c_i P}{RT} \qquad (14.8)$$

where c is the volume correction term. In a usual phase equilibrium calculation, temperature and pressure are the same throughout the system, and the volume correction term cancels out as shown in Equation 4.52. This is not the case in a calculation of the compositional variation with depth. The pressure and possibly the temperature also change with depth, and these changes are related to the fluid density, which property is influenced by a Peneloux correction.

14.1.2 Depth Gradient Calculations for Isothermal Reservoirs

Table 14.2 shows compositions sampled at various depths in a petroleum reservoir (Creek and Schrader 1985). The reservoir temperature is reported to be 88°C, and the paper says nothing about a possible temperature variation with depth. Using Equation 14.5, the compositional variation with depth has been simulated, assuming a constant temperature and starting with the composition at the bottom of the reservoir where pressure is 317 bar (read off from figure in the paper of Creek and Schrader). Simulated compositions are shown in Table 14.3. The SRK equation with Peneloux volume correction has been used (Equation 4.43). The fluid was characterized using the procedure of Pedersen et al. as presented in Chapter 5. Measured and simulated C_1 mole percentages with depth are plotted in Figure 14.2 and agree fairly well.

Table 14.4 shows the composition (Whitson and Belery 1994) of a reservoir fluid sample taken at a depth of 2635 m, referred to in the following as the *reference depth*. At this position, the temperature is 95°C and the pressure 263 bar. Figure 14.3 shows the phase envelope of the mixture in the reference depth calculated using the Soave–Redlich–Kwong (1972) equation with the C_{7+} fraction characterized as suggested by Pedersen et al. It is seen that the mixture is a single-phase liquid at the sampling point. Further upward in the reservoir, the pressure decreases. By considering Figure 14.3, it is seen that for a moderate pressure decrease, the mixture will split into two phases. This can also be seen from Figure 14.4, which shows how the simulated reservoir pressure and mixture saturation pressure develop with depth in an isothermal reservoir. Moving upward in the reservoir from the reference depth, the pressure decreases and the saturation pressure increases. The saturation

TABLE 14.2

Measured Compositions versus Depth in Petroleum Reservoir. The Reservoir Temperature Is 88°C

	Depth (m)									
	1017–1053	1062–1087	1076–1111	1251–1299	1289–1303	1274–1322	1298–1301	1378–1392	1378–1392	1383–1393
	Component Mol%									
N_2	0.02	0.16	0.12	0.13	0.14	0.14	0.10	0.14	0.03	0.13
CO_2	1.42	1.31	1.20	1.42	1.50	1.45	1.28	1.04	1.01	1.19
C_1	71.26	71.85	69.79	67.84	67.37	65.76	65.92	59.67	58.48	58.88
C_2	11.04	10.60	11.63	11.02	11.70	11.27	11.63	11.67	11.11	11.89
C_3	5.66	5.73	5.87	5.83	5.92	6.20	6.36	6.58	6.52	6.79
iC_4	1.39	1.34	1.41	1.45	1.46	1.62	1.63	1.58	1.74	1.82
nC_4	1.79	1.70	1.82	1.90	1.75	2.15	2.10	2.11	2.31	2.46
iC_5	0.73	0.66	0.77	0.80	0.81	0.61	0.88	0.90	1.08	1.06
nC_5	0.66	0.60	0.70	0.76	0.70	0.83	0.81	0.84	1.00	0.99
C_6	0.83	0.72	0.96	0.96	1.25	1.38	1.06	1.17	1.43	1.39
C_{7+}	5.20	5.33	5.73	7.89	7.50	8.59	8.23	14.30	15.29	13.40
C_{7+} M	148	145	145	155	160	158	157	181	190	180
C_{7+} dens	0.782	0.782	0.782	0.800	0.799	0.796	0.791	0.811	0.815	0.803

Source: From Creek, J.L. and Schrader, M.L., East Painter reservoir: An example of a compositional gradient from a gravitational field, SPE 14441, presented at *SPE ATCE*, Las Vegas, NV, September 22–25, 1985.

Note: The C_{7+} density is in g/cm³ and at 1.01 bar and 15°C.

TABLE 14.3

Simulated Compositions versus Depth in Petroleum Reservoir

	Depth (m)								
	1035	1074	1093	1275	1297	1298	1300	1385	1388
	Component Mol%								
N_2	0.19	0.18	0.18	0.15	0.14	0.14	0.14	0.13	0.13
CO_2	1.21	1.22	1.22	1.21	1.21	1.20	1.20	1.19	1.19
C_1	71.33	70.64	70.27	63.56	62.45	62.40	62.31	58.98	58.88
C_2	11.62	11.67	11.70	11.96	11.96	11.96	11.96	11.89	11.89
C_3	5.78	5.86	5.91	6.55	6.62	6.62	6.63	6.79	6.79
iC_4	1.42	1.45	1.47	1.71	1.74	1.75	1.75	1.82	1.82
nC_4	1.82	1.87	1.90	2.28	2.33	2.33	2.33	2.46	2.46
iC_5	0.72	0.74	0.75	0.96	0.98	0.98	0.99	1.06	1.06
nC_5	0.65	0.67	0.68	0.88	0.91	0.91	0.91	0.99	0.99
C_6	0.82	0.85	0.87	1.20	1.25	1.25	1.26	1.39	1.39
C_{7+}	4.45	4.83	5.04	9.54	10.40	10.44	10.52	13.32	13.40

Note: The reservoir temperature is assumed to be 88°C independent of depth. The simulation is started in a depth of 1388 m in which depth the reservoir pressure is 317 bar. Measured compositions are shown in Table 14.2.

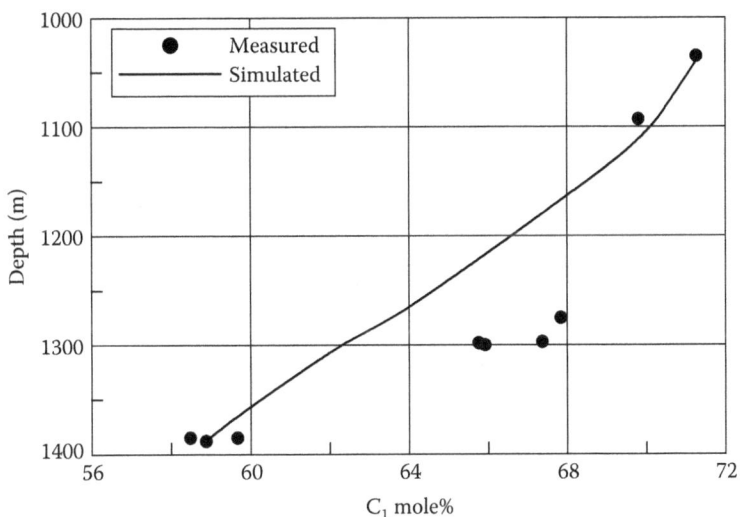

FIGURE 14.2 Measured and simulated C_1 concentrations versus depth in petroleum reservoir. (Measured data from Creek, J.L. and Schrader, M.L., East Painter reservoir: An example of a compositional gradient from a gravitational field, SPE 14441, presented at *SPE ATCE*, Las Vegas, NV, September 22–25, 1985). Tabulated results are shown in Tables 14.2 and 14.3.

pressure (bubble point) increases because the concentrations of heavier components decrease and the concentrations of lighter components increase. At a depth of 2485 m, the reservoir pressure and the saturation pressure coincide, and a separate gas phase is in equilibrium with the oil. This position is called the *gas–oil contact*. Right at the gas–oil contact, the conditions are exactly as in a separator with an oil and a gas in equilibrium. Below the gas–oil contact there is only oil, and further down in the reservoir this oil is gradually enriched in heavy components and stripped of lighter components.

TABLE 14.4

Reservoir Fluid Composition in Depth of 2635 m Where T = 95°C and P = 263 bar

Component	Mol%
N_2	0.27
CO_2	0.79
C_1	46.34
C_2	6.15
C_3	4.46
iC_4	0.87
nC_4	2.27
iC_5	0.96
nC_5	1.41
C_6	2.10
C_{7+}	34.38
C_{7+} M	225.0
C_{7+} density (g/cm³) at 1.01 bar and 15°C	0.870

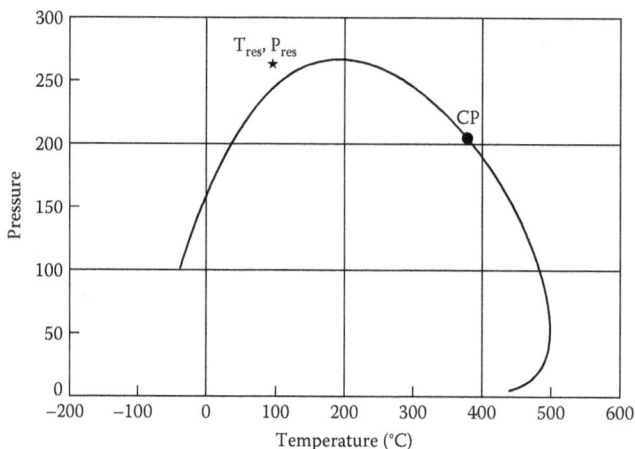

FIGURE 14.3 Phase envelope of oil composition in Table 14.4. The reference temperature and pressure at a depth of 2635 m are indicated. CP stands for critical point.

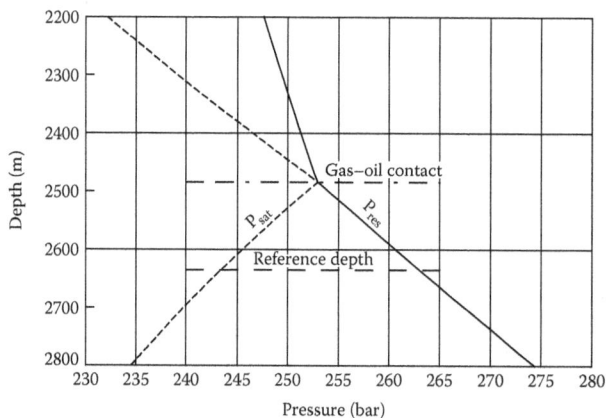

FIGURE 14.4 Development in reservoir pressure (P_{res}) and saturation point (P_{sat}) with depth for composition in Table 14.4.

The gas right at the gas–oil contact is in equilibrium with the oil at this position, whereas the gas further upward contains more of the lighter components and less of the heavy components. The compositional change with depth is also reflected in the plot in Figure 14.5, which shows the variation in gas/oil ratio with depth. The gas/oil ratio has been simulated by a single stage flash to standard conditions (1.01 bar and 15°C) of the reservoir fluid in various depths. In the oil zone below the gas–oil contact, the gas/oil ratio is between 100 and 200 Sm³/Sm³ and slightly decreasing with depth. At the gas/oil contact, a distinct change is seen in gas/oil ratio. Above the gas-oil contact, the gas/oil ratio is 10,000–15,000 Sm³/Sm³ and increasing in upward direction.

Figure 14.6 shows the phase envelopes of the fluids at three different depths. At a depth of 2635 m, the oil is undersaturated (saturation point is lower than the reservoir pressure (upper ▼ in Figure 14.6)). At a depth of 2485 m, the oil is saturated (saturation point and reservoir pressure coincide), and there is a gas phase in equilibrium with the oil. The phase envelope of the gas is also shown in Figure 14.6. The phase envelopes of the oil and the gas at the gas–oil contact intersect at the reservoir P and T (mid ▼ in Figure 14.6), expressing that both fluids are right at the saturation point.

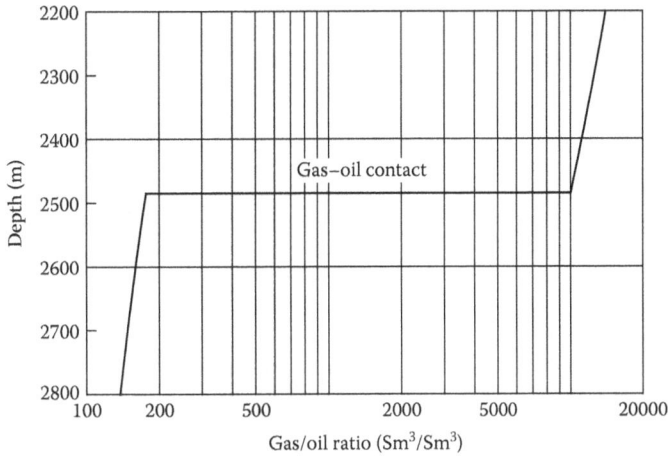

FIGURE 14.5 Development in gas/oil ratio with depth for reservoir fluid. The composition at a depth of 2635 m is given in Table 14.4.

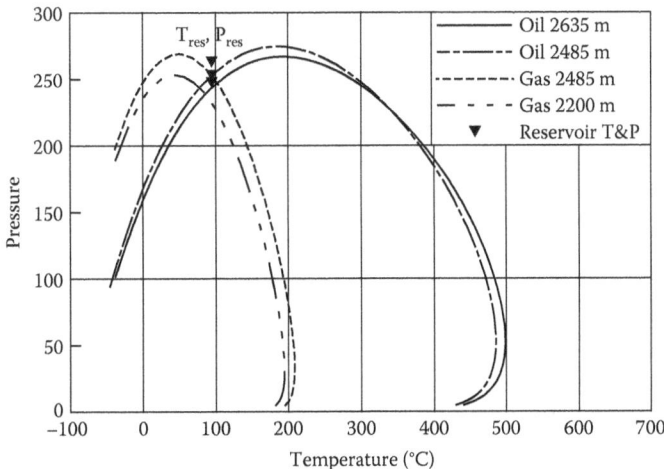

FIGURE 14.6 Phase envelopes of reservoir fluid in depths of 2635 m, 2485 m, and 2200 m. The ▼s indicate the reservoir pressure in the same depths. The composition at a depth of 2635 m is given in Table 14.4.

Finally, Figure 14.6 also shows the phase envelope of the gas at a depth of 2200 m. At this point, the reservoir pressure (lower ▼ in Figure 14.6) is lower than the reservoir pressure at the gas–oil contact, but higher than the saturation pressure of the fluid in the actual depth. This gas is less rich in heavy components than the gas at the gas–oil contact. This is why the dew point pressure of the gas in this depth is lower than that of the equilibrium gas at the gas/oil contact.

Figure 14.7 shows how the reservoir pressure and saturation pressure develop with height for the reservoir fluid in Table 14.5. The figure has been generated using the same models as in the calculations carried out on the mixture in Table 14.4. It is seen that the reservoir pressure and saturation pressure never coincide. Over some distance, the saturation pressure approaches the reservoir pressure, but it then bends off and drifts away from the surrounding reservoir pressure. Figure 14.8 shows simulated phase envelopes of the reservoir fluid at three different depths. The corresponding phase compositions are shown in Table 14.6. At a depth of 4400 m, the critical temperature of the

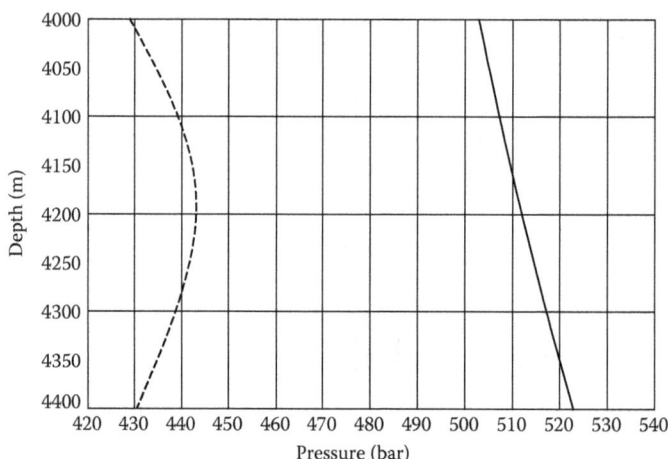

FIGURE 14.7 Development in reservoir pressure (full-drawn line) and saturation point (dashed line) with depth for reservoir fluid. The composition at a depth of 4050 m is given in Table 14.5.

TABLE 14.5
Reservoir Fluid Composition in Depth of 4050 m Where T = 162°C and P = 505 bar

Component	Mol%	M (g/mol)	Density (g/cm³)
N_2	0.504		
CO_2	5.439		
C_1	63.725		
C_2	9.396		
C_3	5.265		
iC_4	0.853		
nC_4	1.894		
iC_5	0.645		
nC_5	0.838		
C_6	0.999		
C_7	1.584	90.6	0.743
C_8	1.648	102.7	0.774
C_9	0.990	116.2	0.790
C_{10+}	6.220	244.1	0.860

Note: The density is at 1.01 bar and 15°C.

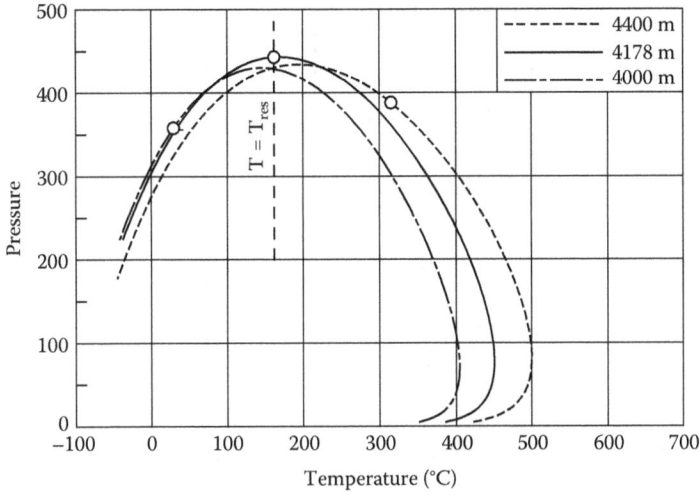

FIGURE 14.8 Phase envelopes of reservoir fluid in various depths. The composition at a depth of 4050 m is given in Table 14.5. Simulated compositions in three different depths are shown in Table 14.6. The circles mark the critical points. The reservoir temperature is shown as a vertical dashed line.

TABLE 14.6
Simulated Reservoir Fluid Compositions (Mol%) in Three Different Depths

Component	4000 m	4178 m	4400 m
N_2	0.518	0.459	0.394
CO_2	5.455	5.352	5.161
C_1	64.715	60.368	54.796
C_2	9.401	9.320	9.020
C_3	5.218	5.371	5.387
iC_4	0.840	0.885	0.903
nC_4	1.859	1.987	2.065
iC_5	0.629	0.689	0.730
nC_5	0.816	0.900	0.962
C_6	0.966	1.095	1.194
C_7	1.516	1.797	2.078
C_8	1.567	1.906	2.257
C_9	0.936	1.163	1.401
C_{10+}	5.562	8.708	13.650

Note: The simulations are based on the composition at a depth of 4050 m, as shown in Table 14.5. Phase envelopes of the fluid compositions are plotted in Figure 14.8.

mixture is higher than the reservoir temperature. Therefore, at these conditions, the reservoir fluid is classified as an oil. At a depth of 4178 m, the mixture-critical temperature is almost identical to the reservoir temperature, and the mixture is classified as a near-critical mixture. At 4000 m, the mixture-critical temperature is considerably lower than the reservoir temperature, and the mixture is classified as a gas condensate. This development in the location of the critical point indicates that the fluid at the bottom of the reservoir behaves like an oil. At a position higher up, it becomes near-critical, and above the near-critical zone, the fluid is a gas condensate.

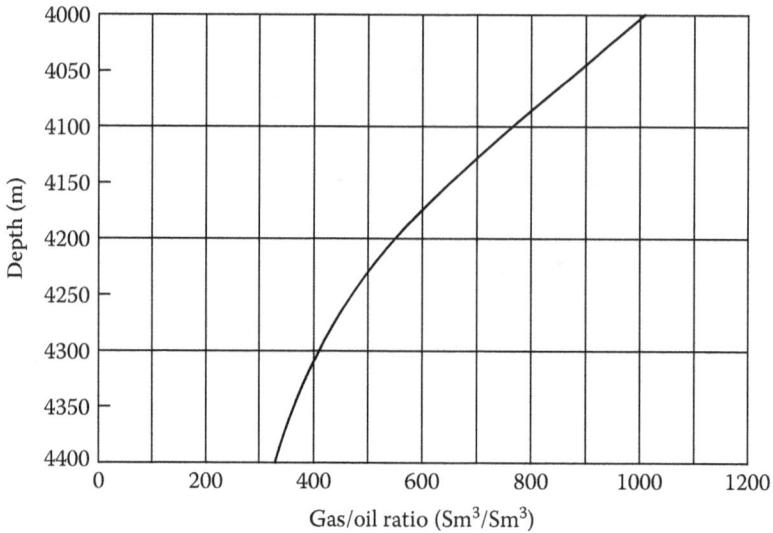

FIGURE 14.9 Development in gas/oil ratio with depth for reservoir fluid. The composition at a depth of 4050 m is given in Table 14.5.

The change from oil to gas condensate does not take place through a gas–oil contact with a discontinuity in the fluid composition, as was the case for the mixture in Table 14.4. As shown in Figure 14.9, the gas/oil ratio gradually increases in the upward direction and not through a distinct change as was seen in Figure 14.5.

14.2 NON-ISOTHERMAL RESERVOIRS

Field data has shown that only part of the compositional variation with depth can be explained by gravity segregation (Montel and Gouel 1985; Pedersen and Lindeloff 2003; Pedersen and Hjermstad 2006, 2015). The temperature in most petroleum reservoirs increases with depth, and a reservoir with a vertical temperature gradient is not at an equilibrium state. There will be a heat and component flux from the deeper (warmer) layers to the colder (upper) layers. As is illustrated in Figure 14.10, a positive vertical temperature gradient will increase the compositional variation with depth. Also, there is evidence of large compositional variations for reservoir fluids close to the saturation point (Fujisawa et al. 2004). The theory presented in Section 14.1.1 is not valid for reservoirs with a temperature gradient. In most petroleum reservoirs, the temperature increases with depth (~0.02°C per m).

14.2.1 THEORY OF NON-ISOTHERMAL RESERVOIRS

The equations needed to solve for the molar compositions in a reservoir with a thermal gradient can be set up using the terminology of irreversible thermodynamics. To simplify the problem, one may assume that the system is in a stationary state. That is, the net component fluxes are zero, and the compositional gradient is assumed to be constant in time. Relative to the equilibrium situation addressed by Schulte (1980), this constitutes a dynamically stabilized system balanced by the gravity and heat flow effects.

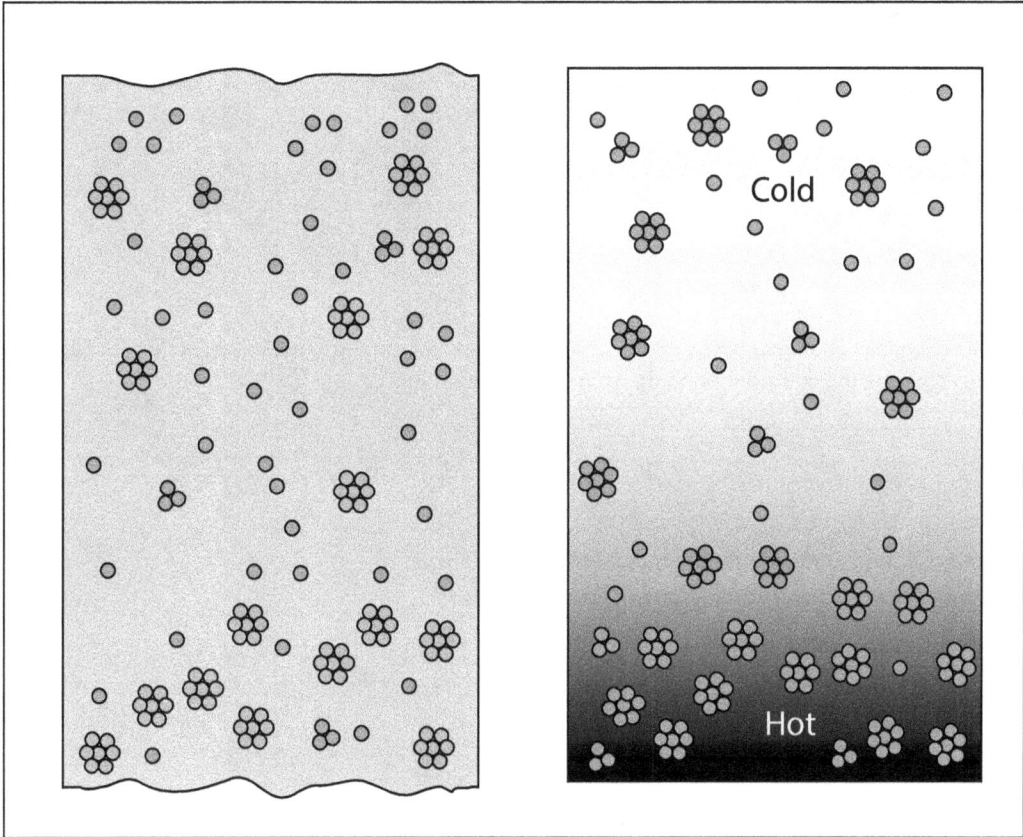

FIGURE 14.10 Schematic illustration of compositional variation with depth in isothermal reservoir (left) and in a reservoir with a positive vertical temperature gradient (right). The temperature gradient makes it more favorable for the high molecular weight hydrocarbons to be in the warmer (lower) part and for the lighter components to be in the upper (cooler) part.

In a petroleum reservoir with no horizontal gradients, the entropy production, σ, per unit time and volume can be written (de Groot and Mazur 1984) as

$$\sigma = -\frac{1}{T^2}\vec{J}_q\frac{dT}{dh} - \frac{1}{T}\sum_{i=1}^{N}\vec{J}_i\left(\frac{T}{M_i}\frac{d\left(\frac{\mu_i}{T}\right)}{dh} - \vec{g}\right)$$ (14.9)

where
T = temperature
$\vec{J}_q$ = heat flux
$\vec{J}_i$ = molar diffusion flux of component i relative to the center of mass velocity
The chemical potential term may be differentiated to

$$T\frac{d\left(\frac{\mu_i}{T}\right)}{dh} = \frac{d\mu_i}{dh} - \frac{\mu_i}{T}\frac{dT}{dh}$$ (14.10)

The following fundamental thermodynamic relation applies:

$$d\mu_i = -\tilde{S}_i dT + \tilde{V}_i dP + \sum_{j=1}^{N}\left(\frac{d\mu_i}{dz_j}\right)z_j \qquad (14.11)$$

where

$\tilde{S}_i$ = partial molar entropy of component i
$\tilde{V}_i$ = partial molar volume of component i
z_j = mole fraction of component j
The term *partial molar property* is explained in Appendix A.

The change in chemical potential may be divided into a term $(d\mu_i)_T$ evaluated at constant T and a term expressing the contribution to $d\mu_i$ from the variation in T:

$$(d\mu_i)_T = \tilde{V}_i dP + \sum_{j=1}^{N}\left(\frac{d\mu_i}{dz_j}\right)z_j \Rightarrow d\mu_i = (d\mu_i)_T - \tilde{S}_i dT \qquad (14.12)$$

Insertion of this expression for $d\mu_i$ into Equation 14.10 gives

$$T\frac{d\left(\frac{\mu_i}{T}\right)}{dh} = \left(\frac{\partial \mu_i}{\partial h}\right)_T - \left(\frac{\mu_i}{T} + \tilde{S}_i\right)\frac{dT}{dh} \qquad (14.13)$$

By use of the general thermodynamic relation

$$\tilde{H}_i = \mu_i + T\tilde{S}_i \qquad (14.14)$$

where $\tilde{H}_i$ is the partial molar enthalpy of component i. Equation 14.13 can be simplified to

$$T\frac{d\left(\frac{\mu_i}{T}\right)}{dh} = \left(\frac{\partial \mu_i}{\partial h}\right)_T - \frac{\tilde{H}_i}{T}\frac{dT}{dh} \qquad (14.15)$$

This expression is inserted into Equation 14.9 to give

$$\sigma = -\frac{1}{T^2}\left(\vec{J}_q - \sum_{i=1}^{N}\frac{\tilde{H}_i}{M_i}\vec{J}_i\right)\frac{dT}{dh} - \frac{1}{T}\sum_{i=1}^{N}\vec{J}_i\left[\left(\frac{\partial \frac{\mu_i}{M_i}}{\partial h}\right)_T - \vec{g}\right] \qquad (14.16)$$

The thermodynamic diffusion force is here defined as

$$F_i = \vec{g} - \left(\frac{\partial \frac{\mu_i}{M_i}}{\partial h}\right)_T \qquad (14.17)$$

From Equation 14.5, it is seen that F_i is the force driving the compositional gradient in the absence of a temperature gradient. Introducing the heat transferred by heat conduction (total heat transfer minus heat transfer from component flow)

$$\vec{J}_q' = \vec{J}_q - \sum_{i=1}^{N}\frac{\tilde{H}_i}{M_i}\vec{J}_i \qquad (14.18)$$

gives the following simplified expression for the entropy production

$$\sigma = -\frac{1}{T^2}\vec{J}'_q\frac{dT}{dh} - \frac{1}{T}\sum_{i=1}^{N}\vec{J}_i\left[\left(\frac{\partial\frac{\mu_i}{M_i}}{\partial h}\right)_T - \vec{g}\right] \qquad (14.19)$$

The molar diffusion flux $\vec{J}_i$ is relative to the center of mass velocity

$$\sum_{i=1}^{N}\vec{J}_i = 0 \Rightarrow \vec{J}_N = -\sum_{i=1}^{N-1}\vec{J}_i \qquad (14.20)$$

which allows the last summation in Equation 14.19 to be rewritten as

$$\sum_{i=1}^{N}\vec{J}_i\left[\left(\frac{\partial\frac{\mu_i}{M_i}}{\partial h}\right)_T - \vec{g}\right] = \sum_{i=1}^{N-1}\vec{J}_i\left[\left(\frac{\partial\frac{\mu_i}{M_i}}{\partial h}\right)_T - \vec{g}\right] + \vec{J}_N\left[\left(\frac{\partial\frac{\mu_N}{M_N}}{\partial h}\right)_T - \vec{g}\right]$$

$$= \sum_{i=1}^{N-1}\vec{J}_i\left[\left(\frac{\partial\frac{\mu_i}{M_i}}{\partial h}\right)_T - \vec{g}\right] - \sum_{i=1}^{N-1}\vec{J}_i\left[\left(\frac{\partial\frac{\mu_N}{M_N}}{\partial h}\right)_T - \vec{g}\right] \qquad (14.21)$$

$$= \sum_{i=1}^{N-1}\vec{J}_i\left(\frac{\partial\left(\frac{\mu_i}{M_i} - \frac{\mu_N}{M_N}\right)}{\partial h}\right)_T$$

This enables the gravitational term, $\vec{g}$, to be eliminated from Equation 14.19:

$$\sigma = -\frac{1}{T^2}\vec{J}'_q\frac{dT}{dh} - \frac{1}{T}\sum_{i=1}^{N-1}\vec{J}_i\left(\frac{\partial\left(\frac{\mu_i}{M_i} - \frac{\mu_N}{M_N}\right)}{\partial h}\right)_T \qquad (14.22)$$

The following phenomenological relations exist for heat and component fluxes (de Groot and Mazur 1984):

$$\vec{J}_q = -L'_{qq}\frac{\frac{dT}{dh}}{T^2} - \frac{1}{T}\sum_{i=1}^{N-1}L'_{qi}\left(\frac{\partial\left(\frac{\mu_i}{M_i} - \frac{\mu_N}{M_N}\right)}{\partial h}\right)_T \qquad (14.23)$$

$$\vec{J}_j = -L'_{jq}\frac{\frac{dT}{dh}}{T^2} - \frac{1}{T}\sum_{i=1}^{N-1}L_{ji}\left(\frac{\partial\left(\frac{\mu_i}{M_i} - \frac{\mu_N}{M_N}\right)}{\partial h}\right)_T \qquad (14.24)$$

where L'_{qq}, L'_{qi}, L'_{jq}, and L_{ji} are the phenomenological coefficients (Onsager 1931a, 1931b).

The theoretical derivations in Equations 14.9 through 14.24 essentially follow the paper of Ghorayeb and Firoozabadi (2000), but the paper gives no guidelines for how to determine the phenomenological coefficients. To make use of the outlined theoretical framework, it is desirable to have Equations 14.23 and 14.24 rewritten in terms of quantities that may be easier to evaluate. Defining the vectors

$$[J] = \begin{pmatrix} \vec{J}_1 \\ \vec{J}_2 \\ .. \\ \vec{J}_{N-1} \end{pmatrix} \qquad (14.25)$$

$$[\nabla_T] = \begin{pmatrix} \left(\dfrac{\partial \left(\frac{\mu_1}{M_1} - \frac{\mu_N}{M_N} \right)}{\partial h} \right)_T \\[2em] \left(\dfrac{\partial \left(\frac{\mu_2}{M_2} - \frac{\mu_N}{M_N} \right)}{\partial h} \right)_T \\[2em] \cdots\cdots \\[1em] \left(\dfrac{\partial \left(\frac{\mu_{N-1}}{M_{N-1}} - \frac{\mu_N}{M_N} \right)}{\partial h} \right)_T \end{pmatrix} \qquad (14.26)$$

$$[L'] = \begin{pmatrix} L'_{1q} \\ L'_{2q} \\ \cdots \\ L'_{N-1,q} \end{pmatrix} \qquad (14.27)$$

$$[L']^T = \left(L'_{q1}, L'_{q2}, ..., L'_{q,N-1} \right) \qquad (14.28)$$

and the matrix

$$[L] = \begin{pmatrix} L_{11} & L_{12} & \cdots & L_{1,N-1} \\ L_{21} & L_{22} & \cdots & L_{2,N-1} \\ \cdots & \cdots & \cdots & \cdots \\ L_{N-1,1} & L_{N-1,2} & \cdots & L_{N-1,N-1} \end{pmatrix} \qquad (14.29)$$

Equations 14.23 and 14.24 may be written as

$$\vec{J}'_q = -L'_{qq} \frac{\frac{dT}{dh}}{T^2} - \frac{1}{T}[L']^T[\nabla_T] \qquad (14.30)$$

$$[J] = -[L'] \frac{\frac{dT}{dh}}{T^2} - \frac{1}{T}[L][\nabla_T] \qquad (14.31)$$

Equation 14.31 multiplied by $[L]^{-1}$ gives

$$[L]^{-1}[J] = -[L]^{-1}[L'] \frac{\frac{dT}{dh}}{T^2} - \frac{1}{T}[\nabla_T] \qquad (14.32)$$

Defining [Q] as

$$[Q] = \begin{pmatrix} Q_1 \\ Q_2 \\ \cdots \\ Q_{N-1} \end{pmatrix} = [L]^{-1}[L'] \Rightarrow [L'] = [L][Q] \qquad (14.33)$$

Equation 14.32 may be written

$$[L]^{-1}[J] = -[Q]\frac{\frac{dT}{dh}}{T^2} - \frac{1}{T}[\nabla_T] \tag{14.34}$$

If no component fluxes exist in the reservoir

$$[J] = 0 \Rightarrow -[Q]\frac{\frac{dT}{dh}}{T} - [\nabla_T] = 0 \tag{14.35}$$

To build up an understanding of the physical importance of [Q], the term [L′] in Equation 14.30 is replaced by [L][Q] from Equation 14.33

$$\vec{J}'_q = -L'_{qq}\frac{\frac{dT}{dh}}{T^2} - \frac{1}{T}[Q]^T[L][\nabla_T] \tag{14.36}$$

Multiplying Equation 14.31 by $[Q]^T$ gives

$$[Q]^T[J] = -[Q]^T[L']\frac{\frac{dT}{dh}}{T^2} - \frac{1}{T}[Q]^T[L][\nabla_T] \tag{14.37}$$

which may be combined with Equation 14.36 to give

$$J'_q - [Q]^T[J] = -L'_{qq}\frac{\frac{dT}{dh}}{T^2} + [Q]^T[L']\frac{\frac{dT}{dh}}{T^2} \tag{14.38}$$

From this expression, it can be seen that [Q] for $dT/dh > 0$ is the heat transfer originating from component flow. The equation must also apply in the absence of a temperature gradient

$$\frac{dT}{dh} = 0 \Rightarrow J'_q = [Q]^T[J] = \sum_{i=1}^{N-1} Q_i \vec{J}_i \tag{14.39}$$

From this equation and Equation 14.18, it is possible to derive the following expression for J_q in the absence of a temperature gradient:

$$\frac{dT}{dh} = 0 \Rightarrow J_q = \sum_{i=1}^{N-1} Q_i \vec{J}_i + \sum_{i=1}^{N} \frac{\tilde{H}_i}{M_i} \vec{J}_i = \sum_{i=1}^{N-1} Q_i \vec{J}_i + \sum_{i=1}^{N-1} \frac{\tilde{H}_i}{M_i} \vec{J}_i + \frac{\tilde{H}_N}{M_N} \vec{J}_N \tag{14.40}$$

In a stationary isothermal reservoir, Equation 14.20 applies, and

$$\frac{dT}{dh} = 0 \Rightarrow J_q = \sum_{i=1}^{N-1}\left[\frac{\tilde{H}_i}{M_i} - \frac{\tilde{H}_N}{M_N} + Q_i\right]\vec{J}_i \tag{14.41}$$

If it is further assumed that

$$\frac{dT}{dh} = 0 \Rightarrow J_q = 0 \tag{14.42}$$

then

$$Q_i = \frac{\tilde{H}_N}{M_N} - \frac{\tilde{H}_i}{M_i} \tag{14.43}$$

This assumption was first proposed by Haase (1969) and enables Equation 14.35 to be written as

$$-\left(\frac{\partial \frac{\mu_i}{M_i}}{\partial h} - \frac{\partial \frac{\mu_N}{M_N}}{\partial h}\right)_T = \left(\frac{\tilde{H}_N}{M_N} - \frac{\tilde{H}_i}{M_i}\right)\frac{\frac{dT}{dh}}{T} \tag{14.44}$$

The Gibbs–Duhem equation (see, e.g., Smith et al. 2001), applied to the chemical potential, may be written as

$$-S\frac{dT}{dh} + V\frac{dP}{dh} - \sum_{j=1}^{N} z_i \frac{d\mu_i}{dh} = 0 \tag{14.45}$$

At constant temperature

$$V\frac{dP}{dh} = \sum_{i=1}^{N} z_i \left(\frac{\partial \mu_i}{\partial h}\right)_T \tag{14.46}$$

In a petroleum reservoir, the pressure gradient will be determined by

$$\frac{dP}{dh} = \rho\vec{g} \tag{14.47}$$

Introducing this expression and the thermal diffusion force defined in Equation 14.17 into Equation 14.46 gives

$$V\rho\vec{g} = \sum_{i=1}^{N} z_i (M_i g - M_i F_i) \tag{14.48}$$

Because

$$V\rho\vec{g} = \vec{g}\sum_{i=1}^{N} z_i M_i \tag{14.49}$$

the following must apply to fulfil the hydrostatic condition in Equation 14.47:

$$\sum_{i=1}^{N} z_i M_i F_i = 0 \tag{14.50}$$

Combining Equations 14.17 and 14.44 gives

$$F_i - F_N = \left(\frac{\tilde{H}_N}{M_N} - \frac{\tilde{H}_i}{M_i}\right)\frac{\frac{dT}{dh}}{T} \tag{14.51}$$

Multiplying this equation by $z_i M_i$ and summing over all i gives

$$\sum_{i=1}^{N} z_i M_i F_i - F_N \sum_{i=1}^{N} z_i M_i = \frac{\frac{dT}{dh}}{T}\sum_{i=1}^{N} z_i M_i \left(\frac{\tilde{H}_N}{M_N} - \frac{\tilde{H}_i}{M_i}\right) \tag{14.52}$$

As can be seen from Equation 14.50, the first term in this equation equals zero, giving

$$F_N = \frac{\frac{dT}{dh} \sum\limits_{i=1}^{N} z_i M_i \left(\frac{\tilde{H}_i}{M_i} - \frac{\tilde{H}_N}{M_N} \right)}{T \sum\limits_{i=1}^{N} z_i M_i} = \left(\frac{H}{M} - \frac{\tilde{H}_N}{M_N} \right) \frac{dT}{dh} \frac{}{T} \qquad (14.53)$$

where H is the molar enthalpy of the mixture and M the average molecular weight. Inserting this into Equation 14.51 gives

$$F_i = \left(\frac{H}{M} - \frac{\tilde{H}_i}{M_i} \right) \frac{\frac{dT}{dh}}{T} a \qquad (14.54)$$

Recalling the definition of F_i (Equation 14.17), this equation may be rewritten as

$$\left(\frac{\partial \mu_i}{\partial h} \right)_T = \vec{g} M_i - M_i \left(\frac{H}{M} - \frac{\tilde{H}_i}{M_i} \right) \frac{\frac{dT}{dh}}{T} \qquad (14.55)$$

which allows the equivalent to Equation 14.5 for the case with dT/dh ≠ 0 to be derived:

$$RT \ln(\varphi_i^h z_i^h P^h) - RT \ln(\varphi_i^{h^0} z_i^{h^0} P^{h^0}) = M_i g \left(h - h^o \right) - M_i \left(\frac{H^{abs}}{M} - \frac{\tilde{H}_i^{abs}}{M_i} \right) \frac{\Delta T}{T} \qquad (14.56)$$

$$i = 1, 2, ..., N$$

The compositional gradient in a reservoir with a known thermal gradient may be determined by solving these N equations together with Equation 14.6.

14.2.2 ABSOLUTE ENTHALPIES

To make use of Equation 14.56, it is necessary to evaluate absolute enthalpies. Textbooks on applied thermodynamics (e.g., Smith et al. 2001) suggest not to evaluate absolute enthalpies but, instead, to evaluate the enthalpy (H) relative to the enthalpy (H^0) in some reference state:

$$H^{abs} = H + H^0 \Rightarrow H = H^{abs} - H^0 \qquad (14.57)$$

The reference state can, for example, be the ideal gas state at 273.15 K. Use of H instead of H^{abs} does not present any problems in heat balance calculations on pure components or mixtures of constant composition. Say a mixture is heated from a state with an enthalpy of H_1^{abs} to a state with an enthalpy of H_2^{abs}, the enthalpy increase can be expressed as

$$\Delta H = H_2^{abs} - H_1^{abs} = (H_2 + H^0) - (H_1 + H^0) = H_2 - H_1 \qquad (14.58)$$

The reference enthalpy is seen to cancel out and does not have to be evaluated.

The situation in a petroleum reservoir is different. The composition changes with depth, and it is necessary to evaluate the absolute partial molar enthalpies. Whether a positive T-gradient with depth according to the Haase model will weaken or strengthen a compositional gradient will depend on the variation in reference enthalpy with component molecular weight.

Pedersen and Hjermstad (2006) have suggested the reference enthalpies in Table 14.7. They used the ideal gas state at 273.15 K as reference state, meaning that the absolute ideal gas enthalpies in Table 14.7 are ideal gas enthalpies at 273.15 K per mass unit.

Figure 14.11 illustrates how two different molecules will react to an external heat source. The smaller molecule has the lower ideal gas enthalpy per mass unit than the larger one and will convert a major part

---

TABLE 14.7
Absolute Ideal Gas Enthalpies per Mass Unit at 273.15 K

Component	$H^{ig}/(M\,R)$ (K/g)
N_2	34.40
CO_2	52.0
C_1	[a]0.0
C_2	40.1
C_3	54.6
iC_4	60.0
nC_4	60.0
iC_5	65.1
nC_5	48.4
C_6	68.4
C_{7+}	83.67–1342/M

Source: Pedersen, K.S. and Hjermstad, H.P., Modeling of compositional variation with depth for five North Sea reservoirs, SPE paper 1075085-MS presented at the SPE ATCE in Houston, Texas, USA, September 28–30, 2015.

[a] Chosen reference value.

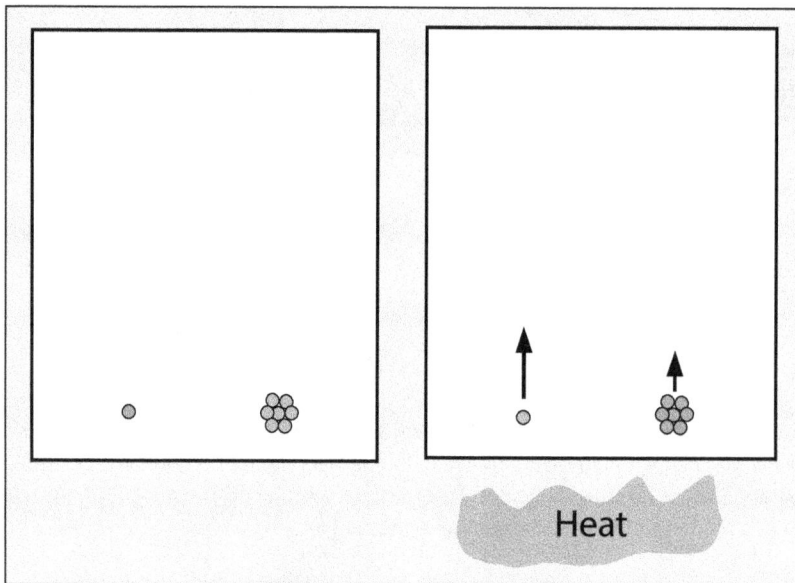

FIGURE 14.11 Schematic illustration of how two different molecules react to being exposed to an external heat source. The smaller molecule has a low number of internal degrees of freedom $\left(\text{low } \dfrac{H_i^{ig}}{M_i} \right)$ and most of the supplied heat is converted to kinetic energy. The larger molecule has a higher $\dfrac{H_i^{ig}}{M_i}$ and will use a large portion of the heat applied to increase the number of molecular vibrations and internal rotations, while only a small portion goes to kinetic energy.

of the added heat to kinetic energy and begin to move. The larger molecule will convert a larger fraction of the added heat to internal energy (intra molecular vibrations and rotations) and less to kinetic energy.

14.2.3 Depth Gradient Calculation for Non-Isothermal Reservoir

Pedersen and Hjermstad (2015) have presented data for the compositional variation with depth in five petroleum reservoirs. A summary of the data for two of the reservoirs is given in Table 14.8 together with viscosity data for the fluid composition at reservoir conditions. Also shown is the fraction of aromatic components in the C_{7+} fraction of each of the reference fluids (Nes and van Westerns 1951). The depths in Table 14.8 are relative to the depth in which the upper fluid sample was taken. Table 14.9

TABLE 14.8
Summary of Depth Gradient and Viscosity Data for Two North Sea Reservoirs.

Reservoir	Depth Interval m	Reference Depth m	Temperature in Reference Depth °C	Pressure in Reference Depth bar	Saturation Pressure in Reference Depth bar	Gas/Oil Ratio Bottom to Top (Sm³/ Sm³)	Approximate d(Psat/dh) in Oil Zone bar/m	Viscosity of Fluid in Reference Depth cP	Aromatic Fraction of C_{7+}
1	0–327	204	94.0	286.0	267.0	228–3,739	−0.17	0.204	0.155
5	0–28	8	79.8	194.0	185.7	115–2,947	−1.57	1.017	0.312

Source: Pedersen, K.S. and Hjermstad, H.P., Modeling of compositional variation with depth for five North Sea reservoirs, SPE 1075085-MS presented at the SPE ATCE in Houston, Texas, USA, September 28–30, 2015.

TABLE 14.9
Compositional Data and Pressure and Temperature Information for Reservoir 1 in Pedersen and Hjermstad (2015). The Depths Are Relative to the Depth of the Upper Fluid Sample. A Gas-Oil Contact Was Found in a Depth of 140 m

	Depths (m)				
	0 m	175 m	204 m	228 m	327 m
	Mol%				
N_2	1.06	0.42	0.42	0.39	0.43
CO_2	0.70	0.69	0.71	0.70	0.77
C_1	75.66	50.04	49.88	48.89	45.66
C_2	7.62	7.85	7.77	7.74	7.60
C_3	5.33	6.77	6.74	6.76	7.13
iC_4	0.73	1.04	1.03	1.04	1.13
nC_4	2.02	3.20	3.21	3.24	3.52
iC_5	0.67	1.16	1.16	1.18	1.31
nC_5	0.84	1.55	1.55	1.58	1.75
C_6	0.91	1.88	1.88	1.93	2.17
C_7	1.29	3.50	3.49	3.62	4.01
C_8	1.06	3.75	3.73	3.85	4.28
C_9	0.54	2.28	2.31	2.36	2.58
C_{10+}	1.57	15.88	16.11	16.70	17.66
C_{7+} Mol Wgt	126	196	197	197	200
C_{7+} Density (g/cm3)	0.785	.846	.848	0.848	0.850
Pressure (bar)	279	284	286	287	293
Temperature (°C)	89	93	94	94	97

Data Source: Pedersen, K.S. and Hjermstad, H.P., Modeling of compositional variation with depth for five North Sea reservoirs, SPE 175085-MS presented at the SPE ATCE in Houston, Texas, USA, September 28–30, 2015.

shows the compositions of the fluid samples from Reservoir 1. Table 14.10 shows the fluids in Table 14.9 characterized to a common EoS model (Section 5.6) for use with the volume-corrected PR equation (Equation 4.48). The binary interaction parameters are found in Table 14.11. A depth gradient simulation is carried out for the fluid in Table 14.10 from a depth of 175 m. Figure 14.12 shows experimental

TABLE 14.10
Common EoS Models for the Fluids in Table 14.9. The EoS Model Is for Use with Volume-Corrected PR Equation. NonZero Binary Interaction Parameters Are Shown in Table 14.11

Component	0 m Mole%	175 m Mole%	204 m Mole%	228 m Mole%	327 m Mole%	Molecular Weight	Crit Temp °C	Crit Pressure bar	Acentric Factor	Volume Shift cm³/mole
N_2	1.060	0.420	0.420	0.390	0.430	28.01	−146.95	33.94	0.0400	−4.23
CO_2	0.700	0.690	0.710	0.700	0.770	44.01	31.05	73.76	0.2250	−1.64
C_1	75.660	50.035	49.885	48.900	45.660	16.04	−82.55	46.00	0.0080	−5.20
C_2	7.620	7.849	7.771	7.742	7.600	30.07	32.25	48.84	0.0980	−5.79
C_3	5.330	6.769	6.741	6.761	7.130	44.10	96.65	42.46	0.1520	−6.35
iC_4	0.730	1.040	1.030	1.040	1.130	58.12	134.95	36.48	0.1760	−7.18
nC_4	2.020	3.200	3.210	3.241	3.520	58.12	152.05	38.00	0.1930	−6.49
iC_5	0.670	1.160	1.160	1.180	1.310	72.15	187.25	33.84	0.2270	−6.20
nC_5	0.840	1.550	1.550	1.580	1.750	72.15	196.45	33.74	0.2510	−5.12
C_6	0.910	1.880	1.880	1.930	2.170	86.18	234.25	29.69	0.2960	1.39
C_7	1.290	3.500	3.490	3.621	4.010	96.00	301.63	29.59	0.3375	15.53
C_8	1.060	3.750	3.730	3.851	4.280	107.00	323.97	27.37	0.3743	19.96
C_9	0.540	2.280	2.310	2.360	2.580	121.00	349.62	25.01	0.4204	25.14
$C_{10}–C_{11}$	0.823	3.264	3.293	3.419	3.430	140.09	381.84	22.59	0.4833	31.89
$C_{12}–C_{13}$	0.392	2.593	2.620	2.720	2.764	167.57	422.96	20.12	0.5703	39.65
$C_{14}–C_{16}$	0.239	2.926	2.961	3.072	3.173	204.75	473.01	17.91	0.6847	46.52
$C_{17}–C_{18}$	0.061	1.459	1.480	1.534	1.611	243.60	519.43	16.40	0.7947	49.62
$C_{19}–C_{21}$	0.037	1.646	1.673	1.733	1.850	275.27	555.64	15.55	0.8805	50.37
$C_{22}–C_{24}$	0.012	1.166	1.187	1.229	1.338	317.02	600.26	14.71	0.9836	48.37
$C_{25}–C_{29}$	0.0051	1.237	1.264	1.308	1.460	370.39	654.86	13.96	1.0983	42.77
$C_{30}–C_{37}$	0.0009	0.958	0.982	1.015	1.180	456.83	737.85	13.16	1.2292	27.92
$C_{38}–C_{80}$	0.00004	0.630	0.652	0.672	0.853	640.76	912.38	12.27	1.1596	−6.04

Source: Vinhal, A., Azeem, J. and Pedersen, K. S., Modeling of compositional grading in heavy oil fields, SPE 205887-MS, presented at SPE ATCE Dubai, UAE, September 21–23, 2021.

TABLE 14.11
NonZero Binary Interaction Parameters (k_{ij}s) Used with the EoS Model in Table 14.10

	N_2	CO_2
CO_2	−0.017	
C_1	0.031	0.120
C_2	0.052	0.120
C_3	0.085	0.120
iC_4	0.103	0.120
nC_4	0.080	0.120
iC_5	0.092	0.120
nC_5	0.100	0.120
C_6	0.080	0.120
C_{7+}	0.080	0.100

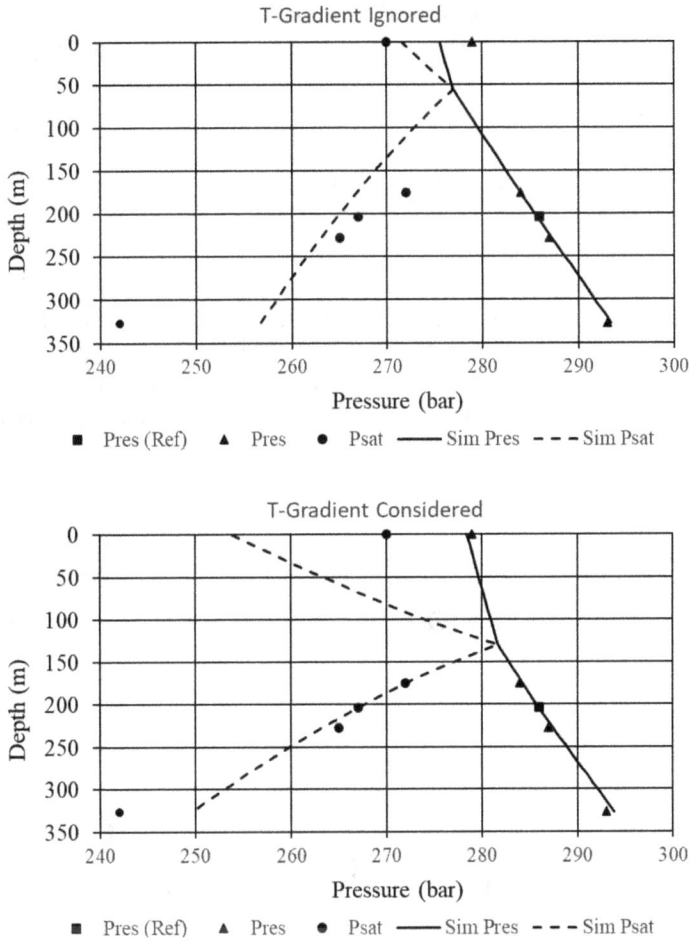

FIGURE 14.12 Experimental and simulated reservoir and saturation pressures versus depth for Reservoir 1 in Pedersen and Hjermstad (2015). Equation 14.56 has been used with fugacity coefficients derived from the volume-corrected PR equation with the EoS model parameters in Table 14.10 and the H_i^{ig} values in Table 14.7. A vertical temperature gradient of 0.025 C/m is used in the simulations.

and simulated data for the development of reservoir pressure and saturation pressure with depth. The simulation results on the upper plot are generated ignoring the vertical temperature gradient, while the lower plot shows the simulations for a vertical temperature gradient of 0.025°C/m.

Espósito et al. (2017) have presented more data for the compositional variation with depth in petroleum reservoirs.

14.3 HEAVY OIL RESERVOIRS

As can be seen in Table 14.8, the viscosity of the reference fluid in Reservoir 5 is about five times higher than the viscosity of the reference fluid in Reservoir 1, and the change in saturation pressure per unit depth is almost ten times higher in Reservoir 5 than in Reservoir 1. One of the assumptions behind the Haase model presented in Section 14.2.1 is that there is an unhindered flux of molecules back and forth between the warmer bottom to the colder top. This assumption will not always hold. In the same medium, a molecule of a high molecular weight molecule will move (diffuse) at a lower rate than a molecule of a lower molecular weight. This applies to both gas (Graham 1846) and liquid phases but is more pronounced in a liquid environment. Hayduk and Minhas (1982) have proposed the following correlation for the diffusion coefficient D (m/s^2) of a component i with a molecular weight of M$_i$ and a density of ρ_i (g/cm^3) in a liquid with a viscosity of η (cP) and at an absolute temperature of T (K).

$$D_i = 13.3 \times 10^{-12} \times T^{1.47} \left(\frac{M_i}{\rho_i} \right)^{-0.71} \eta^{\left[\frac{10.2}{\left(\frac{M_i}{\rho_i} \right)} - 0.791 \right]} \tag{14.59}$$

The exponent in the viscosity term is negative, meaning that the diffusion coefficient for viscosities exceeds 1.0 will decrease with increasing viscosity of the surrounding media. The rate a given heat supply imparts a molecule will therefore be lower in a more viscous medium than in a less viscous medium, and the difference between the speed of movement between a low and a high molecular weight component will increase with the viscosity of the fluid medium. Figure 14.13 illustrates the

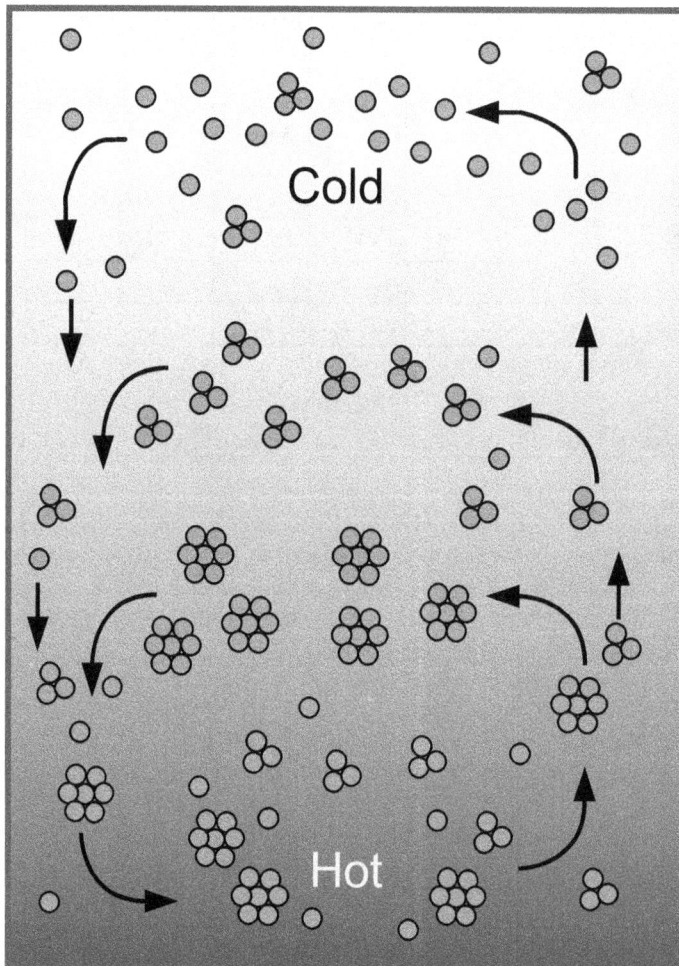

FIGURE 14.13 Schematic illustration of how smaller and larger molecules will diffuse in a reservoir oil zone with a vertical temperature gradient. The diffusion rate decreases with molecular size, and the net result is that the kinetic energy transferred to the heavier molecules will be insufficient to have all of them circulate in the total fluid column as is the case for the lighter molecules. A fraction of the heavier molecules will instead only circulate at the bottom of the reservoir.

impact of high molecular weight components diffusing at a lower rate than low molecular weight components. The high molecular weight components will circulate at the bottom of the reservoir, while the low molecular weight compounds have sufficient kinetic energy to reach the top of the reservoir before the lower temperature at the top makes them seek down again.

Not only does the fluid in Reservoir 5 have a high viscosity; as can be seen in Table 14.8, it also has a high concentration of aromatic components. This suggests that the attenuation of the molecular motion seen in highly viscous reservoirs is further enhanced if the fluid has a high concentration of aromatic components.

14.3.1 Viscosity Correction

To account for the impact of viscosity and aromaticity on the compositional variation with depth, Vinhal et al. 2021) suggested Equation 14.56 was rewritten to

$$RT \, \ln(\varphi_i^h z_i^h P^h) - RT \, \ln(\varphi_i^{h^0} z_i^{h^0} P^{h^0}) = M_i g\left(h - h^o\right) - M_i \left(\frac{H^{abs}}{M} - \frac{\tilde{H}_i^{abs}}{M_i}\right) \qquad (14.60)$$

$$\frac{\Delta T}{T} \times VISFAC \quad i = 1, 2, ..., N$$

VISFAC is a correction factor to take into account that molecules of different sizes move at different rates in a fluid column and that the difference in rate of movement increases with viscosity and with concentration of aromatic components. VISFAC is found from the following expression:

$$VISFAC = 16.875 \times \eta + 5.0 \times AFRAC - 3.375 \qquad (14.61)$$
$$Legal \, range \, for \, VISFRC : 0.5 \leq VISFAC \leq 50 \times AFRAC - 2$$

AFRAC is the fraction of the C_{7+} components that are aromatics, and η is the fluid viscosity in cP.

14.3.2 Depth Gradient Calculation for Heavy Oil Reservoirs

Vinhal et al. (2021) have presented a depth gradient simulation for Reservoir 5 in Pedersen and Hjermstad (2015) using Equation 14.60. The measured and simulated fluid compositions at the sampling depths are shown in Table 14.12 and the experimental and simulated variations in reservoir pressure and saturation pressure with depth can be seen from Figure 14.14. The figure also shows the simulation results of using Equation 14.56, which equation considers the vertical temperature gradient but not the viscosity effect. A comparison of the two plots shows the importance of considering the viscosity effect for heavy aromatic reservoir fluids. Taking that into consideration makes it possible to simulate the very significant compositional variation seen in this fluid column over a depth interval of only 28 m.

Rodriguez et al. (2007) have presented data for the compositional variation with depth in a 970 m fluid column in the Carito-Mulata field in Venezuela. Two of the fluids are shown in Table 14.13 with a common EoS model developed for the volume-corrected SRK equation (Vinhal et al. 2021). Saturation points measured on samples from the field are shown in Table 14.14. A depth gradient calculation was carried out with the fluid in a depth of 3882.5 m as reference fluid assuming a vertical temperature gradient of 0.025°C/m and using Equation 14.60 to represent the compositional grading. Experimental and simulated variations in reservoir pressure and saturation pressure with depth can be seen from Figure 14.15. The figure also shows the simulation results of using Equation 14.56, which ignores the viscosity effect. The viscosity effect must be considered to get a good match of the compositional grading.

TABLE 14.12

Experimental and Simulated Molar Compositions for Reservoir 5 in Table 14.8. The Simulation Results Are for the Volume-Corrected PR Equation Taking the Viscosity Effect into Account (Equation 14.60) and Assuming a Vertical Temperature Gradient of 0.025 °C/m

	0 m		8 m[*1]	21 m		28 m	
Component	Exp Mol%	Sim Mol%	Exp Mol%	Exp Mol%	Sim Mol%	Exp Mol%	Sim Mol%
N_2	2.80	2.61	0.77	0.66	0.72	0.63	0.72
CO_2	0.16	0.12	0.09	0.09	0.09	0.05	0.08
C_1	67.88	69.37	34.41	32.59	31.49	29.47	30.81
C_2	11.06	11.30	10.11	9.92	9.34	9.89	9.04
C_3	8.71	9.06	11.78	11.66	11.02	11.57	10.68
iC_4	1.78	1.55	2.51	2.49	2.39	2.36	2.34
nC_4	3.05	2.42	4.53	4.53	4.30	4.18	4.18
iC_5	1.12	0.82	1.95	1.99	1.88	1.70	1.84
nC_5	0.99	0.76	1.98	2.04	1.90	1.64	1.85
C_6	0.87	0.74	2.52	2.54	2.44	1.95	2.39
C_{7+}	1.58	1.26	29.34	31.49	34.43	36.57	36.07
Saturation Pressure (bar)	–	–	185.7	175.4	–	152.0	–

[*1] Reference depth in which the temperature is 79.8°C.

Source: Vinhal, A., Azeem, J. and Pedersen, K.S., Modeling of compositional grading in heavy oil fields, SPE 205887-MS, presented at *SPE ATCE Dubai*, UAE, September 21–23, 2021.

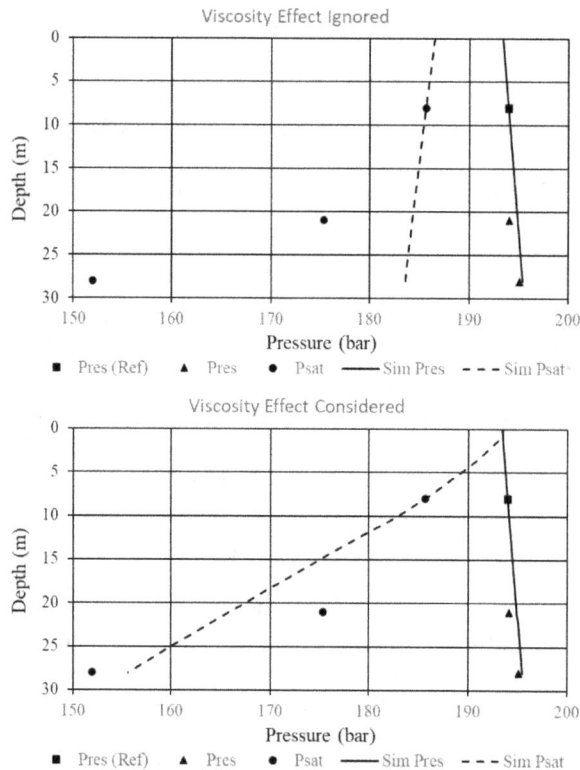

FIGURE 14.14 Experimental and simulated reservoir and saturation pressures versus depth for Reservoir 5 in Pedersen and Hjermstad (2015). The simulation results on the upper plot are generated using Equation 14.56, which does not take the viscosity effect into account. The simulation results on the lower plot are generated accounting for the viscosity effect by using Equation 14.60. The volume-corrected PR equation is used. A vertical temperature gradient of 0.025 °C/m is used in the simulations.

TABLE 14.13

Common EoS Models for Two Fluids (Rodriguez et al. 2007) in a Fluid Column That Extends over a Depth Interval from 3882.5 m to 4853.0 m. The EoS Model Is for Use with the Volume-Corrected SRK Equation. The NonZero Binary Interaction Parameters Are the Same as in Table 14.11

Component	3882.5 m* Mole%	4456.2 m Mole%	Molecular Weight	Crit Temp °C	Crit Pressure bar	Acentric Factor	Volume Shift cm³/mole
N_2	0.17	0.07	28.01	−146.95	33.94	0.0400	−4.23
CO_2	4.73	3.81	44.01	31.05	73.76	0.2250	−1.64
C_1	71.20	46.60	16.04	−82.55	46.00	0.0080	−5.20
C_2	6.67	7.02	30.07	32.25	48.84	0.0980	−5.79
C_3	3.01	5.37	44.10	96.65	42.46	0.1520	−6.35
iC_4	0.64	1.26	58.12	134.95	36.48	0.1760	−7.18
nC_4	0.92	2.43	58.12	152.05	38.00	0.1930	−6.49
iC_5	0.34	1.02	72.15	187.25	33.84	0.2270	−6.20
nC_5	0.30	1.10	72.15	196.45	33.74	0.2510	−5.12
C_6	0.74	2.09	86.18	234.25	29.69	0.2960	1.39
C_7	1.14	2.36	96.00	282.99	29.70	0.3375	11.28
C_8	1.02	2.17	107.00	304.27	27.39	0.3743	14.63
C_9	0.92	2.00	121.00	328.83	24.97	0.4204	18.59
$C_{10}-C_{12}$	2.24	5.07	146.52	369.84	21.82	0.5054	25.29
$C_{13}-C_{15}$	1.63	3.94	189.40	428.25	18.56	0.6387	32.20
$C_{16}-C_{18}$	1.18	3.06	235.80	483.47	16.49	0.7736	34.03
$C_{19}-C_{21}$	0.86	2.38	275.50	526.99	15.36	0.8809	32.35
$C_{22}-C_{25}$	0.79	2.37	323.26	576.19	14.41	0.9982	26.81
$C_{26}-C_{29}$	0.52	1.69	379.16	630.18	13.65	1.1144	16.56
$C_{30}-C_{35}$	0.46	1.68	447.39	693.06	13.00	1.2191	0.38
$C_{36}-C_{45}$	0.34	1.46	552.94	785.57	12.36	1.2801	−29.89
$C_{45}-C_{80}$	0.17	1.05	769.55	971.24	11.65	0.9591	−97.73

[*1] Reference fluid composition sampled at a pressure of 827 bar and a temperature of 152 °C.

Source: Vinhal, A., Azeem, J. and Pedersen, K.S., Modeling of compositional grading in heavy oil fields, SPE 205887-MS, presented at *SPE ATCE Dubai*, UAE, September 21–23, 2021.

TABLE 14.14

Saturation Pressures Measured for Fluid Samples Taken in Different Depths of the Carito-Mulata Field in Venezuela. The Temperature at a Depth of 3882.5 m Is 152°C

Depth M	Saturation Pressure bar
3882.5	574.0
4006.0	561.2
4260.2	490.6
4353.5	402.3
4456.2	307.5
4563.2	251.3
4604.9	237.6
4853.0	218.3

Data Source: Rodriguez, F.D.C., Lopez G., Leonardo, B., Skoreyko, J.A. and Fraser, A., Generation of a compositional model to simulate EOR processes for complex fluid systems of the Caroto-Mulata Field, Venezuela, SPE 107803, presented at *Latin American & Caribbean Petroleum Engineering Conference*, Buenos Aires, Argentina, April 2007.

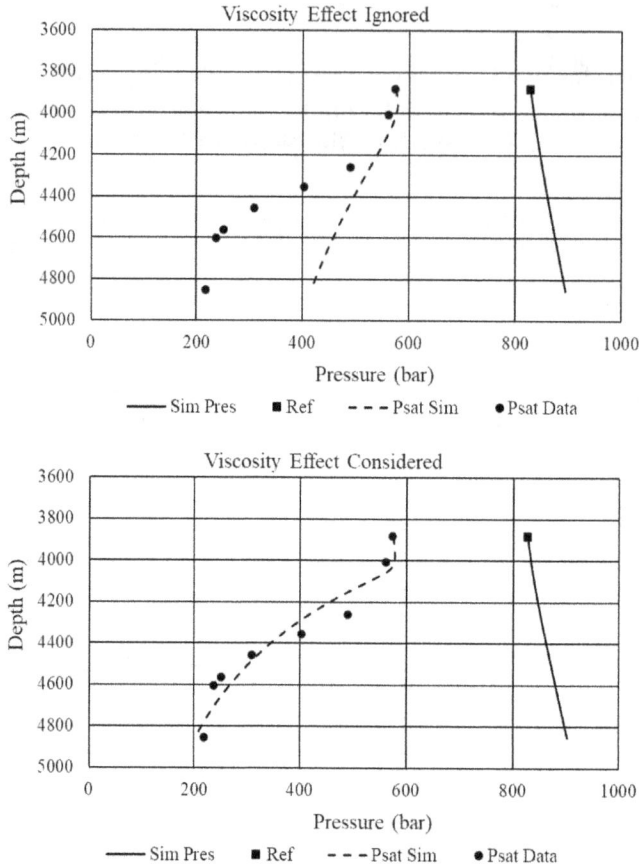

FIGURE 14.15 Experimental and simulated reservoir and saturation pressures versus depth for Carito-Mulata field in Venezuela (Vinhal et al. 2021). The simulation results on the upper plot are generated using Equation 14.56, which does not take the viscosity effect into account. The simulation results on the lower plot are generated accounting for the viscosity effect by using Equation 14.60. The volume-corrected PR equation is used. A vertical temperature gradient of 0.025 °C/m is used in the simulations.

REFERENCES

Creek, J.L. and Schrader, M.L., East Painter reservoir: An example of a compositional gradient from a gravitational field, SPE 14441, presented at the *SPE ATCE*, Las Vegas, NV, September 22–25, 1985.

de Groot, S.R. and Mazur, P., *Non-Equilibrium Thermodynamics*, Dover Edition, New York, 1984.

Espósito, R.O., Alijo, P.H.R., Scilipoti, J.A. and Pavares, F.W., *Compositional Grading in Oil and Gas Reservoirs*, Gulf Professional Publishing, Cambridge, MA, 2017.

Fujisawa, G., Betancourt, S.S., Mullins, O.C., Torgersen, T., O'Keefe, M., Terabayashi, T., Dong, C., and Eriksen, K.O., Large hydrocarbon compositional gradient revealed by in-situ optical spectroscopy, SPE 89704, presented at the *SPE ATCE*, Houston, TX, September 26–29, 2004.

Ghorayeb, K. and Firoozabadi, A., Molecular, pressure, and thermal diffusion in non-ideal multicomponent mixtures, *AICHE J.* 883–891, May 2000.

Graham, T., On the motion of gases, *Philos. Trans. R. Soc. London.* 136, 573–631, 1846.

Haase, R., *Thermodynamics of Irreversible Processes*, Addison-Wesley, Reading, MA, 1969, Chap. 4.

Hayduk, W. and Minhas, B.S., Correlations for molecular diffusivities in liquids, *Can. J. Chem. Eng.* 60, 295–299, 1982.

Montel, F. and Gouel, P.L., Prediction of compositional grading in a reservoir fluid column, SPE 14410, presented at the *SPE ATCE*, Las Vegas, NV, September 22–25, 1985.

Nes, K. and van Westerns, H.A., *Aspects of Constitution of Mineral Oils*, Elsevier, New York, 1951.

Onsager, L., Reciprocal relations in irreversible processes. I, *Phys. Rev.* 37, 405–426, 1931a.

Onsager, L., Reciprocal relations in irreversible processes. II, *Phys. Rev.* 37, 2265–2279, 1931b.

Pedersen, K.S. and Hjermstad, H.P., Modeling of large hydrocarbon compositional gradient, SPE 101275, presented at the *Abu Dhabi International Exhibition and Conference*, Abu Dhabi, November 5–8, 2006.

Pedersen, K.S. and Hjermstad, H.P., Modeling of compositional variation with depth for five North Sea reservoirs, SPE 175085-MS, presented at the *SPE ATCE*, Houston, TX, September 28–30, 2015.

Pedersen, K.S. and Lindeloff, N., Simulations of compositional gradients in hydrocarbon reservoirs under the influence of a temperature gradient, SPE paper 84364, presented at the *SPE ATCE*, Denver, CO, October 5–8, 2003.

Peneloux, A., Rauzy, E., and Fréze, R.A., A consistent correction for Redlich–Kwong–Soave volumes, *Fluid Phase Equililb.* 8, 7–23, 1982.

Peng, D.-Y. and Robinson, D.B., A new two constant equation of state, *Ind. Eng. Chem. Fundam.* 15, 59–64, 1976.

Rodriguez, F.D.C., Lopez, G., Leonardo, B., Skoreyko, J.A., and Fraser, A., Generation of a compositional model to simulate EOR processes for complex fluid systems of the Caroto-Mulata Field, Venezuela, SPE 107803, presented at the *Latin American & Caribbean Petroleum Engineering Conference*, Buenos Aires, Argentina, April 2007.

Schulte, A.M., Compositional variations within a hydrocarbon column due to gravity, SPE 9235, presented at the *SPE ATCE*, Dallas, September 21–24, 1980.

Smith, J.M., Van Ness, H.C., and Abbott, M.M., *Chemical Engineering Thermodynamics*, McGraw-Hill, Boston, MA, 2001.

Soave, G., Equilibrium constants from a modified Redlich–Kwong equation of state, *Chem. Eng. Sci.* 27, 1197–1203, 1972.

Vinhal, A., Azeem, J., and Pedersen, K.S., Modeling of compositional grading in heavy oil fields, SPE 205887-MS, presented at the *SPE ATCE*, Dubai, September 21–23, 2021.

Whitson, C.H. and Belery, P., Compositional gradients in petroleum reservoirs, SPE 28000 presented at the *University of Tulsa/SPE Centennial Petroleum Engineering Symposium*, Tulsa, OK, August 29–31, 1994.

15 Gas Injection EOR

Figure 15.1 shows a sketch of a reservoir produced through natural depletion. The reservoir pressure will fall as production goes on and material is removed from the reservoir. The pressure will eventually reach the saturation pressure at which gas will start to form, as is illustrated in the right-hand sketch of Figure 15.1. From that time on, the majority of the production is likely to come from the gas phase. The gas contains fewer liquid components than the oil, and the final recovery may end up being only a few percent of the total oil in place.

The term *enhanced oil recovery* (EOR) is used to describe various processes applied to increase hydrocarbon recovery from petroleum reservoir fields to a level above that of natural depletion. Injection of water, gas, and water alternating gas are the most applied EOR techniques. For fields with a heavy and highly viscous oil, injection of steam or steam mixed with solvent (expanding solvent—steam assisted gravity drainage (ESSAGD)) can also be applicable.

Injection gases may either be hydrocarbon gases originating from production of oil or gas condensate fluids or the injection gas can be carbon dioxide (CO_2). CO_2 is an effective injection gas, and there may also be environmental reasons in favor of CO_2.

In the upper sketch of Figure 15.2, oil is being displaced by a gas by a piston-like effect. The lower sketch shows an injection gas penetrating the oil phase and making its way to the production well where it will cause an early gas breakthrough. None of the scenarios in Figure 15.2 gives a true picture of the displacement mechanisms.

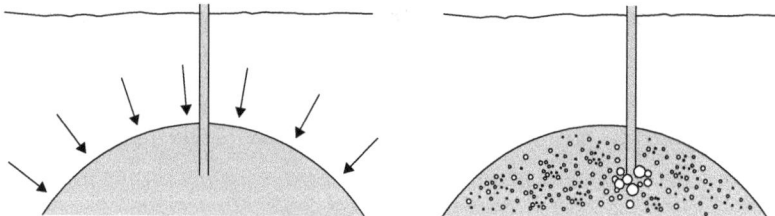

FIGURE 15.1   Reservoir oil produced by natural depletion.

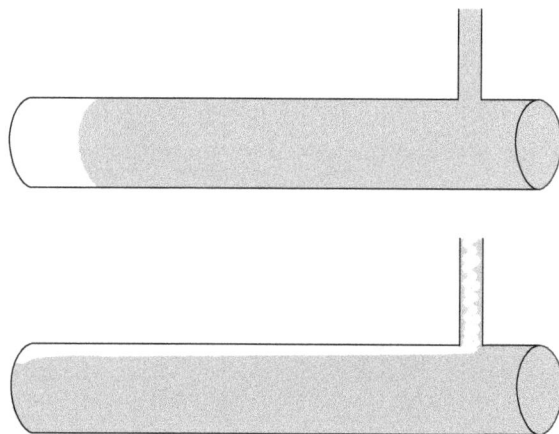

FIGURE 15.2   Piston-like gas displacement (upper) and early gas breakthrough (lower). None of those processes correspond with reality.

DOI: 10.1201/9780429457418-15

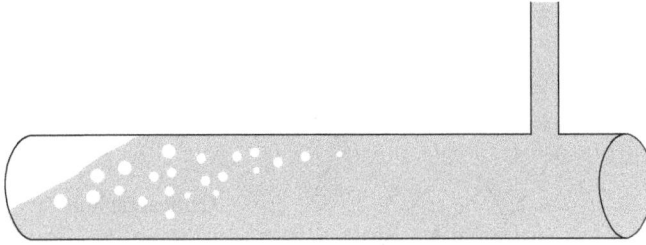

FIGURE 15.3 Transition zone between reservoir oil and injection gas.

In reality, the injected gas will, as illustrated in Figure 15.3, mix with the reservoir fluid and influence the oil–gas phase equilibrium in the reservoir. The gas may selectively take up components from the oil phase (vaporizing mechanism), the oil may take up components from the gas phase (condensing mechanism), or both oil and gas take up components from the other phase (combined vaporizing/condensing drive).

15.1 THREE-COMPONENT MIXTURES

Much of the classical literature on miscible gas displacement is based on consideration of three-component mixtures and ternary diagrams (e.g., Stalkup 1984). Figure 15.4 shows a ternary diagram for a mixture of C_1, C_4, and C_{10}. The diagram is for a fixed pressure and temperature. Each corner in the diagram represents 100% of a given component. Any concentration between 0% and 100% is represented as a proportionate distance between the bottom of the triangle and the opposite corner. Point A in Figure 15.4 corresponds to an oil mixture and point B to a heavy gas mixture. Also shown in Figure 15.4 is the location of the two-phase area. The dashed lines are tie-lines connecting gas and liquid compositions in thermodynamic equilibrium.

The tangent to the two-phase area in the critical point (or plait point) is called the *critical tie-line*. Whether miscibility can be achieved is determined by the position of the initial oil and gas compositions relative to the critical tie-line. One may distinguish between the following cases:

1. *First contact miscibility.* An oil and a gas composition made up of the same three components will be miscible by first contact if the line connecting the two compositions in a ternary diagram does not intersect the two-phase area. This is the situation for oil mixture A and gas mixture B in Figure 15.4. Mixtures of the oil and gas will be single-phase, independent of mixing ratio.
2. *Vaporizing drive.* This principle is illustrated in Figure 15.5. The composition of the injection gas is located on the two-phase side of the critical tie-line and the composition of the original oil on the single-phase side. The line connecting the oil and gas composition intersects the two-phase area. Miscibility between the oil and the injection gas may be achieved through a vaporizing process. When the original oil is mixed with an appropriate amount of injection gas, two phases are formed. The equilibrium phases can be represented as two points on the borderline between the single and two-phase areas in the ternary diagram. The two points are connected by a tie-line in Figure 15.5. The new gas composition may contact original reservoir oil, and a new gas phase will form. It will contain more heavy components and have a composition more like the composition at the critical point. At some stage, the gas composition may become equal to that at the critical point, and this critical phase is miscible with the original oil. This means that only one phase will form, no matter in what proportion

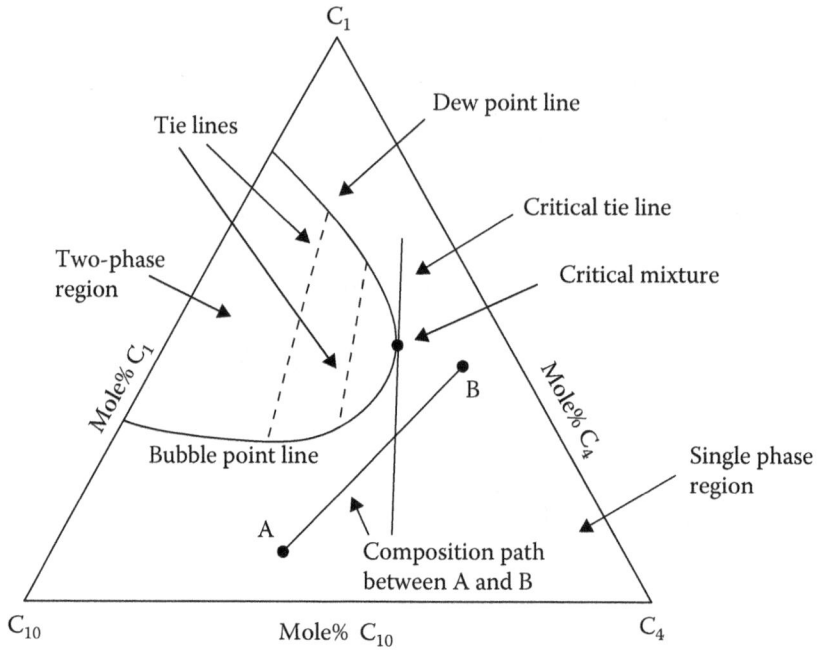

FIGURE 15.4 Ternary diagram for mixtures consisting of C_1, C_4, and C_{10}. Compositions A and B are miscible by first contact.

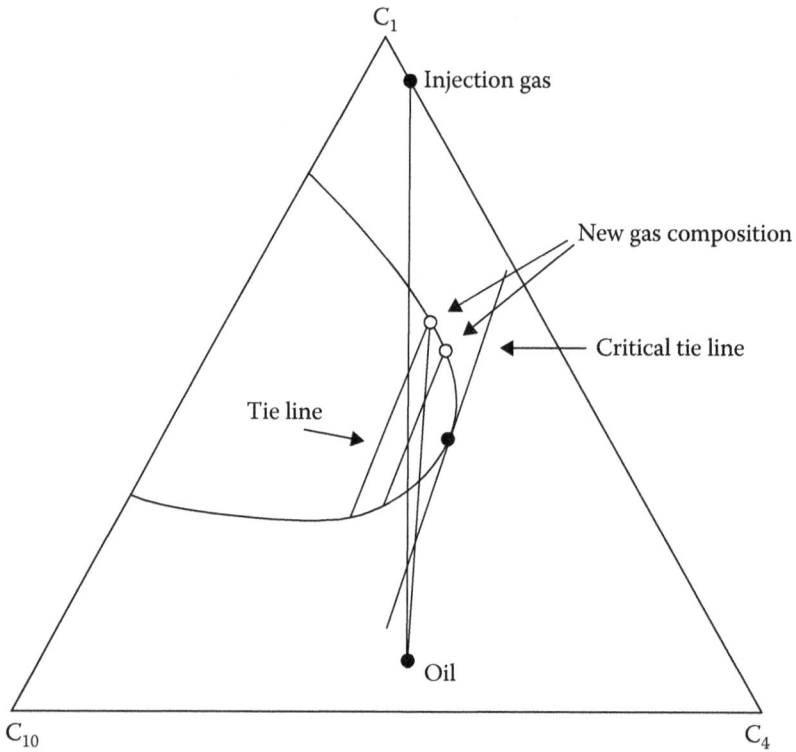

FIGURE 15.5 The principle of a vaporizing gas drive illustrated in a ternary diagram for a mixture of C_1, C_4, and C_{10}.

the critical mixture and the oil are mixed. The oil and the gas are said to be miscible as a result of a vaporizing multicontact process. The term *vaporizing* is used because the gas is gradually enriched with intermediate-molecular-weight components from the oil phase. The vaporizing process is illustrated in a reservoir context in Figure 15.6. The injection gas contacts oil at the injection well, and it takes up intermediate-molecular-weight components from the oil. Because the gas has a higher mobility than the oil, it is pushed away from the injection well by new injection gas, whereas the oil, which is less mobile, is almost stagnant. Away from the injection well, the enriched injection gas contacts fresh oil, takes up more components from the oil, and so on until, after multiple contacts in some distance from the injection well, the enriched gas and the original oil become fully miscible. As can be seen from Figure 15.6, miscibility, for a vaporizing drive, develops at the gas–oil front.

3. *Condensing drive.* The principle is illustrated in Figure 15.7. The composition of the original reservoir fluid (oil) is on the two-phase side of the critical tie-line and the composition of the injection gas on the single-phase side. The line connecting the oil and gas compositions intersects the two-phase area. When the oil is mixed with an appropriate amount of injection gas, two phases are formed. The oil will take up intermediate-molecular-weight components from the gas phase, and its composition will be more like that at the critical point. If the oil that has already taken up gaseous components is contacted by fresh injection gas, some of the heavier components in the gas will again condense into the oil phase, and the composition of the oil will at some stage become equal to that at the critical point. Miscibility has been achieved, because the now-critical "oil" will be miscible with the injection gas in all ratios. The condensing process is illustrated in a reservoir context in Figure 15.8. The injection gas contacts oil at the injection well. The oil takes up intermediate-molecular-weight components from the gas. The stripped gas is pushed away from the injection well by new injection gas. Some of the heavier components in this gas will condense into the oil phase. After a number of contacts, miscibility develops at the injection well.

4. *Miscibility cannot be achieved.* If the injection gas and reservoir oil compositions are both located on the two-phase side of the critical tie-line, miscibility cannot be achieved. This situation is illustrated in Figure 15.9.

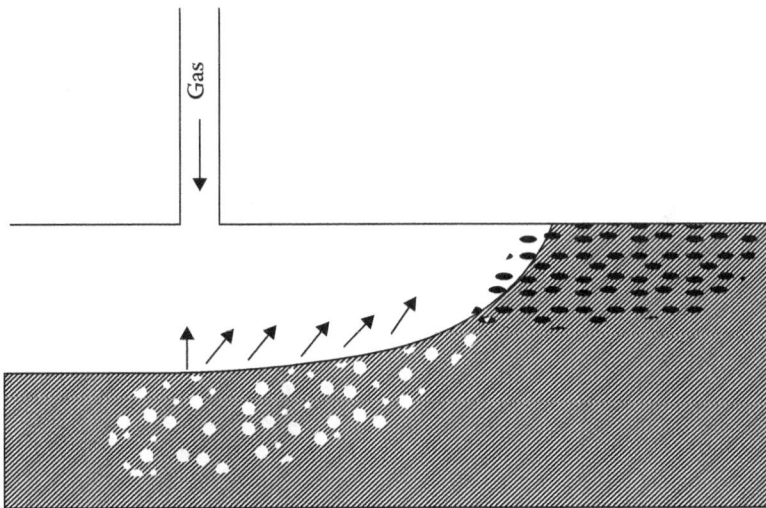

FIGURE 15.6 Miscibility obtained through a vaporizing drive. Passing through the reservoir the gas takes up components from the oil phase and develops miscibility at the gas–oil front.

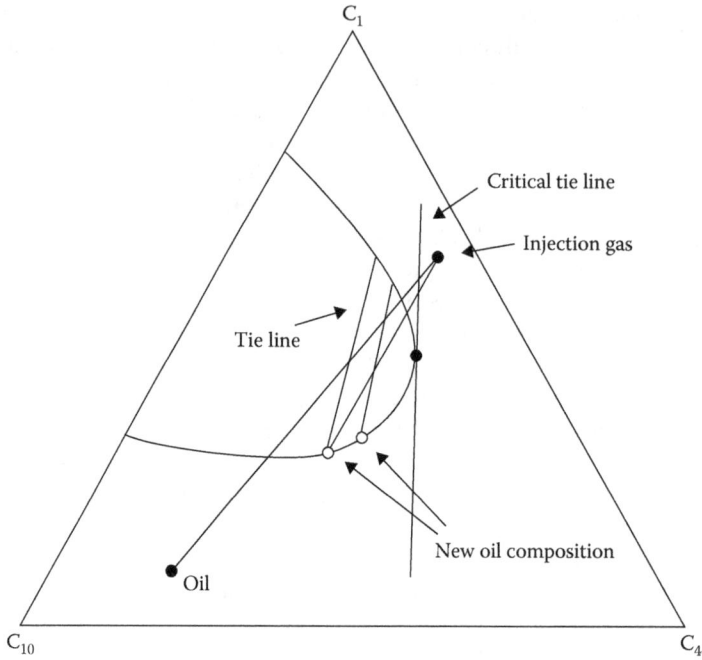

FIGURE 15.7 The principle of a condensing drive illustrated in a ternary diagram for a mixture of C_1, C_4, and C_{10}.

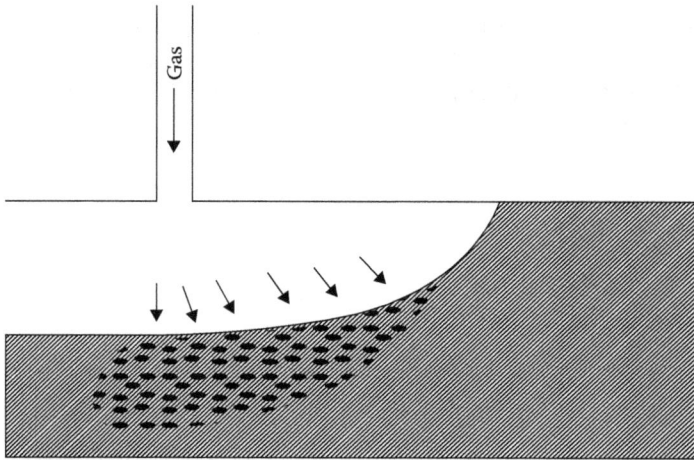

FIGURE 15.8 Miscibility obtained through a condensing drive. Miscibility develops near injection well.

Ternary diagrams represent the phase equilibrium picture at a fixed temperature and pressure. Figure 15.10 illustrates qualitatively the development in the two-phase area with pressure. The size of the two-phase area decreases with increasing pressure, meaning that miscibility is more easily achieved at higher pressures. The lowest pressure, where at a fixed temperature, miscibility can be achieved between a given reservoir fluid and a given injection gas, is called the *minimum miscibility pressure* (MMP). The lowest pressure at which the reservoir oil and the injection gas are miscible by first contact as in Figure 15.4 is called the *first contact minimum miscibility pressure* (FCMMP).

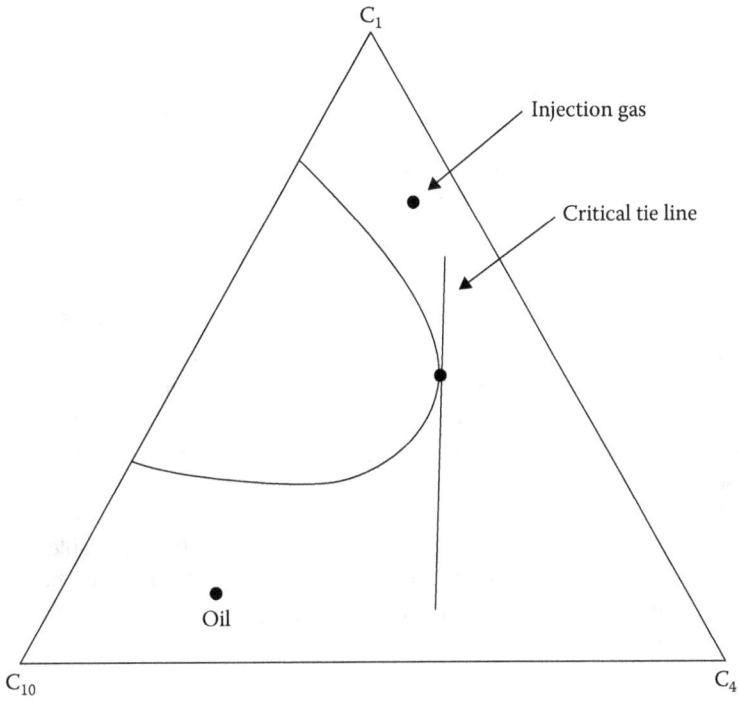

FIGURE 15.9   Oil and gas composition that at the actual pressure are not miscible illustrated for a ternary mixture of C_1, C_4, and C_{10}.

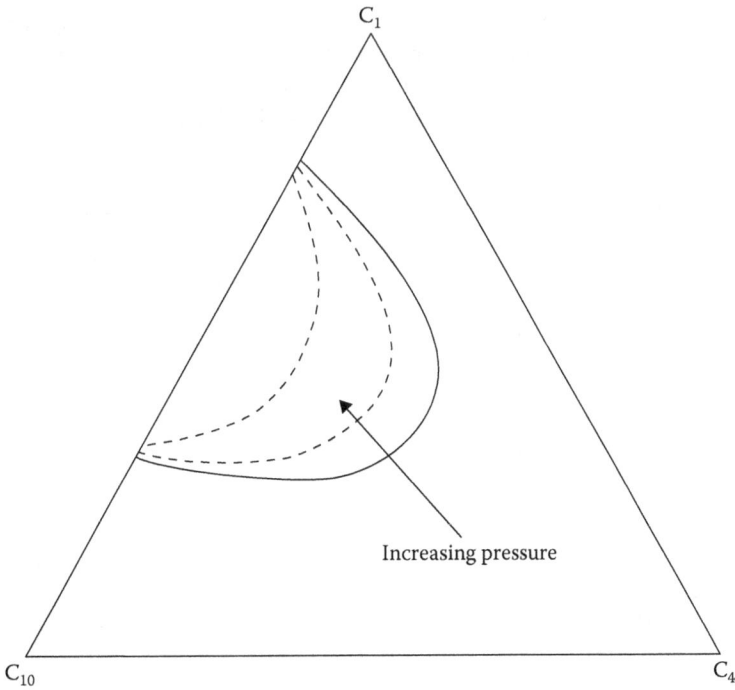

FIGURE 15.10   Development in two-phase area of a petroleum mixture as a function of pressure illustrated for ternary mixture of C_1, C_4, and C_{10}.

15.2   MMP OF MULTICOMPONENT MIXTURES

Ternary diagrams are inapplicable for mixtures with four or more components. As exemplified by Pedersen et al. (1986) and Zick (1986), the two-phase area is undefined in a ternary diagram for a mixture with more than three components. Another deficiency is that a ternary diagram only considers purely vaporizing and purely condensing drives and not the combined vaporizing/condensing mechanism, which for most gas–oil systems determines the miscibility pressure.

15.2.1   First Contact MMP

Figure 15.11 shows the saturation pressure of the reservoir oil in Table 15.1 for a temperature of 73°C, and the influence on the saturation pressure of adding gas of the composition also given in Table 15.1. The simulations have been carried out using the SRK equation of state, and the oil was characterized using the procedure of Pedersen et al. as presented in Chapter 5. The highest saturation pressure is 1074 bar and seen for a gas mole percentage of 92.6. This pressure is the FCMMP. At higher pressures, two phases cannot form, regardless of the ratio in which oil and gas are mixed. Mogensen and Goryachev (2022) have reported a fluid system with high concentrations of H_2S in both reservoir oil and injection gas for which the FCMMP would be achievable in the considered reservoir. Most often the FCMMP is much higher than the reservoir pressure, but, as described in the following, a miscible drive can, for most reservoir fluids, as a result of a multi-contact process, be achieved at a much lower pressure than the FCMMP.

15.2.2   Tie-Line Approach

As was first pointed out by Zick (1986) and Stalkup (1987), the drive for a multicomponent mixture (N > 3) can be combined vaporizing and condensing. This means that the component exchange upon contact between gas and oil is not one-way. The reservoir oil takes up components from the gas and the gas takes up components from the oil. With a combined vaporizing/condensing drive, miscibility develops at a position in the reservoir between the injection well and the gas–oil front. The miscible oil and gas compositions are equal and at their critical point. Miscibility as a result of a combined vaporizing/condensing mechanism is illustrated in Figure 15.12.

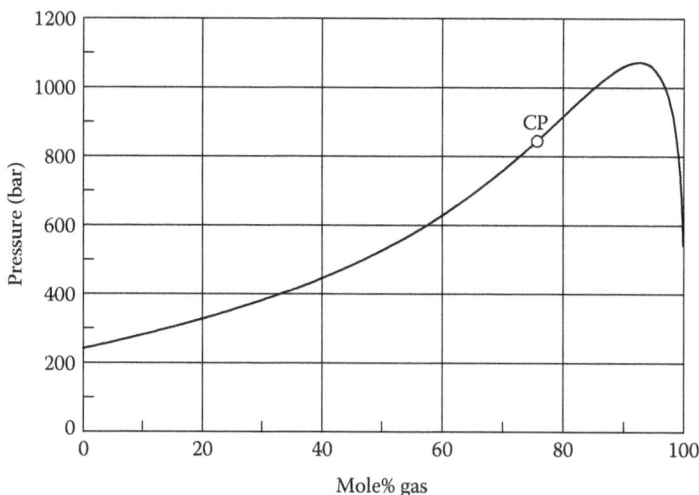

FIGURE 15.11   Saturation pressures at 73°C for mixtures of the oil and gas compositions in Table 15.1 as a function of gas mole percentage of total mixture. CP stands for critical point.

TABLE 15.1

Molar Composition of the Oil and Gas for Which a Simulated Swelling Curve Is Shown in Figure 15.11

| Component | Oil | | | Injection Gas |
	Mol%	Molecular Weight (g/mol)	Density at 1.01 bar, 15°C (g/cm³)	Mol%
N_2	0.53	—	—	1.17
CO_2	1.01	—	—	1.79
C_1	45.30	—	—	85.47
C_2	3.90	—	—	6.93
C_3	1.39	—	—	2.15
iC_4	0.63	—	—	0.77
nC_4	0.81	—	—	0.86
iC_5	0.69	—	—	0.41
nC_5	0.41	—	—	0.17
C_6	1.02	—	—	0.28
C_7	4.22	96	0.733	—
C_8	3.53	107	0.763	—
C_9	3.50	121	0.784	—
C_{10}	3.16	134	0.815	—
C_{11}	2.43	146	0.832	—
C_{12}	2.42	160	0.847	—
C_{13}	2.37	174	0.860	—
C_{14}	2.19	188	0.873	—
C_{15}	1.96	199	0.877	—
C_{16}	1.84	212	0.879	—
C_{17}	1.60	229	0.880	—
C_{18}	1.40	246	0.884	—
C_{19}	1.25	258	0.892	—
C_{20+}	12.43	502	0.933	—

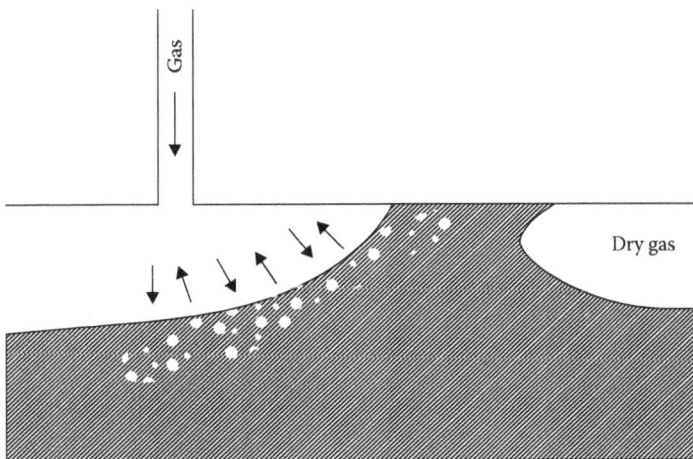

FIGURE 15.12 Miscibility obtained through a combined vaporizing and condensing drive. Miscibility develops at a position between injection well and gas–oil front.

An algorithm that can successfully determine the multicontact MMP for a reservoir oil mixture and an injection gas must consider three different mechanisms for exchange of components between oil and injection gas, these being vaporizing, condensing, and combined vaporizing/condensing. The three mechanisms are illustrated in Figure 15.13. Transfer of intermediate-molecular-weight components (typically C_3–C_6) from the oil phase to the gas phase is the dominant mechanism for a vaporizing drive. With a condensing drive, intermediate-molecular-weight components from the gas phase condense into the oil phase. When the drive is combined vaporizing/condensing, components are transported either way. Some components from the gas phase will condense into the oil, and components from the oil phase will vaporize into the gas phase.

In a reservoir with gas injection and a fluid system represented by N components, N-1 compositional constant composition zones will develop (Monroe et al. 1990) as sketched in Figure 15.14. The zones move with time, leaving gas behind, as is illustrated in the lower sketch of Figure 15.14. A small irreducible oil matrix may persist near the injector (not shown in Figure 15.14). Miscibility develops when the gas and liquid compositions in one of the zones become identical and the fluid composition in that zone is at its critical point. At the gas/oil contact, gas contacts fresh oil. The gas contacting the fresh oil has already contacted oil multiple times. The tie-line connecting the equilibrium gas and oil at the gas/oil front will extend through the fresh oil composition. Mathematically this can be expressed as

$$(1-\beta_1)x_i^1 + \beta_1 y_i^1 = z_i^{oil}; \quad i = 1,2,...,N \tag{15.1}$$

where β can take values from minus infinity to plus infinity. The superscript oil refers to original reservoir oil. If the length of this tie-line becomes zero, the equilibrium gas and oil compositions at the gas/oil front are identical. The fluid is at its critical point and a miscible drive has developed. When a miscible drive develops at the gas–oil front, the drive is 100% vaporizing.

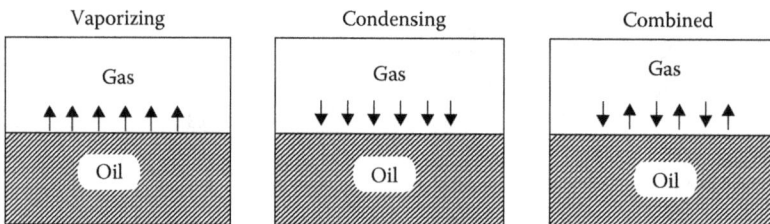

FIGURE 15.13 Mechanisms to be considered when calculating multicontact MMP.

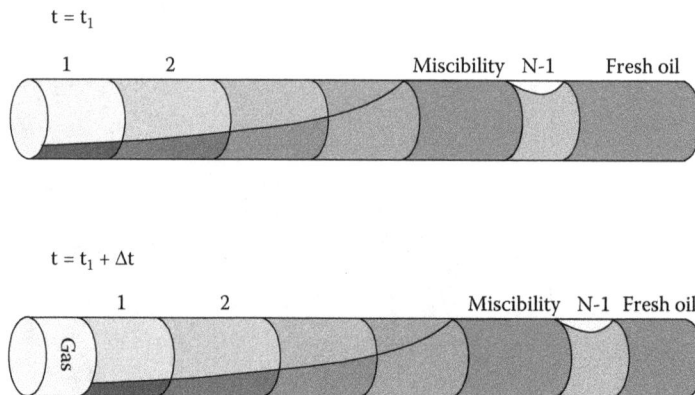

FIGURE 15.14 Compositional zones at time t_1 and $t_1 + \Delta t$ in a reservoir with miscible gas injection. The zones will move with time and leave a gas zone behind.

At the inlet, fresh injection gas contacts oil that after some time has already been contacted many times by fresh gas. The tie-line connecting the equilibrium gas and oil compositions will extend through the injection gas. Mathematically this can be expressed as

$$\left(1-\beta_{N-1}\right)x_i^{N-1}+\beta_{N-1}y_i^{N-1}=z_i^{inj}; \quad i=1,2,...,N \tag{15.2}$$

The superscript inj refers to injection gas. If the length of the tie-line

$$L_n=\sqrt{\sum_{i=1}^{N}\left(y_i^n-x_i^n\right)^2}$$

expressed through the mole fractions in Equation 15.2 becomes zero, the equilibrium gas and oil compositions at the injection well are identical. The fluid is at its critical point and a miscible drive has developed. This drive is 100% condensing. Condensing drives are mainly of theoretical interest. A 100% condensing drive has never been reported in literature for a real case.

At the shift from one compositional zone to the next one in Figure 15.14, the total composition must be some mathematical combination of the phase compositions in each of the two neighboring zones. This is expressed as

$$\beta_j y_i^j+\left(1-\beta_j\right)x_i^j=\beta_{j+1}y_i^{j+1}+\left(1-\beta_{j+1}\right)x_i^{j+1}; \quad i=1,2,...,N, \quad j=1,2,...,N-3 \tag{15.3}$$

These N-3 equations represent N-3 so-called crossover tie-lines connecting the initial oil tine line in Equation 15.1 and the injection gas tie-line in Equation 15.2. Equation 15.3 expresses that each crossover tie-line intersects the tie-line just upstream and the tie-line just downstream.

Johns and Orr (1997) and Wang and Orr (1997, 1998) have presented an algorithm for solving Equations 15.1 through 15.3 starting with a relatively low pressure (could be 50 bar). Once the N-1 key tie-lines have been found for a given pressure, the MMP can be determined by increasing the pressure in increments until one of the key tie-lines becomes a critical tie-line (length of zero). Equation 15.3 must be solved subject to the equilibrium constraint

$$x_i^j\varphi_i^{L,j}=y_i^j\varphi_i^{V,j}; \quad i=1,2,...,N, \quad j=1,2,...,N-1 \tag{15.4}$$

and the summation of mole fraction constraint:

$$\sum_{i=1}^{N}\left[y_i^j-x_i^j\right]=0; \quad j=1,2,...,N-1 \tag{15.5}$$

The multicomponent tie-line calculation procedure can be summarized as follows:

1. Initially solve Equations 15.1 and 15.2.
2. To obtain a set of intersecting tie-lines, Equations 15.3 through 15.5 are solved for a given temperature (typically reservoir temperature) and a (low) pressure.
3. The pressure is increased slightly. The equations in 1 and 2 are solved again using the solution for the previous pressure as initial estimate.
4. Step 3 is repeated until one of the tie-lines shrinks to a point (length L of zero) which condition is fulfilled for tie-line n if

$$L_n=\sqrt{\sum_{i=1}^{N}\left(y_i^n-x_i^n\right)^2}\approx 0$$

The pressure where this happens is the MMP. When the length is zero, the tie-line connects two identical critical compositions.

Figure 15.15 schematically shows the development towards miscibility for an 8-component system. The x-axis represents the distance from the injection well and the y-axis the gas and liquid densities. At a pressure of 150 bar, the constant composition zones have developed, but in all N-1 (=7) zones the liquid density is significantly higher than the gas density. At a pressure of 190 bar the phase densities in zone 6 are almost equal signaling a tie-line length approaching zero and a near critical fluid composition. The fluid has reached miscibility and the MMP is 190 bar. Miscibility develops in a zone between the 1st (injection gas) tie-line and the last (original oil) tie-line. The drive is therefore combined vaporizing-condensing.

Jessen et al. (1998) have presented a modification of the preceding algorithm, with focus on speed and numerical robustness.

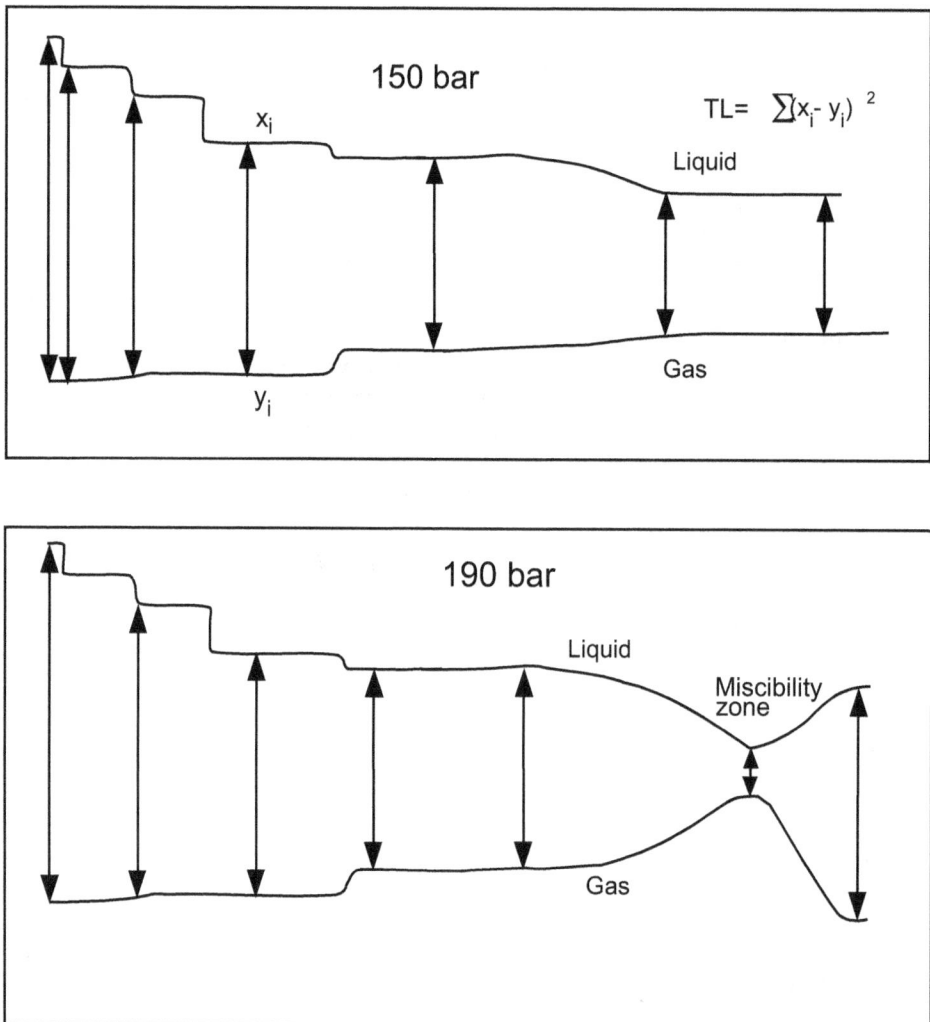

FIGURE 15.15 Density profiles for an 8-component system in a slim tube with gas injection. The MMP is around 190 bar. TL stands for tie-line length. The tie-lines are represented as the lines connecting the liquid and gas densities. x_i and y_i are the liquid and gas mole fractions of component i in the liquid and gas phase, respectively.

Johns et al. (2002) have suggested a method to quantify the displacement mechanisms. On each tie-line, the point for which the vapor mole fraction β equals 0.5 is located. The term d_1 is used for the distance from the $\beta = 0.5$ point on the oil tie-line to the $\beta = 0.5$ point on the second tie-line. The distance is calculated as the square root of the sum of the squared differences between the component mole fractions on the first and second tie-lines for $\beta = 0.5$. The term d_2 is used for the distance from the latter point to the $\beta = 0.5$ point on the third tie-line, and so on. For a four-component mixture, the third tie-line is the last one (the one passing through the injection gas). For a four-component mixture, the fraction of the drive that is vaporizing is given by

$$V_m = \frac{d_2}{d_1 + d_2} \tag{15.6}$$

For a multicomponent system, Johns et al. define the *vaporizing fraction* as the ratio of the total vaporizing (v) path length to the entire composition path

$$V_m = \frac{\sum_{k=1}^{N-2} d_{k,v}}{\sum_{k=1}^{N-2} d_k} \tag{15.7}$$

where $d_{k,v}$ is nonzero for tie-lines for which the displacement mechanism between that tie-line and the next one is vaporizing. This is the case if the tie-lines are longer in the direction toward the gas tie-line than in the direction toward the oil tie-line.

In most cases a slim tube experiment (Section 3.2.4) carried out at the MMP will show a recovery of 90–95%. Often the reservoir conditions do not permit gas injection at a pressure as high as the MMP. Gas injection may still be favorable, and it is of interest to know the recovery for pressures between the saturation pressure of the original oil and the MMP. As an alternative to carrying out a slim tube experiment, an approximate recovery can be simulated using an algorithm of Jessen et al. (2001).

Slim tube data is shown in Table 3.48 for the oil in Table 3.47 with CO_2 as injection gas. As is seen from Figure 3.25, the experimental MMP was 208 bar for a temperature of 85.7°C. The MMP of this system was simulated using a combined vaporizing/condensing MMP algorithm and the SRK equation of state. The characterized fluid composition is shown in Table 15.2 and the nonzero binary interaction coefficients in Table 15.3. The MMP was simulated to 209 bar, which is close to the experimentally determined MMP.

Other experimental slim tube data is shown in Tables 15.4 and 15.5 (Glasø 1985; Firoozabadi and Aziz 1986). Simulated MMPs for these systems are shown in Table 15.6. The simulated MMPs for these systems are generally 5%–15% lower than the experimentally determined slim tube MMPs.

The literature about miscible displacement and determination of MMP by analytical methods is based on theories developed for fluid dynamics and chromatography. Helfferich (1981) has given a good introduction to the terms and assumptions used in the literature about combined vaporizing/condensing gas displacement.

15.2.3 Immiscible Systems

Not all gas–oil systems will have an MMP. Some combinations of gas and oil will after multiple contacts develop a liquid–liquid system, which is immiscible independent of pressure. With CO_2 as injection gas, both phases will be rich in CO_2, but one will have a substantially higher concentration of C_{7+}

TABLE 15.2

Volume-Corrected SRK EoS Model for Oil Composition in Table 3.47. Nonzero Binary Interaction Parameters Can Be Seen in Table 15.3

Component	Mol%	Molecular Weight	Critical Temperature °C	Critical Pressure bar	Acentric Factor	Volume Correction cm³/mol
N_2	1.025	28.01	−146.95	33.94	0.0400	0.92
CO_2	0.251	44.01	31.05	73.76	0.2250	3.03
C_1	17.242	16.04	−82.55	46.00	0.0080	0.63
C_2	5.295	30.07	32.25	48.84	0.0980	2.63
C_2	4.804	44.10	96.65	42.46	0.1520	5.06
C_4	2.592	58.12	145.80	37.44	0.1868	7.65
C_5	0.89	72.15	190.85	33.80	0.2364	11.42
C_6	0.134	86.18	234.25	29.69	0.2960	17.98
C_7–C_{17}	40.524	151.78	349.01	21.03	0.6554	40.42
C_{18}–C_{29}	17.22	311.81	492.09	13.70	1.0185	54.65
C_{30}–C_{80}	10.023	567.94	681.74	12.02	1.2787	−40.39

TABLE 15.3

Nonzero Binary Interaction Parameters for the Volume-Corrected SRK EoS Model in Table 15.2

Component	N_2	CO_2
CO_2	−0.0315	
C_1	0.0278	0.12
C_2	0.0407	0.12
C_2	0.0763	0.12
C_4	0.0789	0.12
C_5	0.0871	0.12
C_6	0.0800	0.12
C_7–C_{17}	0.0800	0.08
C_{18}–C_{29}	0.0800	0.08
C_{30}–C_{80}	0.0800	0.08

TABLE 15.4

Oil and Gas Compositions (in Mol%) Used in Slim Tube Displacement Tests and Experimental Slim Tube MMPs

Component	Reservoir Oil A	Reservoir Oil B	Reservoir Oil C	Injection Gas A	Injection Gas B	Injection Gas C
N_2	0.47	0.92	0.18	0.50	1.40	0.29
CO_2	0.49	0.36	0.44	0.76	1.35	0.76
C_1	42.01	40.60	43.92	72.04	82.17	73.05
C_2	6.05	5.22	10.71	12.41	8.42	13.95
C_3	2.93	3.31	8.81	8.60	4.53	8.17
iC_4	0.61	0.68	1.30	1.19	0.49	0.77
nC_4	0.99	1.89	3.99	2.55	0.95	1.89
iC_5	0.58	0.87	1.36	0.58	0.18	0.29
nC_5	0.42	1.30	1.83	0.65	0.19	0.33

Component	Reservoir Oil A	Reservoir Oil B	Reservoir Oil C	Injection Gas A	Injection Gas B	Injection Gas C
C_6	0.92	1.92	2.55	0.35	0.16	0.24
C_{7+} (C_7 for gases)	44.53	42.93	24.91	0.37	0.16	0.26
C_{7+} M	196.0	215.1	231.0	—	—	—
C_{7+} density (g/cm³)	0.883	0.869	0.855	—	—	—
Reservoir T (°C)	92	79	99	—	—	—
MMP (bar)	390	470	360	—	—	—

The slim tube MMP is here defined as the pressure that gives 95% recovery. The MMPs are shown in the bottom row.
Source: Data from Glasø, Ø., Generalized minimum miscibility correlation, *Soc. Petroleum Eng. J.* 25, 927–934, 1985.

TABLE 15.5
Oil and Gas Compositions (in Mol%) Used in Slim Tube Displacement Tests and Experimental Slim Tube MMPs

Component	Reservoir Oil XA	Reservoir Oil XC	Reservoir Oil XD	Injection Gas XA	Injection Gas XC	Injection Gas XD
N_2	0.25	0.00	0.46	—	2.48	—
CO_2	3.60	0.00	1.34	—	—	—
C_1	56.83	50.39	49.01	100	87.83	100
C_2	9.37	8.82	7.04	—	7.50	—
C_3	5.48	5.91	4.93	—	1.91	—
iC_4	1.46	0.89	0.95	—	—	—
nC_4	2.61	3.28	2.52	—	0.26	—
iC_5	1.20	0.94	1.16	—	—	—
nC_5	1.39	1.29	1.52	—	—	—
C_6	1.26	1.36	3.34	—	—	—
C_{7+}	16.59	27.12	27.73	—	—	—
C_{7+} M	183.3	249.6	250.2	—	—	—
C_{7+} density in g/cm³	0.827	0.900	0.870	—	—	—
Reservoir T (°C)	171	107	151	—	—	—
MMP (bar)	331	414	434	—	—	—

Source: Data from Firoozabadi, A. and Aziz, K., Analysis and correlation of nitrogen and lean-gas miscibility pressure, *SPE Reservoir Eng*. 575–582, November 1986.

TABLE 15.6
Comparison of Experimental and Simulated MMPs

Experimental Compositions in Table	Temperature (°C)	Experimental MMP (bar)	Simulated MMP (bar)	Percentage Deviation
15.2	85.7	208	214	+3
15.4	92	390	319	−22
15.4	79	470	421	−12
15.4	99	360	320	−13
15.5	171	331	315	−5
15.5	107	414	394	−5
15.5	151	434	412	−5

Note: The simulated MMPs are obtained using a combined vaporizing/condensing MMP algorithm.

than the other one. The recovery versus pressure for an immiscible system will not have the distinct bend as shown in Figure 3.24. The recovery curve will be more rounded as the one in Figure 15.16. When no bend is seen on the recovery curve, it is common practice in industry to take the pressure at which the recovery is 90%–95% as the MMP. Many of such systems may in fact be immiscible.

Table 15.7 shows an oil composition sampled from a reservoir with a temperature of 48.6°C (Lindeloff et al. 2013). Slim tube recovery data for the oil with CO_2 as injection gas is shown in Table 15.8 and plotted in Figure 15.17. Tube lengths of 9.1 m and 18.3 m were used. No sharp bend is seen on the recovery versus pressure curve suggesting that miscibility was not achieved.

Two saturation points were experimentally determined for the reservoir fluid mixed with CO_2 in a molar ratio of 1:2. The data points can be seen from Figure 15.18, which also shows the simulated

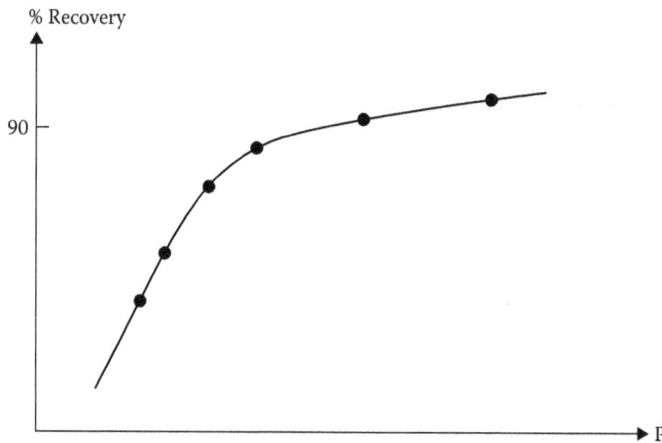

FIGURE 15.16 Stylistic slim tube recovery curve for a system, which could be immiscible as it has no distinct bend as the one in Figure 3.24. Although miscibility is not established, the recovery can still be high because two liquid phases of approximately the same mobility may form.

TABLE 15.7

Molar Composition of Reservoir Oil for Which Slim Tube Data Is Shown in Table 15.8 and Figure 15.17. Saturation Points and Phase Envelope of the Reservoir Oil Mixed with CO_2 in Molar Ratio 1:2 Are Shown in Figure 15.18

Component	Reservoir Oil Mol%
N_2	0.148
CO_2	0.99
C_1	20.429
C_2	1.725
C_3	2.916
iC_4	1.784
nC_4	2.819
iC_5	3.421
nC_5	0.057
C_6	1.989
C_{7+}	63.721
C_{7+} Molecular Weight	306
C_{7+} Density (g/cm³)	0.915

TABLE 15.8
CO$_2$ Slim Tube Data for Oil Composition in Table 15.7 at 48.3°C

9.1 m Slim Tube		18.3 m Slim Tube	
Pressure bar	Recovery Volume%	Pressure bar	Recovery Volume%
173.4	79.05	173.4	73.25
207.9	86.61	207.9	84.58
242.3	91.14	242.3	93.18
276.8	92.79	276.8	96.67
311.3	93.62		

Source: Lindeloff, et al., Investigation of miscibility behavior of CO$_2$ rich hydrocarbon systems—with application for gas
   injection EOR, SPE 166270, presented at *SPE ATCE*, New Orleans, LA, September 30–October 2, 2013.

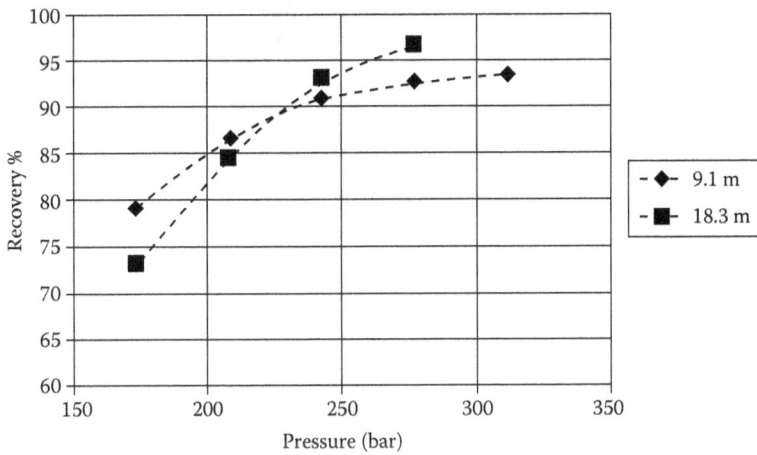

FIGURE 15.17   Recovery versus pressure in a slim tube experiment on the oil in Table 15.7 using CO$_2$ as injection
gas at a temperature of 48.6°C. Tube lengths of 9.1 m and 18.3 m were used. The data can be found in Table 15.8.

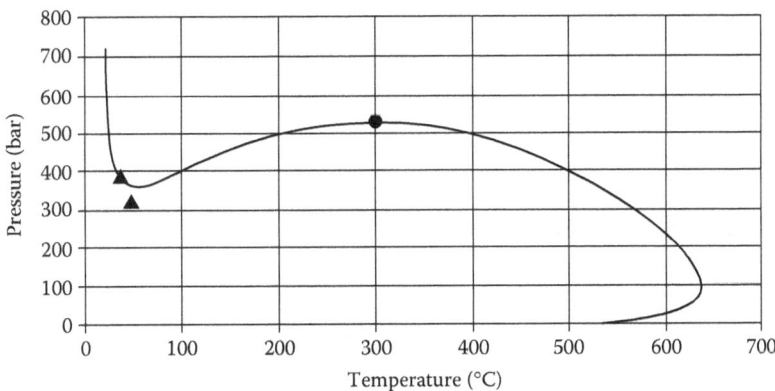

FIGURE 15.18   The triangles show two experimentally determined saturation points (37.8°C, 387.3 bar and
48.6°C, 326.2 bar) for a 2:1 molar mixture of CO$_2$ and the reservoir fluid in Table 15.7. The full-drawn line
shows the phase envelope simulated for the fluid using the SRK equation and the EOS model description in
Tables 15.9 and 15.10. The full circle is the simulated critical point.

phase envelope. Below a temperature of around 50°C, the saturation pressure is seen to increase with decreasing temperature. That is a clear indication of the presence of two liquid phases, which independent of pressure will not develop miscibility. This explains why miscibility did not develop in the slim tube experiment. Because the two liquid phases have approximately the same mobility, the recovery is high despite the system being immiscible. A tie-line algorithm as outlined in Section 15.2.2 will (correctly) fail to find a solution.

The phase envelope in Figure 15.18 was simulated using the SRK equation. The applied eight-component EoS model description is shown in Tables 15.9 and 15.10.

15.2.4 Cell-to-Cell Simulation

A cell-to-cell simulator (slim tube) as first suggested by Metcalfe (1972) is useful to give a detailed picture of the phase behavior with a multi-stage contact between gas and oil.

The starting point is a series of cells of equal volume placed in a row. Each cell is filled with reservoir oil at the reservoir temperature and a fixed pressure above the saturation pressure. A specified amount of gas is added to cell 1. It is assumed that perfect mixing takes place and thermodynamic equilibrium is reached. This means that the conditions in the cell can be found by a PT flash calculation as described in Chapter 6. As the injected gas and the cell fluid mix, the gas plus liquid volume will be larger than the initial cell volume. The excess volume is transferred to

TABLE 15.9

SRK EOS Model for Oil Composition in Table 15.7. Nonzero Binary Interaction Parameters Can Be Seen in Table 15.10

Component	Mol%	Critical Temperature °C	Critical Pressure bar	Acentric Factor
CO_2	0.99	31.05	73.76	0.2250
$N_2 + C_1$	20.577	−83.36	45.85	0.0084
$C_2 + C_3$	4.641	78.14	44.29	0.1365
$C_4 + C_6$	10.072	182.66	34.2	0.2284
$C_7 - C_{16}$	32.426	326.83	20.87	0.6221
$C_{17} - C_{29}$	16.114	497.12	14.67	0.9507
$C_{30} - C_{47}$	9.397	691.57	13.99	1.2461
$C_{48} - C_{80}$	5.783	893.08	14.08	1.0944

TABLE 15.10

Nonzero Binary Interaction Parameters for the SRK EOS Model in Table 15.9

Component	CO_2	$N_2 + C_1$
$N_2 + C_1$	0.1189	
$C_2 + C_3$	0.1200	0.0005
$C_4 + C_6$	0.1200	0.0006
C_7-C_{16}	0.1142	0.0006
C_{17}-C_{29}	0.0952	0.0006
C_{30}-C_{47}	0.0952	0.0006
C_{48}-C_{80}	0.1089	0.0006

cell 2. Metcalfe et al. have, as illustrated in Figure 15.19, used three different criteria for defining the excess volume:

1. *Stagnant oil*: All the gas formed in cell 1 is transferred to cell 2, whereas all the oil remains in cell 1.
2. *Moving excess*: All the gas formed in cell 1 is transferred to cell 2. If the volume of the remaining oil phase exceeds that of the original cell, the excess volume of oil is also transferred to cell 2.
3. *Phase mobility criterion*: The cell volume remains constant throughout the calculation, and the excess volume is transferred to cell 2. If two phases are present, gas and liquid are moved according to their relative phase mobilities:

$$\text{Gas volume moved} = \frac{V_{excess}}{1 + M_{o/g}}$$

$$\text{Oil volume moved} = \frac{V_{excess}\ M_{o/g}}{1 + M_{o/g}}$$

where V_{excess} is the excess cell volume, oil saturation is the volume fraction of oil in the cell, and $M_{o/g}$ is the relative oil/gas mobility:

$$M_{o/g} = \frac{k_{ro}\ \eta_g}{k_{rg}\ \eta_o} \tag{15.8}$$

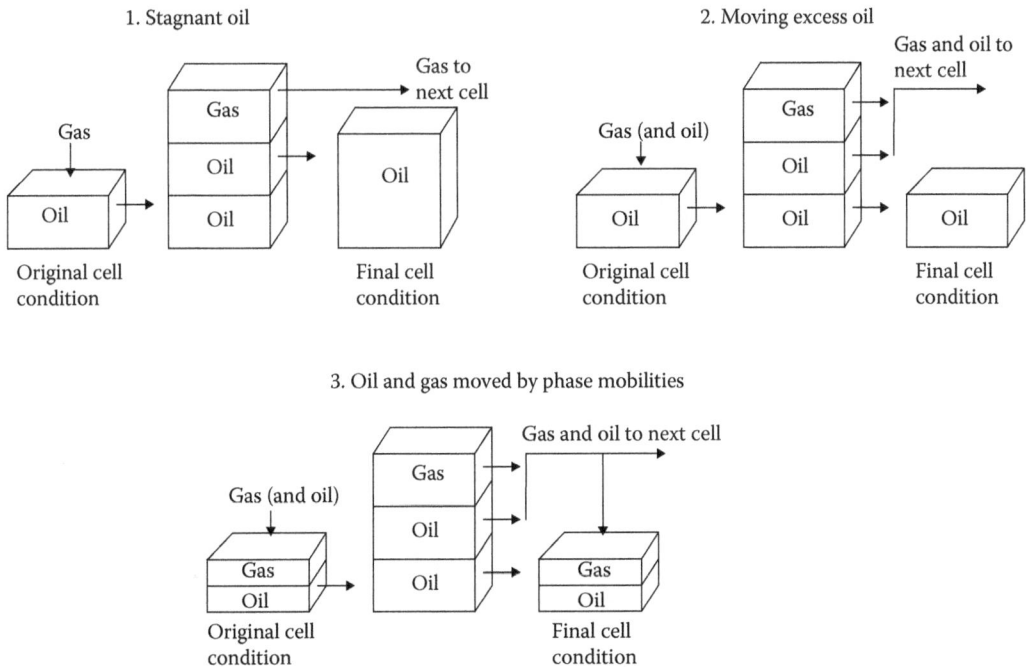

FIGURE 15.19 Criteria for moving excess fluid from one cell to the next one in a cell-to-cell gas injection simulation.

In Equation 15.8, k_{ro} and k_{rg} are the relative oil and gas phase permeabilities, respectively; η_o is the oil viscosity; and η_g the gas viscosity. Relative permeabilities are determined experimentally by gas and oil flooding through core samples (see, e.g., Dandekar 2013). An example of relative permeability curves is given in Figure 15.20. If no relative mobility data exists, a *viscosity mobility criterion* may be used instead. The relative mobility of oil and gas is simply expressed as the ratio of the oil and gas volume fractions times the ratio of the gas and oil viscosities. For liquid–liquid displacement, it may actually be more correct to use the viscosity mobility criterion because any relative permeability data would be measured on a gas–liquid system and not a liquid–liquid system.

The excess volume from cell 1 is transferred to cell 2 and a PT flash calculation is performed on the total cell mixture (phase equilibrium assumed to develop). The excess volume from cell 2 is transferred to cell 3, and so on. The excess volume fluid from the last cell is flashed to standard conditions. When one batch calculation has been completed, a new batch of gas is injected into cell 1 and the cell-to-cell calculation is continued until the injected gas volume at cell conditions equals 1.2 times the volume of all cells. A 100% recovery of the "oil in place" is obtained if the sum of the volumes of stable oil from flashing the excess volume fluid from the last cell to standard conditions equals the volume obtained by flashing the original oil in the cells to standard conditions.

Figure 15.21 shows the recoveries found by a slim tube simulation on the oil composition in Table 3.47 characterized as shown in Table 15.2 for a temperature of 85.7°C with CO_2 as injection gas and using the Peneloux corrected SRK equation. The slim tube was simulated using 2000 cells, and 10,000 time steps (gas injections) were used. The excess volume was moved from one cell to the next one using the *moving excess* criterion. As can be seen from Figure 15.21, the simulated MMP is of the order of 210 bar, which compares well with the experimental MMP of 208 bar (see Figure 3.25).

Cell-to-cell calculations by Zhao et al. (2006) confirm the development of constant composition zones as sketched in Figure 15.15.

Unlike the tie-line method, a cell-to-cell simulation does not provide an analytical solution for the MMP. It is, however, possible to configure a cell-to-cell simulator to detect whether a miscible

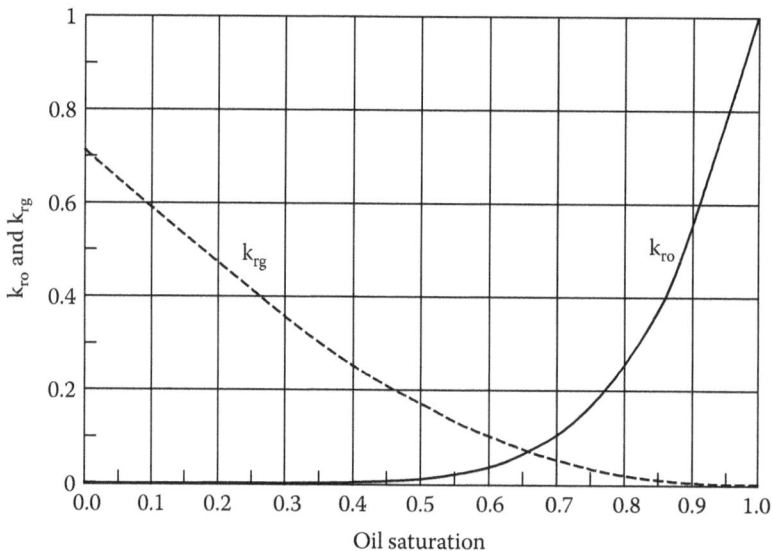

FIGURE 15.20   Typical relative permeability curves for oil and gas.

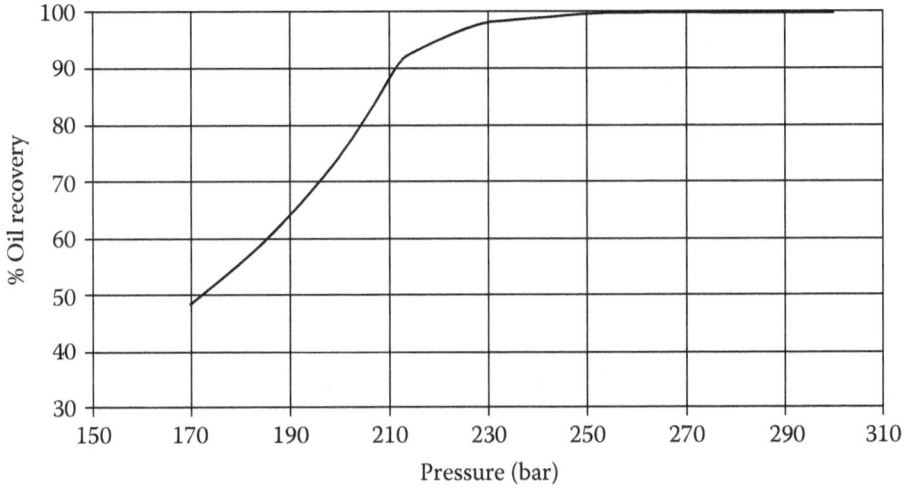

FIGURE 15.21 Slim tube simulation results for a temperature of 85.7°C for oil mixture in Table 3.47 with CO_2 as injection gas. The experimental recoveries can be seen in Table 3.48 and are plotted in Figure 3.25.

drive has developed or not. The fluid in the cell in which miscibility develops is at its critical point. The compositions of the two phases in equilibrium are identical and the K-factors of all components are equal to 1.0. The K-factor of a component is defined as the ratio of the mole fraction of the component in the vapor and liquid phases. The following criteria may be used to decide whether miscibility has developed:

$$\sum_{i=1}^{N} \left(\ln K_i \right)^2 < 1 \tag{15.9}$$

$$\left| T_c \left(\text{mixture} \right) - T \right| < 5 \text{ K and } \left| P_c \left(\text{mixture} \right) - P \right| < 5 \text{ bar} \tag{15.10}$$

where T and P are the cell temperature and pressure. The term mixture is to be understood as an equal molar mixture of the oil and gas compositions in the cell. The mixture critical point (T_c, P_c) may be calculated as outlined by Michelsen and Heideman (1981).

Another indication that miscibility has developed is almost identical phase densities. With the oil composition in Table 15.2, the criterion for miscibility in Equations 15.9 and 15.10 says miscibility develops in cell 1641 in time step 6442. Figure 15.22 shows that the simulated phase densities in the tube for this time step become almost equal consistent with the fact that both phase compositions are at their critical point. It is seen that there are zones of constant density consistent with the constant composition zones sketched in Figure 15.15.

Cell-to-cell simulations are quite time consuming but may possibly be speeded up by taking advantage of the fact that N-1 constant composition zones as sketched in Figure 15.15 develop in tubes or reservoirs with continuous gas injection. If the phase compositions are almost the same in two neighboring cells in the same time step or almost the same in a particular cell in two subsequent time steps, a K-factor flash calculation (Equations 6.22 through 6.24) will provide the correct flash result. Belkadi et al. (2011) have outlined a procedure for detecting when a K-factor flash calculation is sufficient and when a full flash is needed.

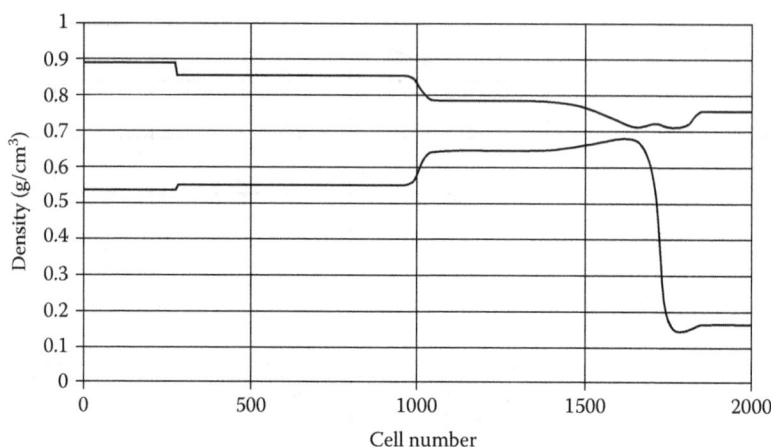

FIGURE 15.22 Simulated oil (upper) and gas (lower) densities in time step 6,442 out of 10,000 in a slim tube simulation for the oil composition in Table 15.2 with CO_2 as injection gas at a temperature of 85.7°C and a pressure of 210 bar. According to the criteria in Equations 15.9 and 15.10, miscibility develops in cell 1,641 in this time step.

REFERENCES

Belkadi, A., Yan, W., Michelsen, M.L., and Stenby, E.H., Comparison of two methods for speeding up flash calculations in compositional simulations, SPE 142132, presented at the *SPE Reservoir Simulation Symposium*, The Woodlands, TX, February 21–23, 2011.

Dandekar, A.Y., *Petroleum Reservoir Rock and Fluid Properties*, 2nd ed., Taylor & Francis, Boca Raton, FL, 2013.

Firoozabadi, A. and Aziz, K., Analysis and correlation of nitrogen and lean-gas miscibility pressure, *SPE Reserv. Eng.* 1, 575–582, November 1986.

Glasø, Ø., Generalized minimum miscibility pressure correlation, *Soc. Pet. Eng. J.* 25, 927–934, 1985.

Helfferich, F.G., Theory of multicomponent, multiphase displacement in porous media, *Soc. Pet. Eng. J.* 21, 51–62, 1981.

Jessen, K., Michelsen, M.L., and Stenby, E., Global approach for calculation of minimum miscibility pressure, *Fluid Phase Equilib.* 153, 251–263, 1998.

Jessen, K., Wang, Y., Ermanov, P., Zhu, P., and Orr, Jr, F.M., Fast approximate solutions for 1D multi-component gas injection problems, *SPE J.* 6, 442–451, 2001.

Johns, R.T. and Orr, F.M., Miscible gas displacement of multicomponent oils, *Soc. Pet. Eng. J.* 2, 268–279, 1997.

Johns, R.T., Yuan, H., and Dindoruk, B., Quantification of displacement mechanisms in multicomponent gas-floods, SPE 77696, presented at the *SPE ATCE*, San Antonio, TX, September 29–October 2, 2002.

Lindeloff, N., Mogensen, K.M., Pedersen, K.S., and Tybjerg, P., Investigation of miscibility behavior of CO_2 rich hydrocarbon systems—with application for gas injection EOR, SPE 166270, presented at the *SPE ATCE*, New Orleans, LA, September 30–October 2, 2013.

Metcalfe, R.S., Fussel, D.D., and Shelton, J.L., A multicell equilibrium separation model for the study of multiple contact miscibility in rich-gas drives, paper presented at the *SPE-AIME 47th Annual Meeting*, San Antonio, TX, October 8–11, 1972.

Michelsen, M.L. and Heideman, R.A., Calculation of critical points from cubic two-constant equations of state, *AICHE J.* 27, 521–523, 1981.

Mogensen, K. and Goryachev, S., First contact miscibility development in a high-temperature, ultra-sour reservoir fluid system offshore Abu Dhabi, SPE 211441-MS, presented at the *Abu Dhabi International Exhibition and Conference*, Abu Dhabi, October 31–November 3, 2022.

Monroe, W.W., Silva, M.K., Larsen, L.L., and Orr, F.M., Jr., Composition paths in four-component systems: Effect of dissolved methane on 1D CO_2 flood performance, *SPE Reserv. Eng.* 5, 423–432, 1990.

Pedersen, K.S., Fjellerup, J., Thomassen, P., and Fredenslund, Aa., Studies of gas injection into oil reservoirs by a cell-to-cell simulation model, SPE 15599, presented at the *SPE ATCE*, New Orleans, LA, October 5–8, 1986.

Stalkup, F.I., *Miscible Displacement, Monograph*, Vol. 8, H.L. Doherty Series, Society of Petroleum Engineers of AIME, New York, 1984.

Stalkup, F.I., Displacement behavior of the condensing/vaporizing gas drive process, paper 16715, presented at the *SPE ATCE*, Dallas, TX, September 27–30, 1987.

Wang, Y. and Orr, F.M., Jr., Analytical calculation of minimum miscibility pressure, *Fluid Phase Equilib.* 139, 101–124, 1997.

Wang, Y. and Orr, F.M., Jr., Calculation of minimum miscibility pressure, SPE paper 39683, presented at the *SPE/DOE Improved Oil Recovery Symposium*, Tulsa, OK, April 19–22, 1998.

Zhao, G.-B., Adidharma, H., Towler, B., and Radosz, M., Using multiple-mixing-cell model to study minimum miscibility pressure controlled by thermodynamic equilibrium tie-lines, *Ind. Eng. Chem. Res.* 45, 7913–7923, 2006.

Zick, A.A., A combined condensing/vaporizing mechanism in the displacement of oil by enriched gases, SPE 15493, presented at the *SPE ATCE*, New Orleans, LA, October 5–8, 1986.

16 Formation Water and Hydrate Inhibitors

Oil and gas are often produced together with water coming from a zone beneath the hydrocarbon zones. The produced formation water will usually contain dissolved salts. The operator may further add methanol or glycol to suppress hydrate formation (Chapter 13). The miscibility between water and oil is quite limited, whereas the water content in gas and gas condensate mixtures can be quite significant. The solubility of hydrocarbons in the aqueous phase is usually small but can nevertheless be quite important. In the arctic and other environmentally sensitive regions, the hydrocarbon concentration in the water phase can, for example, decide the need for cleaning formation water before dumping it into the sea. When storing CO_2 in aquifers, the solubility of CO_2 in water, possibly with dissolved salts, influences how much CO_2 can be stored.

Water, alcohols, and glycols deviate from hydrocarbons by having more polar forces acting between the molecules. Water has a molecular weight (M) of 18, which is close to that of methane (M = 16), but the absolute critical temperature of water is more than three times higher than that of methane, and the critical pressure is almost five times higher. This is because of attractive (association) forces acting between the water molecules, which make water behave as a substance of a somewhat higher molecular weight.

A cubic equation of state with classical a parameter mixing rule (Equation 4.33) is based on the assumption that the molecules are randomly distributed in each phase. This assumption does not hold for water dissolved in a hydrocarbon phase or hydrocarbons dissolved in a water phase. As is illustrated by Figure 16.1, there will be pockets in the hydrocarbon phase with a higher

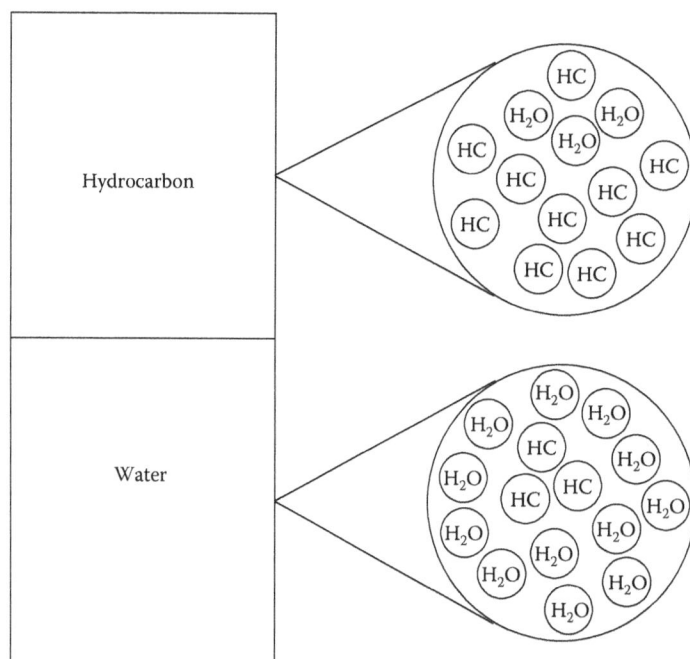

FIGURE 16.1 Close-up of molecules in hydrocarbon phase saturated with water and in water phase saturated with hydrocarbon. The local composition deviates from the overall (or macroscopic) composition.

DOI: 10.1201/9780429457418-16

water concentration than elsewhere, similarly there will be pockets in the water phase with a higher hydrocarbon concentration than in the water phase in general. G^E (or activity coefficient) models such as UNIQUAC (Abrams and Prausnitz 1975) or UNIFAC (Fredenslund et al. 1977) can be used to represent the liquid phase of such systems. As opposed to a cubic equation of state with classical mixing rule, the G^E models can represent liquid phases with local compositions deviating from the overall (or macroscopic) composition. The classical G^E models are, however, limited to liquids and fairly low pressures (≤ 10 bar). An optimum thermodynamic model would be a cubic equation of state extended or modified to also represent mixtures with polar compounds.

16.1 HYDROCARBON-WATER PHASE EQUILIBRIUM MODELS

The simplest way to handle mixtures of hydrocarbons and aqueous components would be to use either the SRK or the PR equation as presented in Chapter 4 (Equations 4.20 and 4.36, respectively). For this to be successful, the equation of state must be able to represent the phase behavior of pure water and of water mixed with hydrocarbons. Figure 16.2 shows the vapor pressure curve of pure water (ASME Steam Tables 1979). Also shown in Figure 16.2 are vapor pressure curves for water calculated using the SRK equation with the classical Soave temperature dependence for the a parameter (Equation 4.21) and using Mathias and Copeman (M & C) temperature dependence (Equations 4.27 and 4.28). Below 50°C, water vapor pressures simulated using the classical Soave temperature dependence are too low, whereas a very good match is seen with the M & C temperature dependence (almost indistinguishable from that of ASME). This is further illuminated in Figure 16.3, which shows a close-up of the vapor pressure curves in Figure 16.2 for the temperature interval from 0 to 50°C. The deviation in Figure 16.3 between the vapor pressures predicted using the classical Soave temperature dependence and using the M & C temperature dependence may not seem very alarming. However, in a situation in which liquid water is in contact with hydrocarbon

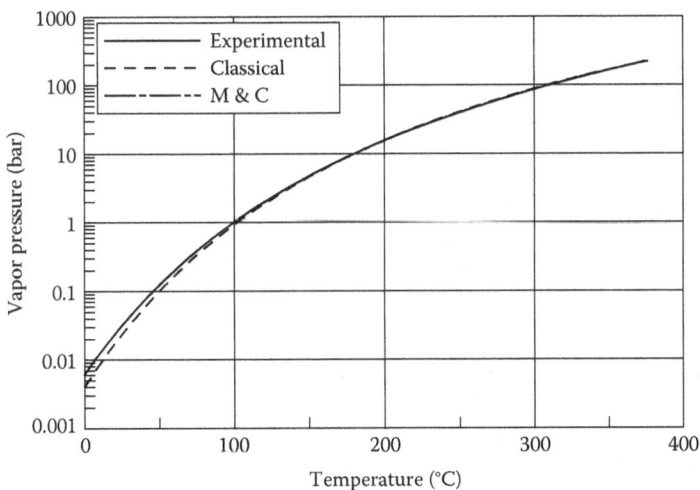

FIGURE 16.2 Experimental (ASME Steam Tables 1979) and simulated vapor pressures of pure water. Simulated results are shown using the classical Soave temperature dependence and using the Mathias and Copeman (M & C) expression. The M & C simulation results are almost indistinguishable from the experimental vapor pressures.

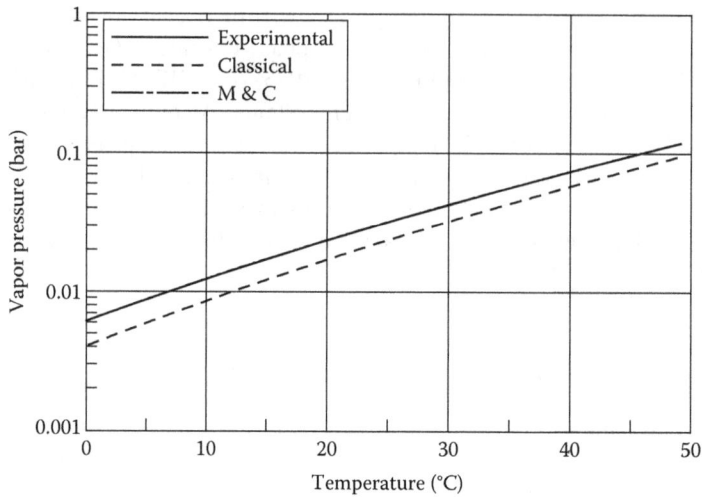

FIGURE 16.3   Close-up of vapor pressure results in Figure 16.2 for temperatures between 0 and 50°C.

gas at a temperature between 0°C and 50°C at low pressure, the simulated concentration of water in the hydrocarbon gas will be approximately proportional with the simulated vapor pressure of water. At 4°C, the actual vapor pressure of water is about 50% higher than that predicted using the classical Soave temperature dependence. In phase equilibrium calculations on mixtures of water and natural gas, use of the classical Soave expression may lead to significant underpredictions of the amount of water contained in the gas. For higher temperatures, satisfactory results are found using both the classical Soave temperature dependence and the M & C temperature dependence.

Table 16.1 shows typical binary interaction coefficients (k_{ij}) for water and hydrocarbon component pairs. The interaction coefficients enter into the mixing rule for the equation of state a parameter as shown in Equations 4.33 and 4.35. Table 16.2 shows data for the mutual solubility of propane (C_3) and water (H_2O) at pressures ranging from 6.9 to 206.9 bar and temperatures of 369.6 and 394.3 K (Kobayashi and Katz 1953). The solubility data is plotted in Figures 16.4 and 16.5. The dashed lines in the two figures show SRK simulation results for the solubility of C_3 in H_2O and the solubility of H_2O in C_3, respectively, using the C_3–H_2O interaction parameters in Table 16.1. The Mathias and Copeman expression (Equations 4.27 and 4.28) was used for the a parameter of water, and the classical Soave temperature dependence (Equation 4.21) was used for C_3. The water concentration in the phase dominated by C_3 is represented well using the SRK equation (Figure 16.5), whereas the C_3 content in the water phase is simulated to be almost negligible and orders of magnitude lower than the experimental results (Figure 16.4). The two figures illustrate the fact that it is not possible with one k_{ij} to get a good match of the solubility of water in a hydrocarbon phase as well as the solubility of hydrocarbons in a water phase. The k_{ij}s in Table 16.1 have been determined to give a fairly good match of the water solubility in a hydrocarbon phase. This is at the expense of the solubility of hydrocarbons in water, which will be simulated to almost zero. Simulations essentially neglecting the solubility of hydrocarbons in water can be satisfactory for many practical purposes, but accurate solubility simulation results are important for some applications. Carbon dioxide (CO_2) and benzene are examples of components with a fairly high solubility in water. The solubility of CO_2 in water is important when assessing the CO_2 storage capacity of an aquifer. Benzene is toxic, and there may be some environmental restrictions on the disposal of formation water

TABLE 16.1

Example of SRK and PR Binary Interaction Coefficients (k_{ij}) for Water–Hydrocarbon Pairs

Component Pair	H_2O
N_2	0.08
CO_2	0.15
H_2S	0.03
C_1	0.45
C_2	0.45
C_3	0.53
iC_4	0.52
nC_4	0.52
iC_5	0.50
nC_5	0.50
C_6	0.50
C_{7+}	0.50

TABLE 16.2

Mutual Solubility of Propane (C_3) and Water (H_2O)

Temperature (K)	Pressure (bar)	x_{C_3} in Water	x_{H_2O} in C_3
369.6	6.9	0.000058	0.13300
369.6	27.6	0.000213	0.03034
369.6	48.2	0.000277	0.00815
369.6	68.9	0.000287	0.00752
369.6	103.5	0.000296	0.00703
369.6	137.9	0.000304	0.00665
369.6	206.9	0.000316	0.00619
394.3	6.9	0.000051	0.29990
394.3	27.6	0.000231	0.07260
394.3	48.2	0.000338	0.03622
394.3	68.9	0.000379	0.01897
394.3	103.5	0.000400	0.01455
394.3	137.9	0.000414	0.01370
394.3	206.9	0.000444	0.01265

Source: Data from Kobayashi, R. and Katz, D.L., Vapor-liquid equilibria for binary hydrocarbon-water systems, *Ind. Eng. Chem.* 45, 440–446, 1953.

Note: x stands for mole fraction.

that has been contacted with a petroleum reservoir fluid with a significant content of benzene. It can therefore be important to have a thermodynamic model that will accurately represent the solubilities of CO_2 and hydrocarbons in water. The Huron and Vidal model, referred to as H & V in Figures 16.4 and 16.5, is an example of such a model and will be dealt with in the following section.

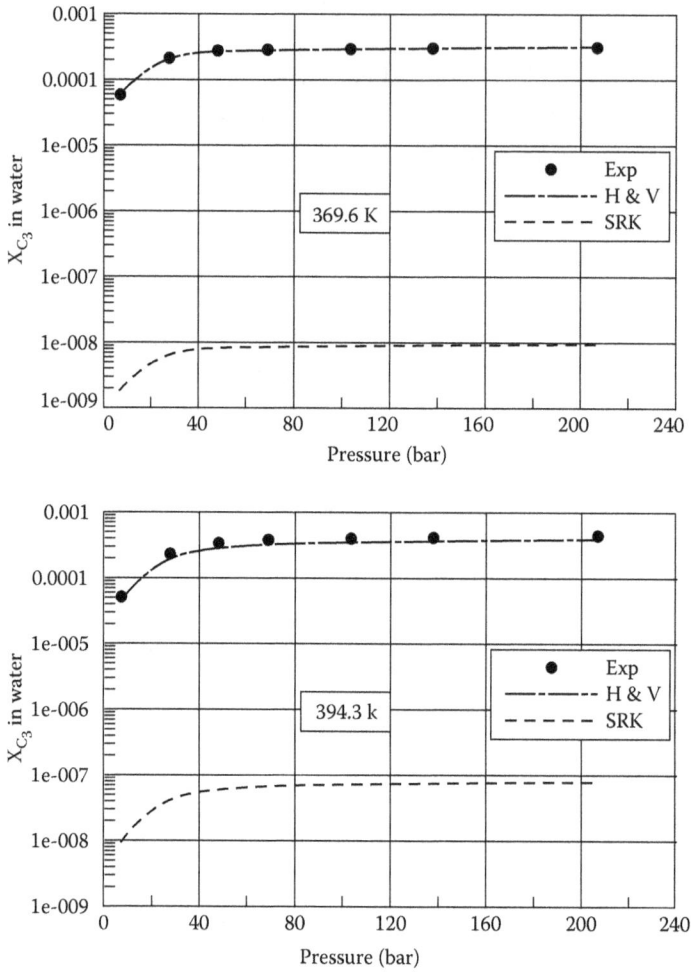

FIGURE 16.4 Experimental and simulated results for the solubility of C₃ in water. The experimental data can be seen in Table 16.2.

16.1.1 Huron and Vidal Mixing Rule

At low pressure, the liquid phase(s) of partially miscible systems, for example, water–hydrocarbon mixtures, can be well-represented using an activity coefficient model (or G^E model). The nonrandom, two-liquid (NRTL) model (Renon and Prausnitz 1968) is one example. It expresses the excess Gibbs energy as

$$\frac{G^E}{RT} = \sum_{i=1}^{N} z_i \frac{\sum_{j=1}^{N} \tau_{ji} z_j \exp(-\alpha_{ji} \ddot{A}_{ji})}{\sum_{k=1}^{N} z_k \exp(-\alpha_{ki} \ddot{A}_{ki})} \tag{16.1}$$

The terms *Gibbs free energy* and *activity coefficient* are further explained in Appendix A. α_{ji} is a non-randomness parameter that takes into account that the mole fraction of molecules of type i around a molecule of type j may deviate from the overall mole fraction of molecules of type

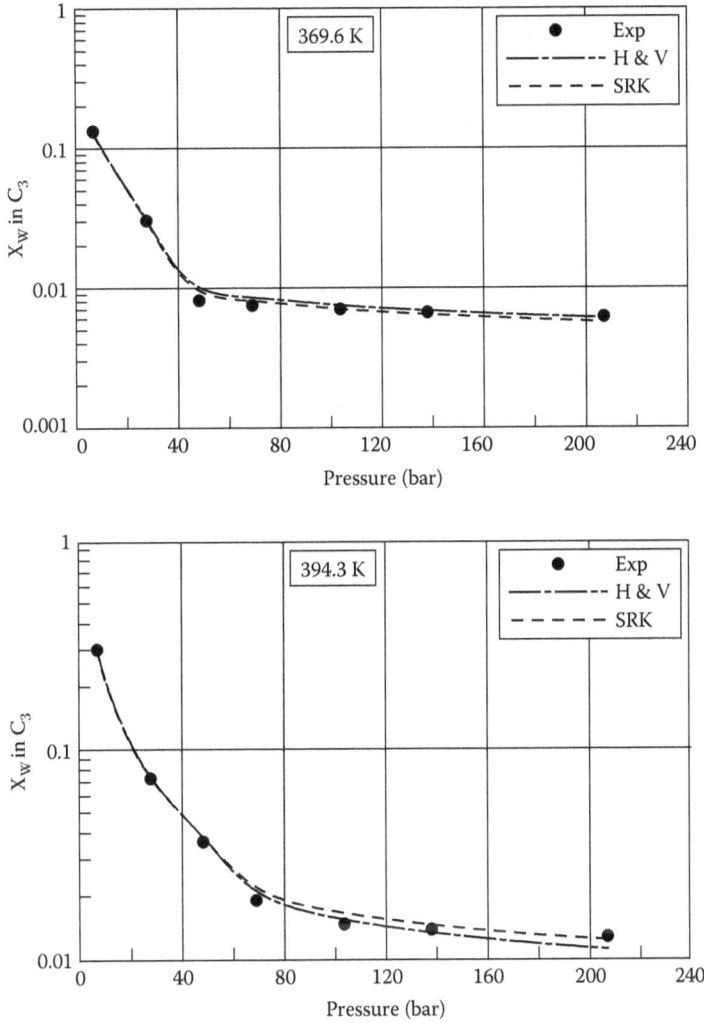

FIGURE 16.5 Experimental and simulated results for the solubility of water in C_3. The experimental data can be seen in Table 16.2.

i in the actual phase. The NRTL model is in other words able to represent phases in which the composition at a microscopic level deviates from the overall (or macroscopic) composition. As is illustrated in Figure 16.1, such deviations are seen for water dissolved in a hydrocarbon phase and for hydrocarbons contained in a water phase. The NRTL model is, therefore, also referred to as a *local composition model*. When α_{ji} in Equation 16.1 is zero, the mixture is completely random, and molecules of the same type will not cluster as shown in Figure 16.1. The parameter τ is defined by

$$\tau_{ji} = \frac{g_{ji} - g_{ii}}{RT} \tag{16.2}$$

where g_{ji} is an energy parameter characteristic of the j–i interaction. τ_{ji} expresses the deviation between the energy interactions between a j and an i molecule and the energy interaction between two i molecules.

The excess Gibbs energy, G^E, of a mixture can be expressed in terms of the mixture fugacity, f, and the pure component fugacities, f_i^*

$$G^E = RT\left(\ln f - \sum_{i=1}^{N} z_i \ln f_i^* \right) \tag{16.3}$$

For the SRK equation, the mixture fugacity, f, and the fugacity, f_i^*, of component i can be found from

$$\ln f = -\ln\left[\frac{P(V-b)}{RT} \right] + \frac{PV}{RT} - 1 - \frac{a}{bRT} \ln\left[\frac{V+b}{V} \right] \tag{16.4}$$

$$\ln f_i^* = -\ln\left[\frac{P\left(\tilde{V}_i^* - b_i\right)}{RT} \right] + \frac{P\tilde{V}_i^*}{RT} - 1 - \frac{a_i}{b_i RT} \ln\left[\frac{\tilde{V}_i^* + b_i}{\tilde{V}_i^*} \right] \tag{16.5}$$

V is the molar volume of the mixture, and $\tilde{V}_i^*$ is partial molar volume of component i. Partial molar properties are introduced in Appendix A. By inserting the fugacity expressions from Equations 16.4 and 16.5 into Equation 16.3, the following expression is obtained for G^E:

$$\frac{G^E}{RT} = -\ln\left[\frac{P(V-b)}{RT} \right] + \sum_{i=1}^{N} z_i \ln\left[\frac{P\left(\tilde{V}_i^* - b_i\right)}{RT} \right] + \frac{PV}{RT} - \sum_{i=1}^{N} z_i \frac{P\tilde{V}_i^*}{RT}$$
$$- \frac{1}{RT}\left[\frac{a}{b} \ln\left[\frac{V+b}{V} \right] - \sum_{i=1}^{N} z_i\left[\frac{a_i}{b_i} \right] \ln\left[\frac{\tilde{V}_i^* + b_i}{\tilde{V}_i^*} \right] \right] \tag{16.6}$$

At infinite pressure, this expression reduces to

$$G_\infty^E = -\left[\frac{a}{b} - \sum_{i=1}^{N} z_i\left[\frac{a_i}{b_i} \right] \right] \gg \tag{16.7}$$

where $\lambda = \ln 2$. By rearranging this equation, the following mixing rule may be obtained for the a parameter of the SRK equation:

$$a = b\left[\sum_{i=1}^{N}\left[z_i \frac{a_i}{b_i} \right] - \frac{G_\infty^E}{\lambda} \right] \tag{16.8}$$

The same mixing rule may be developed for the PR equation with

$$\lambda = \frac{1}{2\sqrt{2}} \ln\left(\frac{\sqrt{2}+1}{\sqrt{2}-1} \right) \tag{16.9}$$

To apply the mixing rule in Equation 16.8 with the G^E expression from Equation 16.5, the interaction parameters α_{ji} and $(g_{ji}-g_{ii})$ must be known for each binary pair contained in the mixture. Because hydrocarbon systems are generally satisfactorily described using the classical mixing rule for the a parameter, it would be desirable not to have to estimate any interaction parameters for hydrocarbon–hydrocarbon pairs. To accomplish this goal, Huron and Vidal (1979) suggested modifying NRTL equation in Equation 16.1 as follows, and to only use the G^E expression at infinite pressure:

$$\frac{G_\infty^E}{RT} = \sum_{i=1}^{N} z_i \frac{\sum_{j=1}^{N} \tau_{ji} b_j z_j \exp(-\alpha_{ji} \tau_{ji})}{\sum_{k=1}^{N} b_k z_k \exp(-\alpha_{ki} \tau_{ki})} \tag{16.10}$$

By comparing this expression with Equation 16.1, it is seen that Huron and Vidal have introduced the b parameter from the SRK equation in the expression for G^E. This modification is practical because, it enables values of the parameters α_{ji}, g_{ii}, and g_{ji} to be determined that will give the same results with the H & V mixing rule as with the classical mixing rule. Huron and Vidal have shown that Equation 16.10 will reduce to (give the same results as) the classical mixing rule in Equation 4.33 if the binary parameters are selected to

$$\alpha_{ji} = 0 \tag{16.11}$$

$$g_{ii} = -\frac{a_i}{b_i} \lambda \tag{16.12}$$

$$g_{ji} = -2 \frac{\sqrt{b_i b_j}}{b_i + b_j} \sqrt{g_{ii} g_{jj}} (1 - k_{ij}) \tag{16.13}$$

where k_{ij} is the binary interaction coefficient in the classical SRK mixing rule (Equations 4.33 and 4.35). For an N component mixture there will be a total of $\frac{N \times (N-1)}{2}$ values of $(g_{ji}-g_{jj})$. If only one component (often H_2O) in a mixture contributes to a non-random molecular order (α-value deviating from zero) then $\frac{(N-1)(N-2)}{2}$ of the interaction parameters can be determined from Equations 16.11–16.13. The remaining (N-1) interaction parameters must be determined from experimental phase equilibrium data for the special component (e.g. H_2O) and the remaining (N-1) components.

That Equation 16.8 really reduces to the classical SRK mixing rule with the interaction parameters in Equation 16.11–16.13 can be seen by initially inserting g_{ii} and g_{jj} derived from Equation 16.12 into Equation 16.13 to give

$$g_{ji} = -2\lambda \sqrt{a_i a_j} \frac{1}{b_i + b_j} (1 - k_{ij}) \tag{16.14}$$

By combining Equations 16.2 and 16.10 and assuming $\alpha_{ji} = 0$ as in Equation 16.11, the following expression is obtained for the excess Gibbs energy at infinite pressure:

$$\frac{G_\infty^E}{RT} = \sum_{i=1}^{N} z_i \frac{\sum_{j=1}^{N} \frac{g_{ji} - g_{ii}}{RT} b_j z_j}{\sum_{k=1}^{N} b_k z_k} = \sum_{i=1}^{N} z_i \frac{\sum_{j=1}^{N} \frac{g_{ji} - g_{ii}}{RT} b_j z_j}{b} \tag{16.15}$$

Further, by using Equation 16.14 for g_{ji} and Equation 16.12 for g_{ii}, the expression for G^E at infinite pressure will take the form

$$G_\infty^E = -\frac{2\lambda}{b} \sum_{i=1}^{N} \sum_{j=1}^{N} z_i z_j \sqrt{a_i a_j} \frac{b_j}{b_i + b_j} (1 - k_{ij}) + \lambda \sum_{i=1}^{N} z_i \frac{a_i}{b_i} \tag{16.16}$$

With this expression inserted into Equation 16.8, the a parameter may be expressed as

$$a = 2 \sum_{i=1}^{N} \sum_{j=1}^{N} z_i z_j \sqrt{a_i a_j} (1 - k_{ij}) \frac{b_j}{b_i + b_j} \tag{16.17}$$

Because $k_{ij} = k_{ji}$, this expression can rewritten as

$$a = \sum_{i=1}^{N} \sum_{j=1}^{N} \left(z_i z_j \sqrt{a_i a_j} (1 - k_{ij}) \left(\frac{b_j}{b_i + b_j} + \frac{b_i}{b_i + b_j} \right) \right) = \sum_{i=1}^{N} \sum_{j=1}^{N} z_i z_j \sqrt{a_i a_j} (1 - k_{ij}) \tag{16.18}$$

which is identical to the classical mixing rule in Equation 4.33.

The Huron and Vidal mixing rule is attractive for use on essentially nonpolar systems with some content of polar compounds. Specific Huron and Vidal interaction parameters are only needed for component pairs with polar compounds. For binaries of nonpolar compounds, the Huron and Vidal parameters can be found from Equations 16.11 through 16.13, which only make use of parameters of the classical SRK/PR mixing rules. The classical mixing rule is further used for binaries of an aqueous component and C_{7+}. The solubility of C_{7+} in an aqueous phase is negligible and k_{ij}s can be applied, which will give a reasonable match of the solubility of aqueous components in C_{7+}. The number of parameters to be estimated for the mixing rule of Huron and Vidal is lower than for other similar nonclassical mixing rules. With, for example, the MHV-2 model (Michelsen 1990) and the Wong and Sandler model (1992), binary interaction parameters would have to be estimated for all pairs of non-identical components. The Huron and Vidal mixing rule has the further advantage that the calculation results are consistent with those obtained with classical SRK mixing rules in the limit with no polar compounds present. Kristensen et al. (1993) used the SRK equation with the Huron and Vidal mixing rule on a gas condensate mixture to represent the distribution of methanol between a water phase, a hydrocarbon liquid phase, and a hydrocarbon gas phase. Water–methanol and water–hydrocarbon pairs were represented using the classical SRK mixing rule. In a related work, Pedersen et al. (1996) further applied the mixing rule of Huron and Vidal to water–methanol and water–hydrocarbon pairs.

Pedersen et al. (2001) have proposed a linear temperature dependence for the interaction parameters, g_{ji} and g_{ij}, where i is H_2O and j one of the components N_2, CO_2, C_1, C_2, C_3, or nC_4:

$$g_{ij} - g_{jj} = (g'_{ij} - g'_{jj}) + T(g''_{ij} - g''_{jj}) \tag{16.19}$$

$$g_{ji} - g_{ii} = (g'_{ij} - g'_{ii}) + T(g''_{ji} - g''_{ii}) \tag{16.20}$$

T is the absolute temperature.

Recommended Huron and Vidal interaction parameters for mixtures with water and methanol can be seen in Tables 16.3 and 6.4, respectively (Pedersen et al. 2001). Boesen et al. (2017) have published H & V parameters MEG-CO_2 and MEG-H_2S.

Figures 16.4 and 16.5 show H & V simulation results for the mutual solubilities of C_3 and H_2O at the pressure and temperature conditions in Table 16.2.

TABLE 16.3
Huron and Vidal Interaction Parameters for Binary Component Pairs of Water (H_2O) and the Indicated Second Component

Soave–Redlich–Kwong Huron–Vidal Interaction Parameters

Second component	$\dfrac{g_{12}-g_{22}}{R}(K)$	$\dfrac{g_{21}-g_{11}}{R}(K)$	$\dfrac{g'_{12}-g'_{22}}{R}(-)$	$\dfrac{g'_{21}-g'_{11}}{R}(-)$	α_{12}
		H$_2$O (1)			
MeOH	−359	352	−1.31	2.44	0.25
MEG	2673	−958	0.00	0.00	0.04
TEG	−2914	2497	0.00	0.00	0.20
N$_2$	73	5194	1.85	−6.05	0.15
CO$_2$	−2131	4158	2.93	−4.20	0.06
H$_2$S	156	1965	0.80	−2.32	0.24
C$_1$	−1218	4525	3.40	−5.16	0.09
C$_2$	−1214	4672	2.42	−5.00	0.09
C$_3$	−2026	6065	−3.82	−3.92	0.05
iC$_4$	−451	4190	−5.00	−1.15	0.08
nC$_4$	−451	4190	−5.00	−1.15	0.08
iC$_5$	681	2507	0.00	0.00	0.14
nC$_5$	681	2507	0.00	0.00	0.14
nC$_6$	681	2507	0.00	0.00	0.14
		Peng–Robinson Huron–Vidal Interaction Parameters			
MeOH	−407	550	−1.02	1.54	0.25
MEG	218	−72	0.00	0.00	0.40
TEG	−2726	2393	0.00	0.00	0.20
N$_2$	120	5330	1.53	−6.10	0.15
CO$_2$	−1907	4047	2.68	−4.15	0.07
H$_2$S	381	2652	0.39	−3.82	0.27
C$_1$	−1072	4541	3.02	−5.06	0.10
C$_2$	−1517	4904	2.28	−5.00	0.08
C$_3$	−1927	5768	−3.02	−3.86	0.06
iC$_4$	−348	4138	−5.00	−1.09	0.08
nC$_4$	−348	4138	−5.00	−1.09	0.08
iC$_5$	728	2514	0.00	0.00	0.15
nC$_5$	726	2514	0.00	0.00	0.15
nC$_6$	720	2514	0.00	0.00	0.15

Source: Pedersen, K.S., Milter, J., and Rasmusssen, C.P., Mutual solubility of water and a reservoir fluid at high temperatures and pressures: Experimental and simulated data, *Fluid Phase Equilib.* 189, 85–97, 2001.

TABLE 16.4
Huron and Vidal Interaction Parameters for Binary Component Pairs of Methanol (MeOH) and the Indicated Second Component

Soave–Redlich–Kwong Huron–Vidal Interaction Parameters

Second component	$\dfrac{g_{12}-g_{22}}{R}(K)$	$\dfrac{g_{21}-g_{11}}{R}(K)$	$\dfrac{g'_{12}-g'_{22}}{R}(-)$	$\dfrac{g'_{21}-g'_{11}}{R}(-)$	α_{12}
		MeOH (1)			
H$_2$O	352	−359	2.44	−1.31	0.25
MEG	49	473	0.00	0.00	−0.05

(Continued)

TABLE 16.4 (*Continued*)

Huron and Vidal Interaction Parameters for Binary Component Pairs of Methanol (MeOH) and the Indicated Second Component

	Soave–Redlich–Kwong Huron–Vidal Interaction Parameters				
	MeOH (1)				
Second component	$\dfrac{g_{12} - g_{22}}{R}(K)$	$\dfrac{g_{21} - g_{11}}{R}(K)$	$\dfrac{g'_{12} - g'_{22}}{R}(-)$	$\dfrac{g'_{21} - g'_{11}}{R}(-)$	α_{12}
TEG	3880	−1130	0.00	0.00	−0.05
N_2	139	−66	−0.50	4.74	0.10
CO_2	−92	−546	0.97	5.00	0.68
H_2S	45	965	0.00	0.00	0.40
C_1	164	1752	0.42	−2.26	0.55
C_2	138	451	1.10	5.00	0.41
C_3	183	848	0.83	0.86	0.42
iC_4	1191	1148	0.00	0.00	0.42
nC_4	1191	1148	0.00	0.00	0.42
iC_5	1216	1059	0.00	0.00	0.37
nC_5	689	1183	0.00	0.00	0.39
nC_6	1385	288	−2.00	2.50	0.38
	Peng–Robinson Huron–Vidal Interaction Parameters				
H_2O	550	−407	1.54	−1.02	0.25
MEG	1960	−957	0.00	0.00	0.05
TEG	6460	−168	0.00	0.00	0.05
N_2	231	−236	−3.53	10.50	0.03
CO_2	−22	−66	0.75	4.33	0.60
H_2S	6	1003	0.00	0.00	0.37
C_1	185	1683	0.36	−1.81	0.55
C_2	141	424	1.10	5.17	0.41
C_3	206	833	0.74	0.94	0.42
iC_4	1204	1155	0.00	0.00	0.42
nC_4	1194	1155	0.00	0.00	0.42
iC_5	1202	1043	0.00	0.00	0.37
nC_5	593	1219	0.00	0.00	0.41
nC_6	1381	280	−2.00	2.50	0.38

Source: Pedersen, K.S., Milter, J., and Rasmusssen, C.P., Mutual solubility of water and a reservoir fluid at high temperatures and pressures: Experimental and simulated data, *Fluid Phase Equilib.* 189, 85–97, 2001.

16.1.2 HURON AND VIDAL MIXING RULE EXTENDED TO HYDROCARBON-SALT WATER SYSTEMS

Formation water produced together with petroleum reservoir fluids will usually contain dissolved salts. These salts will somewhat affect the mutual solubility of hydrocarbons and water.

Sørensen et al. (2002) have proposed to use the SRK equation of state with Huron and Vidal mixing rule for systems of hydrocarbons and salt water. The salts are treated as "ordinary" components with hypothetical critical properties. In the water phase, each salt molecule is assumed to split into as many "molecules" as the number of ions formed when the salt dissociates in aqueous solution. NaCl will, for example, split into a Na$^+$ ion and a Cl$^-$ ion and be treated as two hypothetical molecules (one Na$^+$ and one Cl$^-$). The ions of the salts most common in formation water are assigned the properties in Table 16.5. These properties are close to those of triethylene glycol (TEG). TEG has

TABLE 16.5
Hypothetical Properties of Salts

Salt	T_c (K)	P_c (bar)	Acentric Factor	No. of Ions	No. of Crystal Water
NaCl	700	35.5	1.0	2	0
KCl	700	35.5	1.0	2	0
CaCl$_2$	800	35.5	1.0	3	6

Source: Data from Sørensen, H., et al. Modeling of gas solubility in brine, *Org. Chem.* 33, 635–642, 2002.

a similar effect on water as salts in terms of freezing point depression and lowering of gas solubility. Furthermore, TEG has a low volatility and low solubility in liquid hydrocarbons. To make sure the salts in phase equilibrium simulations stay out of the hydrocarbon phases, large fugacity coefficients are assigned to the salt components in the gas and oil phases.

The presence of salts will lower the mole fraction of water in the water phase. How much the water concentration is lowered depends on the amount of salt and on the number of ions formed when the salt dissociates. As an example, NaCl will split into Na$^+$ and Cl$^-$, and the dilution effect of NaCl will be twice the number of undissociated NaCl molecules. CaCl$_2$ is represented as three molecules. Each Ca^{2+} ion is further assumed to have six H$_2$O molecules associated with it, which reduces the number of free H$_2$O molecules. Imagine a system consisting initially of 100 moles H$_2$O. Add one mole of CaCl$_2$. According to the model of Sørensen et al. CaCl$_2$ will associate with six moles of H$_2$O and split into three moles of salt irrespective of temperature and pressure. The model sees 94 moles of free H$_2$O and 3 moles of salt. The Ca^{2+}–H$_2$O association is similar to the CaCl$_2$ crystal–water bonds acting in the solid state. Figure 16.6 illustrates how NaCl and CaCl$_2$ are handled in aqueous solution.

Salt dissolved in a water phase will make the gas solubility in the water phase decrease. Pedersen and Milter (2004) have estimated both SRK and PR Huron and Vidal interaction parameters that account for this decrease. These parameters are listed in Table 16.6.

16.1.3 Cubic Plus Association

Aqueous compounds tend to self-associate (adhere to each other). This is the primary reason the phase behavior of mixtures with aqueous components is more difficult to model than the phase behavior of a hydrocarbon mixture. The critical point of water is at 647.3 K and 220.9 bar. Had no self-association taken place between the water molecules, the critical point of water would have been more like C$_1$ (190.6 K and 46 bar). Kontogeorgis et al. (1996) have proposed to take the (imaginary) properties of the single water molecules as the starting point and let the model account for the water–water self-association. The model approach is called *cubic plus association* (CPA). The association term enters as a correction to the pressure from a classical cubic equation

$$P = P_{Cubic} + P_{Association} \tag{16.21}$$

The cubic equation can be SRK (Equation 4.20) or any other cubic equation of state. The association term may be expressed as (Michelsen and Hendriks 2001)

$$P_{Association} = -\frac{1}{2}\frac{RT}{V}\left(1 - \frac{1}{V}\frac{\partial \ln g}{\partial\left(\frac{1}{V}\right)}\right)\sum_{i=1}^{N}\left(x_i\sum_{A_i}^{NS_i}\left(1 - X_{A_i}\right)\right) \tag{16.22}$$

N is number of components and NS$_i$ is the number of association sites (A$_i$) for component i. X_{A_i} is the fraction of molecules i not bonded at site A

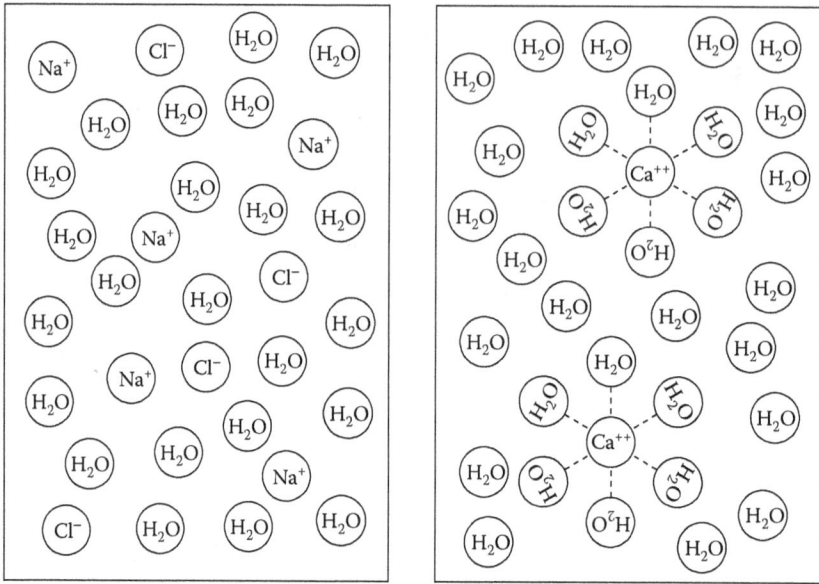

FIGURE 16.6 NaCl and CaCl$_2$ in aqueous solution. NaCl is assumed to dissociate into a Na$^+$ ion and a Cl$^-$ ion and to interact with all water (H$_2$O) molecules in the same manner. CaCl$_2$ is assumed to dissociate into one Ca^{++} and two Cl$^-$ ions. Each Ca^{++} binds six H$_2$O molecules.

TABLE 16.6
Huron and Vidal Interaction Parameters between Gas (1) and Salt Compounds (2) and between Water (1) and Salt Compounds (2)

	α_{12}	$(g_{12}-g_{22})/R(K)$	$(g_{12}-g_{11})/R(K)$
		SRK	
		N$_2$ (1)	
NaCl (2)	0.007	−461.0	46394
KCl (2)	0.008	2335	41212
CaCl$_2$ (2)	0.000	1306	8873
		CO$_2$ (1)	
NaCl (2)	0.023	2463	0.0
KCl (2)	0.000	1539	−972.5
CaCl$_2$ (2)	−0.034	1903	889.5
		C$_1$ (1)	
NaCl (2)	0.099	3674	6493
KCl (2)	0.000	1306	8650
CaCl$_2$ (2)	0.092	2810	9284
		C$_2$ (1)	
NaCl (2)	0.104	3674	6493
KCl (2)	0.000	1360	8650
CaCl$_2$ (2)	0.082	2810	9284
		H$_2$O (1)	
NaCl (2)	−0.734	−11.7	95.12
KCl (2)	−0.826	650.8	92.26

	α_{12}	$(g_{12}\text{-}g_{22})/R(K)$	$(g_{12}\text{-}g_{11})/R(K)$
		SRK	
		N_2 (1)	
CaCl2 (2)	-2.104	170.5	66.88
NaHCO$_3$ (2)	-0.791	-26.6	114
		PR	
		N_2 (1)	
NaCl (2)	0.006	-957.9	46151
KCl (2)	0.008	2293	36188
CaCl$_2$ (2)	0.000	1489	8740
		CO_2 (1)	
NaCl (2)	0.016	2447	0.90
KCl (2)	0.000	1517	47.10
CaCl$_2$ (2)	-0.016	1932	-686
		C_1 (1)	
NaCl (2)	0.104	3676	6886
KCl (2)	0.000	1481	8440
CaCl$_2$ (2)	0.082	2648	7670
		C_2 (1)	
NaCl (2)	0.000	606.0	5506
KCl (2)	0.000	1862	8280
CaCl$_2$ (2)	0.000	1509	7927
		H_2O (1)	
NaCl (2)	-0.781	-19.72	94.70
KCl (2)	-0.563	1009	92.36
CaCl$_2$ (2)	-0.785	-9.70	96.68
NaHCO$_3$ (2)	0.000	4123	361.3

Source: Data from Pedersen, K.S. and Milter, J., Phase equilibrium between gas condensate and brine at HT/HP conditions, SPE 90309, presented at the *SPE ATCE*, Houston, TX, September 26–29, 2004.

$$X_{A_i} = \cfrac{1}{1+\cfrac{1}{V}\sum_{j=1}^{N}\sum_{B_j}^{NS_j} x_j X_{B_j}\Delta^{A_iB_j}} \qquad (16.23)$$

where

$$\Delta^{A_iB_j} = g(\eta)\left[\exp\left(\frac{\varepsilon^{A_iB_j}}{RT}\right)-1\right]b_{ij}\beta^{A_iB_j} \qquad (16.24)$$

$\varepsilon^{A_iB_j}$ and $\beta^{A_iB_j}$ are parameters expressing association energy and association volume for component pair i-j. NS_j is the number of association sites (B_j) for component j. X_{B_j} is the fraction of molecules j not bonded at site B. The parameter b_{ij} is the average of the b parameters for components i and j. The radial distribution function can be found from (Kontogeorgis et al. 1999)

$$g(\eta) = \frac{1}{1-1.9\eta} \text{ where } \eta = \frac{b}{4V} \qquad (16.25)$$

Two associating components of different species may cross-associate and evaluation of Equation (16.23) requires a cross-association energy and a cross-association volume for each pair of cross-associating components. The combining rules Voutsas et al. (1999), often referred to as CR-1, are commonly used

$$\varepsilon^{A_iB_j} = \frac{\varepsilon^{A_iB_i} + \varepsilon^{A_jB_j}}{2} \tag{16.26}$$

$$\beta^{A_iB_j} = \sqrt{\beta^{A_iB_i}\beta^{A_jB_j}} \tag{16.27}$$

CO_2 and H_2S do not self-associate ($\varepsilon^{A_iB_i}$ and $\beta^{A_iB_i}$ are zero) but may cross-associate with water. They are said to be solvating with water and hydrate inhibitors. For those two components a modified CR-1 combining rule can be used as exemplified in the following for water

$$\varepsilon^{A_iB_j} = \frac{\varepsilon^{A_iB_i} + \varepsilon^{A_{H_2O}B_{H_2O}}}{2} = \frac{\varepsilon^{A_{H_2O}B_{H_2O}}}{2} \tag{16.28}$$

$$\beta^{A_iB_j} = \beta^{A_iB_{H_2O}} \tag{16.29}$$

$\beta^{A_iB_{H_2O}}$ is a parameter specific for each solvating component and water. Similar parameters exist for the two solvating components and other aqueous components. The concept of solvation is described by Folas et al. (2006).

Non-aqueous components, for example, C_1 have no association sites and hence no values for neither self-association nor cross-association parameters. In turn they do not contribute to the association term whether as pure components or as binaries with any other component.

An association scheme is required for each associating or solvating component. It tells the number of positive and negative association sites for each component. Table 16.7 shows CPA association schemes. An association scheme represents a certain way of interaction between positive and negative sites of a component. Figure 16.7 illustrates the association schemes 4C with two positively charged and two negatively charged association sites and 2B with one positively charged and one negatively charged association site defined in Table 16.6. All positively charged sites can associate with all negatively charged sites and vice versa. Sites of same charge, positive or negative, cannot associate with each other. This gives four association sites for the 4C association scheme and two association sites for the 2B association scheme. To make use of the CPA model concept, five pure component parameters (a_0, b, c_1, ε^{AB}, and β^{AB}) must be determined for each associating component. The parameter a_0 is equivalent to a_c in Equations 4.22 and 4.38, b is the b parameter entering into the cubic term, and c_1 is equivalent to m in Equations 4.25, 4.40 and 4.42. For a non-associating component, the association term is zero and there are no ε^{AB} and β^{AB} parameters. The sour gases, CO_2 and H_2S, are assumed to solvate with water. For those two inorganic gases the parameters in the cubic term can be determined as for other non-associating components, but to properly account for the interaction with water, a cross-association volume $\beta^{A_iB_{H_2O}}$ must be determined.

A volume correction is generally not used for associating components, which means the c parameter in Equations 4.43 and 4.48 will be zero.

Unlike a classical cubic equation and the Huron–Vidal modification in Chapter 16.1.1, the CPA model is not bound to match the pure component critical point. The CPA parameters for an associating component can therefore not be determined from the critical temperature and pressure as is the case for a non-associating component. The CPA parameters are instead determined from property data. In general, too high critical temperatures and pressures are simulated for associating components using the CPA model. While the association concept holds true for low to moderate pressures, it is unable to correctly account for the molecular interactions

TABLE 16.7
CPA Association Schemes

Component	No. of Positive Sites	No. of Negative Sites	Association Scheme
Water	2	2	4C
MeOH	1	1	2B
MEG	2	2	4C
TEG	2	2	4C
CO_2	0	1	–
H_2S	0	1	–

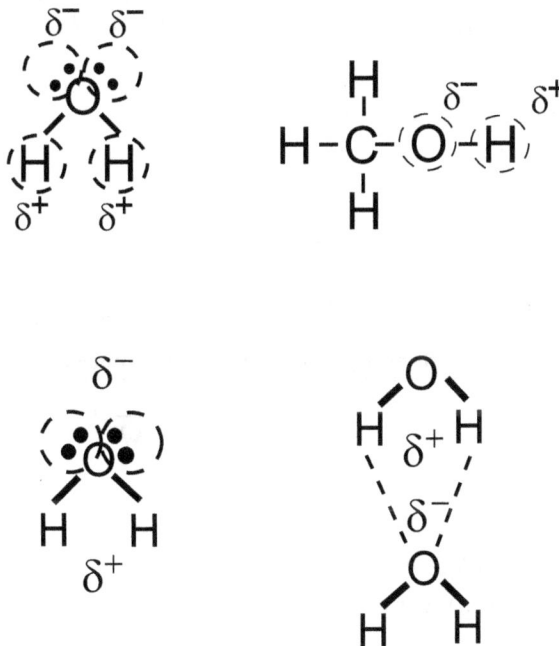

FIGURE 16.7 Illustration of CPA associations sites. The upper left H_2O (water) molecule is assigned association scheme 4C, the lower left H_2O molecule, association scheme 2B, and the CH3OH (methanol) molecule association scheme 2B. The lower right sketch shows two H_2O molecules associated according to scheme 2B. The association schemes are defined in Table 16.7.

at near-critical conditions. While the inability to match the critical point is a deficiency of the CPA model, it has other qualities. Lundstrøm et al. (2006) have shown that liquid isothermal compressibilities (defined in Equation 3.4) and sound velocities (defined in Equation 8.22) of pure water and methanol are better represented using the CPA model than the conventional SRK equation (Equation 4.20).

TABLE 16.8

CPA Pure Component Parameters and Association Schemes for Self-Associating Components

	T_c K	a_0 bar/l²/mol²	b l/mol	c_1 –	ε^{AB} bar l/mol	β^{AB} –	No. of ÷Charge Sites	No. of +Charge Sites
			Soave–Redlich–Kwong CPA Parameters					
Water	647.3	1.2277	1.4515×10^{-2}	0.6736	166.55	6.9200×10^{-2}	2	2
MeOH	512.6	4.0531	3.0978×10^{-2}	0.4310	245.91	1.6100×10^{-2}	1	1
MEG	720.0	10.819	5.1400×10^{-2}	0.6744	197.52	1.4100×10^{-2}	2	2
TEG	769.5	39.126	13.2100×10^{-2}	1.1692	143.37	1.8800×10^{-2}	2	2
			Peng–Robinson CPA Parameters					
Water	647.3	1.5782	1.4788×10^{-2}	0.6736	161.23	6.9662×10^{-2}	2	2
MeOH	512.6	5.3485	3.2112×10^{-2}	0.4310	236.87	1.3239×10^{-2}	1	1
MEG	720.0	12.378	5.1671×10^{-2}	0.6744	194.89	1.4587×10^{-2}	2	2
TEG	769.5	44.349	1.3253×10^{-1}	1.1692	143.37	1.8273×10^{-2}	2	2

Source: Kontogeorgis, G.M., Michelsen, M.L., Folas, G.K., Derawi, S., von Solms, N. and Stenby, E.H., Ten years with CPA (cubic-plus-association) equation of state, Part 1. Pure compounds and self-Associating systems, *Industrial & Engineering Chemistry Research* 45, 4855–4868, 2006.

Sørensen. H., Boesen, R.R. Leekumjorn, S. and Herslund, P.J., Peng–Robinson equation of state extended to handle aqueous components using CPA concept, *J. Natural Gas Eng.* 3, 1–38, 2018.

TABLE 16.9

CPA Binary Interaction Parameters for Self-Associating Components

	MeOH		MEG		TEG	
	k_{ij}^{ref} –	k_{ij}' 1/K	k_{ij}^{ref} –	k_{ij}' 1/K	k_{ij}^{ref} –	k_{ij}' 1/K
		Soave–Redlich–Kwong CPA Binary Interaction Parameters				
H_2O	−0.16526	$−5.738 \times 10^{-4}$	0.0832	0	−0.1953	7.159×10^{-4}
		Peng–Robinson CPA Binary Interaction Parameters				
H_2O	−0.1415	$−4.257 \times 10^{-4}$	−0.0786	0	−0.1961	6.411×10^{-4}

Source: Kontogeorgis, G.M., Michelsen, M.L., Folas, G.K., Derawi, S., von Solms, N. and Stenby, E.H., Ten years with CPA (cubic-plus-association) equation of state, Part 1. Pure compounds and self-Associating systems, *Industrial & Engineering Chemistry Research* 45, 4855–4868, 2006.

Sørensen. H., Boesen, R.R. Leekumjorn, S. and Herslund, P.J., Peng–Robinson equation of state extended to handle aqueous components using CPA concept, *J. Natural Gas Eng.* 3, 1–38, 2018.

Most published work on the CPA model has used SRK (Equation 4.20) as underlying cubic equation (e.g. Kontogeorgis et al. 2006a, 2006b). Sørensen et al. have shown that equally good simulation results can be obtained using the cubic term from PR (Equation 4.36). CPA parameters for use with each of those two equations are shown in Tables 16.8–16.10.

Austegard et al. (2006) found that the Huron and Vidal model with interaction parameters linear in temperature described in mutual solubility of water and $CO_2 + CH_4$ gas slightly better than the CPA model. They considered data for temperatures up to 350°C and pressures up to 3000 bar.

TABLE 16.10

CPA Cross-Association Volume Parameters (β^{AiBj}) and Association Schemes for Solvating Components

		Soave–Redlich–Kwong CPA Parameters For Solvating Components				
	H_2O	MeOH	MEG	TEG	No. of ÷Charge Sites	No. of +Charge Sites
CO_2	0.1413	0.0777	0.1166	0.1987	1	0
H_2S	0.2612	0.1549	/	/	1	0
		Peng–Robinson CPA Parameters For Solvating Components				
β^{AiBj} [–]	H_2O	MeOH	MEG	TEG	No. of ÷Charge Sites	No. of +Charge Sites
CO_2	0.1518	0.0822	0.1316	0.2500	1	0
H_2S	0.2225	0.1309	/	/	1	0

Sources: Folas, G.K., Kontogeorgis, G.M., Michelsen, M.L., and Stenby, E.H., Application of the cubic-plus-association equation to mixtures with polar chemicals and high pressure, *Ind. Eng. Chem. Res.* 45, 1516–1526, 2006.

Sørensen. H., Boesen, R.R. Leekumjorn, S. and Herslund, P.J., Peng–Robinson equation of state extended to handle aqueous components using CPA concept, *J. Natural Gas Eng.* 3, 1–38, 2018.

16.2 EXPERIMENTAL HYDROCARBON-WATER PHASE EQUILIBRIUM DATA

Table 16.11 shows experimental phase equilibrium data for a gas condensate fluid mixed with water and methanol at a pressure of 75.8 bar and temperatures of 10°C and 50°C (Kristensen et al. 1993). The composition of the gas condensate is shown in Table 16.12. Also shown in Table 16.11 are simulation results obtained by using SRK with the Huron and Vidal mixing rule.

Table 16.13 shows experimental phase equilibrium data for a reservoir oil composition mixed with water and methanol at pressures of 120 and 200 bar and temperatures around 7°C (Pedersen et al. 1996). The composition of the reservoir oil (Mixture 1) is shown in Table 16.14. Also shown in Table 16.13 are simulated results obtained using SRK with the Huron and Vidal mixing rule. Similar phase equilibrium data is shown in Table 16.15 for the gas condensate (Mixture 2), the composition of which is also given in Table 16.14.

Table 16.16 shows the composition of a gas condensate mixture. Its mutual solubility with pure water and salt water has been measured at high pressure and temperature. The phase equilibrium data can be seen in Table 16.17 and the saltwater (brine) composition from Table 16.18. Table 16.17 gives a good idea of the mutual solubility of water and hydrocarbons at elevated pressure and temperature, as well as of the influence of dissolved salts on the gas solubility in water at reservoir conditions. Simulated results (Pedersen and Milter 2004) using the H & V model are shown in Table 16.19.

Table 16.20 shows experiment phase equilibrium data (Ng et al. 2001) for an acid gas mixed with water. Also shown are simulation results with SRK Huron and Vidal (H&V), PR H&V, SRK CPA, and PR CPA.

Other phase equilibrium data for petroleum reservoir fluids and water have been presented by Ng and Chen (1995) and by Kokal et al. (2003). The former study further covers phase equilibrium data for model systems of hydrocarbon, water, and methanol and of hydrocarbon, water, and mono ethylene glycol (MEG). Ng et al. (1993) have further presented phase equilibrium data for model systems of hydrocarbon, water, and tri ethylene glycol (TEG) for temperatures ranging from 25°C to 204°C and pressures up to 69 bar.

TABLE 16.11

Experimental (Exp) and Calculated (Calc) Phase Compositions When Mixing Gas Condensate in Table 16.12 with Water and Methanol

	Feed	Vapor		HC Liquid		Aqueous Phase	
		Exp	Calc	Exp	Calc	Exp	Calc
				10°C and 75.8 bar			
Methanol	7.50	0.051	0.047	0.482	0.491	15.7	15.6
Water	40.09	0.034	0.016	1.195	0.014	84.3	84.4
Gas condensate	52.41	99.915	99.937	98.323	98.996	—	0.0
Percentage of feed	100.00	39.4	38.4	13.2	14.1	47.3	47.5
				50°C and 75.8 bar			
Methanol	4.70	0.326	0.263	0.933	0.901	15.0	15.2
Water	24.85	0.213	0.160	0.314	0.103	85.0	84.8
Gas condensate	70.45	99.461	99.577	98.753	98.996	—	—
Percentage of feed	100.00	57.4	56.7	13.5	14.1	29.1	29.2

Source: Data from Kristensen, J.N., et al. A combined Soave–Redlich–Kwong and NRTL equation for calculating the distribution of methanol between water and hydrocarbon phases, *Fluid Phase Equilib.* 82, 199–206, 1993.

Note: HC stands for hydrocarbon.

TABLE 16.12

Molar Composition of Gas Condensate Mixture for Which Phase Equilibrium Data for Mixtures with Water and Methanol Is Shown in Table 16.11

Component	Mole Percentage
N_2	2.280
CO_2	0.373
C_1	71.274
C_2	7.979
C_3	4.725
iC_4	0.675
nC_4	1.741
iC_5	0.530
nC_5	0.684
C_6	1.268
C_7	1.275
C_8	0.884
C_9	0.827
C_{10+}	5.488

Source: Kristensen, J.N., et al. A combined Soave–Redlich–Kwong and NRTL equation for calculating the distribution of methanol between water and hydrocarbon phases, *Fluid Phase Equilib.* 82, 199–206, 1993.

Note: The C_{7+} density is 0.82 g/cm³ and the C_{7+} molecular weight is 212.

TABLE 16.13

Experimental (Exp) and Calculated (Calc) Phase Compositions for Oil (Mixture 1 in Table 16.14) Mixed with Water and Methanol

	Feed	Hydrocarbon Phase		Aqueous Phase	
		Exp	Calc	Exp	Calc
		P = 120 bar and T = 6.5°C			
Res. fluid	60.48	99.815	99.731	—	—
Methanol	5.97	0.185	0.231	15.08	14.76
Water	33.55	—	—	84.92	84.89
		P = 200 bar T = 7.9°C			
Res. fluid	63.85	99.804	99.734	—	—
Methanol	5.46	0.196	0.228	15.08	14.71
Water	30.69	—	—	84.92	84.91

Source: Pedersen, K.S., et al. Phase equilibrium calculations for unprocessed well streams containing hydrate inhibitors, *Fluid Phase Equilib.* 126, 13–28, 1996.

TABLE 16.14

Molar Composition of Reservoir Fluid Compositions for Which Phase Equilibrium Data for Mixtures with Methanol and Water Is Presented in Table 16.13

Component	Mixture 1	Mixture 2
N_2	0.15	0.64
CO_2	2.05	3.10
C_1	25.52	72.74
C_2	8.06	8.01
C_3	7.69	4.26
iC_4	1.78	0.73
nC_4	3.95	1.49
iC_5	1.82	0.53
nC_5	2.39	0.64
C_6	4.29	0.81
C_7	6.71	1.08
C_8	7.85	1.20
C_9	6.31	1.08
C_{10+}	21.43	3.70
C_{7+} M	158	169
C_{7+} density (g/cm^3)	0.83	0.82

Source: Pedersen, K.S., et al. Phase equilibrium calculations for unprocessed well streams containing hydrate inhibitors, *Fluid Phase Equilib.* 126, 13–28, 1996.

Note: The density is at 1.01 bar and 15°C.

TABLE 16.15

Experimental (Exp) and Calculated (Calc) Phase Compositions (Mole Percentage) for Gas Condensate (Mixture 2 in Table 16.14) Mixed with Water and Methanol

	Feed	HC Liquid		HC Vapor		Aqueous	
		Exp	Calc	Exp	Calc	Exp	Calc
			$P = 60.3$ bar and $T = 3.6°C$				
Hydrocarbons	84.76	99.799	99.675	99.957	99.936	—	—
Methanol	2.99	0.201	0.288	0.0429	0.0441	18.68	18.21
Water	7.32	—	—	—	—	81.32	81.40
			$P = 149.9$ bar and $T = 7.7°C$				
Hydrocarbons	64.04	99.812	99.741	99.931	99.909	—	—
Methanol	6.72	0.188	0.214	0.0687	0.0636	18.68	18.44
Water	29.22	—	—	—	—	81.32	80.93

Source: Data from Pedersen, K.S., et al. Phase equilibrium calculations for unprocessed well streams containing hydrate inhibitors, *Fluid Phase Equilib.* 126, 13–28, 1996.

Note: HC stands for hydrocarbon.

TABLE 16.16

Molar Composition of Reservoir Fluid. Phase Equilibrium Data for the Fluid Mixed with Pure Water and Salt Water Is Shown in Table 16.17

Component	Mole Percentage	Molecular Weight	Density (g/cm³)
N_2	0.369	—	—
CO_2	4.113	—	—
C_1	69.243	—	—
C_2	8.732	—	—
C_3	4.270	—	—
iC_4	0.877	—	—
nC_4	1.641	—	—
iC_5	0.625	—	—
nC_5	0.720	—	—
C_6	0.972	—	—
C_7	2.499	94.7	0.738
C_8	0.732	114.2	0.748
C_9	0.637	128.3	0.769
C_{10+}	4.571	229.5	0.961

Source: Pedersen, K.S. and Milter, J., Phase equilibrium between gas condensate and brine at HT/HP conditions, SPE 90309, presented at the *SPE ATCE*, Houston, TX, September 26–29, 2004.

Note: The density is at 1.01 bar and 15°C.

TABLE 16.17

Measured Equilibrium Compositions for the Gas Condensate in Table 16.16 Mixed with Pure Water and Mixed with Salt Water of the Composition in Table 16.18

	Pure Water		Salt Water	
	Vapor	Water	Vapor	Salt-Free Water
1000 bar and 35°C. Feed Consists of 0.69 mol Gas Condensate per Mole Salt-Free Water				
H_2O	0.055	99.4331	0.054	99.5845
N_2	0.337	0.0068	0.345	0.0000
CO_2	3.761	0.2130	4.323	0.1798
C_1	69.183	0.3420	68.934	0.2274
C_2	8.751	0.0043	8.580	0.0068
C_3	4.321	0.0002	4.258	0.0008
iC_4	0.898	0.0000	0.883	0.0001
nC_4	1.707	0.0001	1.637	0.0001
iC_5	0.688	0.0003	0.642	0.0000
nC_5	0.787	0.0001	0.740	0.0000
C_6	1.002	0.0000	0.989	0.0000
C_{7+}	8.511	0.0005	8.615	0.0000
1000 bar and 120°C. Feed Consists of 0.74 mol Gas Condensate per Mole Salt-Free Water				
H_2O	0.753	99.3724	0.663	99.5699
N_2	0.343	0.0274	0.357	0.0000
CO_2	4.068	0.1465	4.413	0.0940
C_1	68.583	0.4306	68.328	0.3204
C_2	8.586	0.0164	8.564	0.0127
C_3	4.244	0.0024	4.254	0.0019
iC_4	0.880	0.0001	0.872	0.0001
nC_4	1.666	0.0002	1.604	0.0003
iC_5	0.660	0.0008	0.629	0.0000
nC_5	0.754	0.0002	0.729	0.0000
C_6	0.982	0.0001	0.973	0.0000
C_{7+}	8.481	0.0029	8.614	0.0007
1000 bar and 200°C. Feed Consists of 0.61 mol Gas Condensate per Mole Salt-Free Water				
H_2O	4.683	98.9982	4.235	99.4011
N_2	0.391	0.0063	0.366	0.0000
CO_2	3.816	0.2113	4.188	0.1040
C_1	65.207	0.7455	65.249	0.4652
C_2	8.184	0.0286	8.244	0.0217
C_3	4.048	0.0040	4.102	0.0041
iC_4	0.845	0.0003	0.841	0.0003
nC_4	1.621	0.0005	1.549	0.0006
iC_5	0.663	0.0005	0.635	0.0004
nC_5	0.765	0.0004	0.746	0.0002
C_6	1.010	0.0001	1.029	0.0001
C_{7+}	8.765	0.0045	8.816	0.0024

Data Source: Pedersen, K.S. and Milter, J., Phase equilibrium between gas condensate and brine at HT/HP conditions, SPE 90309, presented at the *SPE ATCE*, Houston, TX, September 26–29, 2004.

TABLE 16.18
Salt Water (Brine) Composition Used in Phase
Equilibrium Study Reported in Table 16.17

Component	Mole Percentage
H_2O	97.347
$NaHCO_3$	0.035
NaCl	2.404
KCl	0.094
$CaCl_2$	0.075
$MgCl_2$	0.008
$SrCl_2$	0.014
$BaCl_2$	0.024

TABLE 16.19
Simulation Results for CO_2 and C_1 Concentration in Water Phase and Water Concentration in Hydrocarbon Vapor Phase

	Pure Water		Salt Water	
	SRK	PR	SRK	PR
P = 1000 bar and T = 35°C				
CO_2 water	0.215	0.218	0.188	0.189
C_1 water	0.421	0.408	0.306	0.297
H_2O vapor	0.052	0.051	0.049	0.049
P = 1000 bar and T = 120°C				
CO_2 water	0.171	0.175	0.134	0.137
C_1 water	0.396	0.391	0.293	0.290
H_2O vapor	0.843	0.815	0.788	0.762
P = 1000 bar and T = 200°C				
CO_2 water	0.208	0.213	0.155	0.161
C_1 water	0.759	0.757	0.508	0.509
H_2O vapor	4.868	4.689	4.479	4.318
P = 700 bar and T = 200°C				
CO_2 water	0.182	0.186	0.137	0.141
C_1 water	0.657	0.652	0.448	0.446
H_2O vapor	5.876	5.668	5.386	5.201

Source: Pedersen, K.S. and Milter, J., Phase equilibrium between gas condensate and brine at HT/HP conditions, SPE 90309, presented at the *SPE ATCE*, Houston, TX, September 26–29, 2004.

Note: The SRK and PR equations with the Huron and Vidal mixing rule were used. The experimentally determined phase compositions can be seen in Table 16.17.

TABLE 16.20

Phase Equilibrium Data for Acid gas Mixed with Water with Simulation Results for SRK H&V, PR H&V, SRK CPA, and PR CPA

103.4 bar and 49°C											
	Experimental			**SRK H&V**		**PR H&V**		**SRK CPA**		**PR CPA**	
Component	**Feed**	**Gas**	**Aqueous**	**Gas**	**Aqueous**	**Gas**	**Aqueous**	**Gas**	**Aqueous**	**Gas**	**Aqueous**
H_2O	78.380	0.341	97.749	0.256	97.649	0.234	97.715	0.257	97.615	0.243	97.801
CO_2	6.485	29.706	0.679	29.907	0.708	29.890	0.694	29.815	0.741	29.697	0.716
H_2S	6.479	26.962	1.479	26.398	1.566	26.537	1.516	26.492	1.551	26.949	1.391
C_1	8.222	40.821	0.092	41.252	0.076	41.157	0.074	41.247	0.091	40.938	0.091
C_3	0.434	2.171	0.002	2.187	0.001	2.182	0.001	2.189	0.002	2.173	0.001
344.7 bar and 49°C											
H_2O	65.069	0.591	97.451	0.322	97.310	0.265	97.335	0.457	97.182	0.397	97.413
CO_2	10.4775	29.576	0.880	29.589	0.961	29.594	0.960	29.483	1.031	29.453	0.987
H_2S	10.4678	28.212	1.486	28.369	1.554	28.413	1.533	28.372	1.569	28.623	1.388
C_1	13.285	39.515	0.181	39.614	0.174	39.622	0.172	39.579	0.216	39.426	0.211
C_3	0.7008	2.107	0.001	2.106	0.001	2.107	0.001	2.109	0.001	2.100	0.001
103.4 bar and 93°C											
H_2O	83.040	1.273	98.367	1.294	98.256	1.251	98.318	1.228	98.171	1.189	98.113
CO_2	5.087	29.743	0.386	29.856	0.477	29.791	0.472	29.742	0.527	29.812	0.534
H_2S	5.083	25.885	1.189	25.920	1.204	26.150	1.147	25.950	1.223	25.798	1.268
C_1	6.450	40.913	0.057	40.769	0.062	40.652	0.061	40.909	0.077	41.021	0.084
C_3	0.340	2.186	0.001	2.161	0.001	2.155	0.001	2.171	0.002	2.179	0.001
344.7 bar and 93°C											
H_2O	73.833	1.190	97.429	1.147	97.286	1.006	97.403	1.207	97.067	1.091	97.013
CO_2	7.849	29.287	0.730	29.510	0.860	29.498	0.842	29.329	0.977	29.424	0.974
H_2S	7.842	26.845	1.676	26.920	1.686	27.149	1.593	26.915	1.740	26.847	1.785
C_1	9.952	40.493	0.164	40.275	0.168	40.203	0.161	40.388	0.215	40.472	0.226
C_3	0.525	2.185	0.002	2.148	0.001	2.143	0.001	2.161	0.002	2.167	0.002

Data Source: Ng, H.-J., C.-J. Chen, and Schroeder, H., *Water Content of Natural Gas Systems Containing Acid Gas, Research Report RR-174*, Gas Processors Association, Tulsa, OK, 2001.

16.3 WATER PROPERTIES

The density of pure water may be calculated with good accuracy using the SRK or the PR equation of state with Peneloux volume correction (Equations 4.43 and 4.48). The latter correction can possibly be made temperature dependent (Equation 5.9). Accurate results for derived properties for water such as the sound velocity (Equation 8.22) requires a model that is dedicated to water, for example, the model of Keyes et al. (1968).

A model that takes into consideration the attractive forces acting between the water molecules is also required to simulate the viscosity of water (e.g., Meyer et al. 1967; Schmidt 1969). The same applies for the thermal conductivity of water, for which property a model of Sengers and Keyes (1971) may be used.

The surface tension of liquid water (in mN/m) may be calculated from the formula

$$\sigma = 235.8 \left(1 - \frac{T}{T_c}\right)^{1.256} \left(1 - 0.625\left(1 - \frac{T}{T_c}\right)\right) \tag{16.30}$$

where T is the absolute temperature and T_c the critical temperature of water.

The interfacial tension, σ, between a water phase and a hydrocarbon phase (gas or oil) can be calculated from (Firoozabadi and Ramey 1988)

$$\sigma^{1/4} = \frac{a_1 \Delta\rho^{b_1}}{T_r^{0.3125}} \tag{16.31}$$

where

$$\Delta\rho = \left|\rho_w - \rho_{HC}\right| \tag{16.32}$$

In this equation, ρ_w is the density of the water phase and ρ_{HC} the density of the hydrocarbon phase. The densities are in g/cm³. Values of the constants a_1 and b_1 with σ in dyn/cm (= 1 mN/m) can be seen in Table 16.21.

16.3.1 Viscosity of Water-Inhibitor Mixtures

Alder (1966) and van Velzen et al. (1972) have proposed the correlations in Table 16.22 for how the viscosities of saturated liquid methanol, ethanol, and glycols vary with temperature.

TABLE 16.21

Constants in Expression for Interfacial Tension between Hydrocarbon and Water (Equation 16.31)

Dr (g/cm³)	a_1	b_1
< 0.2	2.2062	−0.94716
0.2–0.5	2.9150	−0.76852
≥ 0.5	3.3858	−0.62590

TABLE 16.22

Correlations for Viscosity (η) in cP of Saturated Liquid of Commonly Used Hydrate Inhibitors. T Is the Temperature in K and t the Temperature in °C

Component	Expression	A	B	C	D
Methanol	$\ln\eta = A + \frac{B}{T} + CT + DT^2$	−39.4 × 10	4.83 × 10³	1.13 × 10⁻¹	−1.09 × 10⁻⁴
Ethanol	$\ln\eta = A + \frac{B}{T} + CT + DT^2$	−6.21	1.614 × 10³	6.18 × 10⁻³	−1.132 × 10⁻⁵
MEG	$\ln\eta = A + \frac{B}{t+C}$	−3.61359	986.519	127.861	
TEG	$\ln\eta = A + \frac{B}{t+C}$	−3.11771	914.766	110.068	

Source: Alder, B.J., *Prediction of Transport Properties of Dense Gases and Liquids*, UCRL 14891-T, University of California, Berkeley, CA, May 1966.

These viscosities may be pressure corrected using the following expression suggested by Lucas (1981).

$$\frac{\eta}{\eta_{SL}} = \frac{1 + D\left(\dfrac{\Delta P_r}{2.118}\right)^A}{1 + C\omega P_r} \tag{16.33}$$

where

$$P_r = \frac{P}{P_c}$$

$$\Delta P_r = \frac{P - P^{sat}}{P_c}$$

η = Viscosity of liquid at actual temperature and pressure
η_{SL} = Viscosity of saturated liquid at actual T
ω = Acentric factor

$$A = -0.9991 - \frac{4.674 \times 10^{-4}}{1.0523 T_r^{-0.03877} - 1.0513}$$

$$D = \frac{0.3257}{\left(1.0039 - T_r^{2.573} - 0.208616\right)^{0.2906}} - 0.208616$$

$$C = \sum_{I=0}^{7} c_i T_r^i$$

P_c is the critical pressure and T_r the reduced temperature, T/T_c, where T_c is the critical temperature. The coefficients c_0–c_7 are given in Table 16.23.

The mixing rules of Grunberg and Nissan (1949) may be used to find the viscosity of water–hydrate inhibitor mixtures.

$$\ln \eta_{mix} = \sum_{i=1}^{N} w_i \ln \eta_i + \sum_{i=1}^{N} \sum_{j=1}^{N} w_i w_j G_{ij} \tag{16.34}$$

where N is the number of components in the aqueous phase covering water and inhibitors and G_{ij} is an interaction parameter that can be seen in Table 16.24 for the relevant binaries. The following temperature dependence is used for G_{ij}

$$G_{ij}(T) = 1 - (1 - G_{ij})\frac{573 - T}{275} \tag{16.35}$$

where T is the temperature in K.

TABLE 16.23

Coefficients in Expression for C Entering in Equation 16.33

i	0	1	2	3	4	5	6	7
c_i	−0.07921	2.1616	−13.4040	44.1706	−84.8291	96.1209	59.8127	15.6719

TABLE 16.24

G_{ij} Values for Binaries with Water in Equation 16.34

	H_2O
MeOH	2.5324
EtOH	3.3838
MEG	−1.3209
TEG	−0.2239

16.3.2 THERMAL CONDUCTIVITY OF WATER-INHIBITOR MIXTURES

Alcohols and glycols have lower thermal conductivities than water. A correct representation of the thermal conductivity of a hydrate inhibited water phase therefore requires a correlation that takes the content of inhibitor into account. The thermal conductivities of alcohols and glycols vary slightly with temperature and a linear correlation in temperature may be used for the thermal conductivity (λ_{L0}) at atmospheric pressure

$$\lambda_{L0} = A + B \times T \tag{16.36}$$

The constants A and B for commonly used hydrate inhibitors are shown in Table 16.25.

A pressure correction proposed by Missenard (1970) can be used for methanol and ethanol, while for glycols, the pressure dependence is modest and can be ignored.

The thermal conductivity of a mixture (mix) of water (H_2O) and a hydrate inhibitor (inh) can be calculated from

$$ln\left(\lambda_{mix}\right) = w_{H_2O} ln\left(\lambda_{H_2O}\right) + w_{inh} ln\left(\lambda_{inh}\right) \tag{16.37}$$

where w stands for weight fraction.

16.3.3 PROPERTIES OF SALT WATER

Numbere et al. (1977) have proposed correlations for the density and the viscosity of a water phase with dissolved salts. The density correlation takes the form:

$$\frac{\rho_s}{\rho_w} - 1 = C_S[7.65 \times 10^{-3} - 1.09 \times 10^{-7} P + C_S\left(2.16 \times 10^{-5} + 1.74 \times 10^{-9} P\right)$$
$$- \left(1.07 \times 10^{-5} - 3.24 \times 10^{-10} P\right)T + \left(3.76 \times 10^{-8} - 1.0 \times 10^{-12} P\right)T^2] \tag{16.38}$$

where ρ_s is the salt-water density in g/cm³, ρ_w the density of salt-free water in g/cm³ at the same temperature and pressure, C_s is the salt concentration in weight percentage, T the temperature in °F, and P the pressure in psia. The following is the proposed viscosity correlation for water phase with dissolved salts:

$$\frac{\eta_s}{\eta_w} - 1 = -1.87 \times 10^{-3} C_s^{0.5} + 2.18 \times 10^{-4} C_s^{2.5}$$
$$+ (T^{0.5} - 1.35 \times 10^{-2} T)\left(2.76 \times 10^{-3} C_s - 3.44 \times 10^{-4} C_s^{1.5}\right) \tag{16.39}$$

TABLE 16.25

Constant in Expression for Temperature Dependence of the Thermal Conductivity of Hydrate Inhibitors in W/(m K)

Component	A	B
Methanol	0.2874	−2.897E-4
Ethanol	0.2384	−2.448E-4
MEG	0.2463	2.950E-5
DEG	0.2271	−9.492E-5
TEG	0.2156	−3.348E-5

where η_s is the salt water viscosity and η_w the viscosity of pure water at the same T and P. The temperature T is in °F.

Laliberté and Cooper (2004) have proposed a different salt-water density correlation that distinguishes between different salts

$$\rho_{brine} = \cfrac{1}{\cfrac{w_{H_2O}}{\rho_{H_2O}} + \sum_{i=1}^{NSALT} \cfrac{w_i}{\rho_{App,i}}} \tag{16.40}$$

where
ρ_{brine} Density of salt water (brine)
w_{H_2O} Weight fraction of water
ρ_{H_2O} Density of pure water
w_i Weight fraction of salt number i
$\rho_{App,i}$ Density of salt number i when dissolved in water (*app*arent salt density)

The apparent density (in kg/m³) of a salt dissolved in water is calculated from

$$\rho_{App,i} = \frac{\left((c_{0,i}w_i + c_{1,i}) \times e^{(0.000001(t+c_{4,i})^2)} \right)}{w_i + c_{2,i} + c_{3,i}t} \tag{16.41}$$

where t is the temperature in °C. Table 16.26 shows values of the constants c_0–c_4 for three common salts. Figure 16.8 shows experimental and simulated data for the density of NaCl salt water at 100 bar at three different temperatures and varying salt concentrations.

16.3.4 OIL–WATER EMULSION VISCOSITIES

In a production well or a pipeline transporting both oil and water, water-in-oil or oil-in-water emulsions may form. The viscosity of a water–oil emulsion may exceed the viscosities of the separate phases by several orders of magnitudes.

The maximum viscosity of the emulsion is seen at the water/oil ratio where the emulsion changes from a water-in-oil to an oil-in-water emulsion (inversion point). The following equation (Rønningsen 1995) may be used to predict the viscosity of a water-in-oil emulsion:

$$\ln \eta r = -0.06671 - 0.000775\,t + 0.03484\phi + 0.0000500\,t\phi \tag{16.42}$$

TABLE 16.26
Constant in Expression of Laliberté and Cooper (2004) for Salt Water Density

Constant	NaCl	KCl	CaCl$_2$
c_0	−0.00433	−0.46782	−0.63254
c_1	0.06471	4.30800	0.93995
c_2	1.0166	2.3780	4.2785
c_3	0.014624	0.022044	0.048319
c_4	3315.6	2714.0	3180.9

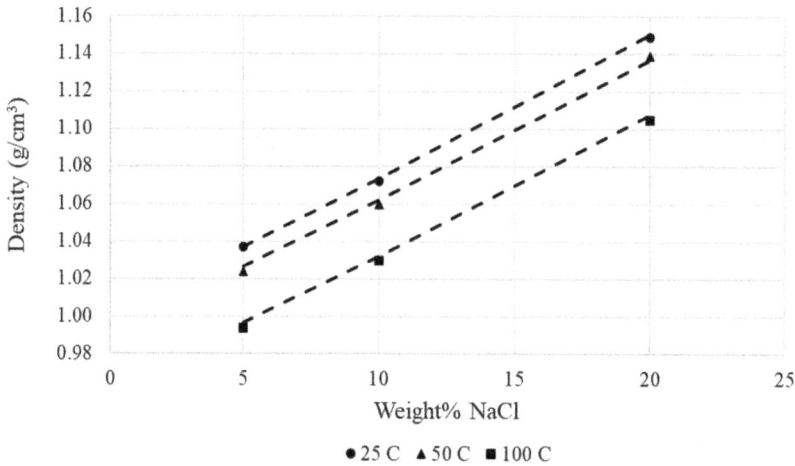

FIGURE 16.8 Experimental and simulated data for the density of NaCl salt water at 100 bar at three different temperature as a function of salt concentrations.

where η_r is the relative viscosity (emulsion/oil), ϕ the volume percentage of water, and t the temperature in °C.

If an experimental viscosity data point exists (η_r at some ϕ), the correlation of Pal and Rhodes (1989) may be used:

$$\eta_{r,h} = \left[1 + \frac{\dfrac{\phi_w}{(\phi)_{\eta_r=100}}}{1.19 - \dfrac{\phi_w}{(\phi)_{\eta_r=100}}} \right]^{2.5} \quad \text{if } \phi_w < \phi_{Inv} \tag{16.43}$$

$$\eta_{r,w} = \left[1 + \frac{\dfrac{\phi_h}{(\phi)_{\eta_r=100}}}{1.19 - \dfrac{\phi_h}{(\phi)_{\eta_r=100}}} \right]^{2.5} \quad \text{if } \phi_w > \phi_{Inv} \tag{16.44}$$

where $\eta_{r,h}$ means the ratio of the water–in–oil emulsion viscosity and the oil viscosity. $\eta_{r,w}$ is the ratio of the oil–in–water emulsion viscosity and the water viscosity. The experimental values of ϕ and η_r are used to calculate $\phi_{\eta_r=100}$ from the following equation:

$$(\phi)_{\eta_r=100} = \frac{\phi}{1.19\left(1-\eta_r^{-0.4}\right)} \tag{16.45}$$

where η_r is $\eta_{r,h}$ in Equation (16.43) and $\eta_{r,w}$ in Equation (16.44).

16.4 PHASE ENVELOPES OF HYDROCARBON-AQUEOUS MIXTURES

Lindeloff and Michelsen (2003) have outlined a procedure for calculating phase envelopes of mixtures of hydrocarbons and aqueous mixtures. Table 16.27 shows a gas condensate mixture (Fluid B in article of Lindeloff and Michelsen). This fluid was characterized for the PR equation of state. The simulated phase envelope is seen as a dashed line in Figure 16.9. Also shown in the same

TABLE 16.27
Molar Composition of Gas Condensate Mixture for Which Phase Envelope Is Shown in Figures 16.9 and 16.10

Component	Mole Percentage
CO_2	2.79
C_1	71.51
C_2	5.77
C_3	4.10
iC_4	1.32
nC_4	1.60
iC_5	0.82
nC_5	0.64
C_6	1.05
C_{7+}	10.40

Note: The C_{7+} density is 0.82 g/cm³ at 1.01 bar and 15°C and the C_{7+} molecular weight is 191. Fluid is the same as Fluid B in Lindeloff and Michelsen (2003).

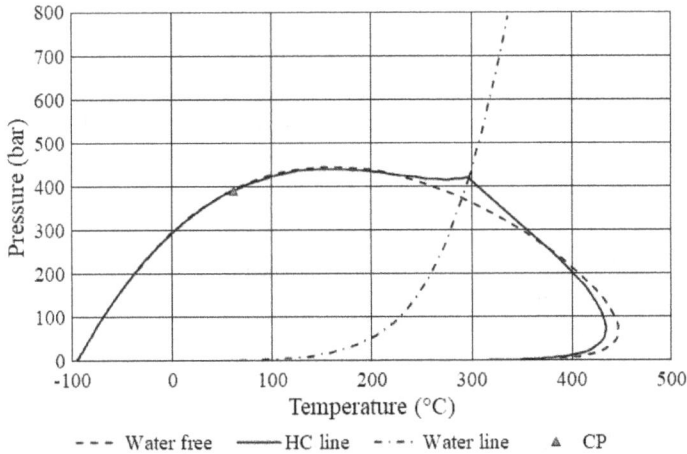

FIGURE 16.9 Simulated phase envelope for fluid in Table 16.27 mixed with water in a molar ratio of 2.09:1.00. PR equation of state with Huron and Vidal mixing rule is used as thermodynamic model. CP stands for critical point and HC for hydrocarbon.

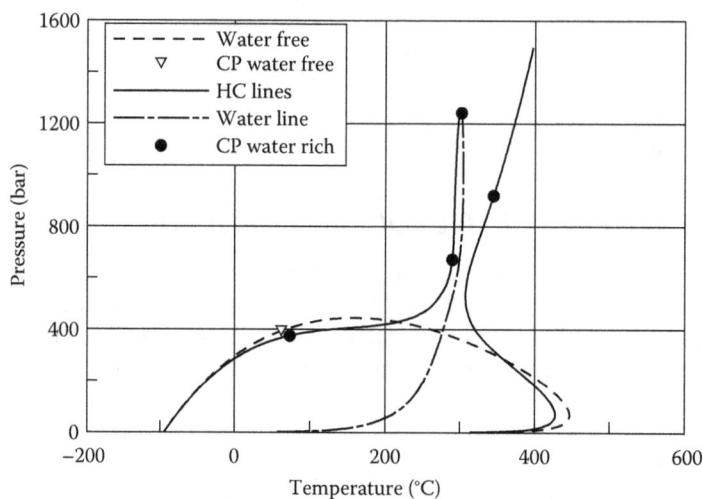

FIGURE 16.10 Simulated phase envelope for fluid in Table 16.27 mixed with an aqueous phase consisting of 71.4 mole percentage water and 28.6 mole percentage methanol. The mixing ratio is 1.49 mole reservoir fluid per mole aqueous phase. PR equation of state is used as thermodynamic model. CP stands for critical point and HC for hydrocarbon.

figure is the phase envelope of a mixture consisting of gas condensate and water mixed in a molar ratio of 2.09:1.00. The Huron and Vidal mixing rule was used for binaries of water and hydrocarbon gas constituents. Figure 16.10 shows the phase envelope for the gas condensate in Table 16.27 mixed with an aqueous phase consisting of 71.4 mole percentage water and 28.6 mole percentage methanol. The mixing ratio was 1.49 mole reservoir fluid per mole aqueous phase. The Huron and Vidal mixing rule was used for component pairs consisting of a hydrocarbon gas constituent and either water or methanol. Four critical points were found for this mixture, as can be seen from Figure 16.10. Figures 16.9 and 16.10 illustrate that aqueous components can have a marked influence on the phase behavior of the hydrocarbon phases of petroleum reservoir fluids.

REFERENCES

Abrams, D.S. and Prausnitz, J.M., Statistical thermodynamics of liquid mixtures: A new expression for the excess Gibbs energy of partly or completely miscible systems, *AICHE J.* 21, 116–128, 1975.

Alder, B.J., *Prediction of Transport Properties of Dense Gases and Liquids*, UCRL 14891-T, University of California, Berkeley, CA, May 1966.

ASME SteamTables, 4th ed., App. I, The American Society of Mechanical Engineers, New York, 11–29, 1979.

Austegard, A., Solbraa, E., De Koeijer, G., and Mølnvik, M.J., Thermodynamic models for calculating mutual solubilities in $H_2O–CO_2–CH_4$ mixtures, *Chem. Eng. Res. Des.* 84, 781–794, 2006.

Boesen, R.R., Herslund, P.J., and Sørensen, H., Loss of monoethylene glycol to CO2—and H2S-rich fluids: Modeled using Soave–Redlich–Kwong with the Huron and Vidal mixing rule and Cubic-Plus-Association equations of state, *Energy Fuels.* 31, 3417–3426, 2017.

Firoozabadi, A. and Ramey, H.J., Surface tension of water-hydrocarbon systems at reservoir conditions, *J. Can. Pet. Technol.* 27, 41–48, 1988.

Folas, G.K., Kontogeorgis, G.M., Michelsen, M.L., and Stenby, E.H., Application of the Cubic-Plus-Association equation to mixtures with polar chemicals and high pressure, *Ind. Eng. Chem. Res.* 45, 1516–1526, 2006.

Fredenslund, Aa., Gmehling, J., and Rasmussen, P., *Vapor-Liquid Equilibria Using UNIFAC*, Elsevier, Amsterdam, North-Holland, 1977.

Grunberg, L. and Nissan, A.H., Mixture law for viscosity, *Nature.* 164, 799–800, 1949.

Huron, M.J. and Vidal, J., New mixing rules in simple equations of state for representing vapor-liquid equilibria of strongly non ideal mixtures, *Fluid Phase Equilib.* 3, 255–271, 1979.

Keyes, F.G., Keenan, J.H., Hill, P.G., and Moore, J.G., A fundamental equation for liquid and vapor water, presented at the *Seventh International Conference on the Properties of Steam*, Tokyo, Japan, September 1968.

Kobayashi, R. and Katz, D.L., Vapor-liquid equilibria for binary hydrocarbon-water systems, *Ind. Eng. Chem.* 45, 440–446, 1953.

Kokal, S., Al-Dokhi, M., and Sayegh, S., Phase behavior of gas condensate/water system, SPE Res Eval & Eng 6, 412–420, 2003.

Kontogeorgis, G.M., Michelsen, M.L., Folas, G.K., Derawi, S., von Solms, N., and Stenby, E.H., Ten years with CPA (Cubic-Plus-Association) equation of state, Part 1. Pure compounds and self-associating systems, *Ind. Eng. Chem. Res.* 45, 4855–4868, 2006a.

Kontogeorgis, G.M., Michelsen, M.L., Folas, G.K., Derawi, S., von Solms, N., and Stenby, E.H., Ten years with CPA (Cubic-Plus-Association) equation of state, Part 2. Cross-associating and multicomponent systems, *Ind. Eng. Chem. Res.* 45, 4869–4878, 2006b.

Kontogeorgis, G.M., Voutsas, E.C., Yakoumis, I.V., and Tassios, D.P., An equation of state for associating fluids, *Ind. Eng. Chem. Res.* 35, 4310–4318, 1996.

Kontogeorgis, G.M., Yakoumis, I.V., Meijer, H., Hendriks, E.M., and Moorwood, T., Multicomponent phase equilibrium calculations for water–methanol–alkane mixtures, *Fluid Phase Equilib.* 158, 201–209, 1999.

Kristensen, J.N., Christensen, P.L., Pedersen, K.S., and Skovborg, P., A combined Soave-Redlich–Kwong and NRTL equation for calculating the distribution of methanol between water and hydrocarbon phases, *Fluid Phase Equilib.* 82, 199–206, 1993.

Laliberté, M. and Cooper, W.E., Model for calculating the density of aqueous electrolyte solutions, *J. Chem. Eng. Data.* 49, 1141–1151, 2004.

Lindeloff, N. and Michelsen, M.L., Phase envelope calculations for hydrocarbon-water mixtures, SPE 85971, *SPE J.* 8, 298–303, 2003.

Lucas, K., Die Druckabhängigkeit der Viskosität von Flüssigkeiten, *Chem. Ing. Tech.* 53, 959–960, 1981 (in German).

Lundstrøm, C., Michelsen, M.L., Kontogeorgis, G.M., Pedersen, K.S., and Sørensen, H., Comparison of the SRK and CPA equations of state for physical properties of water and methanol, *Fluid Phase Equilib.* 247, 149–157, 2006.

Meyer, C.A., McClintock, R.B., Silverstri, G.J., and Spencer, R.C., Jr., Thermodynamic and transport properties of steam, *ASME Steam Tables*, 2nd ed., ASME, New York, 1967.

Michelsen, M.L., A modified Huron–Vidal mixing rule for cubic equations of state, *Fluid Phase Equilib.* 60, 213–219, 1990.

Michelsen, M.L. and Hendriks, E.M., Physical properties from association models, *Fluid Phase Equilib.* 180, 165–174, 2001.

Missenard, F.A., Thermal conductivity of organic liquids of a series or a group of liquids, *Rev. Gen. Thermodyn.* 101, 649–660, 1970.

Ng, H.-J. and Chen, C.-J., Vapor-liquid and vapor-liquid-liquid equilibria for H2S, CO_2, selected light hydrocarbons and a gas condensate in aqueous methanol or ethylene glycol solutions, Research Report RR-149, presented at the *Gas Processors Association*, Tulsa, OK, 1995.

Ng, H.-J., Chen, C.-J., and Razzaghi, M., Vapor-Liquid equilibria of selected aromatic hydrocarbons in triethylene glycol, *Fluid Phase Equilib.* 82, 207–214, 1993.

Ng, H.-J., Chen, C.-J., and Schroeder, H., Water content of natural gas systems containing acid gas, *Research Report RR-174*, Gas Processors Association, Tulsa, OK, 2001.

Numbere, D., Bringham, W.E., and Standing, M.B., Correlations for physical properties of petroleum reservoir brines, *Petroleum Research Institute*, Stanford University, City of Palo Alto, CA, 1977.

Pal, R. and Rhodes, E., Viscosity/concentration relationships for emulsions, *J. Rheol.* 33, 1021–1045, 1989.

Pedersen, K.S., Michelsen, M.L., and Fredheim, A.O., Phase equilibrium calculations for unprocessed well streams containing hydrate inhibitors, *Fluid Phase Equilib.* 126, 13–28, 1996.

Pedersen, K.S. and Milter, J., Phase equilibrium between gas condensate and brine at HT/HP conditions, SPE 90309, presented at the *SPE ATCE*, Houston, TX, September 26–29, 2004.

Pedersen, K.S., Milter, J., and Rasmusssen, C.P., Mutual solubility of water and a reservoir fluid at high temperatures and pressures: Experimental and simulated data, *Fluid Phase Equilib.* 189, 85–97, 2001.

Renon, H. and Prausnitz, J.M., Local composition in thermodynamic excess functions for liquid mixtures, *AICHE J.* 14, 135–144, 1968.

Rønningsen, H.P., Conditions for predicting viscosity of W/O emulsions based on North Sea crude oils, SPE 28968, presented at the *SPE International Symposium on Oilfield Chemistry*, San Antonio, TX, February 14–17, 1995.

Schmidt, E., *Properties of Water and Steam in SI-Units*, Springer-Verlag, New York, 1969.

Sengers, J.V. and Keyes, P.H., Scaling of the thermal conductivity near the gas-liquid critical point, *Tech. Rep.* 71–061, University of Maryland, 1971.

Sørensen, H., Boesen, R.R., Leekumjorn, S., and Herslund, P.J., Peng–Robinson equation of state extended to handle aqueous components using CPA concept, *J. Nat. Gas Eng.* 3, 1–38, 2018.

Sørensen, H., Pedersen, K.S., and Christensen, P.L., Modeling of gas solubility in brine, *Org. Chem.* 33, 635–642, 2002.

van Velzen, D., Cardozo, R.L., and Langekamp, H., A liquid viscosity-temperature-chemical constitution relation for organic compounds, *Ind. Eng. Chem. Fundam.* 11, 20–25, 1972.

Voutsas, E.C., Yakoumis, I.V., and Tassios, D.P., Prediction of phase equilibria in water/alcohol/alkane systems, *Fluid Phase Equilib.* 158, 151–163, 1999.

Wong, D.S.H. and Sandler, S.I., A theoretically correct mixing rule for cubic equations of state, *AICHE J.* 38, 671–680, 1992.

17 Scale Precipitation

Produced petroleum reservoir fluid well streams often carry formation water from an underlying water zone. The formation water may have a considerable content of salts, most of which are lightly soluble in water, for example, NaCl, KCl, and $CaCl_2$. The water may further carry salts of low solubility, such as $BaSO_4$, $CaCO_3$, and $CaSO_4$, which under certain conditions can precipitate as solid salt. Salt deposition is often referred to as *scaling*, and is a potential problem in pipelines transporting unprocessed well streams containing formation water. In the reservoir, scale precipitation can be seen when seawater is injected to obtain enhanced recovery and contacts formation water.

17.1 CRITERIA FOR SALT PRECIPITATION

A particular salt may precipitate if its solubility product, K_{sp}, is exceeded. The stoichiometric solubility product of a salt is defined as the product of the molalities of the salt ions in a water solution saturated with salt. The stoichiometric solubility product of $CaSO_4$ is, for example, defined as:

$$K_{sp}(CaSO_4) = m_{Ca^{++}}\, m_{SO_4^-} \tag{17.1}$$

where m stands for molality (mol/l). The thermodynamic solubility product K_{sp}^o of $CaSO_4$ is defined as the product of the activities of Ca^{++} and SO_4^- in a saturated solution in water:

$$K_{sp}^o(CaSO_4) = a_{Ca^{++}}\, a_{SO_4^-} \tag{17.2}$$

where

$$a_{Ca^{++}} = m_{Ca^{++}} \gamma_{Ca^{++}}; \quad a_{SO_4^-} = m_{SO_4^-} \gamma_{SO^-} \tag{17.3}$$

The term γ stands for activity coefficient and is defined in Equation A.34 in Appendix A. By combining Equations 17.1 and 17.3, it can be seen that the stoichiometric and thermodynamic solubility products are related as follows (again exemplified through $CaSO_4$):

$$K_{sp} = \frac{K_{sp}^o}{\gamma_{Ca^{++}} \gamma_{SO_4^-}} \tag{17.4}$$

It can be seen from this equation that the stoichiometric and thermodynamic solubility products become equal if all activity coefficients are 1.0.

The salts that are most likely to precipitate from formation water are the following:

Calcium sulfate ($CaSO_4$)
Barium sulfate ($BaSO_4$)
Strontium sulfate ($SrSO_4$)
Calcium carbonate ($CaCO_3$)
Iron carbonate ($FeCO_3$)
Iron sulfide (FeS)

Had O_2 been present, iron would have precipitated as ferrihydroxide. The assumption here is that the formation water is free of O_2. Whether a given salt will precipitate or not depends on a number

DOI: 10.1201/9780429457418-17

of factors. Obviously, the concentration of the ions making up the salt is important, but also the acidity (pH = $-\log_{10}[H^+]$) where $[H^+]$ is the hydrogen ion concentration in mol/l), amounts of CO_2 and H_2S dissolved in the water phase and the concentration of other ions, for example, Na^+, K^+, and Cl^-, will influence the scale potential. Whereas the molality is uniquely given by the concentration of salt ions, the activity coefficients are influenced by all the remaining factors.

The equilibria of interest are the following:

Acid equilibria

$$H_2O \leftrightarrow H^+ + OH^-$$

$$H_2O + CO_2 \leftrightarrow H^+ + HCO_3^-$$

$$HCO_3^- \leftrightarrow H^+ + CO_3^{--}$$

$$HA \leftrightarrow H^+ + A^-$$

$$H_2S \leftrightarrow H^+ + HS^-$$

Here, HA is a general organic acid.

Sulfate mineral precipitation reactions:

$$Ca^{++} + SO_4^{--} \leftrightarrow CaSO_4(s)$$

$$Ba^{++} + SO_4^{--} \leftrightarrow BaSO_4(s)$$

$$Sr^{++} + SO_4^{--} \leftrightarrow SrSO_4(s)$$

where s represents solid or precipitate.

Ferrous iron mineral precipitation reactions:

$$Fe^{++} + CO_3^{--} \leftrightarrow FeCO_3(s)$$

$$Fe^{++} + HS^- \leftrightarrow H^+ + FeS(s)$$

Calcium carbonate precipitation reaction:

$$Ca^{++} + CO_3^{--} \leftrightarrow CaCO_3(s)$$

The equilibrium constants and thermodynamic solubility products for these reactions are as follows:

$$K_{H_2O}^o = m_{H^+} \cdot m_{OH^-} \frac{\gamma_{H^+} \gamma_{OH^-}}{a_{H_2O}} \tag{17.5}$$

$$K_{CO_2,1}^o = \frac{m_{H^+} m_{HCO_3^-}}{m_{CO_2}} \frac{\gamma_{H^+} \gamma_{HCO_3^-}}{\gamma_{CO_2} a_{H_2O}} \tag{17.6}$$

$$K_{CO_2,2}^o = \frac{m_{H^+} m_{CO_3^-}}{m_{HCO_3^-}} \frac{\gamma_{H^+} \gamma_{CO_3^-}}{\gamma_{HCO_3^-}} \tag{17.7}$$

$$K_{HA}^o = \frac{m_{H^+} m_{A^-}}{m_{HA}} \frac{\gamma_{H^+} \gamma_{A^-}}{\gamma_{HA}} \tag{17.8}$$

$$K_{H_2S}^o = \frac{m_{H^+} m_{HS^-}}{m_{H_2S}} \frac{\gamma_{H^+} \gamma_{HS^-}}{\gamma_{H_2S}} \tag{17.9}$$

$$K^o_{CaSO_4} = m_{Ca^{++}} m_{SO_4^-} \gamma_{Ca^{++}} \gamma_{SO_4^-} \tag{17.10}$$

$$K^o_{BaSO_4} = m_{Ba^{++}} m_{SO_4^-} \gamma_{Ba^{++}} \gamma_{SO_4^-} \tag{17.11}$$

$$K^o_{SrSO_4} = m_{Sr^{++}} m_{SO_4^-} \gamma_{Sr^{++}} \gamma_{SO_4^-} \tag{17.12}$$

$$K^o_{FeCO_3} = m_{Fe^{++}} m_{CO_3^-} \gamma_{Fe^{++}} \gamma_{CO_3^-} \tag{17.13}$$

$$K^o_{CaCO_3} = m_{Ca^{++}} m_{CO_3^-} \gamma_{Ca^{++}} \gamma_{CO_3^-} \tag{17.14}$$

$$K^o_{FeS} = \frac{m_{Fe^{++}} m_{HS^-}}{m_{H^+}} \frac{\gamma_{Fe^{++}} \gamma_{HS^-}}{\gamma_{H^+}} \tag{17.15}$$

By comparing Equations 17.10 through 17.14 with Equation 17.2, it is seen that the equilibrium constants for $CaSO_4$, $BaSO_4$, $SrSO_4$, $FeCO_3$, and $CaCO_3$ equal the thermodynamic solubility products of the same salts.

To summarize, precipitation of salts is determined by the following:

Molality of ions
Acidity
Equilibrium constants (some of which equal the thermodynamic solubility product of a salt that may precipitate)
Activity coefficients of all ions

17.2 EQUILIBRIUM CONSTANTS

The temperature dependence of the equilibrium constant for the self-ionization of water (Equation 17.5) may for the temperature, T, in K be described by (Olofsson and Hepler 1975):

$$-\log_{10}\left(K^o_{H_2O}(T)\right) = \frac{142613.6}{T} + 4229.195 \log_{10}T - 9.7384\,T + 0.0129638\,T^2$$
$$-1.15068 \times 10^{-5}T^3 + 4.602 \times 10^{-9}T^4 - 8908.483 \tag{17.16}$$

The temperature dependence of the remaining equilibrium constants at a pressure of 1 bar may for the temperature in K be described as:

$$\ln K^o(T) = A + \frac{B}{T} + C\ln T + D\,T + \frac{E}{T^2} \tag{17.17}$$

Values of the constants A to E are given in Table 17.1. The organic acid pool is often taken to be acetic acid, which is usually the main organic acid constituent. The alkalinity (A_T) is defined as

$$A_T = m_{HCO_3^-} + 2m_{CO_3^-} + m_{A^-} + m_{OH^-} + m_{H^+} \tag{17.18}$$

The sum on the right-hand side of this equation is independent of pH.

This is a convenient definition because it keeps the alkalinity constant during pH changes. The pressure dependence of the equilibrium constants is given by:

$$\frac{\partial \ln K}{\partial P} = \frac{P\Delta\kappa - \Delta V}{RT} \tag{17.19}$$

TABLE 17.1

Coefficients in Equation 17.17 for the Temperature Dependence of Equilibrium Constants

	A	B	C	1000 × D	E	Reference
$K_{CO_2,1}$	−820.4327	50275.5	126.8339	−140.2727	−3879660	Haarberg (1989)
$K_{CO_2,2}$	−248.4192	11862.4	38.92561	−74.8996	−1297999	Haarberg (1989)
K_{HA}	−10.937	0	0	0	0	Haarberg (1989)
K_{H_2S}	−16.1121	0	0	0	0	Kaasa and Østvold (1998)
K_{CaSO_4}	11.6592	−2234.4	0	−48.2309	0	Haarberg (1989)
$K_{CaSO_4 2H_2O}$	815.978	−26309.9	−138.361	167.863	18.6143	Haarberg (1989)
K_{BaSO_4}	208.839	−13084.5	−32.4716	−9.58318	2.58594	Haarberg (1989)
K_{SrSO_4}	89.6687	−4033.9	−16.0305	−1.34671	31402.1	Haarberg (1989)
K_{FeCO_3}	21.804	−56.448	−16.8397	0.02298	0	Kaasa and Østvold (1998)
K_{FeS}	−8.3102	0	0	0	0	Kaasa and Østvold (1998)
K_{CaCO_3}	−395.448	6461.5	71.558	−180.28	24847	Haarberg (1989)

Note: T is in K. For T < 373.15 K, the coefficients are for a pressure of 1 atm. For T > 373.15 K, the coefficients are for a pressure equal to the vapor pressure of water.

where R is the universal gas constant, and ΔV is the change in molar volume as a result of the reaction. The compressibility change, Δκ, is given by:

$$\Delta\kappa = \left(\frac{\partial \Delta V}{\partial P}\right)_T \tag{17.20}$$

Equation 17.19 can be integrated to give:

$$\ln\left(\frac{K_i^\circ(P)}{K_i^\circ(P^{Ref})}\right) = \frac{-\Delta V(P - P^{Ref}) + \frac{1}{2}\Delta\kappa(P - P^{Ref})^2}{RT} \tag{17.21}$$

Δκ for the sulfate precipitation reactions is expressed by a third-degree polynomial in temperature t in °C:

$$10^{-3}\Delta\kappa = a + bt + ct^2 + dt^3 \tag{17.22}$$

The coefficients a, b, c, and d for each of the sulfate precipitation reactions are listed in Table 17.2. Haarberg (1989) expresses the compressibility changes associated with both CO_2 acid equilibria (Equations 17.6 and 17.7) as:

$$10^3\left(\Delta\frac{\partial V}{\partial P}\right)_{CO_2,1} = 10^3\left(\Delta\frac{\partial V}{\partial P}\right)_{CO_2,2} = -39.3 + 0.233\,T - 0.000371\,T^2 \tag{17.23}$$

where T is the temperature in K and ∂V/∂P is in cm³/(mol bar). For the calcium carbonate and ferrous carbonate precipitation reactions, Haarberg et al. (1990) have found that the compressibility changes are 0.015 cm³/(mol bar) independent of temperature.

For all reactions other than those explicitly mentioned in the preceding text, the effect of pressure on the equilibrium constants can be neglected.

TABLE 17.2
Coefficients in Equation 17.22 Expressing the Change in Compressibility as a Result of Sulfate Mineral Precipitation Reactions

	A	$100 \times b$	$1000 \times c$	$10^6 \times d$
$BaSO_4$	17.54	−1.159	−17.77	17.06
$SrSO_4$	17.83	−1.159	−17.77	17.06
$CaSO_4$	16.13	−0.944	−16.52	16.71
$CaSO_4\text{–}2H_2O$	17.83	−1.543	−16.01	16.84

Note: Units: t in °C and $\Delta\kappa$ in $cm^3/(mol\ bar)$.

Source: Data from Atkinson, A. and Mecik, M., The chemistry of scale prediction, *J. Petroleum Sci. Eng.* 17, 113–121, 1997.

TABLE 17.3
Coefficients in Equation 17.24 Expressing the Change in Volume as a Result of Sulfate Mineral Precipitation Reactions

	A	B	$1000 \times C$	D	E
$BaSO_4$	−343.6	1.746	−2.567	11.9	−4.0
$SrSO_4$	−306.9	1.574	−2.394	20.0	−8.2
$CaSO_4$	−282.3	1.438	−2.222	21.7	−9.8
$CaSO_4\text{–}2H_2O$	−263.8	1.358	−2.077	21.7	−9.8

Note: Units: T in Kelvin, I in mol/l solvent, and ΔV in cm^3/mol.

Source: Data from Haarberg, T., Mineral Deposition during Oil Recovery, Ph.D. thesis, Department of Inorganic Chemistry, University of Trondheim, Norway, 1989.

The following is a convenient expression for the partial molar volume changes of the sulfate precipitation reactions (Haarberg 1989):

$$\Delta V = A + BT + CT^2 + DI + EI^2 \qquad (17.24)$$

In this expression, I is the ionic strength (mol/l solvent):

$$I = \frac{1}{2} \sum_i m_i \left| z_i \right|^2 \qquad (17.25)$$

The term z is the charge of the ion considered (positive or negative), m is molality, and the index i runs over all ions. The constants A through E for the sulfate mineral precipitation reactions are listed in Table 17.3.

For the calcium carbonate and ferrous carbonate precipitation reactions, Haarberg (1989) described the partial molar volume change by:

$$\Delta V_{CaCO_3} = \Delta V_{FeCO_3} = -328.7 + 1.738\ T - 0.002794\ T^2 \qquad (17.26)$$

Haarberg (1989) expresses the partial molar volume changes of both acid equilibria of CO_2 (Equations 17.6 and 17.7) through:

$$\Delta V_{CO_2,1} = \Delta V_{CO_2,2} = -141.4 + 0.735\,T - 0.001190\,T^2 \tag{17.27}$$

In Equations 17.26 and 17.27, ΔV is in cm³/mol and T in K.

17.3 ACTIVITY COEFFICIENTS

The activity coefficients may be calculated using the Pitzer model (Pitzer 1973, 1975, 1979, 1986; Pitzer et al. 1984). Pitzer expresses the activity of the water in terms of the osmotic coefficient ϕ as:

$$\ln a_{H_2O} = -\phi\,M_{H_2O}\sum_i m_i \tag{17.28}$$

where M is molecular weight, m is molality, and the index i runs over all ions.

The osmotic coefficient is found from:

$$(\phi-1)\sum_i m_i = -\frac{2A_\varphi I^{3/2}}{1+bI^{1/2}} + \sum_c\sum_a m_c m_a\left(B_{ca}^\phi + ZC_{ca}\right)$$
$$+\sum_{c>c'}\sum_{c'} m_c m_{c'}\left(\Phi_{cc'}^\phi + \sum_a m_a\psi_{cc'a}\right) + \sum_{a>a'}\sum_{a'} m_a m_{a'}\left(\Phi_{aa'}^\phi + \sum_c m_c\psi_{ca'a}\right) \tag{17.29}$$

For cations (positive charge), Pitzer expresses the activity coefficients as:

$$\ln\gamma_M = z_M^2 F + \sum_a m_a(2B_{Ma} + ZC_{Ma}) + \sum_c m_c\left(2\Phi_{Mc} + \sum_a m_a\psi_{Mca}\right)$$
$$+\sum_{a>a'}\sum_{a'} m_a m_{a'}\Phi_{Maa'} + |z_M|\sum_c\sum_a m_c m_a C_{ca} \tag{17.30}$$

and for anions (negative charge) as:

$$\ln\gamma_X = z_X^2 F + \sum_c m_c(2B_{cX} + ZC_{cX}) + \sum_a m_a\left(2\Phi_{Xa} + \sum_c m_c\Psi_{cXa}\right)$$
$$+\sum_{c>c'}\sum_{c'} m_c m_{c'}\Psi_{cc'X} + |z_X|\sum_c\sum_a m_c m_a C_{ca} \tag{17.31}$$

In these expressions, a and a' denote anion species, c and c' the cation species, m the molality (mol/l solvent), and I the ionic strength (mol/l solvent) as defined in Equation 17.25. ψ_{ijk} is a model parameter specific for each cation–cation–anion triplet and each cation–anion–anion triplet. Values of ψ_{ijk} can be seen in Table 17.4. The remaining quantities in Equations 17.29 through 17.31 are as follows:

$$F = -A_\phi\left\{\frac{\sqrt{I}}{1+b\sqrt{I}} + \frac{2}{b}\ln\left(1+b\sqrt{I}\right)\right\} + \sum_c\sum_a m_c m_a B_{ca}'$$
$$+\sum_{c>c'}\sum_{c'} m_c m_{c'}\Phi_{cc'}' + \sum_{a>a'}\sum_{a'} m_a m_{a'}\Phi_{aa'}' \tag{17.32}$$

TABLE 17.4
Pitzer Parameters at 25°C

$\beta^{(0)}$ Parameters at 25°C

	H+	Na+	K+	Mg++	Ca++	Sr++	Ba++	Fe++
OH−	0.00000	0.08640	0.12980	0.00000	−0.17470	0.00000	0.17175	0.00000
Cl−	0.17750	0.07650	0.04810	0.35090	0.30530	0.28340	0.26280	0.44790
SO4−−	0.02980	0.01810	0.00000	0.21500	0.20000	0.20000	0.20000	−4.7050
HCO3−	0.00000	0.02800	−0.01070	0.32900	−1.49800	0.00000	0.00000	0.00000
CO3−	0.00000	0.03620	0.12880	0.00000	−0.40000	0.00000	0.00000	1.91900
HS−	0.00000	−0.10300	−0.33700	0.46600	0.069000	0.00000	0.00000	0.00000

$\beta^{(1)}$ Parameters at 25°C

	H+	Na+	K+	Mg++	Ca++	Sr++	Ba++	Fe++
OH−	0.00000	0.25300	0.32000	0.00000	−0.23030	0.00000	1.20000	0.00000
Cl−	0.29450	0.26640	0.21870	1.65100	1.70800	1.62600	1.49630	2.0430
SO4−−	0.00000	1.05590	1.10230	3.36360	3.19730	3.19730	3.19730	17.000
HCO3−	0.00000	0.04400	0.04780	0.60720	7.89900	0.00000	0.00000	14.76000
CO3−−	0.00000	1.51000	1.43300	0.00000	−5.30000	0.00000	0.00000	−5.13400
HS−	0.00000	0.88400	0.88400	2.26400	2.26400	0.00000	0.00000	0.00000

$\beta^{(2)}$ Parameters at 25°C

	H+	Na+	K+	Mg++	Ca++	Sr++	Ba++	Fe++
OH−	0.00000	0.00000	0.00000	0.00000	0.00000	0.00000	0.00000	0.00000
Cl−	0.00000	0.00000	0.00000	0.00000	0.00000	0.00000	0.00000	0.00000
SO4−−	0.00000	0.00000	0.00000	−32.7400	−54.2400	−54.2400	−54.2400	0.00000
HCO3−	0.00000	0.00000	0.00000	0.00000	0.00000	0.00000	0.00000	0.00000
CO3−−	0.00000	0.00000	0.00000	0.00000	−879.200	0.00000	0.00000	0.00000
HS−	0.00000	0.00000	0.00000	0.00000	0.00000	0.00000	0.00000	0.00000

C^{ϕ} Parameters

	H+	Na+	K+	Mg++	Ca++	Sr++	Ba++	Fe++
OH−	0.00000	0.00410	0.00410	0.00000	0.00000	0.00000	0.00000	0.00000
Cl−	0.00080	0.00127	−0.00079	0.00651	0.00215	−0.00089	−0.01938	−0.00861
SO4−−	0.04380	0.00571	0.01880	0.02797	0.00000	0.00000	0.00000	0.02090
HCO3−	0.00000	0.00000	0.00000	0.00000	0.00000	0.00000	0.00000	0.00000
CO3−−	0.00000	0.00520	0.00050	0.00000	0.00000	0.00000	0.00000	0.00000
HS−	0.00000	0.00000	0.00000	0.00000	0.00000	0.00000	0.00000	0.00000

$^S\theta$ Parameters

	H+	Na+	K+	Mg++	Ca++	Sr++	Ba++	Fe++
H+	0.00000	—	—	—	—	—	—	—
Na+	0.03600	0.00000	—	—	—	—	—	—
K+	0.00500	−0.01200	0.00000	—	—	—	—	—
Mg++	0.10000	0.07000	0.00000	0.00000	—	—	—	—
Ca++	0.06120	0.07000	0.03200	0.00700	0.00000	—	—	—
Sr++	0.06500	0.05100	0.00000	0.00000	0.00000	0.00000	—	—
Ba++	0.00000	0.06700	0.00000	0.00000	0.00000	0.00000	0.00000	—
Fe++	0.00000	0.00000	0.00000	0.00000	0.00000	0.00000	0.00000	0.00000

	OH−	Cl−	SO4−−	HCO3−	CO3−−	HS−
OH−	0.00000	—	—	—	—	—
Cl−	−0.05000	0.00000	—	—	—	—

(Continued)

TABLE 17.4 (*Continued*)
Pitzer Parameters at 25°C

SO_4^{--}	−0.01300	0.02000	0.00000	—	—	—
HCO_3^-	0.00000	0.03590	0.01000	0.00000	—	—
CO_3^{--}	0.10000	−0.05300	0.02000	0.08900	0.00000	—
HS^-	0.00000	0.00000	0.00000	0.00000	0.00000	0.00000

ψ Parameters at 25°C
Anion 1 Fixed as Cl⁻

	H^+	Na^+	K^+	Mg^{++}	Ca^{++}	Sr^{++}	Ba^{++}	Fe^{++}
H^+	0.00000	—	—	—	—	—	—	—
Na^+	−0.00400	0.00000	—	—	—	—	—	—
K^+	−0.01100	−0.00180	0.00000	—	—	—	—	—
Mg^{++}	−0.01100	−0.01200	−0.02200	0.00000	—	—	—	—
Ca^{++}	−0.01500	−0.00700	−0.02500	−0.01200	0.00000	—	—	—
Sr^{++}	0.00300	−0.00210	0.00000	0.00000	0.00000	0.00000	—	—
Ba^{++}	0.01370	−0.01200	0.00000	0.00000	0.00000	0.00000	0.00000	—
Fe^{++}	0.00000	0.00000	0.00000	0.00000	0.00000	0.00000	0.00000	0.00000

Anion 1 Fixed as SO_4^{--}

	H^+	Na^+	K^+	Mg^{++}	Ca^{++}	Sr^{++}	Ba^{++}	Fe^{++}
H^+	0.00000	—	—	—	—	—	—	—
Na^+	0.00000	0.00000	—	—	—	—	—	—
K^+	0.19700	−0.01000	0.00000	—	—	—	—	—
Mg^{++}	0.00000	−0.01500	−0.04800	0.00000	—	—	—	—
Ca^{++}	0.00000	−0.05500	0.00000	0.02400	0.00000	—	—	—
Sr^{++}	0.00000	0.00000	0.00000	0.00000	0.00000	0.00000	—	—
Ba^{++}	0.00000	0.00000	0.00000	0.00000	0.00000	0.00000	0.00000	—
Fe^{++}	0.00000	0.00000	0.00000	0.00000	0.00000	0.00000	0.00000	0.00000

Anion 1 Fixed as HCO_3^-

	H^+	Na^+	K^+	Mg^{++}	Ca^{++}	Sr^{++}	Ba^{++}	Fe^{++}
H^+	0.00000	—	—	—	—	—	—	—
Na^+	0.00000	0.00000	—	—	—	—	—	—
K^+	0.00000	−0.00300	0.00000	—	—	—	—	—
Mg^{++}	0.00000	0.00000	0.00000	0.00000	—	—	—	—
Ca^{++}	0.00000	0.00000	0.00000	0.00000	0.00000	—	—	—
Sr^{++}	0.00000	0.00000	0.00000	0.00000	0.00000	0.00000	—	—
Ba^{++}	0.00000	0.00000	0.00000	0.00000	0.00000	0.00000	0.00000	—
Fe^{++}	0.00000	0.00000	0.00000	0.00000	0.00000	0.00000	0.00000	0.00000

Anion 1 fixed as CO_3^{--}

	H^+	Na^+	K^+	Mg^{++}	Ca^{++}	Sr^{++}	Ba^{++}	Fe^{++}
H^+	0.00000	—	—	—	—	—	—	—
Na^+	0.00000	0.00000	—	—	—	—	—	—
K^+	0.00000	0.00300	0.00000	—	—	—	—	—
Mg^{++}	0.00000	0.00000	0.00000	0.00000	—	—	—	—
Ca^{++}	0.00000	0.00000	0.00000	0.00000	0.00000	—	—	—
Sr^{++}	0.00000	0.00000	0.00000	0.00000	0.00000	0.00000	—	—
Ba^{++}	0.00000	0.00000	0.00000	0.00000	0.00000	0.00000	0.00000	—
Fe^{++}	0.00000	0.00000	0.00000	0.00000	0.00000	0.00000	0.00000	0.00000

Cation 1 Fixed as Na⁺

	OH^-	Cl^-	SO_4^{--}	HCO_3^-	CO_3^{--}	HS^-
OH^-	0.00000	—	—	—	—	—
Cl^-	−0.00600	0.00000	—	—	—	—
SO_4^{--}	−0.00900	0.00140	0.00000	—	—	—
HCO_3^-	0.00000	−0.01430	−0.00500	0.00000	—	—
CO_3^{--}	−0.01700	0.00000	−0.00500	0.00000	0.00000	—
HS^-	0.00000	0.00000	0.00000	0.00000	0.00000	0.00000

Cation 1 Fixed as K⁺

	OH^-	Cl^-	SO_4^{--}	HCO_3^-	CO_3^{--}	HS^-
OH^-	0.00000	—	—	—	—	—
Cl^-	−0.00800	0.00000	—	—	—	—
SO_4^{--}	−0.05000	0.00000	0.00000	—	—	—
HCO_3^-	0.00000	0.00000	0.00000	0.00000	—	—
CO_3^{--}	−0.01000	0.02400	−0.00900	−0.03600	0.00000	—
HS^-	0.00000	0.00000	0.00000	0.00000	0.00000	0.00000

Cation 1 Fixed as Mg⁺⁺

	OH^-	Cl^-	SO_4^{--}	HCO_3^-	CO_3^{--}	HS^-
OH^-	0.00000	—	—	—	—	—
Cl^-	0.00000	0.00000	—	—	—	—
SO_4^{--}	0.00000	−0.00400	0.00000	—	—	—
HCO_3^-	0.00000	−0.09600	−0.16100	0.00000	—	—
CO_3^{--}	0.00000	0.00000	0.00000	0.00000	0.00000	—
HS^-	0.00000	0.00000	0.00000	0.00000	0.00000	0.00000

Cation 1 Fixed as Ca⁺⁺

	OH^-	Cl^-	SO_4^{--}	HCO_3^-	CO_3^{--}	HS^-
OH^-	0.00000	—	—	—	—	—
Cl^-	−0.02500	0.00000	—	—	—	—
SO_4^{--}	0.00000	−0.01800	0.00000	—	—	—
HCO_3^-	0.00000	0.00000	0.00000	0.00000	—	—
CO_3^{--}	0.00000	0.00000	0.00000	0.00000	0.00000	—
HS^-	0.00000	0.00000	0.00000	0.00000	0.00000	0.00000

Source: Data from Haarberg, T., Mineral deposition during oil recovery, Ph.D. thesis, Department of Inorganic Chemistry, University of Trondheim, Norway, 1989.

---

where b is a universal constant with the value 1.2 $kg^{1/2}/mol^{1/2}$ and

$$A_\phi = \frac{1}{3}\sqrt{2\pi N_0 d_w}\left(\frac{e^2}{4\pi\varepsilon_0 D kT}\right)^{3/2} \tag{17.33}$$

where N_0 is the Avogadro number (6.022×10^{23}), d_w is the water density in kg/m^3, e is the electronic charge (1.602×10^{-19} C), ε_0 is the permeability of vacuum (8.85419×10^{-12} $C^2/(Nm^2)$), k is the Boltzmann constant (1.381×10^{-23}), and D is the dielectric constant of water, which can be found from a third-degree polynomial in the temperature, t, in °C (Mørk 1989):

$$D = 87.740 - 0.4008\,t + 9.398\times10^{-4}\,t^2 - 1.410\times10^{-6}\,t^3 \tag{17.34}$$

$$Z = \sum_i m_i |z_i| \tag{17.35}$$

$$B_{MX}^{\phi} = \beta_{MX}^{(0)} + \beta_{MX}^{(1)} \exp\left(-\alpha_1 \sqrt{I}\right) + \beta_{MX}^{(2)} \exp\left(-\alpha_2 \sqrt{I}\right) \qquad (17.36)$$

$$B_{MX} = \beta_{MX}^{(0)} + \beta_{MX}^{(1)} g\left(\alpha_1 \sqrt{I}\right) + \beta_{MX}^{(2)} g\left(\alpha_2 \sqrt{I}\right) \qquad (17.37)$$

$$B_{MX}' = \frac{\beta_{MX}^{(1)} g'\left(\alpha_1 \sqrt{I}\right) + \beta_{MX}^{(2)} g'\left(\alpha_2 \sqrt{I}\right)}{I} \qquad (17.38)$$

where M stands for a particular cation and X for a particular anion. Values of the parameters $\beta^{(0)}$, $\beta^{(1)}$, and $\beta^{(2)}$ at 25°C can be seen in Table 17.4. α_1 and α_2 are constants, with $\alpha_1 = 2$ kg$^{1/2}$/mol$^{1/2}$ for ions with charge +1 and −1 and $\alpha_1 = 1.4$ kg$^{1/2}$/mol$^{1/2}$ for ions with charge +2 or −2. The term α_2 equals 12 kg$^{1/2}$/mol$^{1/2}$:

$$g(x) = \frac{2\left[1-(1+x)\exp(-x)\right]}{x^2} \qquad (17.39)$$

$$g'(x) = \frac{-2\left[1-\left(1+x+\frac{x^2}{2}\right)\exp(-x)\right]}{x^2} \qquad (17.40)$$

$$C_{MX} = \frac{C_{MX}^{\phi}}{2\sqrt{|z_M z_X|}} \qquad (17.41)$$

Values of C_{MX}^{ϕ} at 25°C are listed in Table 17.4. Φ_{ij}^{ϕ} is found from

$$\Phi_{ij}^{\phi} = {}^{S}\theta_{ij} + {}^{E}\theta_{ij}(I) + I {}^{E}\theta_{ij}'(I) \qquad (17.42)$$

where values of $^{S}\theta$ can be found in Table 17.4, and

$${}^{E}\theta_{ij}(I) = \frac{z_i z_j}{4I}(J(x_{ij}) - \tfrac{1}{2} J(x_{ii}) - \tfrac{1}{2} J(x_{jj})) \qquad (17.43)$$

$${}^{E}\theta_{ij}'(I) = -\frac{{}^{E}\theta_{ij}(I)}{I} + \frac{z_i z_j}{8I^2}(x_{ij} J'(x_{ij}) - \tfrac{1}{2} x_{ii} J'(x_{ii}) - \tfrac{1}{2} x_{jj} J'(x_{jj})) \qquad (17.44)$$

$$x_{ij} = 6 z_i z_j A_{\phi} \sqrt{I} \qquad (17.45)$$

where Aϕ is defined in Equation 17.33. The subscript ij either represents two cations or two anions.

$$J(x) = \frac{x}{4 + \frac{4.581}{x^{0.7237}} \exp(-0.0120 x^{0.528})} \qquad (17.46)$$

J' in Equation 17.44 is the derivative of J. The Pitzer parameters ψ_{ijk} and $^{S}\theta_{ij}$ are independent of temperature, whereas $\beta_{ij}^{(0)}$, $\beta_{ij}^{(1)}$, $\beta_{ij}^{(2)}$, and depend on temperature. The variation with temperature of these parameters (called X here) may be represented as (Haarberg 1989):

$$X(T) = X(298.15) + \left(\frac{\partial X}{\partial T}\right)_{298.15}(T - 298.15) + \tfrac{1}{2}\left(\frac{\partial^2 X}{\partial T^2}\right)_{298.15}(T - 298.15)^2 \qquad (17.47)$$

The temperature coefficients $\left(\dfrac{\partial X}{\partial T}\right)_{298.15}$ and $\left(\dfrac{\partial^2 X}{\partial T^2}\right)_{298.15}$ are listed in Table 17.5.

TABLE 17.5
Temperature Coefficients in Equation 17.47 Expressing the Temperature Dependence of the Pitzer Parameters

First-Order Temperature Derivative of $\beta^{(0)}$ ($\times 100$)

	H^+	Na^+	K^+	Mg^{++}	Ca^{++}	Sr^{++}	Ba^{++}	Fe^{++}
OH^-	0.00000	−0.01879	0.00000	0.00000	0.00000	0.00000	0.00000	0.00000
Cl^-	−0.18133	0.07159	0.03579	−0.05311	0.02124	0.02493	0.06410	0.00000
SO_4^{--}	0.00000	0.16313	0.09475	0.00730	0.00000	0.00000	0.00000	0.00000
HCO_3^-	0.00000	0.10000	0.10000	0.00000	0.00000	0.00000	0.00000	0.00000
CO_3^{--}	0.00000	0.17900	0.11000	0.00000	0.00000	0.00000	0.00000	0.00000
HS^-	0.00000	0.00000	0.00000	0.00000	0.00000	0.00000	0.00000	0.00000

Second-Order Temperature Derivative of $\beta^{(0)}$ ($\times 100$)

	H^+	Na^+	K^+	Mg^{++}	Ca^{++}	Sr^{++}	Ba^{++}	Fe^{++}
OH^-	0.00000	0.00003	0.00000	0.00000	0.00000	0.00000	0.00000	0.00000
Cl^-	0.00376	−0.00150	−0.00025	0.00038	−0.00057	−0.00621	0.00000	0.00000
SO_4^{--}	0.00000	−0.00115	0.00008	0.00094	0.00000	0.00000	0.00000	0.00000
HCO_3^-	0.00000	−0.00192	0.00000	0.00000	0.00000	0.00000	0.00000	0.00000
CO_3^{--}	0.00000	−0.00263	0.00102	0.00000	0.00000	0.00000	0.00000	0.00000
HS^-	0.00000	0.00000	0.00000	0.00000	0.00000	0.00000	0.00000	0.00000

First-Order Temperature Derivative of $\beta^{(1)}$ ($\times 100$)

	H^+	Na^+	K^+	Mg^{++}	Ca^{++}	Sr^{++}	Ba^{++}	Fe^{++}
OH^-	0.00000	0.27642	0.00000	0.00000	0.00000	0.00000	0.00000	0.00000
Cl^-	0.01307	0.07000	0.11557	0.43440	0.36820	0.20490	0.32000	0.00000
SO_4^{--}	0.00000	−0.07881	0.46140	0.64130	5.46000	5.46000	5.46000	0.00000
HCO_3^-	0.00000	0.11000	0.11000	0.00000	0.00000	0.00000	0.00000	0.00000
CO_3^{--}	0.00000	0.20500	0.43600	0.00000	0.00000	0.00000	0.00000	0.00000
HS^-	0.00000	0.00000	0.00000	0.00000	0.00000	0.00000	0.00000	0.00000

Second-Order Temperature Derivative of $\beta^{(1)}$ ($\times 100$)

	H^+	Na^+	K^+	Mg^{++}	Ca^{++}	Sr^{++}	Ba^{++}	Fe^{++}
OH^-	0.00000	−0.00124	0.00000	0.00000	0.00000	0.00000	0.00000	0.00000
Cl^-	−0.00005	0.00021	−0.00004	0.00074	0.00232	0.05000	0.00000	0.00000
SO_4^{--}	0.00000	0.00908	−0.00011	0.00901	0.00000	0.00000	0.00000	0.00000
HCO_3^-	0.00000	0.00263	0.00000	0.00000	0.00000	0.00000	0.00000	0.00000
CO_3^{--}	0.00000	−0.04170	0.00414	0.00000	0.00000	0.00000	0.00000	0.00000
HS^-	0.00000	0.00000	0.00000	0.00000	0.00000	0.00000	0.00000	0.00000

First-Order Temperature Derivative of $\beta^{(2)}$

	H^+	Na^+	K^+	Mg^{++}	Ca^{++}	Sr^{++}	Ba^{++}	Fe^{++}
OH^-	0.00000	0.00000	0.00000	0.00000	0.00000	0.00000	0.00000	0.00000
Cl^-	0.00000	0.00000	0.00000	0.00000	0.00000	0.00000	0.00000	0.00000
SO_4^-	0.00000	0.00000	0.00000	−0.06100	−0.51600	−0.51600	−0.51600	0.00000
HCO_3^-	0.00000	0.00000	0.00000	0.00000	0.00000	0.00000	0.00000	0.00000
CO_3^{--}	0.00000	0.00000	0.00000	0.00000	0.00000	0.00000	0.00000	0.00000
HS^-	0.00000	0.00000	0.00000	0.00000	0.00000	0.00000	0.00000	0.00000

Second-Order Temperature Derivative of $\beta^{(2)}$

	H^+	Na^+	K^+	Mg^{++}	Ca^{++}	Sr^{++}	Ba^{++}	Fe^{++}
OH^-	0.00000	0.00000	0.00000	0.00000	0.00000	0.00000	0.00000	0.00000
Cl^-	0.00000	0.00000	0.00000	0.00000	0.00000	0.00000	0.00000	0.00000

(Continued)

TABLE 17.5 (Continued)
Temperature Coefficients in Equation 17.47 Expressing the Temperature Dependence of the Pitzer Parameters

SO_4^{--}	0.00000	0.00000	0.00000	−0.01300	0.00000	0.00000	0.00000	0.00000
HCO_3^-	0.00000	0.00000	0.00000	0.00000	0.00000	0.00000	0.00000	0.00000
CO_3^{--}	0.00000	0.00000	0.00000	0.00000	0.00000	0.00000	0.00000	0.00000
HS^-	0.00000	0.00000	0.00000	0.00000	0.00000	0.00000	0.00000	0.00000

First-Order Temperature Derivative of C^ϕ (×100)

	H^+	Na^+	K^+	Mg^{++}	Ca^{++}	Sr^{++}	Ba^{++}	Fe^{++}
OH^-	0.00000	−0.00790	0.00000	0.00000	0.00000	0.00000	0.00000	0.00000
Cl^-	0.00590	−0.01050	−0.00400	−0.01990	−0.01300	0.00000	−0.01540	0.00000
SO_4^{--}	0.00000	−0.36300	−0.00625	−0.02950	0.00000	0.00000	0.00000	0.00000
HCO_3^-	0.00000	0.00000	0.00000	0.00000	0.00000	0.00000	0.00000	0.00000
CO_3^{--}	0.00000	0.00000	0.00000	0.00000	0.00000	0.00000	0.00000	0.00000
HS^-	0.00000	0.00000	0.00000	0.00000	0.00000	0.00000	0.00000	0.00000

Second-Order Temperature Derivative of C^ϕ (×100)

	H^+	Na^+	K^+	Mg^{++}	Ca^{++}	Sr^{++}	Ba^{++}	Fe^{++}
OH^-	0.00000	0.00007	0.00000	0.00000	0.00000	0.00000	0.00000	0.00000
Cl^-	−0.00002	0.00015	0.00003	0.00018	0.00005	0.00000	0.00000	0.00000
SO_4^{--}	0.00000	0.00027	−0.00023	−0.00010	0.00000	0.00000	0.00000	0.00000
HCO_3^-	0.00000	0.00000	0.00000	0.00000	0.00000	0.00000	0.00000	0.00000
CO_3^{--}	0.00000	0.00000	0.00000	0.00000	0.00000	0.00000	0.00000	0.00000
HS^-	0.00000	0.00000	0.00000	0.00000	0.00000	0.00000	0.00000	0.00000

Source: Data from Haarberg, T., Mineral deposition during oil recovery, Ph.D. thesis, Department of Inorganic Chemistry, University of Trondheim, Norway, 1989.

TABLE 17.6
Temperature Coefficients in Equation 17.48 Expressing the Temperature Dependence of the Pitzer Parameters for NaCl

	$\beta^{(0)}_{NaCl}$	$\beta^{(1)}_{NaCl}$	X^ϕ_{NaCl}
Q_1	-6.5681518×10^2	1.1931966×10^2	-6.1084589
Q_2	2.486912950×10^1	$-4.8309327 \times 10^{-1}$	4.0217793×10^{-1}
Q_3	$5.381275267 \times 10^{-5}$	0	2.2902837×10^{-5}
Q_4	-4.4640952	0	$-7.5354649 \times 10^{-2}$
Q_5	$1.110991383 \times 10^{-2}$	1.4068095×10^{-3}	$1.531767295 \times 10^{-4}$
Q_6	$-2.657339906 \times 10^{-7}$	0	$-9.0550901 \times 10^{-8}$
Q_7	$-5.309012889 \times 10^{-6}$	0	$-1.53860082 \times 10^{-8}$
Q_8	$8.634023325 \times 10^{-10}$	0	8.69266×10^{-11}
Q_9	-1.579365943	-4.2345814	$3.53104136 \times 10^{-1}$
Q_{10}	$2.202282079 \times 10^{-3}$	0	$-4.3314252 \times 10^{-4}$
Q_{11}	9.706578079	0	$-9.187145529 \times 10^{-2}$
Q_{12}	$-2.686039622 \times 10^{-2}$	0	5.1904777×10^{-4}

Note: The coefficients are for pressure in bar and temperature in K.

Source: Data from Pitzer, K.S., *Thermodynamics*, Appendix X, McGraw Hill, New York, 1995.

Owing to the appearance of Na and Cl in many systems, Pitzer et al. (1984) have developed an alternative and more sophisticated expression for the temperature and pressure dependence of the parameters for these species. It has the functional form:

$$X(T) = \frac{Q_1}{T} + Q_2 + Q_3P + Q_4\ln(T) + (Q_5 + Q_6P)T$$
$$+(Q_7 + Q_8P)T^2 + \frac{Q_9 + Q_{10}P}{T-227} + \frac{Q_{11} + Q_{12}P}{680-T}$$

(17.48)

The pressure, P, is in bar, the temperature, T, in K, and the coefficients $Q_1, Q_2, \ldots, Q_{12}$ are listed in Table 17.6.

17.4   SOLUTION PROCEDURE

An initial evaluation of which salts are likely to precipitate can be made by calculating the saturation ratio, SR. This quantity for $CaSO_4$ is defined as follows:

$$SR(CaSO_4) = \frac{m_{Ca^{++}} m_{SO_4^{--}}}{K_{sp}^0}$$

(17.49)

where the molalities are evaluated not considering precipitation. Salts having a solubility ratio higher than 1.0 are likely to precipitate, but the actual molality of the ions may be lower because one or both ions can possibly take part in another salt precipitating. A related quantity is the solubility index, SI, which for $CaSO_4$ is defined as:

$$SI(CaSO_4) = \log_{10}(SR(CaSO_4)) = \log_{10}\left(\frac{m_{Ca^{++}} m_{SO_4^{--}}}{K_{sp}^0}\right)$$

(17.50)

SI is greater than zero for salts likely to precipitate. If SI is negative, the salt will not precipitate. SR and SI are qualitative measures and insufficient when it comes to quantitative calculations of the amount of salt precipitate.

The amount of minerals that precipitate from an aqueous solution is evaluated by calculating the amount of ions in solution when equilibrium has established. The remaining salt content has precipitated. The thermodynamic equilibrium constants given from Equations 17.10 through 17.15 must be fulfilled for salts precipitating, whereas this is not the case for salts that do not precipitate. For salts not precipitating, the product of the molalities and activity coefficients in Equations 17.10 through 17.15 will be less than the thermodynamic equilibrium constant.

Solving the system of equations is an iterative process. The activity coefficients are initially set equal to 1.0 for all ions. The equilibrium constants at 1 bar are calculated by use of Equations 17.16 and 17.17, where the following iterative procedure is followed:

1. The equilibrium constants are corrected for pressure using Equation 17.21.
2. The stoichiometric equilibrium constants are calculated from the thermodynamic ones (Equation 17.4) and from the activity coefficients (equal to 1.0 in first iteration).

3. The concentration ratio of CO_2(aq) to H_2S(aq) is calculated. The concentrations of CO_2(aq) and H_2S(aq) are fixed (determined by the fugacities of CO_2 to H_2S in the hydrocarbon phase(s) in equilibrium with the water phase).
4. The amount of sulfate precipitation is calculated, with none of the other precipitation reactions taken into account.
5. The ion products of the iron minerals ($FeCO_3$ and FeS) are checked against the solubility products. Only one of the two iron minerals can precipitate because, for a fixed relative concentration of CO_2(aq) to H_2S(aq), the solubility products will not be fulfilled for both salts at the same time.
6. The ion product of calcium carbonate is checked against its solubility product. If the solubility product is not exceeded, go to step 8. Otherwise continue with step 7.
7. A double-loop iteration is applied to calculate simultaneous precipitation of calcium carbonate and iron minerals. The inner loop calculates the amount of $CaCO_3$ precipitation. This precipitation will affect the amount of sulfate precipitate because some Ca^{++} is removed from the solution. The amount of sulfate precipitate is corrected in each inner-loop iteration. In the outer loop, the iteration variable is the amount of ferrous iron mineral precipitate. Convergence is achieved when the ion product of the ferrous mineral matches the thermodynamic solubility product.
8. The activity coefficients are (re)calculated from Pitzer's activity coefficient model.
9. The procedure is then repeated from step 1 until convergence.

17.5   EXAMPLE CALCULATIONS

Table 17.7 shows two water analyses, one for formation water and another one for seawater. Figure 17.1 shows the amount of scale precipitate at a temperature of 80°C and 135 bar as a function of mixing ratio. The water is assumed to be in equilibrium with a gas containing 4.8 mol% CO_2 and no H_2S. $BaSO_4$ and $SrSO_4$ are seen to precipitate.

TABLE 17.7
Formation Water and Seawater Analysis

Ion	Formation Water (mg/l)	Sea Water (mg/l)
Na^+	8442	10680
K^+	159	396
Ca^{++}	671	409
Mg^{++}	25	1279
Ba^{++}	11	0.02
Sr^{++}	150	7.9
Cl^-	14245	19221
SO_4^{--}	4	2689
Alkalinity based on M of HCO_3.	517	141

Note: No organic acid is present.

Source: Data from Kaasa, B., Prediction of pH, mineral precipitation and multiphase equilibria during oil production, Ph.D. thesis, Department of Inorganic Chemistry, University of Trondheim, Norway, 1998.

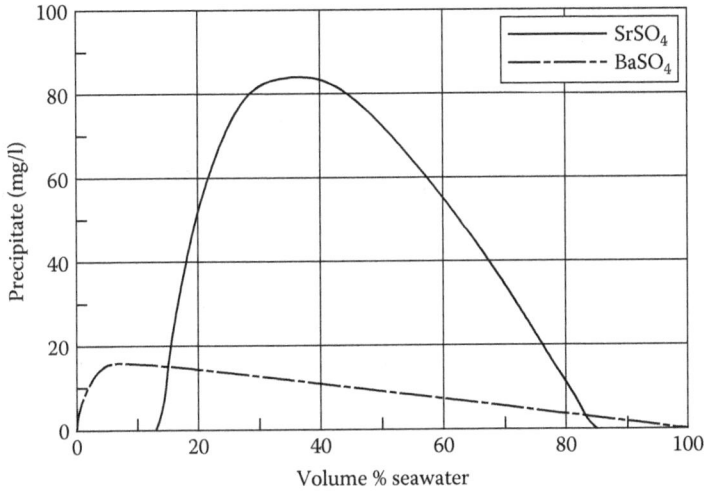

FIGURE 17.1 Scale simulated when mixing formation water and seawater (Table 17.7) at 80°C and 135 bar. The water is assumed to be in equilibrium with a gas containing 4.8 mol% CO_2 and no H_2S.

TABLE 17.8
Formation Water and Seawater Analysis

Ion	Formation Water (mg/l)	Sea Water (mg/l)
Na^+	14834	10680
Ca^{++}	1275	450
Mg^{++}	335	1130
Ba^{++}	50	0
Sr^{++}	335	9
Fe^{++}	30	0
Cl^-	26200	20950
SO_4^{--}	0	3077
Alkalinity based on M of HCO_3^-	415	170

Source: Data from Haarberg, T., et al., The effect of ferrous iron on mineral scaling during oil recovery, *Acta Chem. Scand.* 44, 907–915, 1990.

Note: No organic acid is present.

Similar plots are shown in Figures 17.2 and 17.3 for the combination of formation water and seawater shown in Table 17.8. When the water is in equilibrium with a gas containing CO_2 and no H_2S (Figure 17.2), the iron mineral precipitating is $FeCO_3$. If the water contains H_2S in addition to CO_2, iron precipitates as FeS (Figure 17.3).

$BaSO_4$ has a low solubility in water and the solubility decreases with decreasing pressure. If a water phase is saturated with $BaSO_4$ at high pressure, solid $BaSO_4$ is likely to precipitate at lower pressure. Table 17.9 shows solubility data for $BaSO_4$ in pure water (Bloust 1977). The data is plotted in Figure 17.4 with simulation results using the models presented in this chapter.

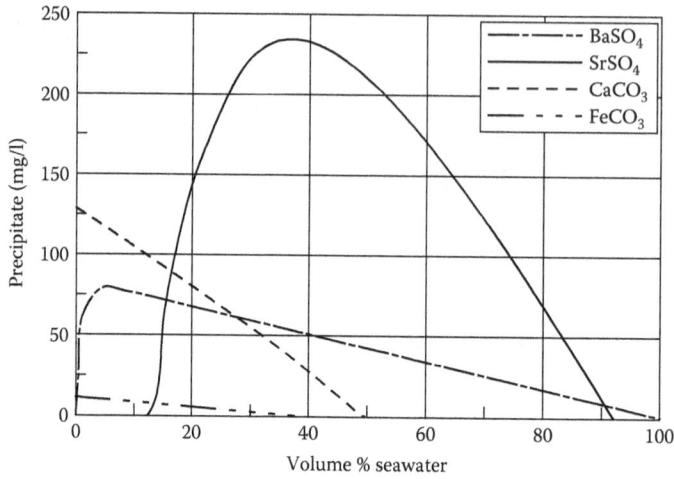

FIGURE 17.2 Scale simulated when mixing formation water and seawater (Table 17.8) at 25°C and 1 bar. The water is assumed to be in equilibrium with a gas containing 3.6 mol% CO_2 and no H_2S.

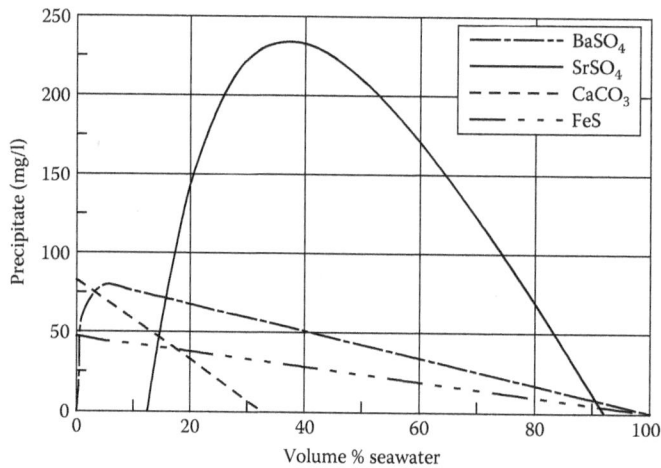

FIGURE 17.3 Scale simulated when mixing formation water and seawater (Table 17.8) at 2°C and 1 bar. The water is assumed to be in equilibrium with a gas containing 3.5 mol% CO_2 and 1.2 mol% H_2S.

TABLE 17.9
Solubility of $BaSO_4$ in Pure Water

Temperature °C	Pressure bar	$BaSO_4$ in millimoles per kg H_2O
96	36	0.0174
96.5	496	0.0243
96.5	862	0.0342
189	100	0.0102
189	402	0.0150
189	804	0.0237

Data source: Bloust, C.W., Barite solubilities and thermodynamic quantities up to 300°C and 1400 bar, *American Mineralogist* 62, 942–057, 1977.

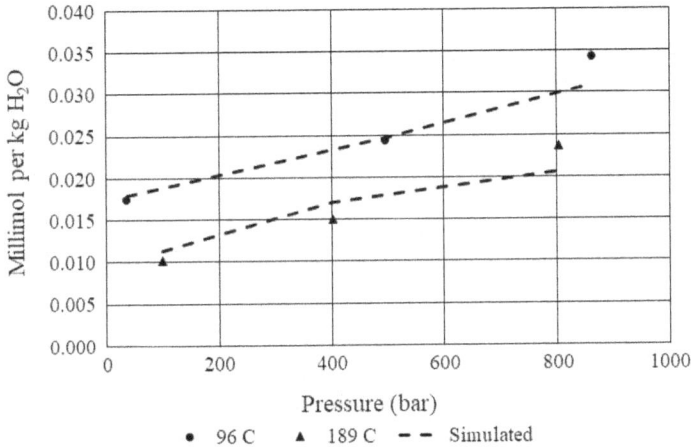

FIGURE 17.4   Experimental (Bloust 1977) and simulated data for the solubility of BaSO$_4$ in water.

REFERENCES

Atkinson, A. and Mecik, M., The chemistry of scale prediction, *J. Pet. Sci. Eng.* 17, 113–121, 1997.

Bloust, C.W., Barite solubilities and thermodynamic quantities up to 300°C and 1400 bar, *Am. Min.* 62, 942–057, 1977.

Haarberg, T., *Mineral Deposition During Oil Recovery*, Ph.D. thesis, Department of Inorganic Chemistry, University of Trondheim, Norway, 1989.

Haarberg, T., Jakobsen, J.E., and Østvold, T., The effect of ferrous iron on mineral scaling during oil recovery, *Acta Chem. Scand.* 44, 907–915, 1990.

Kaasa, B., *Prediction of pH, Mineral Precipitation and Multiphase Equilibria During Oil Production*, Ph.D. thesis, Department of Inorganic Chemistry, University of Trondheim, Norway, 1998.

Kaasa, B. and Østvold, T., Prediction of pH and mineral scaling in waters with varying ionic strength containing CO2 and H$_2$S for 0 < T(°C) < 200 and 1< P(bar) < 500, paper presented at the international conference on *Advances in Solving Oilfield Scaling Aberdeen*, Scotland, January 28–29, 1998.

Mørk, J., *Model til beregning af dielektricitetskonstanter i råolie*, Thesis Project (in Danish), Danish Engineering Academy, 1989.

Olofsson, G. and Hepler, L.G., Thermodynamics of ionization of water over wide ranges of temperature and pressure, *J. Solution Chem.* 4, 127–143, 1975.

Pitzer, K.S., Thermodynamics of electrolytes I. Theoretical basis and general equations, *J. Phys. Chem.* 77, 268–277, 1973.

Pitzer, K.S., Thermodynamics of electrolytes V. Effects of higher-order electrostatic terms, *J. Solution Chem.* 4, 249–265, 1975.

Pitzer, K.S., Theory: Ion interaction approach, *Activity Coefficients in Electrolyte Solutions*, Pytkowicz, R.M., Ed., CRC Press, Boca Raton, FL, 157–208, 1979.

Pitzer, K.S., Theoretical considerations of solubility with emphasis on mixed aqueous electrolytes, *Pure Appl. Chem.* 58, 1599–1610, 1986.

Pitzer, K.S., *Thermodynamics*, Appendix X, McGraw Hill, New York, 1995.

Pitzer, K.S., Peiper, J.C., and Busey, R.H., Thermodynamic properties of aqueous sodium chloride solutions, *J. Phys. Chem. Ref. Data.* 13, 1–102, 1984.

Appendix A: Fundamentals on Phase Equilibrium

This appendix gives an introduction to the fundamental thermodynamic relations that determine the equilibrium state of a multicomponent mixture.

A.1 FIRST AND SECOND LAWS OF THERMODYNAMICS

The first law of thermodynamics is the law of energy conservation, stating that the change in the total energy of a system, dU^t, equals the sum of the heat, dQ, supplied to the system, and the work, dW, done on the system by the surroundings:

$$dU^t = dQ + dW = dQ - pdV^t \qquad (A.1)$$

The superscript t means total. Equation A.1 further states that the work done on the system by the surroundings equals the pressure times the volume decrease.

Before presenting the second law, it is necessary to first introduce the thermodynamic quantity entropy (S). The change in the total entropy of a system is defined as follows:

$$dS^t = \frac{dQ_{rev}}{T} \qquad (A.2)$$

T stands for *temperature* and the subscript rev for *reversible*. The heat dQ is transferred reversibly to the system if the system and the surroundings always are in thermal and mechanical equilibrium during the transfer of heat. In general, it applies that

$$dS^t \geq \frac{dQ}{T} \qquad (A.3)$$

which is the second law of thermodynamics.

A.2 FUNDAMENTAL THERMODYNAMIC RELATIONS

Combining Equations A.1 and A.3 gives

$$dU^t + P\, dV^t \leq T\, dS^t \qquad (A.4)$$

In phase equilibrium calculations, it is convenient to work with Gibbs free energy, G, defined by

$$G^t = H^t - TS^t \qquad (A.5)$$

H^t is the total enthalpy of the system defined by

$$H^t = U^t + PV^t \qquad (A.6)$$

$$G^t = U^t + PV^t - TS^t \qquad (A.7)$$

Differentiation of this equation gives

$$dG^t = dU^t + P \, dV^t + V^t \, dP - T \, dS^t - S^t \, dT \tag{A.8}$$

Equations A.4 and A.8 may be combined to give

$$dG^t \le V^t \, dP - S^t \, dT \tag{A.9}$$

For processes carried out at constant P and T, it will therefore apply that

$$\left(dG^t \right)_{P,T} \le 0 \tag{A.10}$$

For a system only undergoing reversible (equilibrium) processes, Gibbs free energy will remain unchanged. Any irreversible processes occurring at constant P and T will cause a decrease in the Gibbs energy of the system.

This leads to the following definition of the equilibrium state:

The equilibrium state of a closed system at a given P and T is the one for which the Gibbs free energy of the system is at a minimum with respect to all possible changes.

One may ask what changes a closed system is likely to undergo at fixed P and T. Not considering chemical reactions, the system may form one homogeneous phase, or it may separate into two or more unlike phases. The latter will be the case if the system can lower its Gibbs free energy through a phase split as compared to G of the single-phase state.

A.3   PHASE EQUILIBRIUM

Consider a closed cell containing a vapor phase (V) and a liquid phase (L). The two phases can be regarded as two open systems, where mass may be transferred from one system to the other one.

Besides being a function of P and T, the Gibbs free energy is a function of the number of moles of each component n_i in the phase:

$$G = f\left(P, T, n_1, \ldots, n_N \right) \tag{A.11}$$

where N is the total number of components.

Differentiation gives for the change in the total Gibbs free energy of a system:

$$dG = \left(\frac{\partial G}{\partial P} \right)_{T,n} dP + \left(\frac{\partial G}{\partial T} \right)_{P,n} dT + \sum_{i=1}^{N} \left(\frac{\partial G}{\partial n_i} \right)_{P,T,n_{j \ne i}} dn_i \tag{A.12}$$

The partial derivative of G with respect to composition defines the chemical potential

$$\mu_i = \left(\frac{\partial G}{\partial n_i} \right)_{P,T,n_{j \ne i}} \tag{A.13}$$

For P and T constant, this equation may be rewritten to

$$dG = \sum_{i=1}^{N} \mu_i \, dn_i \tag{A.14}$$

Equation A.14 may be applied for each of the two phases present:

$$dG^V = \sum_{i=1}^{N} \mu_i^V \, dn_i^V \qquad (A.15)$$

$$dG^L = \sum_{i=1}^{N} \mu_i^L \, dn_i^L \qquad (A.16)$$

The superscript V stands for vapor and L stands for liquid. For the system as a whole, the following general equilibrium relation applies:

$$(dG^t)_{P,T} = 0 \qquad (A.17)$$

Gibbs free energy of the total system equals the sum of the Gibbs free energies of each phase:

$$(dG^t)_{P,T} = (dG^V)_{P,T} + (dG^L)_{P,T} = 0 \qquad (A.18)$$

$$(dG^t)_{P,T} = \sum_{i=1}^{N} \mu_i^V \, dn_i^V + \sum_{i=1}^{N} \mu_i^L \, dn_i^L = 0 \qquad (A.19)$$

Because the system is closed, the following mass balance must apply:

$$dn_i^V = -dn_i^L \qquad (A.20)$$

and thereby

$$\sum_{i=1}^{N} \left(\mu_i^V - \mu_i^L \right) dn_i^V = 0 \qquad (A.21)$$

Equation A.21 will only be true in general if

$$\mu_i^V = \mu_i^L, \, i = 1, 2, \, ..., \, N \qquad (A.22)$$

This equation states that the chemical potential of any component i, at a fixed P and T, must be the same in two phases in equilibrium. This criterion may be extended to three or more phases in equilibrium.

A.4 FUGACITIES AND FUGACITY COEFFICIENTS

In phase equilibrium calculations, it is practical to work with the term *fugacity* rather than chemical potential. The fugacity f is defined through the following equation:

$$dG = RT \, d \ln f \qquad (A.23)$$

where G is the molar Gibbs free energy and R the gas constant. Equation A.9 applied on a closed system in equilibrium and containing one mole gives

$$dG = V \, dP - S \, dT \qquad (A.24)$$

For T constant, this expression may be reduced to

$$\left(dG \right)_T = V \, dP \qquad (A.25)$$

which, by combining Equations A.23 and A.25, may be used to derive the following pressure dependence of the fugacity:

$$\ln\left(\frac{f(P)}{f(P_{ref})}\right) = \int_{P_{ref}}^{P} \frac{V}{RT} dP \qquad (A.26)$$

where P_{ref} is a reference pressure.

For liquid (L) and solid (S) phases, it is reasonable to assume that the molar volume V is independent of pressure, in which case the pressure dependence of the fugacity becomes

$$f^L(P) = f^L(P_{ref}) \exp\frac{V^L(P-P_{ref})}{RT} \qquad (A.27)$$

$$f^S(P) = f^S(P_{ref}) \exp\frac{V^S(P-P_{ref})}{RT} \qquad (A.28)$$

For an ideal gas, Equation A.25 can be written as

$$(dG)_T = \frac{RT}{P} dP = RT d \ln P \qquad (A.29)$$

Equation A.29 is not generally valid, but it is practical to work with an equally simple relation for real systems. Such a relation is obtained through the definition of the fugacity in Equation A.23. For component i in a mixture, the fugacity is defined through

$$d\mu_i = RT d \ln f_i \qquad (A.30)$$

which may be integrated to give

$$\mu_i = \mu_i^o + RT \ln f_i \qquad (A.31)$$

The reference chemical potential μ_i^o is a function of T. As the two phases in equilibrium are at the same temperature and pressure, Equations A.22 and A.31 may be combined to give the following equilibrium relation:

$$f_i^V = f_i^L \qquad (A.32)$$

The vapor and liquid phase fugacity coefficients of component i, φ_i^V and φ_i^L, are defined through the following relations:

$$f_i^V = y_i \varphi_i^V P; f_i^L = x_i \varphi_i^L P \qquad (A.33)$$

In these equations, y_i is the mole fraction of component i in the vapor phase and x_i the mole fraction of component i in the liquid phase. An alternative way of expressing the fugacity of component i in the liquid phase would be to relate it to the fugacity of pure i, f_i^o, at the same temperature and pressure:

$$f_i^L = x_i \gamma_i f_i^o \qquad (A.34)$$

where g_i is the activity coefficient of component i. The pure component fugacity (f_i^o) is often approximated as the pure component vapor pressure (P_i^{sat}).

The ratio between the vapor- and liquid-phase mole fractions is called the equilibrium ratio or the K-factor:

$$K_i = \frac{y_i}{x_i} \qquad \text{(A.35)}$$

From Equations A.32, A.33, and A.35, it is seen that the equilibrium relation can be expressed in terms of the K-factor:

$$K_i = \frac{y_i}{x_i} = \frac{\varphi_i^L}{\varphi_i^V} \qquad \text{(A.36)}$$

To derive an expression, which makes it possible to calculate the fugacity coefficient of a component in a mixture from an equation of state, the term *partial molar property* is introduced. This term is related to one particular component of a mixture. If this component is called i, a partial molar property of component i is defined as the derivative of that particular property (e.g., G, S, or V) with respect to the mole number of component i at constant temperature and pressure. It is seen from Equation A.13 that the chemical potential equals the partial molar Gibbs free energy:

$$\mu_i = \overline{G}_i \qquad \text{(A.37)}$$

Differentiation of Equation A.24 with respect to mole number leads to the following relation between the partial molar properties of G, V, and S:

$$d\overline{G}_i = \overline{V}_i \, dP - \overline{S}_i \, dT \qquad \text{(A.38)}$$

For T constant, this relation reduces to

$$d\overline{G}_i = \overline{V}_i \, dP; \text{ T constant} \qquad \text{(A.39)}$$

Recalling that the partial molar G equals the chemical potential, it is possible to combine Equations A.23 and A.39 to

$$\overline{V}_i \, dP = RT \, d\ln f_i \qquad \text{(A.40)}$$

For an ideal gas, the partial molar volume equals the pure component molar volume

$$\overline{V}_i = V_i = \frac{RT}{P} \qquad \text{(A.41)}$$

and the fugacity of the component equals its partial pressure:

$$f_i = y_i \, P \qquad \text{(A.42)}$$

where y_i is the mole fraction component i. This gives the following relation for an ideal gas:

$$\frac{RT}{P} dP = RT \, d\ln(y_i P) \qquad \text{(A.43)}$$

Subtraction of Equation A.43 from Equation A.40 gives

$$\left(\overline{V}_i - \frac{RT}{P}\right)dP = RT\,d\ln\frac{f_i}{y_iP} \tag{A.44}$$

Using the definition of the fugacity coefficient in Equation A.33, this relation may be simplified to

$$\left(\overline{V}_i - \frac{RT}{P}\right)dP = RT\,d\,\ln\varphi_i \tag{A.45}$$

$$\ln\varphi_i = \frac{1}{RT}\int_0^P\left(\overline{V}_i - \frac{RT}{P}\right)dP \tag{A.46}$$

which, through some mathematical manipulations, may be rewritten to

$$RT\,\ln\varphi_i = \frac{\partial}{\partial n_i}\left(\int_{V^t}^{\infty}\left(P - \frac{nRT}{V_t}\right)dV^t\right)_{T,V^t,n_{j\neq i}} - RT\,\ln Z \tag{A.47}$$

where n stands for number of moles and Z is the compressibility factor:

$$Z = \frac{PV}{RT} \tag{A.48}$$

Index

For Product Safety Concerns and Information please contact our EU
representative GPSR@taylorandfrancis.com
Taylor & Francis Verlag GmbH, Kaufingerstraße 24, 80331 München, Germany

www.ingramcontent.com/pod-product-compliance
Lightning Source LLC
Chambersburg PA
CBHW080119220326
41598CB00032B/4887

9 781032 642222